Studies in Systems, Decision and Control

Volume 194

The series "Studies in Systems, Decision and Control" (SSDC) covers both new developments and advances, as well as the state of the art, in the various areas of broadly perceived systems, decision making and control–quickly, up to date and with a high quality. The intent is to cover the theory, applications, and perspectives on the state of the art and future developments relevant to systems, decision making, control, complex processes and related areas, as embedded in the fields of engineering, computer science, physics, economics, social and life sciences, as well as the paradigms and methodologies behind them. The series contains monographs, textbooks, lecture notes and edited volumes in systems, decision making and control spanning the areas of Cyber-Physical Systems, Autonomous Systems, Sensor Networks, Control Systems, Energy Systems, Automotive Systems, Biological Systems, Vehicular Networking and Connected Vehicles, Aerospace Systems, Automation, Manufacturing, Smart Grids, Nonlinear Systems, Power Systems, Robotics, Social Systems, Economic Systems and other. Of particular value to both the contributors and the readership are the short publication timeframe and the world-wide distribution and exposure which enable both a wide and rapid dissemination of research output.

More information about this series at http://www.springer.com/series/13304

José Francisco Gómez ·
Lizeth Torres · Ricardo Fabricio Escobar
Editors

Fractional Derivatives with Mittag-Leffler Kernel

Trends and Applications in Science and Engineering

Springer

Editors
José Francisco Gómez
CONACYT-Tecnológico Nacional de México
Centro Nacional de Investigación y Desarrollo Tecnológico
Cuernavaca, Morelos, Mexico

Lizeth Torres
CONACYT-Instituto de Ingeniería
Universidad Nacional Autónoma de México
Mexico City, Mexico

Ricardo Fabricio Escobar
Tecnológico Nacional de México
Centro Nacional de Investigación y Desarrollo Tecnológico
Cuernavaca, Morelos, Mexico

ISSN 2198-4182 ISSN 2198-4190 (electronic)
Studies in Systems, Decision and Control
ISBN 978-3-030-11661-3 ISBN 978-3-030-11662-0 (eBook)
https://doi.org/10.1007/978-3-030-11662-0

Library of Congress Control Number: 2018967421

This Springer imprint is published by the registered company Springer Nature Switzerland AG
The registered company address is: Gewerbestrasse 11, 6330 Cham, Switzerland

Preface

In recent years, fractional calculus has allowed describing several complex problems in the fields of mathematics, physics, biology, and engineering. The complexity of these problems has led to researchers to develop mathematical theories to model the complexities of nature taking into account the fractional calculus. The mathematical models are powerful tools used for describing real-world problems; to develop mathematical models, differential equations and differential operators are required. The differential operators can be local or non-local. The non-local can further be divided into three types: differential operators with a power-law kernel, differential operators with exponential decay law, and finally, differential operators with Mittag-Leffler law. The operators with non-singular kernel have the following features: They do not impose artificial singularities on any model, they have at the same time Markovian and non-Markovian properties, they are at the same time power law, stretched exponential and Brownian motion, the mean square displacement is a crossover from usual diffusion to sub-diffusion, the derivative probability distribution is at the same time Gaussian and non-Gaussian, and it can cross over from Gaussian to non-Gaussian even without passing through the steady state. It means that the fractional derivatives with non-singular kernel are at the same time deterministic and stochastic.

The aim of this book is to present novel developments, trends, and applications of fractional-order derivatives with a non-singular and non-local kernel in the areas of chemistry, mechanics, chaos, epidemiology, fluid mechanics, modeling, and engineering. Non-singular and non-local fractional-order derivatives have been applied in the different chapters to describe complex problems. These 18 contributed chapters, which were put together upon a rigorous review process, have been written by both young and established researchers, who are specialists in their topic.

Cuernavaca, Mexico — José Francisco Gómez
Mexico City, Mexico — Lizeth Torres
Cuernavaca, Mexico — Ricardo Fabricio Escobar
February 2019

Preface

Contents

Reproducing Kernel Method for Fractional Derivative with Non-local and Non-singular Kernel

Ali Akgül

Abstract Atangana and Baleanu introduced a derivative with fractional order to answer some outstanding questions that were posed by many investigators within the field of fractional calculus. Their derivative has a non-singular and nonlocal kernel. Therefore, we apply the reproducing kernel method to fractional differential equations with non-local and non-singular kernel. In this work, a new method has been developed for the newly established fractional differentiation. Examples are given to illustrate the numerical effectiveness of the reproducing kernel method when properly applied in the reproducing kernel space. The comparison of approximate and exact solutions leaves no doubt believing that the reproducing kernel method is very efficient and converges toward exact solution very rapidly.

Keywords Fractional calculus · Atangana–Baleanu fractional derivative · Reproducing kernel method

1 Introduction

Caputo and Fabrizio introduced a new operator that was called the Caputo–Fabrizio derivative with fractional order [1, 2]. Due to the novelty of their results, many researchers applied their derivative in few real world problems with great success [3–17]. The novelty in their operator is that the derivative has no singular kernel and finds applications in many problems in the field of groundwater and thermal science. Atangana with Goufo enhanced the version based upon the Riemann–Liouville approach, and the results were also acquired in the work by Caputo and Fabrizio [18–21]. Apart from the real world applications done with this novel idea, many theoretical works were also given. A couple of issues were pointed out against both derivatives, including the one in Caputo sense and the one in Riemann–Liouville sense.

A. Akgül (✉)
Siirt University Art and Science Faculty Department of Mathematics,
56100 Siirt, Turkey
e-mail: aliakgul00727@gmail.com

J. F. Gómez et al. (eds.), *Fractional Derivatives with Mittag-Leffler Kernel*,
Studies in Systems, Decision and Control 194,
https://doi.org/10.1007/978-3-030-11662-0_1

(1) The kernel was not nonlocal.
(2) The integral associate is not a fractional operator but the average of the function and its integral.
(3) The solution of the following equation $\frac{d^\alpha y}{dx^\alpha} = -ay$ is an exponential equation not a non-local function.

It is therefore concluded by some researchers that the operator was not a derivative with fractional order instead it is a filter with fractional parameter. The fractional parameter can then be viewed as filter regulator. The well known Caputo and Riemann–Liouville also have a big problem; their kernel is nonlocal, but is singular. This weakness has effect when modeling real world problems.

In order to solve the above problems, Atangana and Baleanu suggested a new operator with fractional order based upon the Mittag-Leffler function [22–24]. Their operators have all the benefits of that of Caputo and Fabrizio; in addition, the kernel used is nonlocal. The operators have all the benefits of those of Riemann–Liouville fractional integral of the given function and the function itself. In addition to the above benefits, the derivative was found very useful in thermal science and material sciences [22–24]. These new derivatives with fractional orders are at the same time filters and fractional derivatives [25].

Numerical methods have been known as strong mathematical tools to solve nonlinear ordinary differential equations with local and non-local operators. They have been utilized in many models to predict the behavior of the dynamical system for which the model was enhanced for. They are normally utilized when all the implemented analytical techniques fail. Due to the problems posed by the fractional derivative with power-law kernel, a new fractional differentiation was given. The new fractional differentiation, in order to complete the new fractional calculus, was utilized to produced a new fractional integration. The new fractional differentiation has therefore produced a new class of linear and non-linear ordinary differential equations [26]. We will apply the reproducing kernel method to solve fractional ordinary differential equations that have exact solutions.

When modeling physical processes, clear advantages accrue if the problem can be formulated in a Hilbert space H of differentiable functions on a set E. An important class of such spaces - the reproducing kernel Hilbert spaces - arose in the twentieth century and has become increasingly prominent in the twenty-first [27].

Reproducing kernels were used for the first time at the beginning of the twentieth century by Zaremba in his work on boundary value problems for harmonic and biharmonic functions [28, 29]. The general theory of reproducing kernel Hilbert spaces was established simultaneously and independently by Aronszajn [30] and Bergman [31] in 1950. The introduction of the reproducing kernel Hilbert spaces $W_2^m[a, b]$ by Cui and Lin [32, 33] in the 1980's led to an explosion in applications of reproducing kernel Hilbert space methods to many areas of mathematics: ordinary and partial differential equations [33–37], fractional differential equations [38, 39], nonlinear oscillators with discontinuities [40], nonlinear two-point boundary value problems [41–43], difference equations [44], and integral equations [33, 45–51].

In this paper, we solve the fractional differential equations with fractional derivative with non-local and non-singular kernel by the reproducing kernel method. To do this we take into consideration the following fractional ordinary equation:

$$\begin{cases} {}_0^{ABC}Du(x) = H(x, u(x)), \\ u(0) = u_0. \end{cases} \tag{1}$$

The above equation can be transformed to a fractional integral equation by applying the fundamental theorem of fractional calculus:

$$u(x) - u(0) = \frac{(1-\alpha)}{ABC(\alpha)} H(x, u(x)) + \frac{\alpha}{\Gamma(\alpha)\times ABC(\alpha)} \int_0^x H(\tau, u(\tau))(x-\tau)^{\alpha-1} d\tau. \tag{2}$$

This paper is organized as follows. Section 2 introduces fractional derivative with non-local and non-singular kernel. Some useful reproducing kernel functions are presented in Sect. 3. Main results are given in Sect. 4. Examples are given in Sect. 5 illustrating the numerical effectiveness of the reproducing kernel method. A summary of the results of this research is given in Sect. 6.

2 Fractional Derivative with Non-local and Non-singular Kernel

In this section we present the basic definitions of the new derivatives with fractional order proposed by Atangana and Baleanu.

Definition 1 Let $f \in H^1(a, b), b > a, \alpha \in [0, 1]$ then, Atangana–Baleanu fractional derivative in Caputo sense is given as:

$${}_a^{ABC}D_t^\alpha(f(t)) = \frac{B(\alpha)}{1-\alpha} \int_a^t f'(x) E_\alpha\left[-\alpha\frac{(t-x)^\alpha}{1-\alpha}\right] dx. \tag{3}$$

Of course B has the same properties as in Caputo and Fabrizio case. The above definition will be helpful to real world problem and also will have great advantage when using Laplace transform to solve some physical problem with initial condition [18].

Definition 2 Let $f \in H^1(a, b), b > a, \alpha \in [0, 1]$ and not necessary differentiable then, the definition of the Atangana–Baleanu fractional derivative in Riemann–Liouville sense is given as:

$${}_a^{ABC}D_t^\alpha(f(t)) = \frac{B(\alpha)}{1-\alpha} \frac{d}{dt} \int_a^t f(x) E_\alpha\left[-\alpha\frac{(t-x)^\alpha}{1-\alpha}\right] dx. \tag{4}$$

Definition 3 The fractional integral associate to the new fractional derivative with non-local kernel is defined as:

$$ {}_{a}^{AB}I_{t}^{\alpha}\{f(t)\} = \frac{1-\alpha}{B(\alpha)} f(t) + \frac{\alpha}{B(\alpha)\Gamma(\alpha)} \int_{a}^{t} f(y)(t-y)^{\alpha-1} dy. \tag{5} $$

3 Reproducing Kernel Hilbert Spaces

In this section, we define the notion of a reproducing kernel Hilbert space, exhibit some particular instances of these spaces which will play roles in this work, and describe some well-known properties of these spaces. Convenient references for this section are [30, 33, 52].

Definition 4 Let $(H, \langle \cdot, \cdot \rangle)$ be a Hilbert space of real functions defined on a nonempty set E. A function $K : E \times E \to \mathbb{R}$ is called a *reproducing kernel* for H if and only if

(a) $K(\cdot, z) \in H$ for each fixed $z \in E$,
(b) $\langle \varphi, K(\cdot, z) \rangle = \varphi(z)$ for all $z \in E$ and all $\varphi \in H$.

We will refer to such a Hilbert space H for which there exists a reproducing kernel function K as a *reproducing kernel Hilbert space*.

Condition (b) is called “the reproducing property” of the kernel K because the value of an arbitrary function $\varphi \in H$ at an arbitrary point $z \in E$ is reproduced by the inner product of φ with $K(\cdot, z)$. For brevity we will freely use RKHS instead of the term reproducing kernel Hilbert space. Furthermore, we will usually adhere to the standard convention $K_z(\cdot) = K(\cdot, z)$ for denoting the reproducing kernel in such spaces.

We now list some RKHSs on the closed, bounded interval $[a, b]$. In what follows we will use the symbol $AC[a, b]$ to denote the vector space of real, absolutely continuous functions on the interval $[a, b]$ and $L^2[a, b]$ to denote the real, Lebesgue square integrable functions on [a, b] [27].

Definition 5 Let m be a positive integer. The space $W_2^m[a, b]$ consists of the functions $f : [a, b] \to \mathbb{R}$ such that $f^{(m-1)} \in AC[a, b]$ and $f^{(m)} \in L^2[a, b]$. Equip $W_2^m[a, b]$ with the inner product [33]

$$ \langle f, g \rangle_{W_2^m} = \sum_{i=0}^{m-1} f^{(i)}(a) g^{(i)}(a) + \int_{a}^{b} f^{(m)}(x) g^{(m)}(x) dx. $$

Lemma 1 ([33], p. 8) *If m is a positive integer then* $W_2^m[a,b]$ *is a RKHS.*

One particular instance is

$$W_2^1[0,1] = \{f \in AC[0,1] : \ f' \in L^2[0,1]\},$$

equipped with the inner product

$$\langle f, g\rangle_{W_2^1} = f(0)g(0) + \int_0^1 f'(x)g'(x)dx.$$

The reproducing kernel Q_y of $W_2^1[0,1]$ is given by ([33], pp. 10 and 17)

$$Q_y(x) = \begin{cases} 1+x, & 0 \le x \le y \le 1, \\ 1+y, & 0 \le y < x \le 1. \end{cases} \tag{6}$$

Definition 6 The space $V_2^2[0,a]$ is given by

$$V_2^2[0,a] = \{u \in AC[0,a] : \ u' \in AC[0,a], \ u'' \in L^2[0,a], \ u(0) = 0\},$$

$$\langle u, v\rangle_{V_2^2[0,a]} = u(0)v(0) + u'(0)v'(0) + \int_0^a u''(x)v''(x)\mathrm{d}x, \quad u, v \in \ V_2^2[0,a],$$

and

$$\|u\|_{V_2^2[0,a]} = \sqrt{\langle u, u\rangle_{V_2^2[0,a]}}, \quad u \in V_2^2[0,a],$$

are the inner product and the norm in $V_2^2[0,a]$ respectively.

Theorem 1 *Reproducing kernel function* $R_y(\xi)$ *of* $V_2^2[0,a]$ *is obtained as:*

$$\mathscr{R}_y(\xi) = \begin{cases} \xi y + \frac{1}{2}y\xi^2 - \frac{\xi^3}{6}, & 0 \le \xi \le y \le a, \\ \\ \xi y + \frac{1}{2}y^2\xi - \frac{y^3}{6}, & 0 \le y < \xi \le a. \end{cases} \tag{7}$$

Proof We have

$$\mathscr{R}_y(\xi) = \begin{cases} \sum_{i=1}^4 c_i(y)\xi^{i-1}, & 0 \le \xi \le y \le a, \\ \\ \sum_{i=1}^4 d_i(y)\xi^{i-1}, & 0 \le y < \xi \le a. \end{cases} \tag{8}$$

Then, we obtain

$$\begin{aligned}\langle u, \mathscr{R}_y(\xi)\rangle_{V_2^2[0,a]} &= u(0)\mathscr{R}_y(0) + u'(0)\mathscr{R}_y'(0) + \int_0^a u''(\xi)\frac{\partial''\mathscr{R}_y(\xi)}{\partial\xi''}\mathrm{d}\xi,\\ &= u'(a)\frac{\partial''\mathscr{R}_y(a)}{\partial\xi''} - u'(0)\frac{\partial''\mathscr{R}_y(0)}{\partial\xi''} - \int_0^a u'(\xi)\frac{\partial^3\mathscr{R}_y(\xi)}{\partial\xi^3}\mathrm{d}\xi.\end{aligned}$$

We get

$$\langle u, \mathscr{R}_y(\xi)\rangle_{V_2^2[0,a]} = u(y).$$

This completes the proof.

4 Main Results

The solution of (1) has been obtained in the reproducing kernel space $V_2^2[0, a]$ in this section. On defining the linear operator

$$A : V_2^2[0, a] \to W_2^1[0, 1],$$

as

$$Au(x) =_0^{ABC} Du(x), \tag{9}$$

problem (1) converts the form:

$$\begin{cases} Au(x) = H(x, u(x)), & x \in [0, a],\\ u'(0) = u_0. \end{cases} \tag{10}$$

Lemma 2 *The operator A is a bounded linear operator [39].*

Let us put $\varphi_i(\xi) = Q_{\xi_i}(\xi)$ and $\psi_i(\xi) = A^*\varphi_i(\xi)$, where A^* is conjugate operator of A. The orthonormal system $\{\hat{\psi}_i(\xi)\}_{i=1}^{\infty} \subseteq A_2^2[0, a]$ can be obtained from the well-known Gram-Schmidt orthogonalization process of $\{\psi_i(\xi)\}_{i=1}^{\infty}$:

$$\hat{\psi}_i(\xi) = \sum_{k=1}^{i} \beta_{ik}\psi_k(\xi),\ (\beta_{ii} > 0, i = 1, 2, ...). \tag{11}$$

Lemma 3 *Suppose $\{\xi_i\}_{i=1}^{\infty}$ be dense in $[0, a]$ and $\psi_i(\xi) = A_y\mathscr{R}_\xi(y)|_{y=\xi_i}$. Then the sequence $\{\psi_i(\xi)\}_{i=1}^{\infty}$ is a complete system in $V_2^2[0, a]$ [39].*

Theorem 2 *If $u(\xi)$ is the exact solution of* (1), *then*

$$u(\xi) = \sum_{i=1}^{\infty}\sum_{k=1}^{i} \beta_{ik} H(\xi_k, u_k)\hat{\psi}_i(\xi), \tag{12}$$

where $\{\xi_i\}_{i=1}^{\infty}$ is dense in $[0, a]$.

Proof We obtain

$$\begin{aligned}
u(\xi) &= \sum_{i=1}^{\infty} \left\langle u(\xi), \hat{\psi}_i(\xi) \right\rangle_{V_2^2} \hat{\psi}_i(\xi), \\
&= \sum_{i=1}^{\infty}\sum_{k=1}^{i} \beta_{ik} \left\langle u(\xi), \psi_k(\xi) \right\rangle_{V_2^2} \hat{\psi}_i(\xi), \\
&= \sum_{i=1}^{\infty}\sum_{k=1}^{i} \beta_{ik} \left\langle u(\xi), A^*\varphi_k(\xi) \right\rangle_{V_2^2} \hat{\psi}_i(\xi), \\
&= \sum_{i=1}^{\infty}\sum_{k=1}^{i} \beta_{ik} \left\langle Au(\xi), \varphi_k(\xi) \right\rangle_{W_2^1} \hat{\psi}_i(\xi), \\
&= \sum_{i=1}^{\infty}\sum_{k=1}^{i} \beta_{ik} \left\langle H(\xi, u), Q_{\xi_k} \right\rangle_{W_2^1} \hat{\psi}_i(\xi), \\
&= \sum_{i=1}^{\infty}\sum_{k=1}^{i} \beta_{ik} H(\xi_k, u_k)\hat{\psi}_i(\xi).
\end{aligned}$$

Finite terms of this concludes the approximate solution:

$$u_n(\xi) = \sum_{i=1}^{n}\sum_{k=1}^{i} \beta_{ik} H(\xi_k, u_k)\hat{\psi}_i(\xi). \tag{13}$$

5 Applications

Example 1 We take into consideration the following fractional differential equation, where the fractional derivative is that of Atangana–Baleanu in the Caputo sense:

$$\begin{cases} {}_0^{ABC}D^{\alpha} u(x) = x^2, \\ u(0) = 0. \end{cases} \tag{14}$$

Table 1 Absolute errors in Example 1 using the $V_2^2[0,4]$ kernel and different points near the origin in [0, 4]

x $(\alpha = 0.75)$	$m = 36$	$m = 64$	$m = 81$
0.0	0.000000000	0.000000000	0.000000000
0.5	3.907×10^{-7}	9.300×10^{-9}	5.300×10^{-9}
1.0	9.373×10^{-7}	5.270×10^{-8}	1.270×10^{-8}
1.5	1.922×10^{-6}	7.800×10^{-8}	1.800×10^{-8}
2.0	3.085×10^{-6}	2.150×10^{-7}	1.500×10^{-8}
2.5	5.289×10^{-6}	1.100×10^{-8}	3.100×10^{-8}
3.0	8.980×10^{-6}	3.800×10^{-7}	4.000×10^{-8}
3.5	1.138×10^{-5}	5.800×10^{-7}	5.000×10^{-8}
4.0	1.963×10^{-5}	1.430×10^{-6}	4.000×10^{-8}

The exact solution of Eq. (1) is obtained by applying the Atangana–Baleanu fractional integral on both sides to obtain [26].

$$u(x) = \frac{1-\alpha}{ABC(\alpha)}x^2 + \frac{1}{\Gamma(\alpha)\times ABC(\alpha)}\frac{2x^{\alpha+2}}{(\alpha^2+3\alpha+2)},$$
$$ABC(\alpha) = 1 - \alpha + \frac{\alpha}{\Gamma(\alpha)}. \tag{15}$$

Applying the reproducing kernel method, we demonstrate the numerical solution for $\alpha = \frac{3}{4}$. The comparison of exact solution and numerical solution is given in Tables 1 and 2. The numerical representation let us conclude that the reproducing kernel method is highly accurate and converges very quickly to the exact solution. The numerical method does not depend on the fractional order. For all values of α, we get good agreement between exact solution and approximate solution. More importantly, the results showed that the convergence does not depend on the fractional order.

Example 2 Let us consider the following oscillatory fractional differential equation. Here the fractional derivative is also in Atangana–Baleanu in the Caputo sense:

$$\begin{cases} {}_0^{ABC}D^{\alpha}u(x) = \sin(x), \\ u(0) = 0. \end{cases} \tag{16}$$

By applying, on both sides, the Atangana–Baleanu fractional integral, we acquire the exact solution with LommelS1 as a special function [26]. We give the absolute errors of Example 1 for different values of α in Table 2. Table 3 shows the numerical simulation obtained via the new method for Example 2 for different values of α.

Table 2 Absolute errors in Example 1 using the $V_2^2[0, 4]$ kernel and 100 points near the origin in $[0, 4]$ for different values of α

x	$\alpha = 0.8$	$\alpha = 0.95$	$\alpha = 0.3$	$\alpha = 0.5$
0.0	0.000000000	0.000000000	0.000000000	0.000000000
0.5	4.800×10^{-9}	1.217×10^{-8}	4.000×10^{-9}	5.400×10^{-9}
1.0	1.200×10^{-8}	1.280×10^{-8}	4.000×10^{-9}	1.700×10^{-8}
1.5	1.500×10^{-8}	7.000×10^{-9}	3.000×10^{-9}	5.000×10^{-9}
2.0	9.000×10^{-9}	1.900×10^{-8}	1.000×10^{-8}	3.000×10^{-8}
2.5	3.300×10^{-8}	2.400×10^{-8}	3.000×10^{-8}	5.300×10^{-8}
3.0	4.000×10^{-8}	3.700×10^{-8}	4.000×10^{-8}	4.000×10^{-8}
3.5	6.000×10^{-8}	4.000×10^{-8}	4.000×10^{-8}	6.000×10^{-8}
4.0	5.000×10^{-8}	4.000×10^{-8}	4.000×10^{-8}	5.000×10^{-8}

Table 3 Absolute errors in Example 2 using the $V_2^2[0, 10]$ kernel and 100 points near the origin in $[0, 10]$ for different values of α

x	$\alpha = 0.3$	$\alpha = 0.5$	$\alpha = 0.7$	$\alpha = 0.8$
0.0	0.000000000	0.000000000	0.000000000	0.000000000
1.0	1.700×10^{-8}	5.590×10^{-8}	1.089×10^{-7}	1.007×10^{-7}
2.0	6.300×10^{-8}	1.070×10^{-7}	2.460×10^{-7}	2.310×10^{-7}
3.0	6.760×10^{-8}	1.507×10^{-7}	3.870×10^{-7}	4.110×10^{-7}
4.0	2.390×10^{-8}	1.826×10^{-7}	5.578×10^{-7}	5.731×10^{-7}
5.0	1.300×10^{-8}	4.010×10^{-7}	7.808×10^{-7}	7.525×10^{-7}
6.0	3.780×10^{-8}	4.279×10^{-7}	1.050×10^{-6}	1.093×10^{-6}
7.0	1.580×10^{-7}	3.179×10^{-7}	1.232×10^{-6}	1.283×10^{-6}
8.0	5.160×10^{-7}	1.001×10^{-6}	1.833×10^{-6}	1.951×10^{-6}
9.0	8.392×10^{-7}	9.998×10^{-7}	1.637×10^{-6}	1.572×10^{-6}
10.0	3.275×10^{-7}	3.950×10^{-7}	1.219×10^{-6}	1.884×10^{-7}

After applying the RKM with the reproducing kernel for the space $V_2^2[0, a]$ and an enumeration (x_i, t_i) of the 36, 64, 81, 100 points of

$$\left\{\frac{1}{10}, \frac{1}{5}, \frac{3}{10}, \frac{2}{5}, \frac{1}{2}, \frac{3}{5}, \frac{7}{10}, \frac{4}{5}, \frac{9}{10}, 1\right\}^2,$$

we obtain Tables 1, 2 and 3.

6 Conclusions

Atangana and Baleanu introduced a new kernel based upon the generalized Mittag-Leffler function. In this paper, we applied the reproducing kernel method to the fractional differential equations with novel fractional derivatives. The numerical simulations were performed for different values of fractional order. Numerical examples were given to illustrate the computational effectiveness of the reproducing kernel method.

References

1. Caputo, M., Fabrizio, M.: A new definition of fractional derivative without singular kernel. Prog. Fract. Differ. Appl. **1**, 73–85 (2015)
2. Losada, J., Nieto, J.J.: Properties of a new fractional derivative without singular kernel. Prog. Fract. Differ. Appl. **1**, 87–92 (2015)
3. Atangana, A.: On the new fractional derivative and application to nonlinear Fisher's reaction-diffusion equation. Appl. Math. Comput. **1**(273), 948–956 (2016)
4. Morales-Delgado, V.F., Taneco-Hernández, M.A., Gómez-Aguilar, J.F.: On the solutions of fractional order of evolution equations. Eur. Phys. J. Plus **132**(1), 1–17 (2017)
5. Hristov, J.: Transient heat diffusion with a non-singular fading memory: from the Cattaneo constitutive equation with Jeffrey's kernel to the Caputo-Fabrizio time-fractional derivative. Therm. Sci. **20**(2), 757–762 (2016)
6. Yépez-Martínez, H., Gómez-Aguilar, J.F.: A new modified definition of Caputo-Fabrizio fractional-order derivative and their applications to the multi step homotopy analysis method (MHAM). J. Comput. Appl. Math. **346**, 247–260 (2019)
7. Gómez-Aguilar, J.F., López-López, M.G., Alvarado-Martínez, V.M., Baleanu, D., Khan, H.: Chaos in a cancer model via fractional derivatives with exponential decay and Mittag-Leffler law. Entropy **19**(12), 1–21 (2017)
8. Doungmo Goufo, E.F., Pene, M.K., Jeanine, N.: Duplication in a model of rock fracture with fractional derivative without singular kernel. Open Math. **13**, 839–846 (2015)
9. Gómez-Aguilar, J.F., Escobar-Jiménez, R.F., López-López, M.G., Alvarado-Martínez, V.M.: Atangana-Baleanu fractional derivative applied to electromagnetic waves in dielectric media. J. Electromagn. Waves Appl. **30**(15), 1937–1952 (2016)
10. Brzezinski, D.W.: Accuracy problems of numerical calculation of fractional order derivatives and integrals applying the Riemann-Liouville/Caputo formulas. Appl. Math. Nonlinear Sci. **1**, 23–43 (2016)
11. Jiang, J., Cao, D., Chen, H.: Boundary value problems for fractional differential equation with causal operators. Appl. Math. Nonlinear Sci. **1**, 11–22 (2016)
12. Kumar, S.: A new analytical modelling for telegraph equation via laplace transform. Appl. Math. Model **38**(13), 3154–63 (2014)
13. Coronel-Escamilla, A., Gómez-Aguilar, J.F., Alvarado-Méndez, E., Guerrero-Ramírez, G.V., Escobar-Jiménez, R.F.: Fractional dynamics of charged particles in magnetic fields. Int. J. Mod. Phys. C **27**(08), 1–16 (2016)
14. Gómez-Aguilar, J.F., Yépez-Martínez, H., Escobar-Jiménez, R.F., Astorga-Zaragoza, C.M., Morales-Mendoza, L.J., González-Lee, M.: Universal character of the fractional space-time electromagnetic waves in dielectric media. J. Electromagn. Waves Appl. **29**(6), 727–740 (2015)
15. Kumar, S., Rashidi, M.M.: New analytical method for gas dynamics equation arising in shock fronts. Comput. Phys. Commun. **185**(7), 1947–1954 (2014)

16. Kumar, S., Yao, J.J., Kumar, A.: A fractioanal model to describing the Brownian motion of particles and its analytical solution. Adv. Mech. Eng. **7**(12), 1–11 (2015)
17. Kumar, S., Yin, X.B., Kumar, D.: A modified homotopy analysis method for solution of fractional wave equations. Adv. Mech. Eng. **7**(12), 1–8 (2015)
18. Caputo, M., Fabrizio, M.: Applications of new time and spatial fractional derivatives with exponential kernels. Prog. Fract. Differ. Appl. **2**, 1–11 (2016)
19. Alsaedi, A., Baleanu, D., Etemad, S., Rezapour, S.: On coupled systems of time-fractional differential problems by using a new fractional derivative. J. Funct. Spaces **1**, 1–8 (2016)
20. Gómez-Aguilar, J.F.: Behavior characteristics of a cap-resistor, memcapacitor, and a memristor from the response obtained of RC and RL electrical circuits described by fractional differential equations. Turk. J. Electr. Eng. Comput. Sci. **24**(3), 1–16 (2016)
21. Atangana, A., Baleanu, D.: Caputo-Fabrizio derivative applied to groundwater flow within a confined aquifer. J. Eng. Mech. **1**, 1–16 (2016)
22. Atangana, A., Baleanu, D.: New fractional derivatives with nonlocal and non-singular kernel: theory and application to heat transfer model. Therm. Sci. **18**, 1–10 (2016)
23. Coronel-Escamilla, A., Gómez-Aguilar, J.F., Baleanu, D., Córdova-Fraga, T., Escobar-Jiménez, R.F., Olivares-Peregrino, V.H., Qurashi, M.M.A.: Bateman-Feshbach tikochinsky and Caldirola-Kanai oscillators with new fractional differentiation. Entropy **19**(2), 1–21 (2017)
24. Atangana, A., Gómez-Aguilar, J.F.: Decolonisation of fractional calculus rules: breaking commutativity and associativity to capture more natural phenomena. Eur. Phys. J. Plus **133**, 1–22 (2018)
25. Atangana, A., Koca, I.: Chaos in a simple nonlinear system with Atangana-Baleanu derivatives with fractional order. Chaos Solitons Fractals **89**, 447–454 (2016)
26. Toufik, M., Atangana, A.: New numerical approximation of fractional derivative with non-local and non-singular kernel: application to chaotic models. Eur. Phys. J. Plus **132**, 1–14 (2017)
27. Akgül, A., Grow, D.: Existence of solutions to the telegraph equation in binary reproducing kernel Hilbert spaces (2017)
28. Zaremba, S.: L'équation biharmonique et une classe remarquable de fonctions fondamentales harmoniques. Bulletin International l'Académia des Sciences de Cracovie **1**, 147–196 (1907)
29. Zaremba, S.: Sur le calcul numérique des fonctions demandées dan le probléme de dirichlet et le probleme hydrodynamique. Bulletin International l'Académia des Sciences de Cracovie **1**, 125–195 (1908)
30. Aronszajn, N.: Theory of reproducing kernels. Trans. Am. Math. Soc. **68**, 337–404 (1950)
31. Bergman, S.: The Kernel Function and Conformal Mapping. American Mathematical Society, New York (1950)
32. Cui, M., Zhongxing, D.: On the best operator of interpolation. Math. Numer. Sin. **8**, 209–216 (1986)
33. Cui, M., Yingzhen, L.: Nonlinear Numerical Analysis in the Reproducing Kernel Space. Nova Science Publishers Inc., New York (2009)
34. Mustafa, I., Akgül, A., Kilicman, A.: On solving KdV equation using reproducing kernel Hilbert space method. Abstr. Appl. Anal. **1**, 1–11 (2013)
35. Wang, Y.-L., Chao, L.: Using reproducing kernel for solving a class of partial differential equation with variable coefficients. Appl. Math. Mech. **29**, 129–137 (2008)
36. Wu, B.Y., Li, X.Y.: A new algorithm for a class of linear nonlocal boundary value problems based on the reproducing kernel method. Appl. Math. Lett. **24**, 156–159 (2011)
37. Huanmin, Y., Lin, Y.: Solving singular boundary value problems of higher even order. J. Comput. Appl. Math. **223**, 703–713 (2009)
38. Akgül, A., Mustafa, I., Esra, K., Baleanu, D.: Numerical solutions of fractional differential equations of lane-emden type by an accurate technique. Adv. Differ. Equ.S **1**, 1–20 (2015)
39. Geng, F., Minggen, C.: A reproducing kernel method for solving nonlocal fractional boundary value problems. Appl. Math. Lett. **25**, 818–823 (2012)
40. Wu, B.Y., Li, X.Y.: Iterative reproducing kernel method for nonlinear oscillator with discontinuity. Appl. Math. Lett. **23**, 1301–1304 (2010)

41. Geng, F., Minggen, C.: Solving a nonlinear system of second order boundary value problems. J. Math. Anal. Appl. **327**, 1167–1181 (2007)
42. Mustafa, I., Akgül, A.: Approximate solutions for MHD squeezing fluid flow by a novel method. Bound. Value Probl. **1**, 1–18 (2014)
43. Mustafa, I., Akgül, A., Geng, F.: Reproducing kernel Hilbert space method for solving Bratu's problem. Bull. Malays. Math. Sci. Soc. **38**, 271–287 (2015)
44. Akgül, A., Mustafa, I., Esra, K.: Reproducing kernel functions for difference equations. Discret. Contin. Dyn. Syst. Ser. S **8**, 1055–1064 (2015)
45. Geng, F.: Solving integral equations of the third kind in the reproducing kernel space. Bull. Iran. Math. Soc. **38**, 543–551 (2012)
46. Lorenz, E.N.: Deterministic non-periodic flow. J. Atmos. Sci. **20**(2), 130–141 (1963)
47. Ivancevic Vladimir, G., Tijana, T.I.: Complex Nonlinearity: Chaos, Phase Transitions, Topology Change, and Path Integrals. Springer, Berlin (2008)
48. Safonov Leonid, A., Tomer, E., Strygin Vadim, V., Ashkenazy, Y., Havlin, S.: Multifractal chaotic attractors in a system of delay-differential equations modeling road traffic. Chaos **12**, 1–11 (2002)
49. Vellekoop, M., Berglund, R.: On intervals, transitivity = chaos. Am. Math. Mon. **101**(4), 353–355 (1994)
50. Mustafa, I., Akgül, A.: The reproducing kernel Hilbert space method for solving Troesch's problem. J. Assoc. Arab. Univ. Basic Appl. Sci. **14**, 19–27 (2013)
51. Mustafa, I., Akgül, A., Kilicman, A.: Numerical solutions of the second order one-dimensional telegraph equation based on reproducing kernel Hilbert space method. Abstr. Appl. Anal. **1**, 1–13 (2013)
52. Šremr, J.: Absolutely continuous functions of two variables in the sense of Carathéodory. Electron. J. Differ. Equ. **1**, 1–11 (2010)

Necessary and Sufficient Optimality Conditions for Fractional Problems Involving Atangana–Baleanu's Derivatives

G. M. Bahaa and A. Atangana

Abstract Recently, Atangana and Baleanu proposed a derivative with fractional order to answer some outstanding questions that were posed by many researchers within the field of fractional calculus. Their derivative has a non-singular and nonlocal kernel. In this chapter, the necessary and sufficient optimality conditions for systems involving Atangana–Baleanu's derivatives are discussed. The fractional Euler–Lagrange equations of fractional Lagrangians for constrained systems that contains a fractional Atangana–Baleanu's derivatives are investigated. The fractional contains both the fractional derivatives and the fractional integrals in the sense of Atangana–Baleanu. We present a general formulation and a solution scheme for a class of Fractional Optimal Control Problems (FOCPs) for those systems. The calculus of variations, the Lagrange multiplier, and the formula for fractional integration by parts are used to obtain Euler–Lagrange equations for the FOCP.

Keywords Fractional calculus · Atangana–Baleanu fractional derivative · Fractional optimal control problems

G. M. Bahaa (✉)
Department of Mathematics and Computer Science, Faculty of Science, Beni-Suef University, Beni-Suef, Egypt
e-mail: Bahaa_gm@yahoo.com

G. M. Bahaa
Department of Mathematics, Faculty of Science, Taibah University, Al-Madinah Al-Munawarah, Saudi Arabia

A. Atangana
Faculty of Natural and Agricultural Sciences, Institute of Groundwater Studies, University of Free State, 9301 Bloemfontein, South Africa

J. F. Gómez et al. (eds.), *Fractional Derivatives with Mittag-Leffler Kernel*, Studies in Systems, Decision and Control 194,
https://doi.org/10.1007/978-3-030-11662-0_2

1 Introduction

In recent years, numerous works have been dedicated to the fractional calculus of variations. Most of them deal with Riemann–Liouville fractional derivatives (see [2–5, 15, 16, 23, 33–35, 37, 39], and the references therein) and a few with Caputo or Riesz derivatives [7–14, 24]. Depending on the type of the functional being considered, different fractional Euler–Lagrange type equations are obtained.

Here we propose a new kind of Atangana–Baleanu's fractional derivative, Atangana–Baleanu's fractional integrals [6, 17–22, 25–32, 36, 38, 42, 43] and the functional with a Lagrangian that containing not only Atangana–Baleanu's fractional derivative (ABFD) but also a Atangana–Baleanu's fractional integral (ABFI). We prove the necessary conditions of Euler–Lagrange type for the fundamental fractional problem of the calculus of variations and for the fractional isoperimetric problem. Sufficient optimality conditions are also obtained under appropriate convexity assumptions.

Fractional calculus represents a generalization of ordinary differentiation and integration to arbitrary order [1–3]. During the last decades the fractional calculus started to be used in various fields, e.g. physics, engineering, biology, and many important results were obtained [32, 33].

Lagrangian mechanics and Hamiltonian mechanics are alternative formulations of classical Newtonian mechanics. Their importance is represented by the fact that any of them could be used to solve a problem in classical mechanics. We emphasize that the Newtonian mechanics requires the concept of force, while Lagrangian and Hamiltonian systems are expressed in terms of energy.

The first attempt to find the fractional Lagrangian and Hamiltonian for a given dissipative system is due to Riewe [40, 41]. Important contributions were obtained in the field of variational principles by Agrawal [2–5], Baleanu and Muslih in [16], Baleanu and Avkar in [15], and Baleanu and Agrawal in [14]. Agrawal and Baleanu presented a Hamiltonian formulation and a direct numerical scheme for fractional optimal control problems [4].

For these reasons, the properties of the Lagrangian and Hamiltonian formalism of the fractional extension of constrained systems are considered in this study. Within the classical picture, using only the integer order time derivative, the canonical coordinates of such systems do not remain linearly independent and certain constraints appear among them.

In this study, by using the Atangana–Baleanu's derivative we propose to generalize the notion of equivalent Lagrangian for the fractional case. For a given classical Lagrangian, there are several proposed methods to obtain the fractional Euler–Lagrange equations and the corresponding Hamiltonians. However, the fractional dynamics depends on the fractional derivatives used to construct the Lagrangian to start with; therefore, the existence of several options can be used to treat a specific physical system. In this respect, application of the Atangana–Baleanu's derivative to the fractional dynamics will bring new opportunities in studying the constrained systems mainly because the Atangana–Baleanu's derivative contains both the left and the right derivatives. In addition, the fractional derivative of a function is given by a definite integral, which depend on the values of the function over the entire interval. Therefore, the fractional derivatives are suitable to model systems with long range interactions in space and/or time (memory) and process with many scales of space and/or time involved.

In this chapter we compare the results of fractional Euler–Lagrange equations for the classical Riemann–Liouville and Caputo fractional derivatives stated in [2–5] corresponding to Atangana–Baleanu's fractional derivatives defined in [6].

The plan of this chapter is as follows: In Sect. 2 we collect notations, definitions, and we state the integration by parts involving the Atangana–Baleanu fractional time derivative. Section 3 contains a briefly review of the fractional Lagrangian and Hamiltonian approaches of discrete systems based on fractional Atangana–Baleanu's derivatives and some examples are investigated in detail. In Sect. 4, we calculate the fractional canonical momenta, and we generalize the fractional Lagrangian to n state equations. In Sect. 5 some generalizations for the problem are stated. In Sect. 6, the constrained system within Atangana–Baleanu's derivatives are also discussed and some examples are investigated in detail. In Sect. 7, we discuss the sufficient conditions for optimality. In Sect. 8, Fractional Optimal Control Problem FOCP involving Atangana–Baleanu's derivatives are presented. We state two FOCP, one for time invariant FOCP and the other for time variant FOCP. Section 9 is dedicated to our conclusions.

2 Preliminaries

This section presents the basic definitions and properties of the new Atangana–Baleanu's derivatives in the Caputo and Riemann–Liouville sense. These include the classical Riemann–Liouville, Grünwald–Letnikov, Weyl, Caputo, Marchaud, and Riesz fractional derivatives (see [1] and the references therein). We will formulate the problem in terms of the left and right Atangana–Baleanu's derivatives in the Riemann–Liouville and Caputo senses which will be given later. We also define the classical Riemann–Liouville and Caputo derivatives as they are linked to Atangana–Baleanu's derivatives. We begin with the left and the right classical Riemann–Liouville fractional integrals of order $\alpha > 0$ of a function $x(t)$ which are defined as [6].

Definition 2.1 Let $x : [a, b] \to R$ be a continuous function on $[a, b]$ and $\alpha > 0$ be a real number, and $n = [\alpha]$, where $[\alpha]$ denotes the smallest integer greater than or equal to α. The left (left RLFI) and the right (right RLFI) Riemann–Liouville fractional integrals of order α are defined by

$$ {}_aI_t^{\alpha}x(t) = \frac{1}{\Gamma(\alpha)}\int_a^t (t-s)^{\alpha-1}x(s)ds \ \text{ (left RLFI)}, \tag{1} $$

$$
{}_t I_b^\alpha x(t) = \frac{1}{\Gamma(\alpha)} \int_t^b (s-t)^{\alpha-1} x(s) ds \text{ (right RLFI)}, \tag{2}
$$

where

$$
\Gamma(\alpha) = \int_0^\infty e^{-t} u^{\alpha-1} du, \; {}_a I_t^0 x(t) =_t I_b^0 x(t) = x(t). \tag{3}
$$

In the case of $\alpha = 1$, the fractional integral reduces to the classical integral.

The left (left RLFD) and the right (right RLFD) Riemann–Liouville fractional derivatives of order α are defined by

$$
{}_a D_t^\alpha x(t) = \frac{1}{\Gamma(n-\alpha)} \frac{d^n}{dt^n} \int_a^t (t-s)^{n-\alpha-1} x(s) ds \text{ (left RLFD)}, \tag{4}
$$

$$
{}_t D_b^\alpha x(t) = \frac{(-1)^n}{\Gamma(n-\alpha)} \frac{d^n}{dt^n} \int_t^b (s-t)^{n-\alpha-1} x(s) ds \text{ (right RLFD)}, \tag{5}
$$

where $\alpha \in (n-1, n), n \in N$.

Moreover, the left (left CFD) and the right (right CFD) Caputo fractional derivatives of order α are defined by

$$
{}_a^C D_t^\alpha x(t) = \frac{1}{\Gamma(n-\alpha)} \int_a^t (t-s)^{n-\alpha-1} x^{(n)}(s) ds \text{ (left CFD)}, \tag{6}
$$

provided that the integral is defined.

$$
{}_t^C D_b^\alpha x(t) = \frac{(-1)^n}{\Gamma(n-\alpha)} \int_t^b (s-t)^{n-\alpha-1} x^{(n)}(s) ds \text{ (right CFD)}, \tag{7}
$$

provided that the integral is defined.

The relation between the right RLFD and the right CFD is as follows:

$$
{}_t^C D_b^\alpha x(t) = {}_t D_b^\alpha x(t) - \sum_{k=0}^{n-1} \frac{x^{(k)}(b)}{\Gamma(k-\alpha+1)} (b-t)^{(k-\alpha)}. \tag{8}
$$

If x and $x^{(i)}, i = 1, \ldots, n-1$, vanish at $t = a$, then ${}_a D_t^\alpha x(t) = {}_a^C D_t^\alpha x(t)$, and if they vanish at $t = b$, then ${}_t D_b^\alpha x(t) = {}_t^C D_b^\alpha x(t)$.

Further, it holds

$$
{}_0^C D_t^\alpha c = 0, \text{ where c is a constant}, \tag{9}
$$

and

$$
{}_0^C D_t^\alpha t^n = \begin{cases} 0, & \text{for } n \in \mathbb{N}_0 \text{ and } n < [\alpha]; \\ \frac{\Gamma(n+1)}{\Gamma(n+1-\alpha)} t^{n-\alpha}, & \text{for } n \in \mathbb{N}_0 \text{ and } n \geq [\alpha], \end{cases} \tag{10}
$$

where $\mathbb{N}_0 = 0, 1, 2, \ldots$. We recall that for $\alpha \in \mathbb{N}$ the Caputo differential operator coincides with the usual differential operator of integer order.

Lemma 2.1 *Let* $T > 0, u \in C^m([0, T]), p \in (m-1, m), m \in N$ *and* $v \in C^1([0, T])$. *Then for* $t \in [0, T]$, *the following properties hold*

$$ {}_aD_t^p v(t) = \frac{d}{dt}\, {}_aI_t^{1-p} v(t), \quad m = 1, \tag{11}$$

$$ {}_aD_t^p\, {}_aI_t^p v(t) = v(t); \tag{12}$$

$$ {}_0I_t^p\, {}_0D_t^p u(t) = u(t) - \sum_{k=0}^{m-1} \frac{t^k}{k!} u^{(k)}(0); \tag{13}$$

$$ \lim_{t\to 0^+} {}_0^CD_t^p u(t) = \lim_{t\to 0^+} {}_0I_t^p u(t) = 0. \tag{14}$$

Note also that when $T = +\infty$, ${}_0^CD_t^\alpha f(t)$ *is the Weyl fractional integral of order* α *of* f.

An important tool is the integration by parts formula for Caputo fractional derivatives, which is stated in the following lemma.

Lemma 2.2 *Let* $\alpha \in (0, 1)$, *and* $x, y : [a, b] \to R$ *be two functions of class* C^1. *Then the following integration by parts formula holds:*

$$ \int_a^b y(t)\, {}_a^CD_t x(t) dt = \left[{}_tI_b^{1-\alpha} y(t) x(t) \right]_a^b + \int_a^b x(t)\, {}_tD_b^\alpha y(t) dt. \tag{15}$$

Let us recall some useful definitions of fractional derivatives in the sense of Atangana–Baleanu [6].

Definition 2.2 For a given function $x(t) \in H^1(a, b), b > a, \alpha \in (0, 1)$, the left Atangana–Baleanu fractional derivative (AB derivative) of $x(t)$ of order α in Caputo sense ${}^{ABC}_{\ a}D_t^\alpha x(t)$ (where A denotes Atangana, B denotes Baleanu and C denotes Caputo type) with base point a is defined at a point $t \in (a, b)$ by

$$ {}^{ABC}_{\ a}D_t^\alpha x(t) = \frac{B(\alpha)}{1-\alpha} \int_a^t \frac{d}{ds} x(s) E_\alpha[-\gamma(t-s)^\alpha] ds, \quad \text{(left ABCD)}, \tag{16}$$

where $\gamma = \frac{\alpha}{(1-\alpha)}$, and $B(\alpha)$ being a normalization function satisfying

$$ B(\alpha) = (1-\alpha) + \frac{\alpha}{\Gamma(\alpha)}, \tag{17}$$

where $B(0) = B(1) = 1$, $E_\alpha(.)$ stands for the Mittag-Leffler function defined by

$$E_{\alpha,\beta}(z)=\sum_{k=0}^{\infty}\frac{z^k}{\Gamma(k\alpha+\beta)},\quad E_{\alpha,1}(z)=E_\alpha(z)\quad z\in\mathscr{C}, \tag{18}$$

which is an entire function on the complex plane and $\Gamma(.)$ denotes the Euler's gamma function defined as

$$\Gamma(z)=\int_0^\infty t^{z-1}e^{-t}dt,\quad \mathscr{R}(z)>0.$$

The Mittag-Leffler function $E_{\alpha,\beta}(z)$ is a two-parameter family of entire functions of z of order α. The exponential function is a particular case of the Mittag-Leffler function, namely

$$E_{1,1}(z)=e^z,\quad E_{2,1}(z)=\cosh\sqrt{z},\quad E_{1,2}(z)=\frac{e^z-1}{z},\quad E_{2,2}(z)=\frac{\sinh\sqrt{z}}{\sqrt{z}}.$$

The left Atangana–Baleanu fractional derivative in Riemann–Liouville sense defined with:

$${}^{ABR}_{\ a}D_t^\alpha x(t)=\frac{B(\alpha)}{1-\alpha}\frac{d}{dt}\int_a^t x(s)E_\alpha[-\gamma(t-s)^\alpha]ds,\qquad \text{(left ABRD)}. \tag{19}$$

For $\alpha=1$ in (19) we consider the usual classical derivative ∂_t.

The associated left AB fractional integral is also defined as

$$\begin{aligned}{}^{AB}_{\ a}I_t^\alpha x(t)&=\frac{1-\alpha}{B(\alpha)}x(t)+\frac{\alpha}{B(\alpha)\Gamma(\alpha)}\int_a^t x(s)(t-s)^{\alpha-1}ds,\qquad \text{(left ABI)},\\&=\frac{1-\alpha}{B(\alpha)}x(t)+\frac{\alpha}{B(\alpha)}\,{}_aI_t^\alpha x(t).\end{aligned} \tag{20}$$

Notice that if $\alpha=0$ in Eq. (20) we recover the initial function and if $\alpha=1$ in Eq. (20) we consider the usual ordinary integral.

The left AB Caputo fractional derivatives and the left AB Riemann–Liouville derivative are related by the identity:

$${}^{ABC}_{\ 0}D_t^\alpha x(t)={}^{ABR}_{\ 0}D_t^\alpha x(t)-\frac{B(\alpha)}{1-\alpha}x(0)E_\alpha[-\gamma(t)^\alpha]. \tag{21}$$

Some recent results and properties concerning this operator can be found in [6] and papers therein. So from the definition in [6] we recall the following definition

Definition 2.3 For a given function $x(t)\in H^1(a,b)$, $b>t>a$, the right Atangana–Baleanu fractional derivative of $x(t)$ of order α in Caputo sense with base point T is defined at a point $t\in(a,T)$ by

$${}^{ABC}_{\ t}D_b^\alpha x(t)=-\frac{B(\alpha)}{1-\alpha}\int_t^b E_\alpha\left[-\gamma(s-t)^\alpha\frac{d}{ds}x(s)\right]ds,\qquad \text{(right ABCD)}, \tag{22}$$

and in Riemann–Liouville sense with:

$$ {}^{ABR}_{\ \ t}D^{\alpha}_{b}x(t) = -\frac{B(\alpha)}{1-\alpha}\frac{d}{dt}\int_t^b x(s)E_{\alpha}[-\gamma(s-t)^{\alpha}]ds, \qquad \text{(right ABRD)}. \quad (23) $$

The associated right AB fractional integral is also defined as

$$ \begin{aligned} {}^{AB}_{\ t}I^{\alpha}_{b}x(t) &= \frac{1-\alpha}{B(\alpha)}x(t) + \frac{\alpha}{B(\alpha)\Gamma(\alpha)}\int_t^b x(s)(s-t)^{\alpha-1}ds, \qquad \text{(right ABI)}, \\ &= \frac{1-\alpha}{B(\alpha)}x(t) + \frac{\alpha}{B(\alpha)}\,{}_bI^{\alpha}_{t}x(t). \end{aligned} \quad (24) $$

The right AB Caputo fractional derivatives and the right AB Riemann–Liouville derivative are related by the identity:

$$ {}^{ABC}_{\ \ t}D^{\alpha}_{b}x(t) = {}^{ABR}_{\ \ t}D^{\alpha}_{b}x(t) - \frac{B(\alpha)}{1-\alpha}x(b)E_{\alpha}[-\gamma(b-t)^{\alpha}]. \quad (25) $$

There are useful relations between the left and right AB FDs in the Riemann–Liouville and Caputo senses and the associated AB fractional integrals as the following formulas state.

$$ {}^{AB}_{\ a}I^{\alpha}_{t}\{{}^{ABR}_{\ \ a}D^{\alpha}_{t}x(t)\} = {}^{AB}_{\ t}I^{\alpha}_{b}\{{}^{ABR}_{\ \ t}D^{\alpha}_{b}x(t)\} = x(t), \quad (26) $$

$$ {}^{AB}_{\ a}I^{\alpha}_{t}\{{}^{ABC}_{\ \ a}D^{\alpha}_{t}x(t)\} = x(t) - x(a), \quad (27) $$

$$ {}^{AB}_{\ t}I^{\alpha}_{b}\{{}^{ABC}_{\ \ t}D^{\alpha}_{b}u(t)\} = x(t) - x(b), \quad (28) $$

$$ {}^{ABC}_{\ \ b}D^{\alpha}_{t}x(t) = -\,{}^{ABC}_{\ \ 0}D^{\alpha}_{t}x(t). \quad (29) $$

As a consequence, the backwards in time with the fractional-time derivative with nonsingular Mittag-Leffler kernel at the based point T is equivalently written as a forward in time operator with the fractional-time derivative with nonsingular Mittag-Leffler kernel ${}^{ABC}_{\ \ 0}D^{\alpha}_{t}$.

Note also that when $T \to +\infty$(Large enough), ${}^{ABC}_{\ \ 0}D^{\alpha}_{t}f(t)$ is the Caputo fractional derivative of order α of f and ${}^{ABR}_{\ \ 0}D^{\alpha}_{t}f(t)$ is the Riemann–Liouville derivative of order α of f.

Next we state the following proposition which gives the integration by parts [1].

Proposition 2.3 (Integration by parts) (see [1]) *Let* $\alpha > 0$, $p \geq 1$, $q \geq 1$, *and* $\frac{1}{p} + \frac{1}{q} \leq 1 + \alpha (p \neq 1$ *and* $q \neq 1$ *in the case* $\frac{1}{p} + \frac{1}{q} = 1 + \alpha)$. *Then for any* $\phi(x) \in L^p(a, b)$, $\psi(x) \in L^q(a, b)$, *we have*

$$\int_a^b \phi(x)\, {}^{ABC}_{\ a}D_t^\alpha \psi(x)dx = \int_a^b \psi(x)\, {}^{ABR}_{\ \ t}D_b^\alpha \phi(x)dx + \frac{B(\alpha)}{1-\alpha}\psi(t)E^1_{\alpha,1,\frac{-\alpha}{1-\alpha},b}\phi(t)|_a^b, \tag{30}$$

$$\int_a^b \phi(x)\, {}^{ABC}_{\ \ t}D_b^\alpha \psi(x)dx = \int_a^b \psi(x)\, {}^{ABR}_{\ \ a}D_t^\alpha \phi(x)dx - \frac{B(\alpha)}{1-\alpha}\psi(t)E^1_{\alpha,1,\frac{-\alpha}{1-\alpha},a}\phi(t)|_a^b, \tag{31}$$

$$\int_a^b \phi(x)\, {}^{ABR}_{\ \ a}D_t^\alpha \psi(x)dx = \int_a^b \psi(x)\, {}^{ABR}_{\ \ t}D_b^\alpha \phi(x)dx, \tag{32}$$

$$\int_a^b \phi(x)\, {}^{AB}{}_aI_t^\alpha \psi(x)dx = \int_a^b \psi(x)\, {}^{AB}{}_tI_b^\alpha \phi(x)dx, \tag{33}$$

$$\int_a^b \phi(x)\, {}^{AB}{}_tI_b^\alpha \psi(x)dx = \int_a^b \psi(x)\, {}^{AB}{}_aI_t^\alpha \phi(x)dx, \tag{34}$$

where the left generalized fractional integral operator

$$E^\alpha_{\gamma,\mu,\omega,a}x(t) = \int_a^t (t-\tau)^{\mu-1}E^\alpha_{\gamma,\mu}[\omega(t-\tau)^\gamma]x(\tau)d\tau, \quad t > a,$$

and the right generalized fractional integral operator

$$E^\alpha_{\gamma,\mu,\omega,b}x(t) = \int_t^b (\tau-t)^{\mu-1}E^\alpha_{\gamma,\mu}[\omega(\tau-t)^\gamma]x(\tau)d\tau, \quad t < b.$$

3 Fractional Variational Principles Within Fractional Atangana–Baleanu's Derivatives

The fractional Euler–Lagrange and fractional Hamilton equations within Atangana–Baleanu's derivatives are briefly presented in the following.

Theorem 3.1 *Let $J[x]$ be a functional of the form*

$$J[x] = \int_a^b L(t,\ {}^{AB}{}_aI_t^{\alpha-1}x(t),\ {}^{ABC}_{\ \ a}D_t^\beta x(t))dt, \tag{35}$$

defined by the set of functions which have continuous Atangana–Baleanu fractional derivative and integral in the Caputo sense on the set of order α in $[a, b]$ and satisfy the boundary conditions $x(a) = x_a$ and $x(b) = x_b$. Then a necessary condition for $J[x]$ to have a maximum for given function $x(t)$ is that $x(t)$ must satisfy the following Euler–Lagrange equation

$$ {}^{AB}_{t}I_{b}^{\alpha-1}\frac{\partial L}{\partial({}^{AB}_{a}I_{t}^{\alpha-1}x(t))}+{}^{ABR}_{t}D_{b}^{\beta}\left(\frac{\partial L}{\partial({}^{ABC}_{a}D_{t}^{\beta}x(t))}\right)=0. \tag{36}$$

Proof To obtain the necessary conditions for the extremum, assume that $x^*(t)$ is the desired function. Let $\varepsilon \in R$, and define a family of curves

$$x(t)=x^*(t)+\varepsilon\eta(t), \tag{37}$$

where $\eta(t)$ is an arbitrary curve except that it satisfies the boundary conditions, i.e. we require that

$$\eta(a)=\eta(b)=0. \tag{38}$$

To obtain the Euler–Lagrange equation, we substitute equation (37) into (35), differentiate the resulting equation with respect to ε and set the result to 0. This leads to the following condition for extremum:

$$\int_a^b\left[{}^{AB}_{a}I_{t}^{\alpha-1}\eta(t)\cdot\frac{\partial L}{\partial({}^{AB}_{a}I_{t}^{\alpha-1}(x(t))}+{}^{ABC}_{a}D_{t}^{\beta}\eta(t)\cdot\frac{\partial L}{\partial({}^{ABC}_{a}D_{t}^{\beta}x(t))}\right]dt=0, \tag{39}$$

using Eqs. (31) and (34), we obtain

$$\int_a^b{}^{AB}_{a}I_{t}^{\alpha-1}\eta(t)\cdot\frac{\partial L}{\partial({}^{AB}_{a}I_{t}^{\alpha-1}x(t))}dt=\int_a^b\eta(t)\cdot{}^{AB}_{t}I_{b}^{\alpha-1}\left(\frac{\partial L}{\partial({}^{AB}_{a}I_{t}^{\alpha-1}(x(t))}\right)dt,$$

$$\begin{aligned}\int_a^b{}^{ABC}_{a}D_{t}^{\beta}\eta(t)\cdot\frac{\partial L}{\partial({}^{ABC}_{a}D_{t}^{\beta}x(t))}dt=&\int_a^b\eta(t)\cdot{}^{ABR}_{t}D_{b}^{\beta}\left(\frac{\partial L}{\partial({}^{ABC}_{a}D_{t}^{\beta}x(t))}\right)dt\\&+\eta(t)\cdot\frac{B(\beta)}{1-\beta}E^{1}_{\beta,1,\frac{-\beta}{1-\beta},b}\left(\frac{\partial L}{\partial({}^{ABC}_{a}D_{t}^{\beta}x(t))}\Bigg|_a^b\right),\end{aligned}$$

then Eq. (39) can be written as

$$\begin{aligned}\int_a^b\left[{}^{AB}_{t}I_{b}^{\alpha-1}\left(\frac{\partial L}{\partial({}^{AB}_{a}I_{t}^{\alpha-1}x(t))}\right)+{}^{ABR}_{t}D_{b}^{\beta}\left(\frac{\partial L}{\partial({}^{ABC}_{a}D_{t}^{\beta}x(t)}\right)\right]\eta(t)dt\\+\eta(t)\frac{B(\beta)}{1-\beta}E^{1}_{\beta,1,\frac{-\beta}{1-\beta},b}\left(\frac{\partial L}{\partial({}^{ABC}_{a}D_{t}^{\beta}x(t))}\Bigg|_a^b\right)=0,\end{aligned} \tag{40}$$

we call $E^{1}_{\beta,1,\frac{-\beta}{1-\beta},b}(\frac{\partial L}{\partial({}^{ABC}_{a}D_{t}^{\beta}x(t))}|_a^b)=0$, the natural boundary conditions, since $\eta(t)$ is arbitrary, it follows from a well established result in calculus of variations that

$$ {}^{AB}_{\ t}I_b^{\alpha-1}\left(\frac{\partial L}{\partial({}^{AB}_{\ a}I_t^{\alpha-1}x(t))}\right)+{}^{ABR}_{\ \ t}D_b^{\beta}\left(\frac{\partial L}{\partial({}^{ABC}_{\ \ a}D_t^{\beta}x(t)}\right)=0. \tag{41}$$

Equation (41) is the Generalized Euler–Lagrange Equation GELE for the Fractional Calculus Variation (FCV) problem defined in terms of the Atangana–Baleanu Fractional Derivatives ABFD. Note that the Atangana–Baleanu derivatives in the Caputo and Riemann–Liouville sense appear in the resulting differential equations.

As α and β goes to 1, the fractional Euler–Lagrange equations (41) becomes the classical Euler–Lagrange equation

$$\frac{\partial L}{\partial x}-\frac{d}{dt}\frac{\partial L}{\partial x}=0. \tag{42}$$

Example Let us consider the following fractional Lagrangian is given by:

$$L=\frac{1}{2}\left[{}^{AB}_{\ a}I_t^{\alpha-1}x(t)+{}^{ABC}_{\ \ a}D_t^{\beta}x(t)\right]^2, \tag{43}$$

then independent fractional Euler–Lagrange equation (41) is given by

$$ {}^{AB}_{\ t}I_b^{\alpha-1}\left[{}^{AB}_{\ a}I_t^{\alpha-1}x(t)+{}^{ABC}_{\ \ a}D_t^{\beta}x(t)\right]+{}^{ABR}_{\ \ t}D_b^{\beta}\left[{}^{AB}_{\ a}I_t^{\alpha-1}x(t)+{}^{ABC}_{\ \ a}D_t^{\beta}x(t)\right]=0. \tag{44}$$

4 The Fractional Canonical Momenta

For a given fractional Lagrangian

$$L=L(t,\ {}^{AB}_{\ a}I_t^{\alpha-1}x(t),\ {}^{ABC}_{\ \ a}D_t^{\beta}x(t)), \tag{45}$$

the fractional canonical momenta are defined as

$$P_\alpha=\frac{\partial L}{\partial({}^{AB}_{\ a}I_t^{\alpha-1}x(t))},\ P_\beta=\frac{\partial L}{\partial({}^{ABC}_{\ \ a}D_t^{\beta}x(t))}. \tag{46}$$

Therefore, we construct the corresponding fractional Hamiltonian as follows,

$$H(x,P_\alpha,P_\beta)=P_\alpha\ {}^{AB}_{\ a}I_t^{\alpha-1}x(t)+P_\beta\ {}^{ABC}_{\ \ a}D_t^{\beta}x(t)-L. \tag{47}$$

Then we have

$$dH = P_\beta d({}^{ABC}_{\ a}D_t^\beta x(t)) + {}^{ABC}_{\ a}D_b^\beta x(t) dP_\beta + P_\alpha d({}^{AB}_{\ a}I_t^{\alpha-1}x(t)) + {}^{AB}_{\ a}I_t^{\alpha-1}x(t)dP_\alpha - \frac{\partial L}{\partial t}dt - \frac{\partial L}{\partial({}^{AB}_{\ a}I_t^{\alpha-1}x(t))}d({}^{AB}_{\ a}I_t^{\alpha-1}x(t)) - \frac{\partial L}{\partial({}^{ABC}_{\ a}D_t^\beta x(t))}d({}^{ABC}_{\ a}D_t^\beta x(t)). \tag{48}$$

This suggests that H is a function of t, P_α, P_β only. Therefore, we can write

$$dH = \frac{\partial H}{\partial t}dt + \frac{\partial H}{\partial P_\alpha}dP_\alpha + \frac{\partial H}{\partial P_\beta}dP_\beta. \tag{49}$$

By using (41), (45) and (46) we obtain the fractional Hamilton's equations

$${}^{ABC}_{\ a}D_b^\beta x(t) = \frac{\partial H}{\partial P_\beta}, \quad {}^{AB}_{\ a}I_t^{\alpha-1}x(t) = \frac{\partial H}{\partial P_\alpha}, \tag{50}$$

and

$$\frac{\partial H}{\partial t} = -\frac{\partial L}{\partial t}. \tag{51}$$

Equation (50) represents two fractional deferential and integral equations of order β, α respectively for the system which is equivalent to the system (41). Because of their similarity with the canonical Euler equations for integer order systems, we call Eqs. (50) the fractional canonical system of Euler equations or simply the fractional canonical Euler equations.

Example Function L in Eq. (35) can be thought of as a function containing both the left and the right Atangana–Baleanu in the Caputo sense Fractional Derivatives ABCFDs

$$L = L(x, {}^{ABC}_{\ a}D_t^\alpha x(t), {}^{ABC}_{\ t}D_b^\alpha x(t),)$$

for which the GELE is given as [21]

$$\frac{\partial L}{\partial x} + {}^{ABR}_{\ t}D_b^\alpha \frac{\partial L}{\partial\, {}^{ABC}_{\ a}D_t^\alpha x(t)} + {}^{ABR}_{\ a}D_t^\alpha \frac{\partial L}{\partial\, {}^{ABC}_{\ t}D_b^\alpha x(t)} = 0. \tag{52}$$

Also function L in Eq. (35) can be thought of as a function containing both the left and the right Caputo Fractional Derivatives CFDs

$$L = L(x, {}^C_aD_t^\alpha x(t), {}^C_tD_b^\alpha x(t)),$$

for which the GELE is given as [21]

$$\frac{\partial L}{\partial x} + {}_tD_b^\alpha \frac{\partial L}{\partial\, {}^C_aD_t^\alpha x(t)} + {}_aD_t^\alpha \frac{\partial L}{\partial\, {}^C_tD_b^\alpha x(t)} = 0. \tag{53}$$

For $\alpha = 1$, the Euler–Lagrange equation is given as

$$\frac{\partial L}{\partial x} - \frac{d}{dt}\frac{\partial L}{\partial x} = 0. \tag{54}$$

Equations (41), (51) (and its equivalent in terms of ABRFD) and (52) are similar, and they all contain both forward and backward derivatives. Note that $-d/dt$ is essentially a backward derivative. Thus, backward derivatives in Eqs. (41) and (51) appear explicitly, whereas they appear in Eq. (52) in a disguise form.

5 Some Generalization

We now give some generalizations of Theorem 3.1.

5.1 Extension to Several Dependent Variables

We can generalize in a straight forward manner to problems containing several unknown functions. We denote by $\mathcal{F}_n$ the set of all functions $x_1(t), x_2(t), \ldots, x_n(t)$ which have continuous left ABC fractional derivative of order α and right ABC fractional derivative of order β for $t \in [a, b]$ and satisfy the conditions

$$x_i(a) = x_{ia}, x_i(b) = x_{ib}, i = 1, 2, \ldots, n.$$

The problem can be defined as follows: find the functions $x_1, x_2, \ldots, x_n$ from $\mathcal{F}_n$, for which the functional

$$J[x_1, x_2, \ldots, x_n] =$$

$$\int_a^b L\left[t, {}^{AB}{}_aI_t^{\alpha-1}x_1(t), \ldots, {}^{AB}{}_aI_t^{\alpha-1}x_n(t), {}^{ABC}_{\ a}D_t^{\beta}x_1(t), \ldots, {}^{ABC}_{\ a}D_t^{\beta}x_n(t)\right]dt,$$

has an extremum, where $L(t, u_1, \ldots, u_n, v_1, \ldots, v_n)$ is a function with continuous first and second partial derivatives with respect to all its arguments. A necessary condition for $J[x_1, x_2, \ldots, x_n]$ to admit an extremum is that $x_1(t), x_2(t), \ldots, x_n(t)$ satisfy Euler–Lagrange equations:

$${}^{AB}{}_tI_b^{\alpha-1}\left(\frac{\partial L}{\partial({}^{AB}{}_aI_t^{\alpha-1}x_i(t))}\right) + {}^{ABR}_{\ \ t}D_b^{\beta}\left(\frac{\partial L}{\partial({}^{ABC}_{\ a}D_t^{\beta}x_i(t)}\right) = 0, \quad i = 1, 2, \ldots, n. \tag{55}$$

Example We consider the system of two planar pendula, both of length l and mass m, suspended a same distance apart on a horizontal line so that they move in the same plane. The fractional form of the Lagrangian is given by:

$$L\left(t,\ {}^{AB}_{\ a}I_t^{\alpha-1}x_1(t),\ {}^{AB}_{\ a}I_t^{\alpha-1}x_2(t),\ {}^{ABC}_{\ a}D_t^{\alpha}x_1,\ {}^{ABC}_{\ a}D_t^{\alpha}x_2\right) =$$
$$\frac{1}{2}m\left[\left({}^{ABC}_{\ a}D_t^{\alpha}x_1\right)^2+\left({}^{ABC}_{\ a}D_t^{\alpha}x_2\right)^2\right]-\frac{1}{2}\frac{mg}{l}\left[\left({}^{AB}_{\ a}I_t^{\alpha-1}x_1(t)\right)^2+\left({}^{AB}_{\ a}I_t^{\alpha-1}x_2(t)\right)^2\right]. \tag{56}$$

To obtain the fractional Euler–Lagrange equation, we use

$${}^{AB}_{\ t}I_b^{\alpha-1}\left(\frac{\partial L}{\partial({}^{AB}_{\ a}I_t^{\alpha-1}x_i(t))}\right)+{}^{ABR}_{\ \ t}D_b^{\beta}\left(\frac{\partial L}{\partial({}^{ABC}_{\ a}D_t^{\beta}x_i(t)}\right)=0,\quad i=1,2. \tag{57}$$

It follows

$${}^{AB}_{\ t}I_b^{\alpha-1}\left(-\frac{mg}{l}({}^{AB}_{\ a}I_t^{\alpha-1}x_1(t)\right)+{}^{ABR}_{\ \ t}D_b^{\beta}(m\,{}^{ABC}_{\ a}D_t^{\alpha}x_1)=0, \tag{58}$$

$${}^{AB}_{\ t}I_b^{\alpha-1}\left(-\frac{mg}{l}({}^{AB}_{\ a}I_t^{\alpha-1}x_2(t)\right)+{}^{ABR}_{\ \ t}D_b^{\beta}(m\,{}^{ABC}_{\ a}D_t^{\alpha}x_2)=0. \tag{59}$$

These equations reduce to the equation of motion of the harmonic oscillator when $\alpha\to 1$.

$$x_1+\frac{g}{l}x_1=0,\quad x_2+\frac{g}{l}x_2=0. \tag{60}$$

5.2 Extension to Variational Problems of Non-commensurate Order

We now consider problems of the calculus of variations with Atangana–Baleanu's derivatives and integrals of non commensurate order, i.e., we consider functionals containing ABFI and ABFD of different fractional orders. Let

$$J[x]=\int_a^b L\left[t,\ {}^{AB}_{\ a}I_t^{\alpha_1-1}x(t),\ldots,{}^{AB}_{\ a}I_t^{\alpha_n-1}x(t),\ {}^{ABC}_{\ a}D_t^{\beta_1}x(t),\ldots,\ {}^{ABC}_{\ a}D_t^{\beta_m}x(t)\right]dt,$$

where n and m are two positive integers and $\alpha_i,\beta_j\in(0,1)$, $i=1,\ldots,n$ and $j=1,\ldots m$. Following the proof of Theorem 3.1, we deduce the following result.

$$\sum_{i=1}^{n}{}^{AB}_{\ t}I_b^{\alpha_i-1}\left(\frac{\partial L}{\partial({}^{AB}_{\ a}I_t^{\alpha_i-1}x(t))}\right)+\sum_{j=1}^{m}{}^{ABR}_{\ \ t}D_b^{\beta_j}\left(\frac{\partial L}{\partial({}^{ABC}_{\ a}D_t^{\beta_j}x(t)}\right)=0,\ \text{for all}\ t\in[a,b]. \tag{61}$$

6 Fractional Variational Principles and Constrained Systems Within Atangana–Baleanu's Derivatives

The above fractional canonical equations are valid for the case when no primary constraints exist; namely, all canonical momenta are linearly independent. Many dynamical systems possessing physical interest have constraints. The problem can be defined as follows. Find the extremum of the functional

$$J[x] = \int_a^b L(t,\ {}^{AB}{}_aI_t^{\alpha-1}x(t), {}^{ABC}_{\ a}D_t^{\beta}x(t))dt,$$

subject to the dynamical constraint

$$\int_a^b G(t,\ {}^{AB}{}_aI_t^{\alpha-1}x(t), {}^{ABC}_{\ a}D_t^{\beta}x(t))dt = l,$$

with the boundary conditions

$$x(a) = x_a, \quad x(b) = x_b,$$

where l is a prescribed value. This problem was solved in [8] for functionals containing Caputo fractional derivatives and RLFI. Using similar techniques as the ones discussed in [8], one proves the following. In this case we define the functional

$$S[x] = \int_a^b [\lambda_0 L + \lambda G]dt,$$

where λ, λ_0 are the Lagrange multipliers which are not both zero. Then Eqs. (41) in this case take the form:

$$ {}^{AB}{}_tI_b^{\alpha-1}\left(\frac{\partial S}{\partial({}^{AB}{}_aI_t^{\alpha-1}x(t))}\right) + {}^{ABR}_{\ \ t}D_b^{\beta}\left(\frac{\partial S}{\partial({}^{ABC}_{\ a}D_t^{\beta}x(t))}\right) = 0, \tag{62}$$

which can be written as

$$\begin{aligned} &\lambda_0\left[{}^{AB}{}_tI_b^{\alpha-1}\left(\frac{\partial L}{\partial({}^{AB}{}_aI_t^{\alpha-1}x(t))}\right) + {}^{ABR}_{\ \ t}D_b^{\beta}\left(\frac{\partial L}{\partial({}^{ABC}_{\ a}D_t^{\beta}x(t)}\right)\right] \\ +&\lambda\left[{}^{AB}{}_tI_b^{\alpha-1}\left(\frac{\partial G}{\partial({}^{AB}{}_aI_t^{\alpha-1}x(t))}\right) + {}^{ABR}_{\ \ t}D_b^{\beta}\left(\frac{\partial G}{\partial({}^{ABC}_{\ a}D_t^{\beta}x(t))}\right)\right] = 0. \end{aligned} \tag{63}$$

Example Let take

$$J[x] = \int_0^1 ({}^{ABC}_{\ 0}D_t^{\beta}x(t))^2 dt,$$

with the boundary conditions

$$x(0) = 0, \quad x(1) = 0,$$

$$K_1[x] = \int_0^1 {}^{AB}{}_0I_t^{\alpha-1}x(t)dt = 0, \quad K_2[x] = \int_0^1 t\,{}^{ABC}_{\ 0}D_t^{\beta}x(t)dt = 1.$$

Then we have:

$$S[x] = \int_0^1 \left[\lambda_0({}^{ABC}_{\ 0}D_t^{\beta}x(t))^2 + \lambda_1\,{}^{AB}{}_0I_t^{\alpha-1}x(t) + \lambda_2 t\,{}^{ABC}_{\ 0}D_t^{\beta}x(t)\right]dt,$$

where $\lambda_0, \lambda_1, \lambda_2$ are the Lagrange multipliers. Then Eq. (62) takes the form:

$$2\lambda_0\,{}^{ABR}_{\ \ t}D_1^{\beta}({}^{ABC}_{\ 0}D_t^{\beta}x(t)) + \lambda_1\,{}^{AB}{}_tI_1^{\alpha-1}1 + \lambda_2\,{}^{ABR}_{\ \ t}D_1^{\beta}t = 0, \tag{64}$$

which can be written as

$$2\lambda_0\,{}^{ABR}_{\ \ t}D_1^{\beta}({}^{ABC}_{\ 0}D_t^{\beta}x(t)) = -\lambda_1\,{}^{AB}{}_tI_1^{\alpha-1}1 - \lambda_2\,{}^{ABR}_{\ \ t}D_1^{\beta}t. \tag{65}$$

7 Sufficient Conditions

In this section we prove the sufficient conditions that ensure the existence of minimums. Similarly to what happens in the classical calculus of variations, some conditions of convexity are in order.

Definition 7.1 Given a function L, we say that $L(\underline{x}, u, v)$ is convex in $S \subset R^3$ if $\frac{\partial L}{\partial u}$ and $\frac{\partial L}{\partial v}$ exist and are continuous and verify the following condition:

$$L(x, u + u_1, v + v_1) - L(x, u, v) \geq \frac{\partial L(x, u, v)}{\partial u}u_1 + \frac{\partial L(x, u, v)}{\partial v}v_1,$$

$$\text{for all } (x, u, v), (x, u + u_1, v + v_1) \in S.$$

Similarly, we can define convexity for $L(\underline{x}, \underline{u}, v)$.

Theorem 7.1 . *Let $L(\underline{x}, u, v)$ be a convex function in Ta; $[a, b] \times R^2$ and let x_0 be a curve satisfying the fractional Euler–Lagrange equation (36). Then, x_0 minimizes (35).*

Proof The following holds:

$$
\begin{aligned}
J(x_0+\eta)-J(x_0) &= \int_a^b [L(t,\ {}^{AB}{}_aI_t^{\alpha-1}x_0(t)+{}^{AB}{}_aI_t^{\alpha-1}\eta(t),\ {}^{ABR}_{\ \ a}D_t^{\beta}x_0(t)+{}^{ABR}_{\ \ a}D_t^{\beta}\eta(t)) \\
&\quad - L(t,\ {}^{AB}{}_aI_t^{\alpha-1}x_0(t),\ {}^{ABR}_{\ \ a}D_t^{\beta}x_0(t))]dt \\
&\geq \int_a^b \Bigg[\frac{\partial L(t,\ {}^{AB}{}_aI_t^{\alpha-1}x_0(t),\ {}^{ABR}_{\ \ a}D_t^{\beta}x_0(t))}{\partial({}^{AB}{}_aI_t^{\alpha-1}x_0(t))}\cdot {}^{AB}{}_aI_t^{\alpha-1}\eta(t)+ \\
&\quad \frac{\partial L(t,\ {}^{AB}{}_aI_t^{\alpha-1}x_0(t),\ {}^{ABR}_{\ \ a}D_t^{\beta}x_0(t))}{\partial({}^{ABR}_{\ \ a}D_t^{\beta}x_0(t))}\cdot {}^{ABR}_{\ \ a}D_t^{\beta}\eta(t))\Bigg]dt \\
&\geq \int_a^b \Bigg[{}^{AB}{}_tI_b^{\alpha-1}\left(\frac{\partial L(t,\ {}^{AB}{}_aI_t^{\alpha-1}x_0(t),\ {}^{ABR}_{\ \ a}D_t^{\beta}x_0(t))}{\partial({}^{AB}{}_aI_t^{\alpha-1}x_0(t))}\right)+ \\
&\quad {}^{ABR}_{\ \ t}D_b^{\beta}\left(\frac{\partial L(t,\ {}^{AB}{}_aI_t^{\alpha-1}x_0(t),\ {}^{ABR}_{\ \ a}D_t^{\beta}x_0(t))}{\partial({}^{ABR}_{\ \ a}D_t^{\beta}x_0(t))}\right]\eta(t)dt=0.
\end{aligned}
$$

Thus $J(x_0+\eta)\geq J(x_0)$.

8 Fractional Optimal Control Problem Involving Atangana–Baleanu's Derivatives

Using the above definitions, the Fractional Optimal Control Problem FOCP under consideration can be defined as follows. To find the optimal control $u(t)$ for a FDS that minimizes the performance index

$$J[u]=\int_0^1 F({}^{ABC}_{\ \ 0}D_t^{\alpha}x(t),u,t)dt, \tag{66}$$

subject to the dynamical constraint

$${}^{ABC}_{\ \ 0}I_t^{\alpha-1}x(t)=G(x,u,t), \tag{67}$$

with the boundary conditions

$$x(0)=x_0. \tag{68}$$

where $x(t)$ is the state variable, t represents the time, and F and G are two arbitrary functions. Note that Eq. (66) may also include some additional terms containing state variables at the end point. This term in not considered here for simplicity. When $\alpha=1$, the above problem is reduced to a standard optimal control problem. Here the limits of integration have been taken as 0 and 1. Furthermore, we consider $0<\alpha<1$. These are not the limitations of the approach. Any limits can be considered and the derivative can be of any order. However, these conditions are considered for simplicity. To find the optimal control we follow the traditional approach and define a modified performance index as

In this case we define the functional

$$\mathscr{J}[x] = \int_0^1 [F({}^{ABC}_{\ 0}D_t^{\alpha} x(t), u, t) + \lambda(G(x, u, t) - {}^{ABC}_{\ 0}I_t^{\alpha-1} x(t))]dt, \tag{69}$$

where λ is the Lagrange multiplier also known as a costate or an adjoint variable. Taking variation of Eq. (64), we obtain

$$\begin{aligned}\delta\mathscr{J}[u] = \int_0^1 \Bigg[&\frac{\partial F}{\partial({}^{ABC}_{\ 0}D_t^{\alpha} x(t))} \delta({}^{ABC}_{\ 0}D_t^{\alpha} x(t)) + \frac{\partial F}{\partial u}\delta u + \delta\lambda(G(x, u, t) - {}^{AB}{}_0I_t^{\alpha-1} x(t)) + \\ &\lambda\left(\frac{\partial G}{\partial x}\delta x + \frac{\partial G}{\partial u}\delta u - \delta({}^{ABC}_{\ 0}I_t^{\alpha-1} x(t))\right)\Bigg] dt,\end{aligned} \tag{70}$$

Using Eq. (7), the last integral in Eq. (69) can be written as

$$\int_0^1 \lambda\delta({}^{AB}{}_0I_t^{\alpha-1} x(t))dt = \int_0^1 \delta x(t)({}^{AB}{}_tI_1^{\alpha-1}\lambda)dt, \tag{71}$$

provided $\delta x(0) = 0$ or $\lambda(0) = 0$, and $\lambda x(1) = 0$ or $\lambda(1) = 0$. Because $x(0)$ is specified, we have $\delta x(0) = 0$, and since $x(1)$ is not specified, we require $\lambda(1)$ to be zero. With these assumptions, the identity in Eq. (70) is satisfied. Note that we have assumed that the order of variation and the fractional derivative can be interchanged. Using Eqs. (69) and (70), we obtain

$$\begin{aligned}\delta\mathscr{J}[u] = \int_0^1 \Bigg[&\delta\lambda(G(x, u, t) - {}^{AB}{}_0I_t^{\alpha-1} x(t)) + \delta({}^{ABC}_{\ 0}D_t^{\alpha} x(t))\left(\frac{\partial F}{\partial({}^{ABC}_{\ 0}D_t^{\alpha} x(t))}\right), \\ &+ \delta x\left[\lambda\frac{\partial G}{\partial x} - {}^{AB}{}_tI_1^{\alpha-1}\lambda\right] + \delta u\left[\frac{\partial F}{\partial u} + \lambda\frac{\partial G}{\partial u}\right]\Bigg] dt.\end{aligned} \tag{72}$$

Minimization of $\mathscr{J}[u]$ (and hence minimization of $J(u)$) requires that the coefficients of δx, δu, and $\delta\lambda$ in Eq. (71) be zero. This leads to

$${}^{AB}{}_0I_t^{\alpha-1} x(t) = G(x, u, t), \tag{73}$$

$${}^{AB}{}_tI_1^{\alpha-1}\lambda = \lambda\frac{\partial G}{\partial x}, \tag{74}$$

$$\frac{\partial F}{\partial({}^{ABC}_{\ 0}D_t^{\alpha} x(t))} = 0, \tag{75}$$

$$\frac{\partial F}{\partial u} + \lambda\frac{\partial G}{\partial u} = 0. \tag{76}$$

and

$$x(0) = x_0 \text{ and } \lambda(1) = 0. \tag{77}$$

Equations (72)–(75) represent the Euler–Lagrange equations for the FOCP. These equations give the necessary conditions for the optimality of the FOCP considered here. They are very similar to the Euler–Lagrange equations for classical optimal control problems except that the resulting differential equations contain the left and the right fractional derivatives. Furthermore, the derivation of these equations is very similar to the derivation for an optimal control problem containing integral order derivatives. Determination of the optimal control for the fractional system requires solution of Eqs. (72)–(76).

Observe that Eq. (72) contains left Atangana–Baleanu integral where as Eq. (73) contains right Atangana–Baleanu integral. This clearly indicates that the solution of optimal control problems requires knowledge not only of forward integrals but of backward integrals to account for end conditions. In classical optimal control theories, this issue is either not discussed or they are not clearly stated. This is mainly because the backward integrals of order 1 turns out to be the negative of the forward integrals of order 1.

As a special case, assume that the performance index is an integral of quadratic forms in the state and the control,

$$J[u] = \frac{1}{2}\int_0^1 [q(t)({}^{ABC}_{0}D^{\alpha}_{t}x(t))^2 + r(t)u^2]dt, \tag{78}$$

where $q(t) \geq 0$ and $r(t) > 0$, and the dynamics of the system is described by the following linear fractional differential equation,

$$^{AB}{}_{0}I^{\alpha-1}_{t}x(t) = a(t)x + b(t)u, \tag{79}$$

This linear system for $\alpha = 1$ and $0 < \alpha < 1$ has been studied extensively, and formulations and solution schemes for this system within the classical Riemann–Liouville and Caputo derivatives are well documented in many textbooks and journal articles (see e.g. [2, 3]). Her we consider the new Atangana–Baleanu's derivatives and integers. For $0 < \alpha < 1$, the Euler–Lagrange Eqs. (72)–(75) lead to Eq. (78) and

$$^{AB}{}_{t}I^{\alpha}_{1}\lambda = a(t)\lambda, \tag{80}$$

$$q(t)^{ABC}_{0}D^{\alpha}_{t}x(t) = 0, \tag{81}$$

and

$$r(t)u + b(t)\lambda = 0. \tag{82}$$

From Eqs. (78) and (81), we get

$$^{AB}{}_{0}I^{\alpha-1}_{t}x(t) = a(t)x(t) - r^{-1}(t)b^2(t)\lambda. \tag{83}$$

The state $x(t)$ and the costate $\lambda(t)$ are obtained by solving the fractional differential equations (78)–(80) subject to the terminal conditions given by Eq. (82). Once $\lambda(t)$ is known, the control variable $u(t)$ can be obtained using Eq. (81).

9 Conclusions

A general formulation has been presented for a class of fractional optimal control problems involving the new Atangana–Baleanu's fractional derivatives. The formulation utilized the calculus of variations, the Lagrange multiplier technique, and the formula for fractional integration by parts to obtain the Euler–Lagrange equations for the fractional optimal control problems. The formulation presented and the resulting equations are very similar to those for classical optimal control problems. The formulation is specialized for a system with quadratic performance index subject to a fractional system dynamic constraint. From the above and other literature in the field of fractional calculus, it is clear that many of the ideas of the ordinary calculus can be extended to fractional calculus with only minor changes. The advantage of the new fractional derivative has no singularity, which was not precisely illustrated in the previous definitions.

As a final remark, we note that very little progress has been made in the field of FOCP involving the new Atangana–Baleanu's fractional derivatives. This is mainly due to the fact that the underlying mathematics for fractional derivatives was not well developed. Recent development in the field of fractional derivatives has eliminated this barrier. From the formulation presented above, it is clear that many of the concepts of classical control theory can be directly extended to FOCPs. Although only one class of FOCPs involving the new Atangana–Baleanu's fractional derivatives was considered here, the formulation can be extended to many other FOCPs involving the new Atangana–Baleanu's fractional derivatives. It is hoped that this observation will initiate some interest in the areas of fractional variational calculus and fractional optimal control.

References

1. Abdeljawad, T., Baleanu, D.: Integration by parts and its applications of a new nonlocal fractional derivative with Mittag-Leffler nonsingular kernel. J. Nonlinear Sci. Appl. **10**, 1098–1107 (2017)
2. Agrawal, O.P.: Formulation of Euler-Lagrange equations for fractional variational problems. J. Math. Anal. Appl. **272**, 368–379 (2002)
3. Agrawal, O.P.: A general formulation and solution scheme for fractional optimal control problems. Nonlinear Dyn. **38**, 323–337 (2004)
4. Agrawal, O.P., Baleanu, D.A.: Hamiltonian formulation and direct numerical scheme for fractional optimal control problems. J. Vib. Control. **13**(9–10), 1269–1281 (2007)

5. Agarwal, R.P., Baghli, S., Benchohra, M.: Controllability for semilinear functional and neutral functional evolution equations with infinite delay in Fréchet spaces. Appl. Math. Optim. **60**, 253–274 (2009)
6. Atangana, A., Baleanu, D.: New fractional derivatives with non-local and non-singular kernel: theory and application to heat transfer model. Therm. Sci. **20**(2), 763–769 (2016)
7. Bahaa, G.M.: Fractional optimal control problem for infinite order system with control constraints. Adv. Differ. Equ. **250**, 1–16 (2016)
8. Bahaa, G.M.: Fractional optimal control problem for variational inequalities with control constraints. IMA J. Math. Control. Inf. **33**(3), 1–16 (2016)
9. Bahaa, G.M.: Fractional optimal control problem for differential system with control constraints. Filomat **30**(8), 2177–2189 (2016)
10. Bahaa, G.M.: Fractional optimal control problem for differential system with delay argument. Adv. Differ. Equ. **69**, 1–19 (2017)
11. Bahaa, G.M.: Fractional optimal control problem for variable-order differential systems. Fract. Calc. Appl. Anal. **20**(6), 1–16 (2017)
12. Bahaa, G.M., Tang, Q.: Optimal control problem for coupled time-fractional evolution systems with control constraints. J. Dyn. Differ. Equ. **1**, 1–21 (2017)
13. Bahaa, G.M., Tang, Q.: Optimality conditions for fractional diffusion equations with weak Caputo derivatives and variational formulation. J. Fract. Calc. Appl. **9**(1), 100–119 (2018)
14. Baleanu, D., Agrawal, O.M.P.: Fractional Hamilton formalism within Caputo's derivative. Czechoslov. J. Phys. **56**(10–11), 1087–1092 (2000)
15. Baleanu, D., Avkar, T.: Lagrangian with linear velocities within Riemann-Liouville fractional derivatives. Nuovo Cimento B **119**, 73–79 (2004)
16. Baleanu, D., Muslih, S.I.: Lagrangian formulation on classical fields within Riemann-Liouville fractional derivatives. Phys. Scr. **72**(2–3), 119–121 (2005)
17. Baleanu, D., Jajarmi, A., Hajipour, M.: A new formulation of the fractional optimal control problems involving Mittag-Leffler nonsingular kernel. J Optim. Theory Appl. **175**, 718–737 (2017)
18. Coronel-Escamilla, A., Gómez-Aguilar, J.F., Alvarado-Méndez, E., Guerrero-Ramírez, G.V., Escobar-Jiménez, R.F.: Fractional dynamics of charged particles in magnetic fields. Int. J. Mod. Phys. C **27**(08), 1–16 (2016)
19. Coronel-Escamilla, A., Gómez-Aguilar, J.F., Baleanu, D., Córdova-Fraga, T., Escobar-Jiménez, R.F., Olivares-Peregrino, V.H., Qurashi, M.M.A.: Bateman-Feshbach tikochinsky and Caldirola-Kanai oscillators with new fractional differentiation. Entropy **19**(2), 1–21 (2017)
20. Cuahutenango-Barro, B., Taneco-Hernández, M.A., Gómez-Aguilar, J.F.: On the solutions of fractional-time wave equation with memory effect involving operators with regular kernel. Chaos Solitons Fractals **115**, 283–299 (2018)
21. Djida, J.D., Atangana, A., Area, I.: Numerical computation of a fractional derivative with non-local and non-singular kernel. Math. Model. Nat. Phenom. **12**(3), 4–13 (2017)
22. Djida, J.D., Mophou, G.M., Area, I.: Optimal control of diffusion equation with fractional time derivative with nonlocal and nonsingular Mittag-Leffler kernel (2017). arXiv:1711.09070
23. El-Sayed, A.M.A.: On the stochastic fractional calculus operators. J. Fract. Calc. Appl. **6**(1), 101–109 (2015)
24. Frederico Gastao, S.F., Torres Delfim, F.M.: Fractional optimal control in the sense of Caputo and the fractional Noether's theorem. Int. Math. Forum **3**(10), 1–17 (2008)
25. Gómez-Aguilar, J.F.: Behavior characteristics of a cap-resistor, memcapacitor, and a memristor from the response obtained of RC and RL electrical circuits described by fractional differential equations. Turk. J. Electr. Eng. Comput. Sci. **24**(3), 1–16 (2016)
26. Gómez-Aguilar, J.F.: Novel analytical solutions of the fractional Drude model. Optik **168**, 728–740 (2018)
27. Gómez-Aguilar, J.F., Dumitru, B.: Fractional transmission line with losses. Zeitschrift für Naturforschung A **69**(10–11), 539–546 (2014)
28. Gómez-Aguilar, J.F., Escobar-Jiménez, R.F., López-López, M.G., Alvarado-Martínez, V.M.: Atangana-Baleanu fractional derivative applied to electromagnetic waves in dielectric media. J. Electromagn. Waves Appl. **30**(15), 1937–1952 (2016)

29. Gómez-Aguilar, J.F., Torres, L., Yépez-Martínez, H., Baleanu, D., Reyes, J.M., Sosa, I.O.: Fractional Liénard type model of a pipeline within the fractional derivative without singular kernel. Adv. Differ. Equ. **2016**(1), 1–17 (2016)
30. Gómez-Aguilar, J.F., Yépez-Martínez, H., Escobar-Jiménez, R.F., Astorga-Zaragoza, C.M., Reyes-Reyes, J.: Analytical and numerical solutions of electrical circuits described by fractional derivatives. Appl. Math. Model. **40**(21–22), 9079–9094 (2016)
31. Gómez-Aguilar, J.F., Yépez-Martínez, H., Torres-Jiménez, J., Córdova-Fraga, T., Escobar-Jiménez, R.F., Olivares-Peregrino, V.H.: Homotopy perturbation transform method for nonlinear differential equations involving to fractional operator with exponential kernel. Adv. Differ. Equ. **2017**(1), 1–18 (2017)
32. Gómez-Aguilar, J.F., Atangana, A., Morales-Delgado, J.F.: Electrical circuits RC, LC, and RL described by Atangana-Baleanu fractional derivatives. Int. J. Circuit Theory Appl. **1**, 1–22 (2017)
33. Hafez, F.M., El-Sayed, A.M.A., El-Tawil, M.A.: On a stochastic fractional calculus. Fract. Calc. Appl. Anal. **4**(1), 81–90 (2001)
34. Jarad, F., Maraba, T., Baleanu, D.: Fractional variational optimal control problems with delayed arguments. Nonlinear Dyn. **62**, 609–614 (2010)
35. Jarad, F., Maraba, T., Baleanu, D.: Higher order fractional variational optimal control problems with delayed arguments. Appl. Math. Comput. **218**, 9234–9240 (2012)
36. Kilbas, A.A., Saigo, M., Saxena, K.: Generalized Mittag-Leffler function and generalized fractional calculus operators. Integr. Transform. Spec. Funct. **15**(1), 1–13 (2004)
37. Mophou, G.M.: Optimal control of fractional diffusion equation with state constraints. Comput. Math. Appl. **62**, 1413–1426 (2011)
38. Morales-Delgado, V.F., Taneco-Hernández, M.A., Gómez-Aguilar, J.F.: On the solutions of fractional order of evolution equations. Eur. Phys. J. Plus **132**(1), 1–17 (2017)
39. Ozdemir, N., Karadeniz, D., Iskender, B.B.: Fractional optimal control problem of a distributed system in cylindrical coordinates. Phys. Lett. A **373**, 221–226 (2009)
40. Riewe, F.: Nonconservative Lagrangian and Hamiltonian mechanics. Phys. Rev. E **53**, 1890–1899 (1996)
41. Riewe, F.: Mechanics with fractional derivatives. Phys. Rev. E **55**, 3581–3592 (1997)
42. Saad, K.M., Gómez-Aguilar, J.F.: Analysis of reaction diffusion system via a new fractional derivative with non-singular kernel. Phys. A Stat. Mech. Appl. **509**, 703–716 (2018)
43. Yépez-Martínez, H., Gómez-Aguilar, J.F., Sosa, I.O., Reyes, J.M., Torres-Jiménez, J.: The Feng's first integral method applied to the nonlinear mKdV space-time fractional partial differential equation. Rev. Mex. Fís **62**(4), 310–316 (2016)

Variable Order Mittag–Leffler Fractional Operators on Isolated Time Scales and Application to the Calculus of Variations

Thabet Abdeljawad, Raziye Mert and Delfim F. M. Torres

Abstract We introduce new fractional operators of variable order in isolated time scales with Mittag–Leffler kernels. This allows a general formulation of a class of fractional variational problems involving variable-order difference operators. Main results give fractional integration by parts formulas and necessary optimality conditions of Euler–Lagrange type.

Keywords Fractional calculus · Atangana–Baleanu fractional derivative · Fractional variational problems

1 Introduction

Fractional calculus is a generalization of ordinary differentiation and integration to an arbitrary non-integer order. It has been used effectively in the modeling of many problems in various fields of science and engineering, reflecting successfully the description of non-local properties of complex systems [12, 42]. For the sake of finding more fractional operators with different kernels, recently several authors have introduced and studied new non-local derivatives with non-singular kernels and have applied them successfully to some real world problems [2, 3, 8–11, 19, 20, 27–29, 33, 34]. What makes those fractional derivatives with Mittag–Leffler kernels

T. Abdeljawad (✉)
Department of Mathematics and General Sciences, Prince Sultan University,
P.O. Box 66833, Riyadh 11586, Saudi Arabia
e-mail: tabdeljawad@psu.edu.sa

R. Mert
Mechatronic Engineering Department, University of Turkish Aeronautical Association,
06790 Ankara, Turkey
e-mail: rmert@thk.edu.tr

D. F. M. Torres
Department of Mathematics, Center for Research and Development in Mathematics
and Applications (CIDMA), University of Aveiro, 3810-193 Aveiro, Portugal
e-mail: delfim@ua.pt

J. F. Gómez et al. (eds.), *Fractional Derivatives with Mittag-Leffler Kernel*,
Studies in Systems, Decision and Control 194,
https://doi.org/10.1007/978-3-030-11662-0_3

more interesting is that their corresponding fractional integrals contain Riemann–Liouville fractional integrals as part of their structure. Moreover, such operators enable numerical analysts to develop more efficient algorithms in solving fractional dynamical systems by concentrating only on the coefficients of the differential equations rather than worrying about the singularity of the kernels, as in the case of classical fractional operators [7].

In 1993, Samko and Ross investigated integrals and derivatives not of a constant but of variable order [39–41]. Afterwards, several pure mathematical and applicational papers contributed to the theory of variable order fractional calculus [6, 21–26, 32, 37, 38, 44, 50]. Here we continue this line of research.

The article is organized as follows. In Sect. 2, we introduce new definitions of two different types of left and right nabla fractional sums of variable order, two different types of discrete versions of the left and right generalized fractional integral operators, together with two different types of fractional sums and differences of variable order in the sense of Atangana–Baleanu. Afterwards, in Sect. 3, we prove integration by parts formulas for Atangana–Baleanu fractional sums and differences with variable order. We end with Sect. 4, applying our results to the calculus of variations.

2 Fractional Sums and Differences of Variable Order

The study of fractional calculus on time scales was initiated with the papers [13–15] and is now under strong development: see, e.g., [16–18, 36, 43]. Here, inspired by the results of [3, 8], we introduce new nabla fractional operators of variable order in isolated time scales. The reader interested on the motivation and importance to consider variable order operators is referred to [46–48] and references therein.

Let $a, b \in \mathbb{R}$ with $b - a$ a positive integer. The sets $\mathbb{N}_a$, ${}_b\mathbb{N}$, and $\mathbb{N}_{a,b}$ are defined by

$$\mathbb{N}_a = \{a, a+1, a+2, \ldots\}, \quad {}_b\mathbb{N} = \{\ldots, b-2, b-1, b\}, \quad \mathbb{N}_{a,b} = \{a, a+1, a+2, \ldots, b\},$$

respectively. Our operators use the concepts of *rising function* and *discrete Mittag–Leffler function*.

Definition 1 (*Rising function* [30]) (i) For a natural number m and $t \in \mathbb{R}$, the m rising (ascending) factorial of t is defined by

$$t^{\overline{m}} = \prod_{k=0}^{m-1}(t+k), \quad t^{\overline{0}} = 1.$$

(ii) For any real number α, the (generalized) rising function is defined by

$$t^{\overline{\alpha}} = \frac{\Gamma(t+\alpha)}{\Gamma(t)}, \quad t \in \mathbb{R} \setminus \{\ldots, -2, -1, 0\}, \quad 0^{\overline{\alpha}} = 0.$$

Definition 2 (*Nabla discrete Mittag–Leffler function* [1, 4]) For $\lambda \in \mathbb{R}$, $|\lambda| < 1$ and $\alpha, \beta, z \in \mathbb{C}$ with $Re(\alpha) > 0$, the nabla discrete Mittag–Leffler function is defined by

$$E_{\overline{\alpha,\beta}}(\lambda, z) = \sum_{k=0}^{\infty} \lambda^k \frac{z^{\overline{k\alpha+\beta-1}}}{\Gamma(\alpha k + \beta)}.$$

For $\beta = 1$, we write

$$E_{\overline{\alpha}}(\lambda, z) \triangleq E_{\overline{\alpha,1}}(\lambda, z) = \sum_{k=0}^{\infty} \lambda^k \frac{z^{\overline{k\alpha}}}{\Gamma(\alpha k + 1)}.$$

To start, we define two different types of nabla fractional sums of variable order.

Definition 3 (*Left nabla fractional sums of order $\alpha(t)$ — types I and II*) Let $0 < \alpha(t) \leq 1$ for all $t \in \mathbb{N}_a$. For a function $f : \mathbb{N}_a \to \mathbb{R}$,

1. the type I left nabla fractional sum of order $\alpha(t)$ is defined by

$$_a\nabla^{-\alpha(t)} f(t) = \frac{1}{\Gamma(\alpha(t))} \sum_{s=a+1}^{t} (t - \rho(s))^{\overline{\alpha(t)-1}} f(s), \quad t \in \mathbb{N}_{a+1};$$

2. the type II left nabla fractional sum of order $\alpha(t)$ is defined by

$$^*_a\nabla^{-\alpha(t)} f(t) = \sum_{s=a+1}^{t} (t - \rho(s))^{\overline{\alpha(s)-1}} f(s) \frac{1}{\Gamma(\alpha(s))}, \quad t \in \mathbb{N}_{a+1}.$$

Definition 4 (*Right nabla fractional sums of order $\alpha(t)$ — types I and II*) Let $0 < \alpha(t) \leq 1$ for all $t \in {}_b\mathbb{N}$. For a function $f : {}_b\mathbb{N} \to \mathbb{R}$,

1. the type I right nabla fractional sum of order $\alpha(t)$ is defined by

$$\nabla_b^{-\alpha(t)} f(t) = \frac{1}{\Gamma(\alpha(t))} \sum_{s=t}^{b-1} (s - \rho(t))^{\overline{\alpha(t)-1}} f(s), \quad t \in {}_{b-1}\mathbb{N};$$

2. the type II right nabla fractional sum of order $\alpha(t)$ is defined by

$$^*\nabla_b^{-\alpha(t)} f(t) = \sum_{s=t}^{b-1} (s - \rho(t))^{\overline{\alpha(s)-1}} f(s) \frac{1}{\Gamma(\alpha(s))}, \quad t \in {}_{b-1}\mathbb{N}.$$

Following [3], we now define two different discrete versions of the left and right generalized fractional integral operators.

Definition 5 (*Discrete left generalized fractional integral operators — types I and II*) Let $0 < \alpha(t) < 1/2$ for all $t \in \mathbb{N}_a$. For a function $\varphi : \mathbb{N}_a \to \mathbb{R}$,

1. the type I discrete left generalized fractional integral operator is defined by

$$\mathbf{E}_{\overline{\alpha(t)},1,\frac{-\alpha(t)}{1-\alpha(t)},a^+}\varphi(t) = \frac{B(\alpha(t))}{1-\alpha(t)} \sum_{s=a+1}^{t} E_{\overline{\alpha(t)}}\left[\frac{-\alpha(t)}{1-\alpha(t)}, t-\rho(s)\right]\varphi(s), \quad t \in \mathbb{N}_{a+1}; \quad (1)$$

2. the type II discrete left generalized fractional integral operator is defined by

$$\mathscr{E}_{\overline{\alpha(t)},1,\frac{-\alpha(t)}{1-\alpha(t)},a^+}\varphi(t) = \sum_{s=a+1}^{t} \frac{B(\alpha(s))}{1-\alpha(s)} E_{\overline{\alpha(s)}}\left[\frac{-\alpha(s)}{1-\alpha(s)}, t-\rho(s)\right]\varphi(s), \quad t \in \mathbb{N}_{a+1}. \quad (2)$$

Definition 6 (*Discrete right generalized fractional integral operators — types I and II*) Let $0 < \alpha(t) < 1/2$ for all $t \in {}_b\mathbb{N}$. For a function $\varphi : {}_b\mathbb{N} \to \mathbb{R}$,

1. the type I discrete right generalized fractional integral operator is defined by

$$\mathbf{E}_{\overline{\alpha(t)},1,\frac{-\alpha(t)}{1-\alpha(t)},b^-}\varphi(t) = \frac{B(\alpha(t))}{1-\alpha(t)} \sum_{s=t}^{b-1} E_{\overline{\alpha(t)}}\left[\frac{-\alpha(t)}{1-\alpha(t)}, s-\rho(t)\right]\varphi(s), \quad t \in {}_{b-1}\mathbb{N}; \quad (3)$$

2. the type II discrete right generalized fractional integral operator is defined by

$$\mathscr{E}_{\overline{\alpha(t)},1,\frac{-\alpha(t)}{1-\alpha(t)},b^-}\varphi(t) = \sum_{s=t}^{b-1} \frac{B(\alpha(s))}{1-\alpha(s)} E_{\overline{\alpha(s)}}\left[\frac{-\alpha(s)}{1-\alpha(s)}, s-\rho(t)\right]\varphi(s), \quad t \in {}_{b-1}\mathbb{N}. \quad (4)$$

We now define two different types of fractional sums and differences of variable order in the sense of Atangana–Baleanu [8] (the so-called AB operators).

Definition 7 (*Left AB nabla fractional sums of order $\alpha(t)$ — types I and II*) Let $0 < \alpha(t) \leq 1$ for all $t \in \mathbb{N}_a$. For a function $f : \mathbb{N}_a \to \mathbb{R}$,

1. the type I left AB nabla fractional sum of order $\alpha(t)$ is defined by

$$\begin{aligned} {}^{AB}_{a}\nabla^{-\alpha(t)} f(t) &= \frac{1-\alpha(t)}{B(\alpha(t))} f(t) + \frac{\alpha(t)}{B(\alpha(t))\Gamma(\alpha(t))} \sum_{s=a+1}^{t} (t-\rho(s))^{\overline{\alpha(t)-1}} f(s) \\ &= \frac{1-\alpha(t)}{B(\alpha(t))} f(t) + \frac{\alpha(t)}{B(\alpha(t))} {}_a\nabla^{-\alpha(t)} f(t), \quad t \in \mathbb{N}_{a+1}; \end{aligned} \quad (5)$$

2. the type II left AB nabla fractional sum of order $\alpha(t)$ is defined by

$$\begin{aligned} {}^{*AB}_{a}\nabla^{-\alpha(t)} f(t) &= \frac{1-\alpha(t)}{B(\alpha(t))} f(t) + \sum_{s=a+1}^{t} \frac{\alpha(s)}{B(\alpha(s))\Gamma(\alpha(s))} (t-\rho(s))^{\overline{\alpha(s)-1}} f(s) \\ &= \frac{1-\alpha(t)}{B(\alpha(t))} f(t) +^{*}_{a} \nabla^{-\alpha(t)} \frac{\alpha f}{B \circ \alpha}(t), \quad t \in \mathbb{N}_{a+1}. \end{aligned} \quad (6)$$

Definition 8 (*Right AB nabla fractional sums of order $\alpha(t)$ — types I and II*) Let $0 < \alpha(t) \leq 1$ for all $t \in {}_b\mathbb{N}$. For a function $f : {}_b\mathbb{N} \to \mathbb{R}$,

1. the type I right AB nabla fractional sum of order $\alpha(t)$ is defined by

$$\begin{aligned} {}^{AB}\nabla_b^{-\alpha(t)} f(t) &= \frac{1-\alpha(t)}{B(\alpha(t))} f(t) + \frac{\alpha(t)}{B(\alpha(t))\Gamma(\alpha(t))} \sum_{s=t}^{b-1} (s-\rho(t))^{\overline{\alpha(t)-1}} f(s) \\ &= \frac{1-\alpha(t)}{B(\alpha(t))} f(t) + \frac{\alpha(t)}{B(\alpha(t))} \nabla_b^{-\alpha(t)} f(t), \quad t \in {}_{b-1}\mathbb{N}; \end{aligned} \tag{7}$$

2. the type II right AB nabla fractional sum of order $\alpha(t)$ is defined by

$$\begin{aligned} {}^{*AB}\nabla_b^{-\alpha(t)} f(t) &= \frac{1-\alpha(t)}{B(\alpha(t))} f(t) + \sum_{s=t}^{b-1} \frac{\alpha(s)}{B(\alpha(s))\Gamma(\alpha(s))} (s-\rho(t))^{\overline{\alpha(s)-1}} f(s) \\ &= \frac{1-\alpha(t)}{B(\alpha(t))} f(t) + {}^{*}\nabla_b^{-\alpha(t)} \frac{\alpha f}{B \circ \alpha}(t), \quad t \in {}_{b-1}\mathbb{N}. \end{aligned} \tag{8}$$

Note that in Definitions 7 and 8, if $\alpha(t) \equiv 0$, then we recover the initial function; if $\alpha(t) \equiv 1$, then we recover the ordinary sum.

Definition 9 (*Left Riemann–Liouville AB nabla fractional differences of order $\alpha(t)$ — types I and II*) Let $0 < \alpha(t) < 1/2$ for all $t \in \mathbb{N}_a$. For a function $f : \mathbb{N}_a \to \mathbb{R}$,

1. the type I left Riemann–Liouville AB nabla fractional difference of order $\alpha(t)$ is defined by

$$ {}^{ABR}_{\ a}\nabla^{\alpha(t)} f(t) = \nabla \mathbf{E}_{\overline{\alpha(t),1,\frac{-\alpha(t)}{1-\alpha(t)}},a^+} f(t), \quad t \in \mathbb{N}_{a+1}; \tag{9}$$

2. the type II left Riemann–Liouville AB nabla fractional difference of order $\alpha(t)$ is defined by

$$ {}^{ABR}_{\ a}\widehat{\nabla}^{\alpha(t)} f(t) = \nabla \mathscr{E}_{\overline{\alpha(t),1,\frac{-\alpha(t)}{1-\alpha(t)}},a^+} f(t), \quad t \in \mathbb{N}_{a+1}. \tag{10}$$

Definition 10 (*Right Riemann–Liouville AB nabla fractional differences of order $\alpha(t)$ — types I and II*) Let $0 < \alpha(t) < 1/2$ for all $t \in {}_b\mathbb{N}$. For a function $f : {}_b\mathbb{N} \to \mathbb{R}$,

1. the type I right Riemann–Liouville AB nabla fractional difference of order $\alpha(t)$ is defined by

$$ {}^{ABR}\nabla_b^{\alpha(t)} f(t) = -\Delta \mathbf{E}_{\overline{\alpha(t),1,\frac{-\alpha(t)}{1-\alpha(t)}},b^-} f(t), \quad t \in {}_{b-1}\mathbb{N};$$

2. the type II right Riemann–Liouville AB nabla fractional difference of order $\alpha(t)$ is defined by

$$ {}^{ABR}\widehat{\nabla}_b^{\alpha(t)} f(t) = -\Delta \mathscr{E}_{\overline{\alpha(t),1,\frac{-\alpha(t)}{1-\alpha(t)}},b^-} f(t), \quad t \in {}_{b-1}\mathbb{N}.$$

Definition 11 (*Left Caputo AB nabla fractional differences of order $\alpha(t)$ — types I and II*) Let $0 < \alpha(t) < 1/2$ for all $t \in \mathbb{N}_a$. For a function $f : \mathbb{N}_a \to \mathbb{R}$,

1. the type I left Caputo AB nabla fractional difference of order $\alpha(t)$ is defined by

$$ {}^{ABC}_{\ a}\nabla^{\alpha(t)} f(t) = \mathbf{E}_{\overline{\alpha(t)},1,\frac{-\alpha(t)}{1-\alpha(t)},a^+} \nabla f(t), \quad t \in \mathbb{N}_{a+1}; $$

2. the type II left Caputo AB nabla fractional difference of order $\alpha(t)$ is defined by

$$ {}^{ABC}_{\ a}\widehat{\nabla}^{\alpha(t)} f(t) = \mathscr{E}_{\overline{\alpha(t)},1,\frac{-\alpha(t)}{1-\alpha(t)},a^+} \nabla f(t), \quad t \in \mathbb{N}_{a+1}. $$

Definition 12 (*Right Caputo AB nabla fractional differences of order $\alpha(t)$ — types I and II*) Let $0 < \alpha(t) < 1/2$ for all $t \in {}_b\mathbb{N}$. For a function $f : {}_b\mathbb{N} \to \mathbb{R}$,

1. the type I right Caputo AB nabla fractional difference of order $\alpha(t)$ is defined by

$$ {}^{ABC}\nabla_b^{\alpha(t)} f(t) = -\mathbf{E}_{\overline{\alpha(t)},1,\frac{-\alpha(t)}{1-\alpha(t)},b^-} \Delta f(t), \quad t \in {}_{b-1}\mathbb{N}; $$

2. the type II right Caputo AB nabla fractional difference of order $\alpha(t)$ is defined by

$$ {}^{ABC}\widehat{\nabla}_b^{\alpha(t)} f(t) = -\mathscr{E}_{\overline{\alpha(t)},1,\frac{-\alpha(t)}{1-\alpha(t)},b^-} \Delta f(t), \quad t \in {}_{b-1}\mathbb{N}. $$

Remark 1 If we replace $\alpha(t)$ in (1) and (3) by $\alpha(t-s)$ and replace each $\alpha(s)$ in (2) and (4) by $\alpha(t-s)$, then the ABR and ABC fractional differences with variable order can be expressed in convolution form. Similarly, if we replace $\alpha(t)$ in (5) and (7) by $\alpha(t-s)$ and replace each $\alpha(s)$ in (6) and (8) by $\alpha(t-s)$, then the second part of the AB fractional integrals with variable order can be expressed in convolution form.

3 Summation by Parts for Variable Order Fractional Operators

Summation/integration by parts has a very important role in mathematics: see, e.g., [31, 35, 45]. This is particularly true in the calculus of variations and optimal control, to prove necessary optimality conditions of Euler–Lagrange type (cf. proof of Theorem 3).

Lemma 1 (Integration by parts formula for nabla fractional sums of order $\alpha(t)$) *Let $0 < \alpha(t) \leq 1$ for all $t \in \mathbb{N}_{a,b}$. For functions $f, g : \mathbb{N}_{a,b} \to \mathbb{R}$, we have*

$$ \sum_{t=a+1}^{b-1} f(t)\, {}_a\nabla^{-\alpha(t)} g(t) = \sum_{t=a+1}^{b-1} g(t)\, {}^{*}\nabla_b^{-\alpha(t)} f(t); $$

$$\sum_{t=a+1}^{b-1} f(t)\, \nabla_b^{-\alpha(t)} g(t) = \sum_{t=a+1}^{b-1} g(t)\, {}_a^*\nabla^{-\alpha(t)} f(t).$$

Proof From Definition 3, and by changing the order of summation, we get

$$\begin{aligned}\sum_{t=a+1}^{b-1} f(t)\, {}_a\nabla^{-\alpha(t)} g(t) &= \sum_{t=a+1}^{b-1} f(t) \frac{1}{\Gamma(\alpha(t))} \sum_{s=a+1}^{t} (t-\rho(s))^{\overline{\alpha(t)-1}} g(s)\\ &= \sum_{s=a+1}^{b-1} g(s) \left(\sum_{t=s}^{b-1} (t-\rho(s))^{\overline{\alpha(t)-1}} f(t) \frac{1}{\Gamma(\alpha(t))} \right)\\ &= \sum_{s=a+1}^{b-1} g(s)\, {}^*\nabla_b^{-\alpha(t)} f(s).\end{aligned}$$

The proof of the second assertion follows similarly.

Now, with the help of Lemma 1, we can prove the following integration by parts formula for AB fractional sums of variable order.

Theorem 1 (Integration by parts formula for AB nabla fractional sums of order $\alpha(t)$) *Let* $0 < \alpha(t) \leq 1$ *for all* $t \in \mathbb{N}_{a,b}$. *For functions* $f, g : \mathbb{N}_{a,b} \to \mathbb{R}$, *we have*

$$\sum_{t=a+1}^{b-1} f(t)\, {}_a^{AB}\nabla^{-\alpha(t)} g(t) = \sum_{t=a+1}^{b-1} g(t)\, {}^{*AB}\nabla_b^{-\alpha(t)} f(t);$$

$$\sum_{t=a+1}^{b-1} f(t)\, {}_a^{*AB}\nabla^{-\alpha(t)} g(t) = \sum_{t=a+1}^{b-1} g(t)\, {}^{AB}\nabla_b^{-\alpha(t)} f(t).$$

Proof From Definition 7 and the first part of Lemma 1, we get

$$\begin{aligned}\sum_{t=a+1}^{b-1} f(t)\, {}_a^{AB}\nabla^{-\alpha(t)} g(t) &= \sum_{t=a+1}^{b-1} f(t) \frac{1-\alpha(t)}{B(\alpha(t))} g(t) + \sum_{t=a+1}^{b-1} f(t) \frac{\alpha(t)}{B(\alpha(t))}\, {}_a\nabla^{-\alpha(t)} g(t)\\ &= \sum_{t=a+1}^{b-1} f(t) \frac{1-\alpha(t)}{B(\alpha(t))} g(t) + \sum_{t=a+1}^{b-1} g(t)\, {}^*\nabla_b^{-\alpha(t)} \frac{\alpha f}{B \circ \alpha}(t)\\ &= \sum_{t=a+1}^{b-1} g(t) \left(\frac{1-\alpha(t)}{B(\alpha(t))} f(t) +^* \nabla_b^{-\alpha(t)} \frac{\alpha f}{B \circ \alpha}(t) \right)\\ &= \sum_{t=a+1}^{b-1} g(t)\, {}^{*AB}\nabla_b^{-\alpha(t)} f(t).\end{aligned}$$

The proof of the second assertion is similar to the first one. It follows from Definition 7 and the second part of Lemma 1.

Lemma 2 *Let $0 < \alpha(t) < 1/2$ for all $t \in \mathbb{N}_{a,b}$. For functions $f, g : \mathbb{N}_{a,b} \to \mathbb{R}$, we have*

$$\sum_{t=a+1}^{b-1} f(t)\, \boldsymbol{E}_{\overline{\alpha(t)},1,\frac{-\alpha(t)}{1-\alpha(t)},a^+} g(t) = \sum_{t=a+1}^{b-1} g(t)\, \mathscr{E}_{\overline{\alpha(t)},1,\frac{-\alpha(t)}{1-\alpha(t)},b^-} f(t);$$

$$\sum_{t=a+1}^{b-1} f(t)\, \mathscr{E}_{\overline{\alpha(t)},1,\frac{-\alpha(t)}{1-\alpha(t)},a^+} g(t) = \sum_{t=a+1}^{b-1} g(t)\, \boldsymbol{E}_{\overline{\alpha(t)},1,\frac{-\alpha(t)}{1-\alpha(t)},b^-} f(t).$$

Proof From Definitions 5 and 6, and by changing the order of summation, we have

$$\begin{aligned}
\sum_{t=a+1}^{b-1} f(t)\, \mathbf{E}_{\overline{\alpha(t)},1,\frac{-\alpha(t)}{1-\alpha(t)},a^+} g(t) &= \sum_{t=a+1}^{b-1} f(t) \frac{B(\alpha(t))}{1-\alpha(t)} \sum_{s=a+1}^{t} E_{\overline{\alpha(t)}}\left[\frac{-\alpha(t)}{1-\alpha(t)}, t-\rho(s)\right] g(s)\\
&= \sum_{s=a+1}^{b-1} g(s) \sum_{t=s}^{b-1} \frac{B(\alpha(t))}{1-\alpha(t)}\, E_{\overline{\alpha(t)}}\left[\frac{-\alpha(t)}{1-\alpha(t)}, t-\rho(s)\right] f(t)\\
&= \sum_{s=a+1}^{b-1} g(s)\, \mathscr{E}_{\overline{\alpha(s)},1,\frac{-\alpha(s)}{1-\alpha(s)},b^-} f(s).
\end{aligned}$$

The proof of the second assertion follows similarly.

Theorem 2 *Let $0 < \alpha(t) < 1/2$ for all $t \in \mathbb{N}_{a,b}$. For functions $f, g : \mathbb{N}_{a,b} \to \mathbb{R}$, we have*

$$\sum_{t=a+1}^{b-1} f(t)\, {}^{ABC}_{\ a}\nabla^{\alpha(t)} g(t) = g(t)\, \mathscr{E}_{\overline{\alpha(t)},1,\frac{-\alpha(t)}{1-\alpha(t)},b^-} f(t)\Big|_a^{b-1} + \sum_{t=a+1}^{b-1} g(t-1)\, {}^{ABR}\widehat{\nabla}_b^{\alpha(t)} f(t-1);$$

$$\sum_{t=a+1}^{b-1} f(t)\, {}^{ABC}_{\ a}\widehat{\nabla}^{\alpha(t)} g(t) = g(t)\, \boldsymbol{E}_{\overline{\alpha(t)},1,\frac{-\alpha(t)}{1-\alpha(t)},b^-} f(t)\Big|_a^{b-1} + \sum_{t=a+1}^{b-1} g(t-1)\, {}^{ABR}\nabla_b^{\alpha(t)} f(t-1);$$

$$\sum_{t=a+1}^{b-1} f(t)\, {}^{ABC}\nabla_b^{\alpha(t)} g(t) = -g(t)\, \mathscr{E}_{\overline{\alpha(t)},1,\frac{-\alpha(t)}{1-\alpha(t)},a^+} f(t)\Big|_{a+1}^{b} + \sum_{t=a+1}^{b-1} g(t+1)\, {}^{ABR}_{\ a}\widehat{\nabla}^{\alpha(t)} f(t+1);$$

$$\sum_{t=a+1}^{b-1} f(t)\, {}^{ABC}\widehat{\nabla}_b^{\alpha(t)} g(t) = -g(t)\, \boldsymbol{E}_{\overline{\alpha(t)},1,\frac{-\alpha(t)}{1-\alpha(t)},a^+} f(t)\Big|_{a+1}^{b} + \sum_{t=a+1}^{b-1} g(t+1)\, {}^{ABR}_{\ a}\nabla^{\alpha(t)} f(t+1).$$

Proof We will only prove the first assertion. The proof of the others follow similarly. From Definitions 10 and 11, the first part of Lemma 2 and the summation by parts formula from ordinary difference calculus, we get

$$
\begin{aligned}
&\sum_{t=a+1}^{b-1} f(t)\, {}_a^{ABC}\nabla^{\alpha(t)} g(t) \\
&= \sum_{t=a+1}^{b-1} f(t)\, \mathbf{E}_{\overline{\alpha(t),1},\frac{-\alpha(t)}{1-\alpha(t)},a^+} \nabla g(t) \\
&= \sum_{t=a+1}^{b-1} \nabla g(t)\, \mathscr{E}_{\overline{\alpha(t),1},\frac{-\alpha(t)}{1-\alpha(t)},b^-} f(t) \\
&= g(t)\, \mathscr{E}_{\overline{\alpha(t),1},\frac{-\alpha(t)}{1-\alpha(t)},b^-} f(t)\Big|_a^{b-1} - \sum_{t=a+1}^{b-1} g(t-1) \nabla \mathscr{E}_{\overline{\alpha(t),1},\frac{-\alpha(t)}{1-\alpha(t)},b^-} f(t) \\
&= g(t)\, \mathscr{E}_{\overline{\alpha(t),1},\frac{-\alpha(t)}{1-\alpha(t)},b^-} f(t)\Big|_a^{b-1} - \sum_{t=a+1}^{b-1} g(t-1) \Delta \mathscr{E}_{\overline{\alpha(t),1},\frac{-\alpha(t)}{1-\alpha(t)},b^-} f(t-1) \\
&= g(t)\, \mathscr{E}_{\overline{\alpha(t),1},\frac{-\alpha(t)}{1-\alpha(t)},b^-} f(t)\Big|_a^{b-1} + \sum_{t=a+1}^{b-1} g(t-1)\, {}^{ABR}\widehat{\nabla}_b^{\alpha(t)} f(t-1).
\end{aligned}
$$

The proof is complete.

4 Variable Order Fractional Variational Principles

The fractional calculus of variations of variable-order is a subject under strong current development [5, 49]. However, to the best of our knowledge, available results are only for the continuous time scale $\mathbb{T} = \mathbb{R}$. Here we obtain the main result of a variational calculus, that is, an Euler–Lagrange necessary optimality condition, for the isolated time scale $\mathbb{T} = \mathbb{N}_{a+1,b-1}$.

Let J be a functional of the form

$$
J(f) = \sum_{t=a+1}^{b-1} L(t, f^{\rho}(t), {}_a^{ABC}\nabla^{\alpha(t)} f(t)),
$$

where $0 < \alpha(t) < 1/2$ for all $t \in \mathbb{N}_{a+1,b-1}$, $f : \mathbb{N}_{a,b-1} \to \mathbb{R}$ and $L : \mathbb{N}_{a+1,b-1} \times \mathbb{R} \times \mathbb{R} \to \mathbb{R}$.

Theorem 3 *Let f be a local extremizer of J satisfying the boundary conditions*

$$
f(a) = A, \quad f(b-1) = B.
$$

Then f satisfies the Euler–Lagrange equation

$$
L_1^{\sigma}(t) + {}^{ABR}\widehat{\nabla}_b^{\alpha(t)} L_2(t) = 0, \quad t \in \mathbb{N}_{a+1,b-2},
$$

where $L_1 = \frac{\partial L}{\partial f^\rho}$ *and* $L_2 = \frac{\partial L}{\partial\, {}_a^{ABC}\nabla^{\alpha(t)} f}$.

Proof Let ε be a small real parameter and $\eta : \mathbb{N}_{a,b-1} \to \mathbb{R}$ be a function such that $\eta(a) = \eta(b-1) = 0$. Consider a variation of f, say $f + \varepsilon\eta$. Since the Caputo difference operator ${}_a^{ABC}\nabla^{\alpha(t)}$ is linear, it follows that

$$J(f+\varepsilon\eta) = \sum_{t=a+1}^{b-1} L(t, f^\rho(t) + \varepsilon\eta^\rho(t), {}_a^{ABC}\nabla^{\alpha(t)} f(t) + \varepsilon\, {}_a^{ABC}\nabla^{\alpha(t)}\eta(t)).$$

Define $\hat{J}(\varepsilon) = J(f+\varepsilon\eta)$. Because f is a local extremizer of J, $\hat{J}$ attains a local extremum at $\varepsilon = 0$. Differentiating $\hat{J}(\varepsilon)$ at zero, we get

$$\sum_{t=a+1}^{b-1} \eta^\rho(t) \frac{\partial L}{\partial f^\rho}(t, f^\rho(t), {}_a^{ABC}\nabla^{\alpha(t)} f(t)) + {}_a^{ABC}\nabla^{\alpha(t)}\eta(t) \frac{\partial L}{\partial\, {}_a^{ABC}\nabla^{\alpha(t)} f}(t, f^\rho(t), {}_a^{ABC}\nabla^{\alpha(t)} f(t)) = 0.$$

Using the first integration by parts formula in Theorem 2, we have

$$\sum_{t=a+1}^{b-1} \eta^\rho(t) \Bigg[\frac{\partial L}{\partial f^\rho}(t, f^\rho(t), {}_a^{ABC}\nabla^{\alpha(t)} f(t)) + \left({}^{ABR}\widehat{\nabla}_b^{\alpha(t)} \frac{\partial L}{\partial\, {}_a^{ABC}\nabla^{\alpha(t)} f}(t, f^\rho(t), {}_a^{ABC}\nabla^{\alpha(t)} f(t)) \right)(t-1) \Bigg] + \eta(t) \left(\mathscr{E}_{\overline{\alpha(t),1,\frac{-\alpha(t)}{1-\alpha(t)},b^-}} \frac{\partial L}{\partial\, {}_a^{ABC}\nabla^{\alpha(t)} f}(t, f^\rho(t), {}_a^{ABC}\nabla^{\alpha(t)} f(t)) \right)(t) \Bigg|_a^{b-1} = 0.$$

Since $\eta(a) = \eta(b-1) = 0$ and η is arbitrary, it follows that

$$\frac{\partial L}{\partial f^\rho}(t, f^\rho(t), {}_a^{ABC}\nabla^{\alpha(t)} f(t)) + \left({}^{ABR}\widehat{\nabla}_b^{\alpha(t)} \frac{\partial L}{\partial\, {}_a^{ABC}\nabla^{\alpha(t)} f}(t, f^\rho(t), {}_a^{ABC}\nabla^{\alpha(t)} f(t)) \right)(t-1) = 0$$

for all $t \in \mathbb{N}_{a+2,b-1}$.

Although we only consider here a class of fractional variable order variational problems (FVOVP), our Theorem 3 can be easily extended to many other related FVOVPs involving the new variable-order fractional differences introduced in Sect. 2. We trust that this observation will initiate some interest in further future developments.

Acknowledgements Abdeljawad is grateful to Prince Sultan University for funding this work through research group *Nonlinear Analysis Methods in Applied Mathematics* (NAMAM), number RG-DES-2017-01-17; Torres to the support of FCT within the R&D unit CIDMA, UID/MAT/04106/2019.

References

1. Abdeljawad, T.: On delta and nabla Caputo fractional differences and dual identities. Discret. Dyn. Nat. Soc. **1**, 1–12 (2013)
2. Abdeljawad, T., Baleanu, D.: Discrete fractional differences with nonsingular discrete Mittag-Leffler kernels. Adv. Differ. Equ. **1**, 1–18 (2016)
3. Abdeljawad, T., Baleanu, D.: On fractional derivatives with exponential kernel and their discrete versions. Rep. Math. Phys. **80**(1), 11–27 (2017)
4. Abdeljawad, T., Jarad, F., Baleanu, D.: A semigroup-like property for discrete Mittag-Leffler functions. Adv. Differ. Equ. **72**, 1–7 (2012)
5. Almeida, R., Tavares, D., Torres, D.F.M.: The Variable-Order Fractional Calculus of Variations. Springer Briefs in Applied Sciences and Technology. Springer, Cham (2018)
6. Atanackovic, T.M., Pilipovic, S.: Hamilton's principle with variable order fractional derivatives. Fract. Calc. Appl. Anal. **14**(1), 94–109 (2011)
7. Atangana, A.: Fractional Operators with Constant and Variable Order with Application to Geo-Hydrology. Academic, London (2018)
8. Atangana, A., Baleanu, D.: New fractional derivatives with non-local and non-singular kernel: theory and application to heat transfer model. Therm. Sci. **20**(2), 763–769 (2016)
9. Atangana, A., Gómez-Aguilar, J.F.: A new derivative with normal distribution kernel: theory, methods and applications. Phys. A Stat. Mech. Appl. **476**, 1–14 (2017)
10. Atangana, A., Gómez-Aguilar, J.F.: Decolonisation of fractional calculus rules: breaking commutativity and associativity to capture more natural phenomena. Eur. Phys. J. Plus **133**, 1–22 (2018)
11. Atangana, A., Gómez-Aguilar, J.F.: Fractional derivatives with no-index law property: application to chaos and statistics. Chaos Solitons Fractals **114**, 516–535 (2019)
12. Baleanu, D., Diethelm, K., Scalas, E., Trujillo, J.J.: Fractional Calculus. Series on Complexity, Nonlinearity and Chaos, vol. 3. World Scientific Publishing Co. Pte. Ltd., Hackensack (2012)
13. Bastos, N.R.O., Ferreira, R.A.C., Torres, D.F.M.: Necessary optimality conditions for fractional difference problems of the calculus of variations. Discret. Contin. Dyn. Syst. **29**(2), 417–437 (2011)
14. Bastos, N.R.O., Ferreira, R.A.C., Torres, D.F.M.: Discrete-time fractional variational problems. Signal Process. **91**(3), 513–524 (2011)
15. Bastos, N.R.O., Mozyrska, D., Torres, D.F.M.: Fractional derivatives and integrals on time scales via the inverse generalized Laplace transform. Int. J. Math. Comput. **11**, 1–9 (2011)
16. Bayour, B., Torres, D.F.M.: Complex-valued fractional derivatives on time scales. In: Differential and Difference Equations with Applications. Springer Proceedings in Mathematics and Statistics, vol. 164, pp. 79–87 (2015)
17. Benkhettou, N., Brito da Cruz, A.M.C., Torres, D.F.M.: Nonsymmetric and symmetric fractional calculi on arbitrary nonempty closed sets. Math. Methods Appl. Sci. **39**(2), 261–279 (2016)
18. Benkhettou, N., Hammoudi, A., Torres, D.F.M.: Existence and uniqueness of solution for a fractional Riemann-Liouville initial value problem on time scales. J. King Saud Univ. Sci. **28**(1), 87–92 (2016)
19. Caputo, M., Fabrizio, M.: A new definition of fractional derivative without singular kernel. Prog. Fract. Differ. Appl. **1**(2), 73–85 (2015)
20. Caputo, M., Fabrizio, M.: Applications of new time and spatial fractional derivatives with exponential kernels. Prog. Fract. Differ. Appl. **2**(1), 1–11 (2016)
21. Coimbra, C.F.M.: Mechanics with variable-order differential operators. Ann. Phys. **12**(11–12), 692–703 (2003)
22. Coronel-Escamilla, A., Gómez-Aguilar, J.F., Torres, L., Escobar-Jiménez, R.F., Valtierra-Rodríguez, M.: Synchronization of chaotic systems involving fractional operators of Liouville-Caputo type with variable-order. Phys. A Stat. Mech. Appl. **487**, 1–21 (2017)

23. Coronel-Escamilla, A., Gómez-Aguilar, J.F., Torres, L., Valtierra-Rodríguez, M., Escobar-Jiménez, R.F.: Design of a state observer to approximate signals by using the concept of fractional variable-order derivative. Digit. Signal Process. **69**, 127–139 (2017)
24. Coronel-Escamilla, A., Gómez-Aguilar, J.F., Torres, L., Escobar-Jiménez, R.F.: A numerical solution for a variable-order reaction-diffusion model by using fractional derivatives with non-local and non-singular kernel. Phys. A Stat. Mech. Appl. **491**, 406–424 (2018)
25. Diaz, G., Coimbra, C.F.M.: Nonlinear dynamics and control of a variable order oscillator with application to the van der Pol equation. Nonlinear Dyn. **56**(1–2), 145–157 (2009)
26. Gómez-Aguilar, J.F.: Analytical and Numerical solutions of a nonlinear alcoholism model via variable-order fractional differential equations. Phys. A Stat. Mech. Appl. **494**, 52–75 (2018)
27. Gómez-Aguilar, J.F.: Novel analytical solutions of the fractional Drude model. Optik **168**, 728–740 (2018)
28. Gómez-Aguilar, J.F., Atangana, A.: New insight in fractional differentiation: power, exponential decay and Mittag-Leffler laws and applications. Eur. Phys. J. Plus **132**(1), 1–13 (2017)
29. Gómez-Aguilar, J.F., Escobar-Jiménez, R.F., López-López, M.G., Alvarado-Martínez, V.M.: Analysis of projectile motion: a comparative study using fractional operators with power law, exponential decay and Mittag-Leffler kernel. Eur. Phys. J. Plus **133**(3), 1–26 (2018)
30. Goodrich, C., Peterson, A.C.: Discrete Fractional Calculus. Springer, Cham (2015)
31. Hang, Y., Liu, Y., Xu, X., Chen, Y., Mo, S.: Sensitivity analysis based on Markovian integration by parts formula. Math. Comput. Appl. **22**(4), 1–12 (2017)
32. Lorenzo, C.F., Hartley, T.T.: Variable order and distributed order fractional operators. Nonlinear Dyn. **29**(1–4), 57–98 (2002)
33. Losada, J., Nieto, J.J.: Properties of a new fractional derivative without singular kernel. Prog. Fract. Differ. Appl. **1**(2), 87–92 (2015)
34. Morales-Delgado, V.F., Taneco-Hernández, M.A., Gómez-Aguilar, J.F.: On the solutions of fractional order of evolution equations. Eur. Phys. J. Plus **132**(1), 1–17 (2017)
35. Nelsen, R.B.: Proof without words: integration by parts. Math. Mag. **64**(2), 1–13 (1991)
36. Pei, B., Xu, Y., Yin, G., Zhang, X.: Averaging principles for functional stochastic partial differential equations driven by a fractional Brownian motion modulated by two-time-scale Markovian switching processes. Nonlinear Anal. Hybrid Syst. **27**, 107–124 (2018)
37. Ramirez, L.E.S., Coimbra, C.F.M.: On the selection and meaning of variable order operators for dynamic modeling. Int. J. Differ. Equations **1**, 1–16 (2010)
38. Ramirez, L.E.S., Coimbra, C.F.M.: On the variable order dynamics of the nonlinear wake caused by a sedimenting particle. Phys. D **240**(13), 1111–1118 (2011)
39. Ross, B., Samko, S.: Fractional integration operator of variable order in the Hölder spaces $H^{\lambda(x)}$. Int. J. Math. Math. Sci. **18**(4), 777–788 (1995)
40. Samko, S.G.: Fractional integration and differentiation of variable order. Anal. Math. **21**(3), 213–236 (1995)
41. Samko, S.G., Ross, B.: Integration and differentiation to a variable fractional order. Integr. Transform. Spec. Funct. **1**(4), 277–300 (1993)
42. Samko, S.G., Kilbas, A.A., Marichev, O.I.: Fractional Integrals and Derivatives (translated from the 1987 Russian original). Gordon and Breach Science Publishers, Yverdon (1993)
43. Sheng, K., Zhang, W., Bai, Z.: Positive solutions to fractional boundary-value problems with p-Laplacian on time scales. Bound. Value Probl. **1**(70), 1–17 (2018)
44. Solís-Pérez, J.E., Gómez-Aguilar, J.F., Atangana, A.: Novel numerical method for solving variable-order fractional differential equations with power, exponential and Mittag-Leffler laws. Chaos Solitons Fractals **114**, 175–185 (2018)
45. Tang, J.: Computation of an infinite integral using integration by parts. Math. Methods Appl. Sci. **41**(3), 929–935 (2018)
46. Tavares, D., Almeida, R., Torres, D.F.M.: Caputo derivatives of fractional variable order: numerical approximations. Commun. Nonlinear Sci. Numer. Simul. **3**, 69–87 (2016)
47. Tavares, D., Almeida, R., Torres, D.F.M.: Constrained fractional variational problems of variable order. IEEE/CAA J. Autom. Sin. **4**(1), 80–88 (2017)

48. Tavares, D., Almeida, R., Torres, D.F.M.: Fractional Herglotz variational problems of variable order. Discret. Contin. Dyn. Syst. Ser. S **11**(1), 143–154 (2018)
49. Tavares, D., Almeida, R., Torres, D.F.M.: Combined fractional variational problems of variable order and some computational aspects. J. Comput. Appl. Math. **339**, 374–388 (2018)
50. Zúñiga-Aguilar, C.J., Romero-Ugalde, H.M., Gómez-Aguilar, J.F., Escobar-Jiménez, R.F., Valtierra-Rodríguez, M.: Solving fractional differential equations of variable-order involving operators with Mittag-Leffler kernel using artificial neural networks. Chaos Solitons Fractals **103**, 382–403 (2017)

Modeling and Analysis of Fractional Leptospirosis Model Using Atangana–Baleanu Derivative

Saif Ullah and Muhammad Altaf Khan

Abstract In this chapter, a fractional epidemic model for the leptospirosis disease with Atangana–Baleanu (AB) derivative is formulated. Initially, we present the model equilibria and basic reproduction number. The local stability of disease free equilibrium point is proved using fractional Routh Harwitz criteria. The Picard–Lindelof method is applied to show the existence and uniqueness of solutions for the model. A numerical scheme using Adams–Bashforth method for solving the proposed fractional model involving the AB derivative is presented. Finally, numerical simulations are performed in order to validate the importance of the arbitrary order derivative. The numerical result shows that the fractional order plays an important role to better understand the dynamics of disease.

Keywords Fractional calculus · Atangana–Baleanu fractional derivative · Leptospirosis model

1 Introduction

Leptospirosis is one of the most common bacterial infectious diseases which affects both humans and animals. A spiral-shaped bacteria called Leptospira is responsible for this infection. More than one million of the population are infected with this infection each year around the word. Person to person transmission of this infection is rare. It can be transferred to humans through direct or indirect contact with urine of infected animals, wounds or any other contaminated fluid. Fever, headache, chills, muscle aches, vomiting cough, conjunctival suffusion, jaundice, and sometimes a rash are included in main symptoms of this disease. The incubation period is usually

S. Ullah
Department of Mathematics, University of Peshawar, Peshawar, KP, Pakistan

M. A. Khan (✉)
Department of Mathematics, City University of Science and Information Technology, Peshawar, KP, Pakistan
e-mail: altafdir@gmail.com

J. F. Gómez et al. (eds.), *Fractional Derivatives with Mittag-Leffler Kernel*, Studies in Systems, Decision and Control 194,
https://doi.org/10.1007/978-3-030-11662-0_4

5–14 days, with a range of 2–30 days. The Leptospira bacteria appears in tropical and subtropical regions and is capable to live in urban and suburban areas. The Rodents are main carriers of this bacteria. It has been detected in more than 150 species of mammals, specially in animals like dogs, cats, cattle and pigs. Leptospirosis infection is present in almost all countries around the globe particularly, in South America, Africa, Asian Pacific region, and in Mexico [1–3].

In the last decades, mathematical models have been used to better understand the dynamics of various infectious diseases. In the existing literature a number of such models have been formulated for the transmission dynamics of leptospirosis. In [4], the authors studied the stability and controlling strategies for leptospirosis. A transmission model with possible infection of humans with animal or with free living bacteria has been presented in [5]. Analysis of the leptospirosis infection in the population of Thailand has been carried out in [6]. A model with saturated incidence rate is developed in [7]. An effective controlling strategies of transmission model with time delay have been developed in [8]. Khan et al. developed a number of models for the dynamics of leptospirosis and can be found in [9–12]. The above leptospirosis models are restricted to classical integer-order, delay or stochastic differential equations. In the present paper we extended the leptospirosis model in fractional environment using AB derivative. First we give an overview of fractional calculus (FC) and recent development in fractional mathematical models for such infectious disease.

The FC deals with fractional-order differentiation and integration which are more prominent and helpful than the classical integer order in the modeling of real phenomena due to hereditary properties and description of the memory [13, 14]. Also in the real world explanation, the integer-order derivative does not explore the dynamics between two different points. Different types of the non-local or fractional orders derivatives are suggested in the existing literature to handle the limitation of classical derivative. For example, Riemann and Liouville introduced the concept of fractional orders differentiation in [14] based on power law. In [15] a new fractional derivative using the exponential kernel has been proposed by Caputo and Fabrizio. This derivative has some problems with respect to the locality of its kernel. Recently, to overcome this limitation Atangana and Baleanu (AB) suggest another version of fractional order derivative with the help of generalized Mittag-Leffler (ML) function as non-local and non-singular kernel in [16]. Due to the use of generalized ML function as kernel the, AB derivative provides an excellent description of memory and possess the crossover properties for the mean-square displacement [17–19]. Further, the ML kernel in AB derivative guarantees no singularity, which provides good information at the beginning and at the end of the evolution of the spread. Therefore, this new derivative has been successfully implemented in the modeling of various real phenomena such as [20–39]. Recently in 2018, the AB derivative is also successfully used to model the infectious diseases. For example the transmission model for the Ebola virus with AB derivative is presented in [40]. A fractional model for the dynamics of smoking with local and non-local kernel is proposed in [41]. There is no rich literature available on fractional models of leptospirosis infection. Only few fractional models have been found in recent past but they have been restricted to ordinary fractional derivative [42, 43].

To the best of our knowledge no one has yet considered the fractional order leptospirosis model with AB derivative. Therefore, motivated by the above works, we consider a fractional order leptospirosis model using the new Atangana–Baleanu derivative [16]. From numerical simulation, one can see that the fractional leptospirosis model provides better and more flexible results than the integer-order leptospirosis model. In the next section we present basic definitions of the AB fractional derivative.

2 Preliminaries

We first give the definitions of new fractional Atangana–Baleanu derivatives with non-singular and non-local kernel, [16].

Let $g \in H^1(a,b), b > a, \sigma \in [0,1]$ then the new fractional derivatives in Caputo sense is given below:

$$ {}^{ABC}_{a}D_t^{\sigma}(g(t)) = \frac{B(\sigma)}{1-\sigma}\int_a^t g'(\xi)E_{\sigma}\left[-\sigma\frac{(t-\xi)^{\sigma}}{1-\sigma}\right]d\xi. \quad (1)$$

Let $g \in H^1(a,b)$, $b > a$, $\sigma \in [0,1]$ and not necessarily differentiable then the AB fractional derivative in Riemann-Liouville sense is given as:

$$ {}^{ABR}_{a}D_t^{\sigma}(g(t)) = \frac{B(\sigma)}{1-\sigma}\frac{d}{dt}\int_a^t g(\xi)E_{\sigma}\left[-\sigma\frac{(t-\xi)^{\sigma}}{1-\sigma}\right]d\xi. \quad (2)$$

The fractional integral associate to the new fractional derivative wit non local kernel is defined as:

$$ {}^{ABC}_{a}I_t^{\alpha}(g(t)) = \frac{1-\sigma}{B(\sigma)}g(t) + \frac{\sigma}{B(\sigma)\Gamma(\sigma)}\int_a^t g(y)(t-y)^{\sigma-1}dy. \quad (3)$$

The initial function is recovered when the fractional order turns to zero. Also when the order turns to 1 we have the classical integral.

Theorem 2.1 *Let g be a continuous function on a closed interval [a,b]. Then the following inequality is obtained [16].*

$$ \|{}^{ABC}_{a}D_t^{\sigma}(g(t))\| < \frac{B(\sigma)}{1-\sigma}\|g(x)\|, \text{ where } \|g(x)\| = max_{a\le x\le b}|g(x)|. \quad (4)$$

Theorem 2.2 *Both of (ABC) and (ABR) derivatives satisfy the Lipschitz condition given below, [16]:*

$$ \|{}^{ABC}_{a}D_t^{\sigma}g_1(t) - {}^{ABC}_{a}D_t^{\sigma}g_2(t)\| < \mathscr{K}_1\|g_1(t)-g_2(t)\|, \quad (5)$$

and

$$\|{}_{a}^{ABC}D_t^{\sigma} g_1(t) - {}_{a}^{ABC}D_t^{\sigma} g_2(t)\| < \mathcal{K}_2\|g_1(t) - g_2(t)\|. \tag{6}$$

Theorem 2.3 *The time fractional ordinary differential equation given below:*

$${}_{a}^{ABC}D_t^{\sigma} g(t) = z(t), \tag{7}$$

has a unique solution with applying the inverse Laplace transform and using the convolution theorem below [16]:

$$g(t) = \frac{1-\sigma}{ABC(\sigma)} z(t) + \frac{\sigma}{ABC(\sigma)\Gamma(\sigma)} \int_a^t z(\xi)(t-\xi)^{\sigma-1} d\xi. \tag{8}$$

3 Leptospirosis Model with AB Derivative

In this section we extend the leptospirosis disease model [44], to fractional order using generalized Mittag-Leffler function as kernel. The classical integer-order leptospirosis disease model is formulated by the following nonlinear system of differential equations:

$$\begin{aligned}
\frac{dS_H}{dt} &= b_1 - \beta_1 S_H I_H - \beta_2 S_H I_V + \lambda_H R_H - \mu_H S_H,\\
\frac{dI_H}{dt} &= \beta_1 S_H I_H + \beta_2 S_H I_V - (\mu_H + \delta_H + \gamma_H) I_H,\\
\frac{dR_H}{dt} &= \gamma_H I_H - (\mu_H + \lambda_H) R_H,\\
\frac{dS_V}{dt} &= b_2 - \gamma_V S_V - \beta_3 S_V I_H,\\
\frac{dI_V}{dt} &= \beta_3 S_V I_H - (\gamma_V + \delta_V) I_V.
\end{aligned} \tag{9}$$

In the above model (9), $S_H(t)$, $I_H(t)$ and $R_H(t)$ represent the susceptible, infected and recovered human respectively, while $S_V(t)$ and $I_V(t)$ are susceptible and infected vectors. The detail description of model parameters and numerical values are given in Table 1.

In this paper our aim is to generalize the classical leptospirosis model (9) to a fractional model by replacing the integer-order time derivative by AB derivative and can be written as below:

$$\begin{aligned}
{}_{0}^{ABC}D_t^{\sigma} S_H &= b_1 - \beta_1 S_H I_H - \beta_2 S_H I_V + \lambda_H R_H - \mu_H S_H,\\
{}_{0}^{ABC}D_t^{\sigma} I_H &= \beta_1 S_H I_H + \beta_2 S_H I_V - (\mu_H + \delta_H + \gamma_H) I_H,\\
{}_{0}^{ABC}D_t^{\sigma} R_H &= \gamma_H I_H - (\mu_H + \lambda_H) R_H,\\
{}_{0}^{ABC}D_t^{\sigma} S_V &= b_2 - \gamma_V S_V - \beta_3 S_V I_H,\\
{}_{0}^{ABC}D_t^{\sigma} I_V &= \beta_3 S_V I_H - (\gamma_V + \delta_V) I_V.
\end{aligned} \tag{10}$$

Table 1 Description of the parameters and variables for the leptospirosis model (10)

Parameter	Description	Numerical values
b_1	Human recruitment rate	0.9
b_2	Vector recruitment rate	1.2
λ_H	Recovery rate of infected human	0.066
δ_H	Disease induced death rate in human	0.0001
μ_H	Human natural death rate	0.00046
δ_V	Disease induced death rate in vector	0.0001
γ_V	Natural death rate in vector	0.0018
γ_H	Transfer rate of I_H to R_H	0.027
β_1	Contact rate of susceptible and infected human	0.0013
β_2	Contact rate of infected vector with suspectable human	0.0089
β_3	Contact rate of susceptible vector and infected human	0.0079

The initial conditions involved in (10) are

$$S_H(0) = c_1,\ I_H(0) = c_2,\ R_H(0) = c_3,\ S_V(0) = c_4,\ and\ I_V(0) = c_5. \tag{11}$$

The disease free equilibrium $\mathscr{M}_0$ of system (10) is obtained by solving the following system:

$${}_0^{ABC}D_t^{\sigma} S_H = {}_0^{ABC} D_t^{\sigma} I_H = {}_0^{ABC} D_t^{\sigma} R_H = {}_0^{ABC} D_t^{\sigma} S_V = {}_0^{ABC} D_t^{\sigma} I_V = 0,$$

and is given by $\mathscr{M}_0 = (\frac{b_1}{\mu_H}, 0, 0, \frac{b_2}{\gamma_V}, 0)$. The endemic equilibria denoted by $\mathscr{M}_1$ is given by

$$\begin{cases} S_H^{**} &= \frac{(\delta_V+\gamma_V)(\delta_H+\gamma_H+\mu_H)(\gamma_V+\beta_3 I_H^{**})}{b_2\beta_2\beta_3+\beta_1(\delta_V+\gamma_V)(\gamma_V+\beta_3 I_H)}, \\ S_V^{**} &= \frac{b_2}{\gamma_V+\beta_3 I_H^{**}}, \\ I_V^{**} &= \frac{\beta_3 b_2 I_H^{**}}{(\delta_V+\gamma_V)(\gamma_V+\beta_3 I_H^{**})}, \\ R_H^{**} &= \frac{\gamma_H I_H^{**}}{\mu_H+\lambda_H}. \end{cases}$$

The basic reproduction number $\mathscr{R}_0$, is the number of secondary cases that one case would produce in a completely susceptible population of the model obtained by using the next generation technique [45] and is given below:

$$\mathscr{R}_0 = \mathscr{R}_{01} + \mathscr{R}_{02},$$

where

$$\mathscr{R}_{01} = \frac{\beta_2\beta_3 b_1 b_2}{\gamma_V \mu_H(\delta_V + \gamma_V)(\delta_H + \gamma_H + \mu_H)}, \quad \mathscr{R}_{02} = \frac{\beta_1 b_1}{\mu_H(\delta_H + \gamma_H + \mu_H)}.$$

Next, to proceed with the stability of the DFE, first, we calculate the following Jacobian of the linearized system (10) as below:

$$J_{\mathscr{M}_0} = \begin{pmatrix} -\mu_H & -\beta_1 S_H^0 & \lambda_H & 0 & -\beta_2 S_H^0 \\ 0 & -(\delta_H + \gamma_H + \mu_H) + \beta_1 S_H^0 & 0 & 0 & \beta_2 S_H^0 \\ 0 & \gamma_H & -(\mu_H + \lambda_H) & 0 & 0 \\ 0 & \beta_3 S_V^0 & 0 & -\gamma_V & 0 \\ 0 & \beta_3 S_V^0 & 0 & 0 & -(\delta_V + \gamma_V) \end{pmatrix}.$$

Theorem 3.1 *The DFE point $\mathscr{M}_0$ of the model (10) is locally asymptotically stable if $\mathscr{R}_0 < 1$.*

Proof The characteristic equation of the matrix $J_{\mathscr{M}_0}$ is given by

$$(\Lambda + \mu_H)(\Lambda + (\mu_H + \lambda_H))(\Lambda + \gamma_V) \times (\Lambda^2 + q_1\Lambda + q_2) = 0, \tag{12}$$

where

$$q_1 = (\delta_V + \gamma_V) + (\delta_H + \gamma_H + \mu_H)(1 - \mathscr{R}_{02}),$$

and

$$q_2 = (\delta_V + \gamma_V)(\delta_H + \gamma_H + \mu_H)(1 - \mathscr{R}_0).$$

The arguments of the roots of the equation $\Lambda + \mu_H = 0$, $\Lambda + (\mu_H + \lambda_H) = 0$ and $\Lambda + \gamma_V = 0$ clearly satisfy the conditions $|arg\Lambda_i| > \alpha\frac{\pi}{2}$.

Further, it is clear that if $\mathscr{R}_0 < 1$, then $q_1 > 0$ and $q_2 > 0$. Hence using Ahmed et al. [46], the Routh–Hurwitz conditions are necessary and sufficient for Matignon's conditions to be satisfied. Thus, the disease-free equilibrium is locally asymptotically stable for $\alpha \in (0, 1)$ if $\mathscr{R}_0 < 1$ and unstable otherwise.

4 Existence of Solutions for Fractional Leptospirosis Infection Model

Since the model (10) is nonlinear and nonlocal, there is no particular method to provide the exact solutions of this system. However, the existence of a solution guarantees that under some conditions one will have the exact solution. In this section, we present the existence of the solution of the proposed model in detail via fixed-point theorem Picard–Lindelof approach.

Let us re-write the system (10) in the following convenient form

$$\begin{aligned} {}_0^{ABC}D_t^{\sigma}[S_H(t)] &= K_1(t, S_H),\\ {}_0^{ABC}D_t^{\sigma}[I_H(t)] &= K_2(t, I_H),\\ {}_0^{ABC}D_t^{\sigma}[R_H(t)] &= K_3(t, R_H),\\ {}_0^{ABC}D_t^{\sigma}[S_V(t)] &= K_4(t, S_V),\\ {}_0^{ABC}D_t^{\sigma}[I_V(t)] &= K_5(t, I_V). \end{aligned} \tag{13}$$

Now using theorem 3, the system (13) can be converted to the Volterra type integral equation with the AB fractional integral as below.

$$\begin{aligned} S_H(t) - S_H(0) &= \tfrac{(1-\sigma)}{ABC(\sigma)} K_1(t, S_H) + \tfrac{\sigma}{ABC(\sigma)\Gamma(\sigma)} \int_0^t K_1(y, S_H)(t-y)^{\sigma-1}dy,\\ I_H(t) - I_H(0) &= \tfrac{(1-\sigma)}{ABC(\sigma)} K_2(t, I_H) + \tfrac{\sigma}{ABC(\sigma)\Gamma(\sigma)} \int_0^t K_2(y, I_H)(t-y)^{\sigma-1}dy,\\ R_H(t) - R_H(0) &= \tfrac{(1-\sigma)}{ABC(\sigma)} K_3(t, R_H) + \tfrac{\sigma}{ABC(\sigma)\Gamma(\sigma)} \int_0^t K_3(y, R_H)(t-y)^{\sigma-1}dy, \quad (14)\\ S_V(t) - S_V(0) &= \tfrac{(1-\sigma)}{ABC(\sigma)} K_4(t, S_V) + \tfrac{\sigma}{ABC(\sigma)\Gamma(\sigma)} \int_0^t K_4(y, S_V)(t-y)^{\sigma-1}dy,\\ I_V(t) - I_V(0) &= \tfrac{(1-\sigma)}{ABC(\sigma)} K_5(t, I_V) + \tfrac{\sigma}{ABC(\sigma)\Gamma(\sigma)} \int_0^t K_5(y, I_V)(t-y)^{\sigma-1}dy. \end{aligned}$$

Theorem 4.1 *The kernel K_1 satisfy the Lipchitz condition and contraction if the inequality given below holds*

$$0 \le (\beta_1 a_1 + \beta_2 a_2 + \mu_H) < 1.$$

Proof Let S_H and S_{1H} be two functions, then

$$\begin{aligned} ||K_1(t, S_H) - K_1(t, S_{1H})|| &= ||-(\beta_1 I_H + \beta_2 I_V + \mu_H)(S_H - S_{1H})||\\ &\le (\beta_1\|I_H\| + \beta_2\|I_V\| + \mu_H)\|S_H(t) - S_H(t_1)\|\\ &\le (\beta_1 a_1 + \beta_2 a_2 + \mu_H)\|S_H(t) - S_H(t_1)\|\\ &\le M||S(t) - S(t_1)||. \end{aligned} \tag{15}$$

Where $M = (\beta_1 a_1 + \beta_2 a_2 + \mu_H)$, $||I_H(t)|| \le a_1$ and $||I_V(t)|| \le a_2$. Hence

$$||K_1(t, S_H) - K_1(t, S_{1H})|| \le M||S(t) - S(t_1)||. \tag{16}$$

Hence, for K_1 the Lipschitz condition is obtained. Similarly for the remaining cases the Lipschiz condition can be easily verified.

Theorem 4.2 *The solution of leptospirosis fractional model given in (10) will exist and will be unique under the conditions given below holds.*

$$\frac{(1-\sigma)}{ABC(\sigma)}M + \frac{T^{\sigma}\sigma}{ABC(\sigma)\Gamma(\sigma)}M < 1. \tag{17}$$

Proof To proof the above theorem we apply the Picard–Lindelof approach along with fixed theorem. To proceed let us consider

$$M_1 = sup_{A[c,v_1]}\|K_1(t, S_H)\|,\ M_2 = sup_{A[c,v_2]}\|K_2(t, I_H)\|,\ M_3 = sup_{A[c,v_3]}\|K_3(t, R_H)\|,$$
$$M_3 = sup_{A[c,v_4]}\|K_4(t, S_V)\|,\ M_5 = sup_{A[c,v_5]}\|K_5(t, I_V)\|,$$

where

$$\begin{aligned}
A[c, v_1] &= [t-c, t+c] \times [x - v_1, x + v_1] = C \times V_1,\\
A[c, v_2] &= [t-c, t+c] \times [x - v_2, x + v_2] = C \times V_2,\\
A[c, v_3] &= [t-c, t+c] \times [x - v_3, x + v_3] = C \times V_3,\\
A[c, v_4] &= [t-c, t+c] \times [x - v_4, x + v_4] = C \times V_4,\\
A[c, v_5] &= [t-c, t+c] \times [x - v_5, x + v_5] = C \times V_5.
\end{aligned}$$

We will employ the fixed point theorem using the metric on $C[c, v_i]$ together with uniform norm given below:

$$\|F(t)\|_\infty = sup_{t\in C}|F(t)|. \tag{18}$$

Consider the following Picard's operator defined between two continuous functional spaces.

$$\Psi : A(C, V_1, V_2, V_3, V_4, V_5) \to A(C, V_1, V_2, V_3, V_4, V_5), \tag{19}$$

defined as bellow:

$$\Psi[F(t)] = F_0(t) + K(t, F(t))\frac{(1-\sigma)}{ABC(\sigma)} + \frac{\sigma}{ABC(\sigma)\Gamma(\sigma)}\int_0^t (t-y)^{\sigma-1}K(y, F(y))dy,$$

where

$$F(t) = (S_H(t), I_H(t), R_H(t), S_V(t), I_V(t)),\quad F_0(t) = (h_1, h_2, h_3, h_4, h_5),$$

and

$$K(t, F(t)) = (K_1(t, S_H), K_2(t, I_H), K_3(t, R_H), K_4(t, S_V), K_5(t, I_V)).$$

Now we compute the following

$$\|\Psi[F_1(t)] - \Psi[F_2(t)]\| = sup_{t\in C}|\Psi[F_1(t)] - \Psi[F_2(t)]|. \tag{20}$$

Applying definition of the operator Ψ in Eq. (1), we have

$$\|\Psi[F_1(t)] - \Psi[F_2(t)]\| \leq \left\| \frac{(1-\sigma)}{ABC(\sigma)}(K(t, F_1(t)) - K(t, F_2(t)) + \frac{\sigma}{ABC(\sigma)\Gamma(\sigma)} \times \int_0^t (t-y)^{\sigma-1}(K(y, F_1(y)) - K(y, F_2(y)))dy \right\|. \tag{21}$$

After using the triangular inequality and Lipschiz condition, we obtained

$$\|\Psi[F_1(t)] - \Psi[F_2(t)]\| \leq \left(\frac{(1-\sigma)M}{ABC(\sigma)} + \frac{\sigma}{ABC(\sigma)\Gamma(\sigma)} MT^{\sigma} \right) \|F_1(t) - F_2(t)\|.$$

Thus we obtained

$$\|\Psi[F_1(t)] - \Psi[F_2(t)]\| \leq L\|F_1(t) - F_2(t)\|, \tag{22}$$

where

$$L = \frac{(1-\sigma)M}{ABC(\sigma)} + \frac{\sigma}{ABC(\sigma)\Gamma(\sigma)} MT^{\sigma}.$$

The operator Ψ will be a contraction if condition (17) fulfil. Hence due to Banach fixed point theorem, there exists a unique solution of the model (13).

5 Numerical Results

In this section, we present a numerical solution of the fractional order model (10). Then The numerical simulations are obtained using the proposed scheme. For this purpose we use the fractional Adams Bashforth method [34] to approximate the AB fractional integral operator. To obtain an iterative scheme, we proceed with the first equation of the system (14) as below:

$$S_H(t) - S_H(0) = \frac{(1-\sigma)}{ABC(\sigma)} K_1(t, S_H) + \frac{\sigma}{ABC(\sigma)\Gamma(\sigma)} \int_0^t K_1(y, S_H)(t-y)^{\sigma-1} dy.$$

At $t = t_{n+1}$, $n = 0, 1, 2, \ldots$, we have

$$\begin{aligned} S_H(t_{n+1}) - S_H(0) &= \frac{1-\sigma}{ABC(\sigma)} K_1(t_n, S_H) + \\ &\quad \frac{\sigma}{ABC(\sigma) \times \Gamma(\sigma)} \int_0^{t_{n+1}} K_1(y, S_H)(t_{n+1} - y)^{\sigma-1} dy, \\ &= \frac{1-\sigma}{ABC(\sigma)} K_1(t_n, S_H) + \\ &\quad \frac{\sigma}{ABC(\sigma) \times \Gamma(\sigma)} \sum_{k=0}^{n} \int_{t_k}^{t_{k+1}} K_1(y, S_H)(t_{n+1} - y)^{\sigma-1} dy. \end{aligned} \tag{23}$$

The function $K_1(y, S_H)$ can be approximated over $[t_k, t_{k+1}]$, using the interpolation polynomial

$$K_1(y, S_H) \cong \frac{K_1(t_k, S_H(t_k))}{h}(t - t_{k-1}) - \frac{K_1(t_{k-1}, S_H(t_{k-1})),}{h}(t - t_k) \quad (24)$$

which gives

$$\begin{aligned} S_H(t_{n+1}) &= S_H(0) + \frac{1-\sigma}{ABC(\sigma)}K_1(t_n, S_H) + \\ &\frac{\sigma}{ABC(\sigma)\Gamma(\sigma)}\sum_{k=0}^{n}\Big(\frac{K_1(t_k, S_H(t_k))}{h}\int_{t_k}^{t_{k+1}}(t - t_{k-1})(t_{n+1} - t)^{\sigma-1}dt \\ &-\frac{K_1(t_{k-1}, S_H(t_{k-1}))}{h}\int_{t_k}^{t_{k+1}}(t - t_k)(t_{n+1} - t)^{\sigma-1}dt\Big). \end{aligned} \quad (25)$$

Now

$$I_{\sigma,1} = \int_{t_k}^{t_{k+1}}(t - t_{k-1})(t_{n+1} - t)^{\sigma-1}dt, \quad (26)$$

and

$$I_{\sigma,2} = \int_{t_k}^{t_{k+1}}(t - t_k)(t_{n+1} - t)^{\sigma-1}dt. \quad (27)$$

Calculating these integrals we get

$$I_{\sigma,1} = h^{\sigma+1}\,\frac{(n+1-k)^{\sigma}(n-k+2+\sigma) - (n-k)^{\sigma}(n-k+2+2\sigma)}{\sigma(\sigma+1)}, \quad (28)$$

and

$$I_{\sigma,2} = h^{\sigma+1}\,\frac{(n+1-k)^{\sigma} - (n-k)^{\sigma}(n-k+1+\sigma)}{\sigma(\sigma+1)}. \quad (29)$$

Finally

$$\begin{aligned} S_H(t_{n+1}) &= S_H(t_0) + \frac{1-\sigma}{ABC(\sigma)}K_1(t_n, S_H(t_n)) + \\ &\frac{\sigma}{ABC(\sigma)\Gamma(\sigma)}\sum_{k=0}^{n} \\ &\Big(\frac{h^{\sigma}K_1(t_k, S_H(t_k))}{\Gamma(\sigma+2)}((n+1-k)^{\sigma}(n-k+2+\sigma) - (n-k)^{\sigma}(n-k+2+2\sigma)) \\ &-\frac{h^{\sigma}K_1(t_{k-1}, S_H(t_{k-1}))}{\Gamma(\sigma+2)}((n+1-k)^{\sigma+1} - (n-k)^{\sigma}(n-k+1+\sigma))\Big). \end{aligned} \quad (30)$$

In similar way for the rest of equations of system (10) we obtained the recursive formula as below:

$$
\begin{aligned}
I_H(t_{n+1}) &= I_H(t_0) + \frac{1-\sigma}{ABC(\sigma)} K_2(t_n, I_H(t_n)) + \\
&\frac{\sigma}{ABC(\sigma)\Gamma(\sigma)} \sum_{k=0}^{n} \\
&\left(\frac{h^\sigma K_2(t_k, I_H(t_k))}{\Gamma(\sigma+2)} ((n+1-k)^\sigma (n-k+2+\sigma) - (n-k)^\sigma (n-k+2+2\sigma)) \right. \\
&\left. - \frac{h^\sigma K_2(t_{k-1}, I_H(t_{k-1}))}{\Gamma(\sigma+2)} ((n+1-k)^{\sigma+1} - (n-k)^\sigma (n-k+1+\sigma)) \right), \\
R_H(t_{n+1}) &= R_H(t_0) + \frac{1-\sigma}{ABC(\sigma)} K_3(t_n, R_H(t_n)) + \\
&\frac{\sigma}{ABC(\sigma)\Gamma(\sigma)} \sum_{k=0}^{n} \\
&\left(\frac{h^\sigma K_3(t_k, R_H(t_k))}{\Gamma(\sigma+2)} ((n+1-k)^\sigma (n-k+2+\sigma) - (n-k)^\sigma (n-k+2+2\sigma)) \right. \\
&\left. - \frac{h^\sigma K_3(t_{k-1}, R_H(t_{k-1}))}{\Gamma(\sigma+2)} ((n+1-k)^{\sigma+1} - (n-k)^\sigma (n-k+1+\sigma)) \right), \\
S_V(t_{n+1}) &= S_V(t_0) + \frac{1-\sigma}{ABC(\sigma)} K_4(t_n, S_V(t_n)) + \\
&\frac{\sigma}{ABC(\sigma)\Gamma(\sigma)} \sum_{k=0}^{n} \\
&\left(\frac{h^\sigma K_4(t_k, S_V(t_k))}{\Gamma(\sigma+2)} ((n+1-k)^\sigma (n-k+2+\sigma) - (n-k)^\sigma (n-k+2+2\sigma)) \right. \\
&\left. - \frac{h^\sigma K_4(t_{k-1}, S_V(t_{k-1}))}{\Gamma(\sigma+2)} ((n+1-k)^{\sigma+1} - (n-k)^\sigma (n-k+1+\sigma)) \right), \\
I_V(t_{n+1}) &= I_V(t_0) + \frac{1-\sigma}{ABC(\sigma)} K_5(t_n, I_V(t_n)) + \\
&\frac{\sigma}{ABC(\sigma)\Gamma(\sigma)} \sum_{k=0}^{n} \\
&\left(\frac{h^\sigma K_5(t_k, I_V(t_k))}{\Gamma(\sigma+2)} ((n+1-k)^\sigma (n-k+2+\sigma) - (n-k)^\sigma (n-k+2+2\sigma)) \right. \\
&\left. - \frac{h^\sigma K_5(t_{k-1}, I_V(t_{k-1}))}{\Gamma(\sigma+2)} ((n+1-k)^{\sigma+1} - (n-k)^\sigma (n-k+1+\sigma)) \right). \qquad (31)
\end{aligned}
$$

Now to give the numerical simulations of the fractional order model (10) with AB derivative, we apply the iterative solution given in (30) and (31). The time level is taken up to 100 unit. The numerical values of the parameters used in the simulations are given in Table 1. The behavior of the individuals of the model (10), for $\sigma = 1, 0.90, 0.80, 0.70, 0.60$, is given in Figs. 1, 2, 3, 4 and 5 respectively such that in each figure the solid line represents the model simulations when its order is 1 while

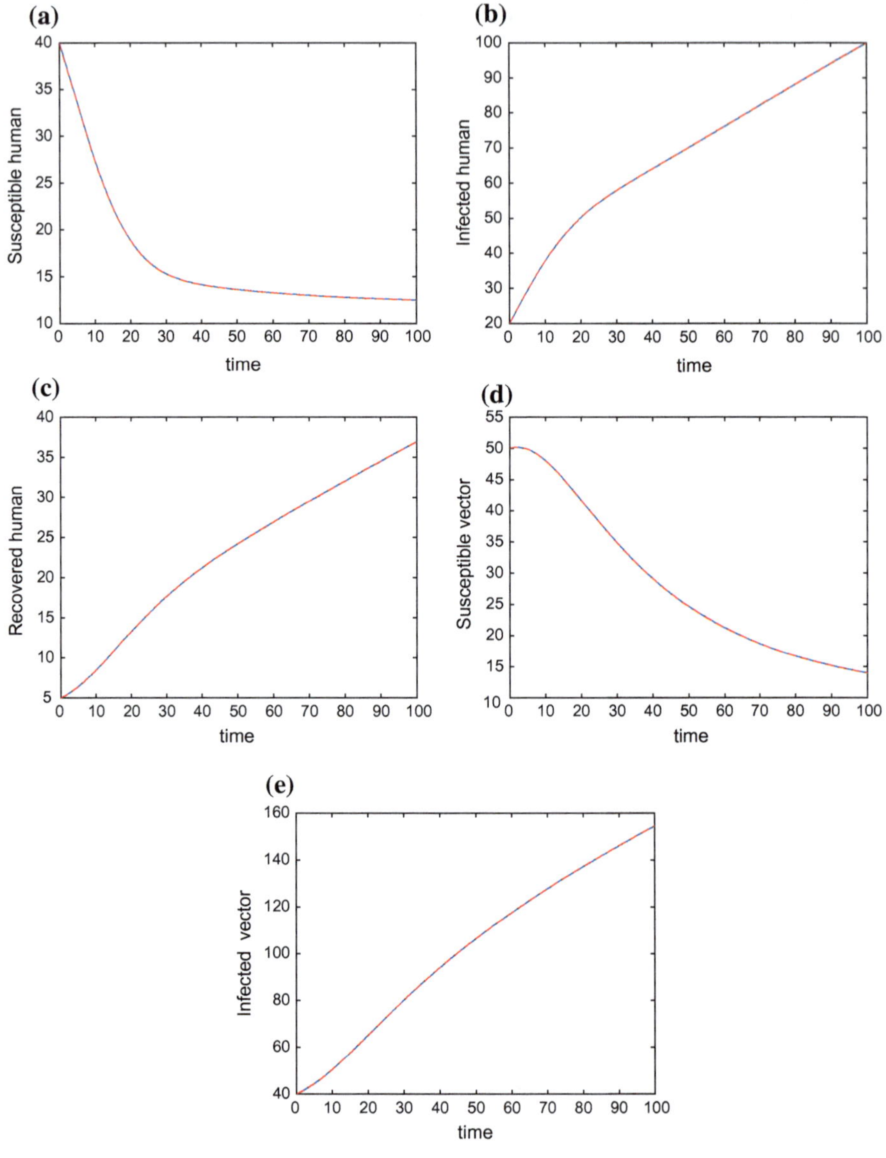

Fig. 1 Simulation of leptospirosis model (10) using AB derivative for $\sigma = 1$

the doted line is the graph of the model for specific values of σ other than 1. The graphical results show that by decreasing the value of fractional parameter σ, the number of susceptible human and vectors increases while the number of infected classes decreases significantly.

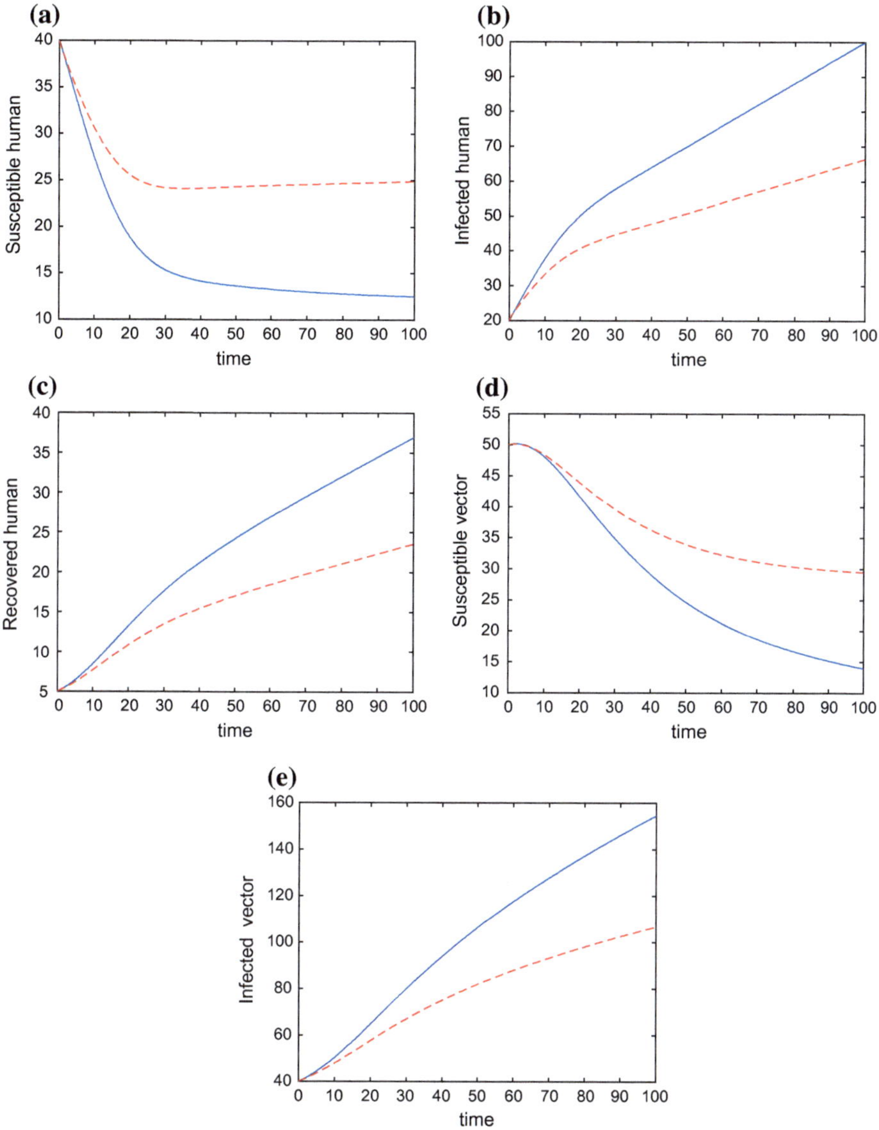

Fig. 2 Simulation of leptospirosis model (10) using AB derivative for $\sigma = 0.90$

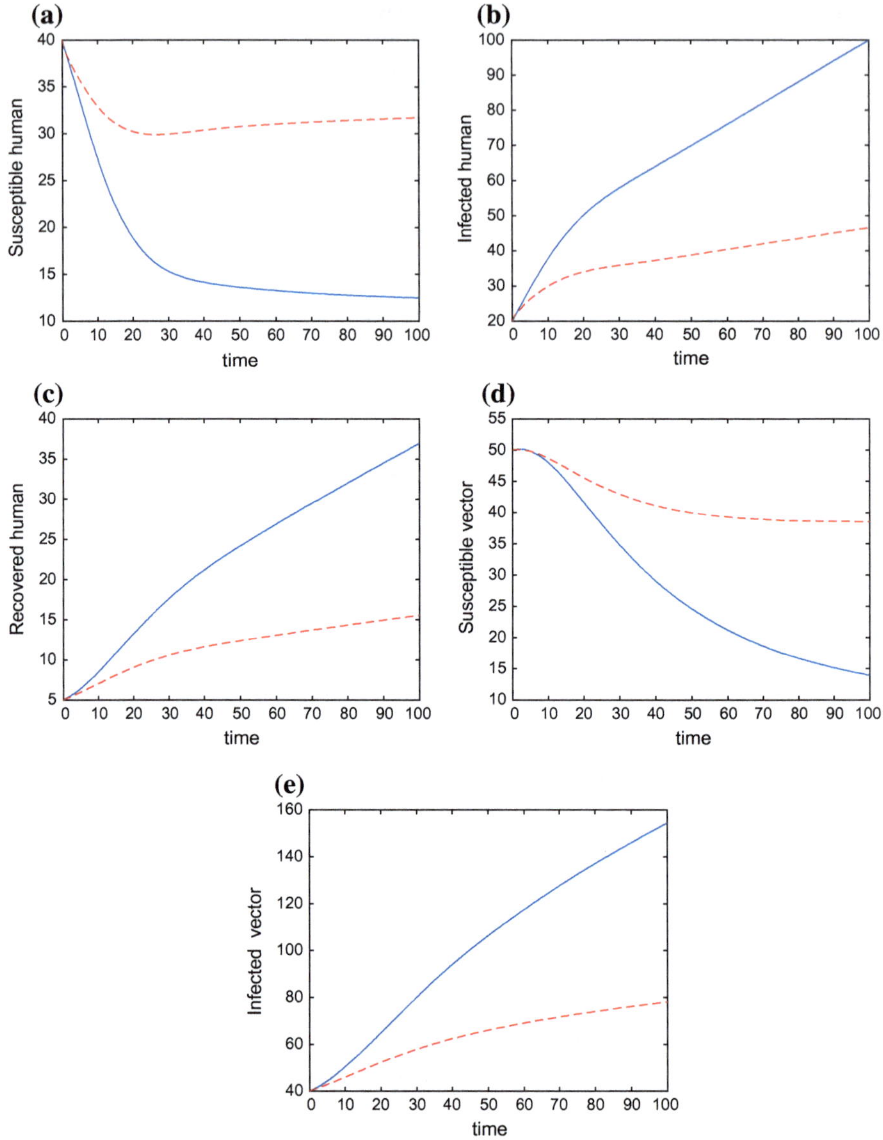

Fig. 3 Simulation of leptospirosis model (10) using AB derivative for $\sigma = 0.80$

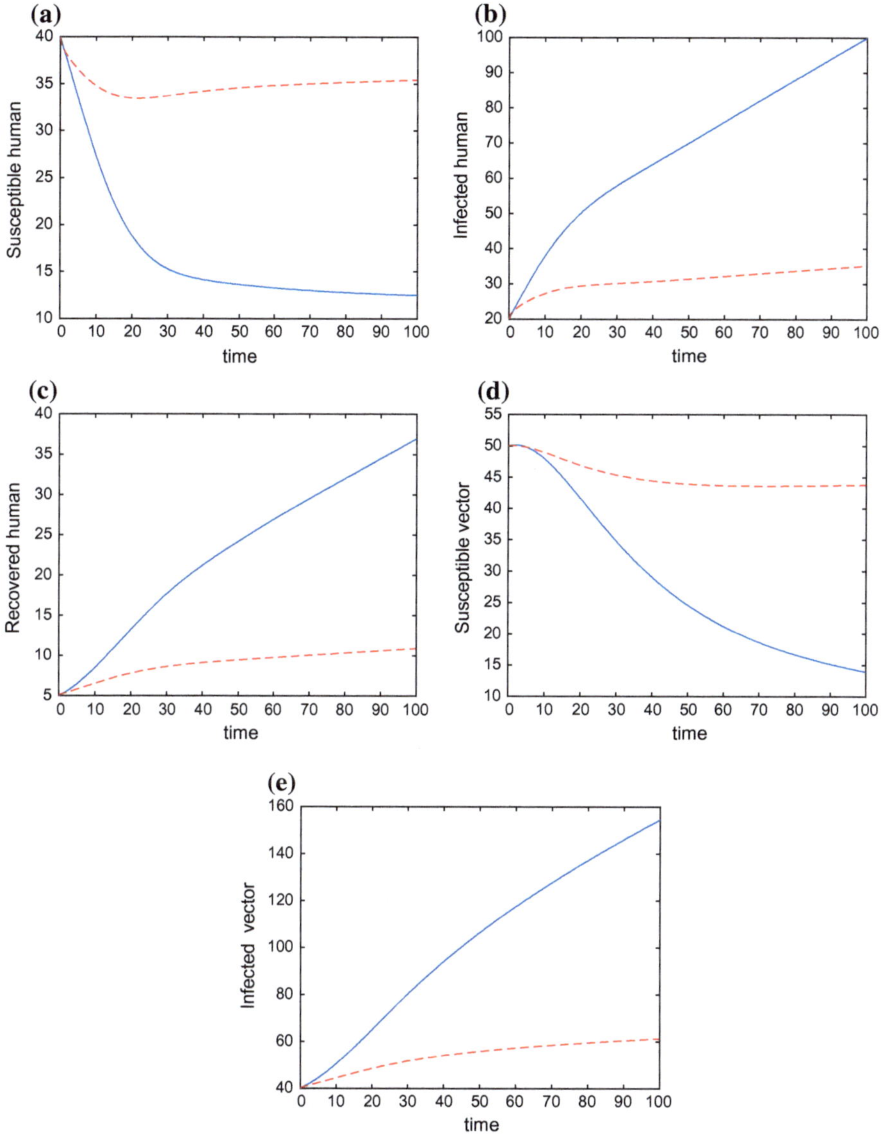

Fig. 4 Simulation of leptospirosis model (10) using AB derivative for $\sigma = 0.70$

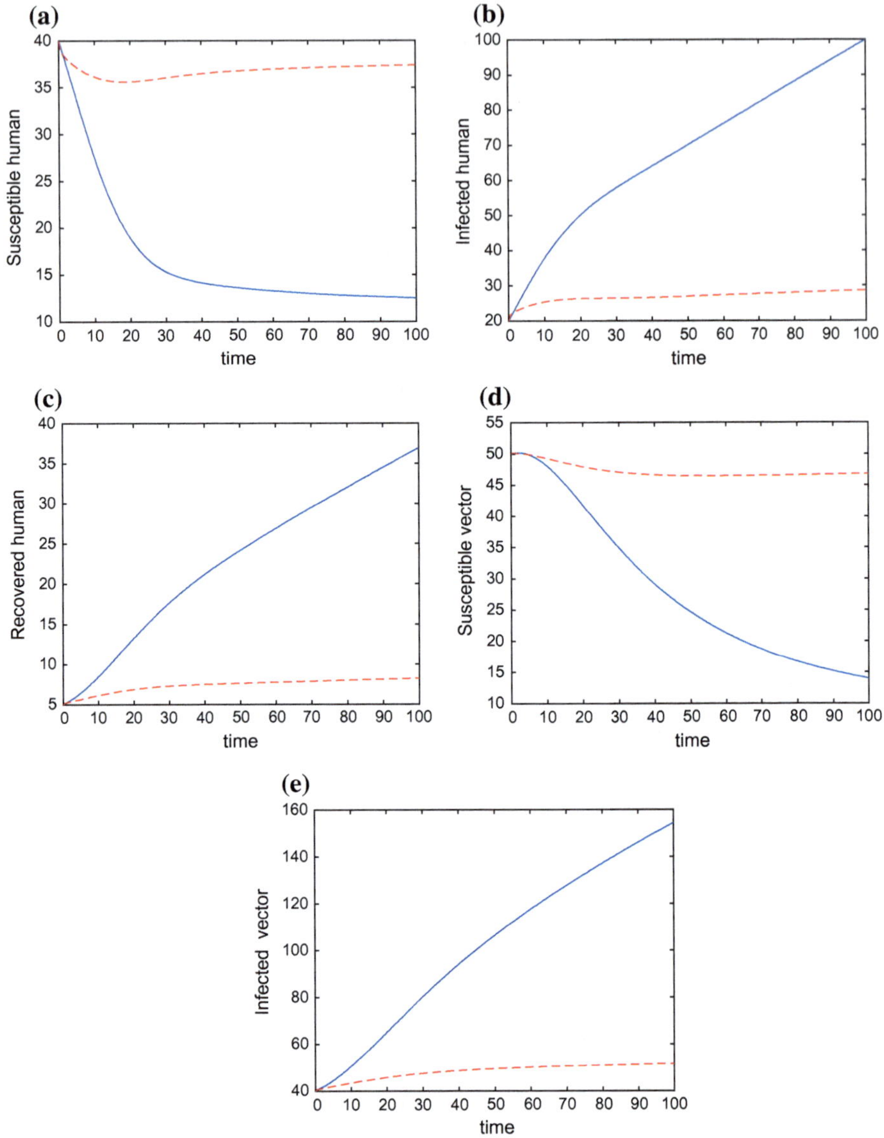

Fig. 5 Simulation of leptospirosis model (10) using AB derivative for $\sigma = 0.60$

6 Conclusion

A fractional order model for leptospirosis disease transmission with AB-derivative is analyzed. The model equilibria, their steady state and basic properties are explored. The local stability analysis is carried out using Matignon's conditions. The existence and uniqueness of solutions for the model with AB derivative is proved in detail using Picard–Lindelof technique. The model numerical solution using two step fractional modified Adams–Bashforth scheme with AB derivative is obtained. By taking different values of the fractional parameter σ, the numerical simulations are obtained and briefly discussed. The graphical results show that the new fractional model continuously depend on the fractional parameter σ and provide better and more flexible information to explore the dynamics of the leptospirosis infection.

References

1. Marr, J.S., Cathey, J.T.: New hypothesis for cause of epidemic among native 236 americans, new england, 1616–1619. Emerg. Infect. Dis. **16**(2), 1–281 (2010)
2. Victoriano, A.F.B., Smythe, L.D., Barzaga, N.G., Cavinta, L.L., Kasai, T., Limpakarnjanarat, K., Ong, B.L., Gongal, G., Hall, J., Coulombe, C.A.: Leptospirosis in the asia pacific region. BMC Infect. Dis. **9**(1), 1–147 (2009)
3. Holt, J., Davis, S., Leirs, H.: A model of leptospirosis infection in an African rodent to determine risk to humans: seasonal fluctuations and the impact of rodent control. Acta Trop. **99**(2–3), 218–225 (2006)
4. Okosun, K.O., Mukamuri, M., Makinde, D.O.: Global stability analysis and control of leptospirosis. Open Math. **14**(1), 567–585 (2016)
5. Baca-Carrasco, D., Olmos, D., Barradas, I.: A mathematical model for human and animal leptospirosis. J. Biol. Syst. **23**(01), 55–65 (2015)
6. Chadsuthi, S., Modchang, C., Lenbury, Y., Iamsirithaworn, S., Triampo, W.: Modeling seasonal leptospirosis transmission and its association with rainfall and temperature in Thailand using time-series and ARIMAX analyses. Asian Pac. J. Trop. Med. **5**(7), 539–546 (2012)
7. Khan, M.A., Saddiq, S.F., Islam, S., Khan, I., Shafie, S.: Dynamic behavior of leptospirosis disease with saturated incidence rate. Int. J. Appl. Comput. Math. **2**(4), 435–452 (2016)
8. Sadiq, S.F., Khan, M.A., Islam, S., Zaman, G., Jung, H., Khan, S.A.: Optimal control of an epidemic model of leptospirosis with nonlinear saturated incidences. Annu. Res. Rev. Biol. **4**(3), 560–576 (2014)
9. Khan, M.A., Zaman, G., Islam, S., Chohan, M.I.: Optimal campaign in leptospirosis epidemic by multiple control variables. Appl. Math. **3**(11), 1655–1663 (2012)
10. Khan, M.A., Islam, S., Khan, S.A., Khan, I., Shafie, S., Gul, T.: Prevention of Leptospirosis infected vector and human population by multiple control variables. Abstr. Appl. Anal. (Hindawi) **1**, 1–14 (2014)
11. Sadiq, S.F., Khan, M.A., Islam, S., Zaman, G., Jung, H., Khan, S.A.: Optimal control of an epidemic model of leptospirosis with nonlinear saturated incidences. Annu. Res. Rev. Biol. **4**(3), 560–576 (2014)
12. Khan, M.A., Islam, S., Khan, S.A.: Mathematical modeling towards the dynamical interaction of leptospirosis. Appl. Math. Inf. Sci. **8**(3), 1–8 (2014)
13. Podlubny, I.: Fractional Differential Equations: An Introduction to Fractional Derivatives, Fractional Differential Equations, to Methods of Their Solution and Some of Their Applications. Academic, San Diego (1999)

14. Samko, S.G., Kilbas, A.A., Marichev, O.I.: Fractional Integrals and Derivatives: Theory and Applications. Gordon and Breach, Yverdon (1993)
15. Caputo, M., Fabrizio, M.: A new definition of fractional derivative without singular kernel. Prog. Fract. Differ. Appl. **1**(2), 73–85 (2015)
16. Atangana, A., Baleanu, D.: New fractional derivatives with non-local and non-singular kernel: theory and application to heat transfer model. Therm. Sci. **20**(2), 763–769 (2016)
17. Atangana, A.: Non validity of index law in fractional calculus: a fractional differential operator with Markovian and non-Markovian properties. Phys. A Stat. Mech. Appl. **505**, 688–706 (2018)
18. Atangana, A., Gómez-Aguilar, J.F.: Fractional derivatives with no-index law property: application to chaos and statistics. Chaos Solitons Fractals **114**, 516–535 (2018)
19. Atangana, A., Gómez-Aguilar, J.F.: Decolonisation of fractional calculus rules: breaking commutativity and associativity to capture more natural phenomena. Eur. Phys. J. Plus **133**, 1–22 (2018)
20. Atangana, A., Koca, I.: Chaos in a simple nonlinear system with Atangana-Baleanu derivatives with fractional order. Chaos Solitons Fractals **89**, 447–454 (2016)
21. Atangana, A., Gómez-Aguilar, J.F.: A new derivative with normal distribution kernel: theory, methods and applications. Phys. A Stat. Mech. Appl. **476**, 1–14 (2017)
22. Gómez-Aguilar, J.F., Dumitru, B.: Fractional transmission line with losses. Zeitschrift für Naturforschung A **69**(10–11), 539–546 (2014)
23. Gómez-Aguilar, J.F., Torres, L., Yépez-Martínez, H., Baleanu, D., Reyes, J.M., Sosa, I.O.: Fractional Liénard type model of a pipeline within the fractional derivative without singular kernel. Adv. Differ. Equ. **2016**(1), 1–17 (2016)
24. Yépez-Martínez, H., Gómez-Aguilar, J.F., Sosa, I.O., Reyes, J.M., Torres-Jiménez, J.: The Feng's first integral method applied to the nonlinear mKdV space-time fractional partial differential equation. Rev. Mex. Fís **62**(4), 310–316 (2016)
25. Saad, K.M., Gómez-Aguilar, J.F.: Analysis of reaction-diffusion system via a new fractional derivative with non-singular kernel. Phys. A Stat. Mech. Appl. **509**, 703–716 (2018)
26. Gómez-Aguilar, J.F., Escobar-Jiménez, R.F., López-López, M.G., Alvarado-Martínez, V.M.: Atangana-Baleanu fractional derivative applied to electromagnetic waves in dielectric media. J. Electromagn. Waves Appl. **30**(15), 1937–1952 (2016)
27. Coronel-Escamilla, A., Gómez-Aguilar, J.F., Baleanu, D., Córdova-Fraga, T., Escobar-Jiménez, R.F., Olivares-Peregrino, V.H., Qurashi, M.M.A.l.: Bateman-Feshbach tikochinsky and Caldirola–Kanai oscillators with new fractional differentiation. Entropy **19**(2), 1–21 (2017)
28. Gómez-Aguilar, J.F., Yépez-Martínez, H., Escobar-Jiménez, R.F., Astorga-Zaragoza, C.M., Morales-Mendoza, L.J., González-Lee, M.: Universal character of the fractional space-time electromagnetic waves in dielectric media. J. Electromagn. Waves Appl. **29**(6), 727–740 (2015)
29. Saad, K.M., Gómez-Aguilar, J.F.: Coupled reaction-diffusion waves in a chemical system via fractional derivatives in Liouville-Caputo sense. Rev. Mex. Fís. **64**(5), 539–547 (2018)
30. Coronel-Escamilla, A., Gómez-Aguilar, J.F., López-López, M.G., Alvarado-Martínez, V.M., Guerrero-Ramírez, G.V.: Triple pendulum model involving fractional derivatives with different kernels. Chaos Solitons Fractals **91**, 248–261 (2016)
31. Gómez-Aguilar, J.F., Atangana, A.: New insight in fractional differentiation: power, exponential decay and Mittag-Leffler laws and applications. Eur. Phys. J. Plus **132**(1), 1–13 (2017)
32. Alkahtani, B.S.T.: Chua's circuit model with Atangana-Baleanu derivative with fractional order. Chaos Solitons Fractals **89**, 547–551 (2016)
33. Gómez-Aguilar, J.F., Yépez-Martínez, H., Escobar-Jiménez, R.F., Astorga-Zaragoza, C.M., Reyes-Reyes, J.: Analytical and numerical solutions of electrical circuits described by fractional derivatives. Appl. Math. Model. **40**(21–22), 9079–9094 (2016)
34. Toufik, M., Atangana, A.: New numerical approximation of fractional derivative with non-local and non-singular kernel: application to chaotic models. Eur. Phys. J. Plus **132**, 444 (2017)
35. Algahtani, O.J.J.: Comparing the Atangana-Baleanu and Caputo-Fabrizio derivative with fractional order: Allen Cahn model. Chaos Solitons Fractals **89**, 552–559 (2016)
36. Gómez-Aguilar, J.F.: Analytical and numerical solutions of a nonlinear alcoholism model via variable-order fractional differential equations. Phys. A Stat. Mech. Appl. **494**, 52–75 (2018)

37. Coronel-Escamilla, A., Gómez-Aguilar, J.F., Torres, L., Valtierra-Rodríguez, M., Escobar-Jiménez, R.F.: Design of a state observer to approximate signals by using the concept of fractional variable-order derivative. Digit. Signal Process. **69**, 127–139 (2017)
38. Solís-Pérez, J.E., Gómez-Aguilar, J.F., Atangana, A.: Novel numerical method for solving variable-order fractional differential equations with power, exponential and Mittag-Leffler laws. Chaos Solitons Fractals **114**, 175–185 (2018)
39. Alkahtani, B.S.T., Atangana, A., Koca, I.: Novel analysis of the fractional Zika model using the Adams type predictor-corrector rule for non-singular and non-local fractional operators. J. Nonlinear Sci. Appl. **10**, 3191–3200 (2017)
40. Koca, I.: Modelling the spread of Ebola virus with Atangana-Baleanu fractional operators. Eur. Phys. J. Plus **133**(3), 1–16 (2018)
41. Morales-Delgado, V.F., Gómez-Aguilar, J.F., Taneco-Hernándeza, M.A., Escobar-Jiménez, R.F., Olivares-Peregrino, V.H.: Mathematical modeling of the smoking dynamics using fractional differential equations with local and nonlocal kernel. J. Nonlinear Sci. Appl. **11**, 994–1014 (2018)
42. Khan, M.A., Saddiq, S.F., Islam, S., Khan, I., Ching, D.L.C.: Epidemic model of leptospirosis containing fractional order. Abstr. Appl. Anal. (Hindawi) **1**, 1–13 (2014)
43. El-Shahed, M.: Fractional order model for the spread of leptospirosis. Int. J. Math. Anal. **8**(54), 2651–2667 (2014)
44. Zaman, G., Khan, M.A., Islam, S., Chohan, M.I., Jung, I.H.: Modeling dynamical interactions between leptospirosis infected vector and human population. Appl. Math. Sci. **6**(26), 1287–1302 (2012)
45. Van den Driessche, P., Watmough, J.: Reproduction numbers and sub-threshold endemic equilibria for compartmental models of disease transmission. Math. Biosci. **180**(1–2), 29–48 (2002)
46. Ahmed, E., El-Sayed, A.M.A., El-Saka, H.A.: On some Routh-Hurwitz conditions for fractional order differential equations and their applications in Lorenz, Rössler, Chua and Chen systems. Phys. Lett. A **358**(1), 1–4 (2006)

Dual Fractional Analysis of Blood Alcohol Model Via Non-integer Order Derivatives

Kashif Ali Abro and J. F. Gómez-Aguilar

Abstract The concentration of alcohol in blood differs with vessel diameter (arterial diameter). In case of arteries having thinner diameter, alcohol concentrates around their walls because of Fahraeus–Lindqvist effect. The fluctuating concentration of alcohol in blood directly affects normal human body functions causing peptic ulcer and hypertension. In this work, we made the comparative analysis of blood alcohol model via Caputo–Fabrizio and Atangana–Baleanu fractional derivatives. The governing ordinary differential equations of blood alcohol model have been converted in terms of non-integers order derivatives. The analytic calculations of the concentrations of alcohol in stomach ($C_1(t)$) and the concentrations of alcohol in the blood ($C_2(t)$) have been investigated by applying Laplace transform method. The general solutions of the concentrations of alcohol in stomach ($C_1(t)$) and the concentrations of alcohol in the blood ($C_2(t)$) are expressed in the terms of wright function $\Phi(a, b; c)$. The graphs of both types of concentrations are depicted on the basis of fractional parameters of Caputo–Fabrizio and Atangana–Baleanu fractional derivatives. Finally, the comparative analysis of both fractional types of concentration of alcohol level in blood decay faster for higher fractional order.

Keywords Fractional calculus · Atangana–Baleanu fractional derivative · Blood alcohol model

K. A. Abro
Department of Basic Sciences and Related Studies, Mehran University of Engineering and Technology, Jamshoro, Pakistan

J. F. Gómez-Aguilar (✉)
CONACYT-Tecnológico Nacional de México, Centro Nacional de Investigación y Desarrollo Tecnológico, Interna del Internado, Palmira, 62490 Cuernavaca, Morelos, México
e-mail: jgomez@cenidet.edu.mx

J. F. Gómez et al. (eds.), *Fractional Derivatives with Mittag-Leffler Kernel*, Studies in Systems, Decision and Control 194,
https://doi.org/10.1007/978-3-030-11662-0_5

1 Introduction

There is no denying fact that alcohol consumption not only effects in the area of disorders but also influences the incidences of chronic diseases, injuries, and few health problems. This is because alcoholic beverages have become a part of many cultures for thousands of years. The impact of alcohol consumption is categorized in three factors, namely (i) the quality of alcohol consumed, (ii) the volume of alcohol consumed and (iii) the consumption pattern, on rare occasions. In context with the above three categories, alcohol consumption has become a detrimental and beneficial health impact. For instance few epidemiological and animal studies suggested that excessive alcohol consumption depresses cardiac function and causes cardiomyopathy or cardiomyopathy injury. The heavy alcohol consumption not only depresses cardiac function, but also it includes being thirsty, tired, sleepy, drowsy, weak and nauseaus as weil as having dry mouth and headache and several other concentration problems [1–8]. On the other hand, the World Health Organization has suggested a terrifying report, the harmful utilization of alcohol causes approximately 5.1% of the global burden of disease is which as attributable to alcohol consumption and 3.3 million deaths every year (or 5.9% of all the global deaths) [9]. The non-integer order derivatives have attracted many researchers and scientists due to its several significant applications in science and engineering; these derivatives model the various dynamical processes and they carry information regarding their present as well as past states (memory effects). In order to characterize memory property of complex systems, one need to employ the non-integer order derivatives because these operators give a complete description of different physical processes with dissipation and long-range interaction. On the basis of non-integer order derivatives, several researchers have utilized these derivatives among different physical aspects. For instance, pharmacokinetics [10], anomalous diffusion [11–14], control theory [15], electromagnetism [16, 17], rheological fluids [18–21], electrical engineering [22, 23], and heat transfer [24, 25]. Ludwin in [26] investigated the blood alcohol content as a function of time by employing direct integration method for the exact solutions of the concentrations of alcohol in stomach and alcohol in the blood respectively. Here, an experimental data was depicted with exact solutions of the concentrations of alcohol in stomach and alcohol in the blood and it was suggested that the average accuracy of the model was found to be 94.4%. Almeida et al. [27] compared integer and fractional models versus real experimental data. Liouville–Caputo fractional derivative was considered to solve the fractional order differential equations that describe the process studied. A numerical optimization approach based on least squares approximation was used to determine the order of the fractional derivative that better describes real experimental data, as well as other related parameters. Aqsa et al. [28] traced out the mathematical modeling of $CD4 + T-$ cells through analytical technique Optimal Variational Iteration Method (OVIM) on the system of governing nonlinear differential equations.

In this paper, our aim is to analyze the blood alcohol model via Caputo–Fabrizio and Atangana–Baleanu fractional derivatives in the Liouville–Caputo sense. The governing ordinary differential equations of blood alcohol model have been converted in terms of non-integers order derivatives. The analytic calculations of the concentrations of alcohol in stomach ($C_1(t)$) and the concentrations of alcohol in the blood ($C_2(t)$) have been investigated by applying Laplace transform method. The general solutions of the concentrations of alcohol in stomach ($C_1(t)$) and the concentrations of alcohol in the blood ($C_2(t)$) are expressed in terms of wright function $\Phi(a, b; c)$.

2 Fractional Modeling of Blood Alcohol Model

The governing ordinary differential equations of blood alcohol model can be written as [27]

$$\frac{dC_1(t)}{dt} + R_1 C_1(t) = 0, \tag{1}$$

$$\frac{dC_2(t)}{dt} - R_1 C_1(t) + R_2 C_2(t) = 0, \tag{2}$$

where $C_1(t)$ and $C_2(t)$ represent the concentrations of alcohol in stomach and alcohol in the blood respectively. R_1 and R_2 are non-zero constants. While the corresponding initial conditions are taken into consideration for Eqs. (1) and (2) which satisfy the analytic solutions for validations which are defined as

$$C_1(0) = C_0, \qquad C_2(0) = 0. \tag{3}$$

Equation (3) is the imposed condition for Eqs. (1) and (2) respectively. Converting Eqs. (1) and (2) in terms of non-integer time derivatives of Atangana–Baleanu–Caputo type, we have [29–31]

$$^{ABC}\left(\frac{d^{\alpha_1} C(t)}{dt^{\alpha_1}}\right) = \int_0^t E_{\alpha_1}\left[-\frac{\alpha_1 (z-t)^{\alpha_1}}{1-\alpha_1}\right]\frac{C'(t)}{1-\alpha_1}dt, \qquad 0 < \alpha_1 \le 1. \tag{4}$$

Meanwhile, in the Caputo–Fabrizio–Caputo sense, we get [32–35]

$$^{CFC}\left(\frac{d^{\beta_1} C(t)}{dt^{\beta_1}}\right) = \int_0^t \exp\left[-\frac{\beta_1 (z-t)^{\beta_1}}{1-\beta_1}\right]\frac{C'(t)}{1-\beta_1}dt, \qquad 0 < \beta_1 \le 1. \tag{5}$$

Using Eqs. (4) and (5), we arrive at the generalized equations in fractional form as

$$\frac{d^{\alpha_1} C_1(t)}{dt^{\alpha_1}} + R_1 C_1(t) = 0,$$

$$\frac{d^{\alpha_2} C_2(t)}{dt^{\alpha_2}} - R_1 C_1(t) + R_2 C_2(t) = 0, \tag{6}$$

and

$$\frac{d^{\beta_1}C_1(t)}{dt^{\beta_1}}+R_1C_1(t)=0,$$

$$\frac{d^{\beta_2}C_2(t)}{dt^{\beta_2}}-R_1C_1(t)+R_2C_2(t)=0, \tag{7}$$

where the fractional derivatives are in Atangana–Baleanu–Caputo and Caputo–Fabrizio–Caputo sense, respectively.

The system of fractional differential equations (6) and (7) can be solved by the technique of Laplace transforms with inversions. Even fractional differential equations (6) and (7) can be explained in principle by enormous methodologies and their effectiveness is usually subjective by the domain of definition.

3 Solution of the Fractional Blood Alcohol Model

Calculation of the Problem via Atangana–Baleanu–Caputo Fractional Operator

Applying Laplace transform on fractional differential equations (6) and using imposed condition for Eq. (3), we obtain

$$\frac{s^{\alpha_1}\eta_1C_1(s)-C_0}{s^{\alpha_1}+\alpha_1\eta_1}+R_1C_1(s)=0,$$

$$\frac{s^{\alpha_2}\eta_2C_2(s)}{s^{\alpha_2}+\alpha_2\eta_2}-R_1C_1(s)+R_2C_2(s)=0, \tag{8}$$

where $\eta_1=\frac{1}{1-\alpha_1}$ and $\eta_2=\frac{1}{1-\alpha_2}$ represent the letting parameters. Simplifying Eq. (8), we get

$$C_1(s)=\frac{C_0\lambda_0\eta_1}{s^{\alpha_1}+\lambda_1},$$

$$C_2(s)=\frac{C_0R_1\lambda_0\lambda_2(s^{\alpha_2}+\alpha_2\eta_2)}{(s^{\alpha_1}+\lambda_1)(s^{\alpha_2}+\lambda_3)}, \tag{9}$$

where $\lambda_0=\frac{1}{\eta_1+R_1}$, $\lambda_1=\frac{\eta_1R_1\alpha_1}{\eta_1+R_1}$, $\lambda_2=\frac{1}{\eta_2+R_2}$ and $\lambda_3=\frac{\eta_2R_2\alpha_2}{\eta_2+R_2}$ are the rheological parameters. Employing the fact of infinite series $\frac{1}{1+x}=\sum_{n=0}^{\infty}(-x)^n$ [36] on Eq. (9), we investigated the equivalent form of Eq. (9) as

$$C_1(s)=\frac{C_0\lambda_0\eta_1}{\lambda_1}\sum_{n=0}^{\infty}\left(-\frac{1}{\lambda_1}\right)^n\frac{1}{s^{-n\alpha_1}},$$

$$C_2(s) = \frac{C_0 R_1 \lambda_0 \lambda_2}{\lambda_1} \sum_{n=0}^{\infty} \Big(-\frac{1}{\lambda_1}\Big)^n \sum_{m=0}^{\infty} \Big(-\lambda_3\Big)^m \frac{1}{s^{-n\alpha_1+m\alpha_2}} +$$

$$+ \frac{C_0 R_1 \lambda_0 \lambda_2 \alpha_2 \eta_2}{\lambda_1 \lambda_3} \sum_{n=0}^{\infty} \Big(-\frac{1}{\lambda_1}\Big)^n \sum_{m=0}^{\infty} \Big(-\frac{1}{\lambda_3}\Big)^m \frac{1}{s^{-n\alpha_1-m\alpha_2}}. \tag{10}$$

Applying inverse Laplace transform on Eq. (10), we have

$$C_1(t) = \frac{C_0 \lambda_0 \eta_1}{\lambda_1} \sum_{n=0}^{\infty} \Big(-\frac{1}{\lambda_1}\Big)^n \frac{t^{-n\alpha_1-1}}{\Gamma(-n\alpha_1)},$$

$$C_2(t) = \frac{C_0 R_1 \lambda_0 \lambda_2}{\lambda_1} \sum_{n=0}^{\infty} \Big(-\frac{1}{\lambda_1}\Big)^n \sum_{m=0}^{\infty} \Big(-\lambda_3\Big)^m \frac{t^{-n\alpha_1+m\alpha_2-1}}{\Gamma(-n\alpha_1+m\alpha_2)} +$$

$$+ \frac{C_0 R_1 \lambda_0 \lambda_2 \alpha_2 \eta_2}{\lambda_1 \lambda_3} \sum_{n=0}^{\infty} \Big(-\frac{1}{\lambda_1}\Big)^n \sum_{m=0}^{\infty} \Big(-\frac{1}{\lambda_3}\Big)^m \frac{t^{-n\alpha_1-m\alpha_2-1}}{\Gamma(-n\alpha_1-m\alpha_2)}. \tag{11}$$

In order to eliminate the Gamma function, Eq. (11) is expressed in terms of wright function $\Phi(a, b; c) = \sum_{n=0}^{\infty} \frac{(c)^n}{n!\Gamma(a-bn)}$, we obtain the final expression of concentrations of alcohol in stomach and concentrations of alcohol in the blood as

$$C_1(t) = \frac{C_0 \lambda_0 \eta_1}{\lambda_1 t} \Phi\Big(0, -\alpha_1; -\frac{1}{\lambda_1 t^{\alpha_1}}\Big),$$

$$C_2(t) = \frac{C_0 R_1 \lambda_0 \lambda_2}{\lambda_1} \sum_{n=0}^{\infty} \Big(-\frac{1}{\lambda_1}\Big)^n \Phi\Big(-\alpha_1 n, \alpha_2; -\lambda_3 t^{\alpha_2}\Big) +$$

$$+ \frac{C_0 R_1 \lambda_0 \lambda_2 \alpha_2 \eta_2}{\lambda_1 \lambda_3} \sum_{n=0}^{\infty} \Big(-\frac{1}{\lambda_1}\Big)^n \Phi\Big(-\alpha_1 n, \alpha_2; -\frac{1}{\lambda_3 t^{\alpha_2}}\Big). \tag{12}$$

Equation (12) represents the final expressions for concentration of alcohol in stomach and concentration of alcohol in the blood in terms of Atangana–Baleanu fractional operator in Liouville–Caputo sense. It is also pointed out that one can retrieve classical concentration of alcohol in stomach and concentration of alcohol in the blood by taking $\alpha_1 = \alpha_2 = 1$ in Eq. (12).

Calculation of the Problem via Caputo–Fabrizio–Caputo Fractional Operator

Applying Laplace transform on fractional differential equations (7) and using imposed condition for Eq. (3), we obtain

$$\frac{s\eta_3 C_1(s) - C_0}{s + \beta_1 \eta_3} + R_1 C_1(s) = 0,$$

$$\frac{s\eta_4 C_2(s)}{s + \beta_2 \eta_4} - R_1 C_1(s) + R_2 C_2(s) = 0, \tag{13}$$

where $\eta_3 = \frac{1}{1-\beta_1}$ and $\eta_4 = \frac{1}{1-\beta_2}$ represent the letting parameters. Simplifying Eq. (13), we get

$$C_1(s) = \frac{C_0 \Lambda_0 \eta_3}{s + \Lambda_1},$$

$$C_2(s) = \frac{C_0 R_1 \Lambda_0 \Lambda_2 (s + \beta_2 \eta_4)}{(s + \Lambda_1)(s + \Lambda_3)}, \tag{14}$$

where $\Lambda_0 = \frac{1}{\eta_3 + R_1}$, $\Lambda_1 = \frac{\eta_3 R_1 \beta_1}{\eta_3 + R_1}$, $\Lambda_2 = \frac{1}{\eta_4 + R_2}$ and $\Lambda_3 = \frac{\eta_4 R_2 \beta_2}{\eta_4 + R_2}$ are the rheological parameters. Employing the fact of infinite series $\frac{1}{1+x} = \sum_{n=0}^{\infty} (-x)^n$ [36] on Eq. (14), we investigated the equivalent form of Eq. (14) as

$$C_1(s) = \frac{C_0 \Lambda_0 \eta_3}{\Lambda_1} \sum_{n=0}^{\infty} \left(-\frac{1}{\Lambda_1} \right)^n \frac{1}{s^{-n}},$$

$$C_2(s) = \frac{C_0 R_1 \Lambda_0 \Lambda_2}{\Lambda_1} \sum_{n=0}^{\infty} \left(-\frac{1}{\Lambda_1} \right)^n \sum_{m=0}^{\infty} \left(-\Lambda_3 \right)^m \frac{1}{s^{-n+m}} +$$

$$+ \frac{C_0 R_1 \Lambda_0 \Lambda_2 \beta_2 \eta_4}{\Lambda_1 \Lambda_3} \sum_{n=0}^{\infty} \left(-\frac{1}{\Lambda_1} \right)^n \sum_{m=0}^{\infty} \left(-\frac{1}{\Lambda_3} \right)^m \frac{1}{s^{-n-m}}. \tag{15}$$

Applying inverse Laplace transform on Eq. (15), we have

$$C_1(t) = \frac{C_0 \Lambda_0 \eta_3}{\Lambda_1} \sum_{n=0}^{\infty} \left(-\frac{1}{\Lambda_1} \right)^n \frac{t^{-n-1}}{\Gamma(-n)},$$

$$C_2(t) = \frac{C_0 R_1 \Lambda_0 \Lambda_2}{\Lambda_1} \sum_{n=0}^{\infty} \left(-\frac{1}{\Lambda_1} \right)^n \sum_{m=0}^{\infty} \left(-\Lambda_3 \right)^m \frac{t^{-n+m-1}}{\Gamma(-n+m)} +$$

$$+ \frac{C_0 R_1 \Lambda_0 \Lambda_2 \beta_2 \eta_4}{\Lambda_1 \Lambda_3} \sum_{n=0}^{\infty} \left(-\frac{1}{\Lambda_1} \right)^n \sum_{m=0}^{\infty} \left(-\frac{1}{\Lambda_3} \right)^m \frac{t^{-n-m-1}}{\Gamma(-n-m)}. \tag{16}$$

In order to eliminate the Gamma function, Eq. (16) is expressed in terms of the wright function $\Phi(a, b; c) = \sum_{n=0}^{\infty} \frac{(c)^n}{n! \Gamma(a-bn)}$, we obtain the final expression of concentrations of alcohol in stomach and concentrations of alcohol in the blood as

$$C_1(t) = \frac{C_0 \Lambda_0 \eta_3}{\lambda_1 t} \Phi\left(0, -1; -\frac{1}{\Lambda_1 t}\right),$$

$$C_2(t) = \frac{C_0 R_1 \Lambda_0 \Lambda_2}{\Lambda_1} \sum_{n=0}^{\infty} \left(-\frac{1}{\Lambda_1}\right)^n \Phi\left(-n, 1; -\lambda_3 t\right) +$$

$$+ \frac{C_0 R_1 \Lambda_0 \Lambda_2 \beta_2 \eta_4}{\Lambda_1 \Lambda_3} \sum_{n=0}^{\infty} \left(-\frac{1}{\Lambda_1}\right)^n \Phi\left(-n, 1; -\frac{1}{\Lambda_3 t}\right). \tag{17}$$

Equation (17) represents the final expressions for concentration of alcohol in stomach and concentration of alcohol in the blood in terms of Atangana–Baleanu fractional operator in Liouville–Caputo sense. It is also pointed out that one can retrieve classical concentration of alcohol in stomach and concentration of alcohol in the blood by taking $\alpha_1 = \alpha_2 = 1$ in Eq. (17).

4 Results and Conclusions

Now, a comparative analysis of blood alcohol model via Caputo–Fabrizio–Caputo and Atangana–Baleanu–Caputo fractional derivatives is studied. The ordinary differential equations of blood alcohol model are generalized via non-integers order derivatives. The analytic calculations of the concentrations of alcohol in stomach $C_1(t)$ and the concentrations of alcohol in the blood $C_2(t)$ have been traced out by implementing a powerful technique namely Laplace transform method. The general solutions of the concentrations of alcohol in stomach $C_1(t)$ and the concentrations of alcohol in the blood $C_2(t)$ are expressed in the wright function $\Phi(a, b; c)$. The graphical illustrations are based on smaller time $t = 0.02$ s and larger time $t = 5$ s, for both types of concentrations. The influence of the fractional derivatives namely Caputo–Fabrizio–Caputo and Atangana–Baleanu–Caputo fractional operators on the concentrations of alcohol in stomach $C_1(t)$ is depicted through Mathcad software as shown in Fig. 1a–f for smaller time $t = 0.02$ s and larger time $t = 5$ s. It can be seen from Fig. 1a–f that the comparative analysis for the concentrations of alcohol in stomach $C_1(t)$ suggested that for smaller time $t = 0.02$ s, the concentration of alcohol in stomach $C_1(t)$ has increasing behavior with Atangana–Baleanu–Caputo approach but decreasing with Caputo–Fabrizio–Caputo approach. While, for larger time $t = 5$ s, the concentration of alcohol in stomach $C_1(t)$ with Caputo–Fabrizio–Caputo and Atangana–Baleanu–Caputo fractional derivatives have reciprocal behavior. On the other hand, for unit time $t = 1$s, the concentrations of alcohol in stomach $C_1(t)$ are identical in both fractional approaches. The rheological impacts of the fractional derivatives of Caputo–Fabrizio–Caputo and Atangana–Baleanu–Caputo fractional type on the concentrations of alcohol in blood $C_2(t)$ are showed in Fig. 2a–f for smaller time $t = 0.02$ s and larger time $t = 5$ s. It is observed that in the

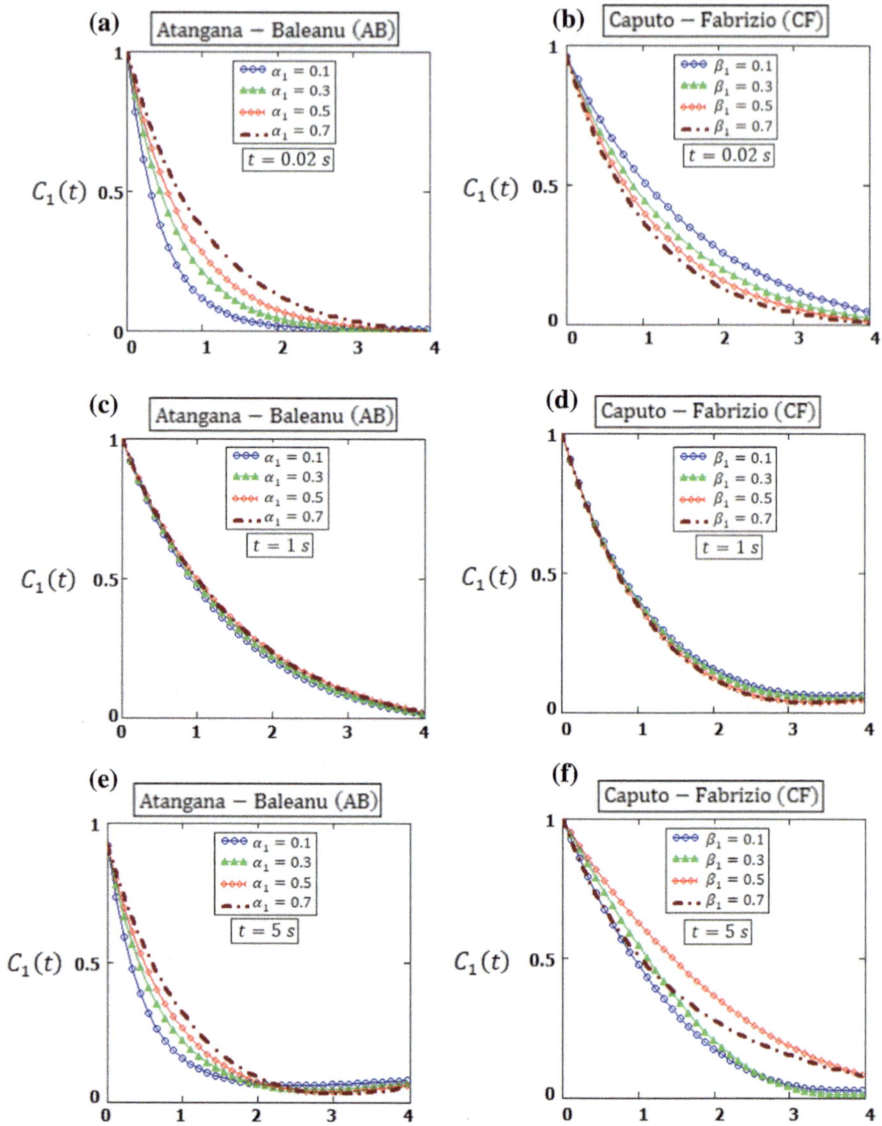

Fig. 1 Comparative analysis of the concentrations of alcohol in the stomach $C_1(t)$ via Atangana–Baleanu–Caputo and Caputo–Fabrizio–Caputo fractional derivatives for smaller and larger time when embedded parameter are $C_0 = 225$, $R_1 = 0.025$, $R_2 = 0.031$, $\alpha_1 = \beta_1 = [0, 1]$

Atangana–Baleanu–Caputo fractional operators decrease the concentration of alcohol in blood $C_2(t)$ for smaller time $t = 0.02$ s but for larger time $t = 5$ s, the concentration of alcohol in blood $C_2(t)$ increases in the Atangana–Baleanu–Caputo fractional derivative. Meanwhile, an opposite trend is perceived in the Caputo–Fabrizio–Caputo sense for the concentration of alcohol in blood $C_2(t)$. Finally, the comparative

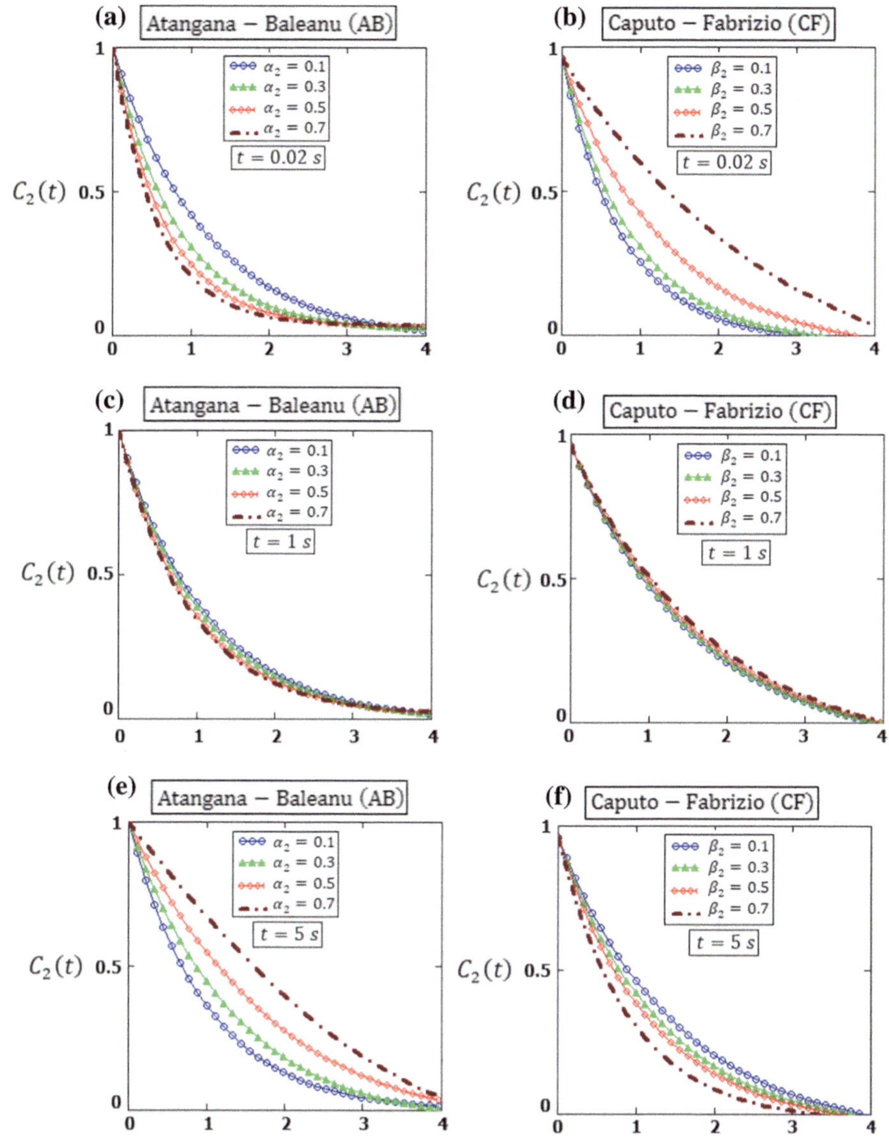

Fig. 2 Comparative analysis of the concentrations of alcohol in the stomach $C_2(t)$ via Atangana–Baleanu–Caputo and Caputo–Fabrizio–Caputo fractional derivatives for smaller and larger time when embedded parameter are $C_0 = 225$, $R_1 = 0.025$, $R_2 = 0.031$, $\alpha_1 = \beta_1 = [0, 1]$

analysis of both fractional types of concentration of alcohol level in blood decay faster for higher fractional order. It is also pointed out that the comparison of fractional model verses ordinary/classical model can also be depicted as well for both concentrations of alcohol in stomach $C_1(t)$ and in blood $C_2(t)$.

Acknowledgements The authors are highly thankful and grateful to Mehran university of Engineering and Technology, Jamshoro, Pakistan for generous support and facilities of this research work. José Francisco Gómez Aguilar acknowledges the support provided by CONACyT: Cátedras CONACyT para jóvenes investigadores 2014 and SNI-CONACyT.

References

1. National Health and Family Planning Commission. China Health and Family Planning Statistics Yearbook 2014. Peking Union Medical College Press, Beijing (2014)
2. Jokinen, E.: Obesity and cardiovascular disease. Minerva Pediatr. **67**(1), 1–25 (2015)
3. O'Keefe, J.H., Bybee, K.A., Lavie, C.J.: Alcohol and cardiovascular health: the razor-sharp double-edged sword. J. Am. Coll. Cardiol. **50**(11), 1009–1014 (2007)
4. Larsson, S.C., Wallin, A., Wolk, A.: Alcohol consumption and risk of heart failure: meta-analysis of 13 prospective studies. Clin. Nutr. **37**(4), 1247–1251 (2018)
5. Friedman, H.S.: Cardiovascular effects of alcohol with particular reference to the heart. Alcohol **1**(4), 333–339 (1984)
6. Global strategy to reduce the harmful use of alcohol. World Health Organization, Switzerland, Geneva (2010)
7. Penning, R., McKinney, A., Verster, J.C.: Alcohol hangover symptoms and their contribution to the overall hangover severity. Alcohol Alcohol. **47**(3), 248–252 (2012)
8. McKinney, A.: A review of the next day effects of alcohol on subjective mood ratings. Curr. Drug Abus. Rev. **3**, 88–91 (2010)
9. Global Strategy to Reduce the Harmful Use of Alcohol. World Health Organization, Geneva, Switzerland (2014)
10. Sopasakis, P., Sarimveis, H., Macheras, P., Dokoumetzidis, A.: Fractional calculus in pharmacokinetics. J. Pharmacokinet. Pharmacodyn. **45**(1), 107–125 (2018)
11. Hristov, J.: Derivatives with non-singular kernels from the Caputo–Fabrizio definition and beyond: appraising analysis with emphasis on diffusion models. Front. Fract. Calc. **1**, 270–342 (2017)
12. Atangana, A., Koca, I.: On the new fractional derivative and application to nonlinear Baggs and Freedman model. J. Nonlinear Sci. Appl. **9**, 2467–2480 (2016)
13. Atangana, A., Koca, I.: Chaos in a simple nonlinear system with Atangana–Baleanu derivatives with fractional order. Chaos Solitons Fractals **89**, 447–454 (2016)
14. Atangana, A.: On the new fractional derivative and application to nonlinear Fisher's reaction-diffusion equation. Appl. Math. Comput. **273**, 948–956 (2016)
15. Zhuo, L., Liu, L., Dehghan, S., Yang, Q.C., Xue, D.: A review and evaluation of numerical tools for fractional calculus and fractional order controls. Int. J. Control. **90**(6), 1165–1181 (2016)
16. Kashif, A.A., Mukarrum, H., Mirza, M.B.: A mathematical analysis of magnetohydrodynamic generalized Burger fluid for permeable oscillating plate. Punjab Univ. J. Math. **50**(2), 97–111 (2018)
17. Ali, F., Saqib, M., Khan, I., Sheikh, N.A.: Application of Caputo–Fabrizio derivatives to MHD free convection flow of generalized Walters'-B fluid model. Eur. Phys. J. Plus **131**(10), 1–10 (2016)
18. Sheikh, N.A., Ali, F., Khan, I., Gohar, M., Saqib, M.: On the applications of nanofluids to enhance the performance of solar collectors: a comparative analysis of Atangana–Baleanu and Caputo–Fabrizio fractional models. Eur. Phys. J. Plus **132**(12), 1–11 (2017)
19. Muhammad, J., Kashif, A.A., Najeeb, A.K.: Helices of fractionalized Maxwell fluid. Nonlinear Eng. **4**(4), 191–201 (2015)
20. Sheikh, N.A., Ali, F., Saqib, M., Khan, I., Jan, S.A.A., Alshomrani, A.S., Alghamdi, M.S.: Comparison and analysis of the Atangana–Baleanu and Caputo–Fabrizio fractional derivatives

for generalized Casson fluid model with heat generation and chemical reaction. Results Phys. **7**, 789–800 (2017)
21. Ali, F., Jan, S.A.A., Khan, I., Gohar, M., Sheikh, N.A.: Solutions with special functions for time fractional free convection flow of Brinkman-type fluid. Eur. Phys. J. Plus **131**(9), 1–11 (2016)
22. Gómez-Aguilar, J.F., Morales-Delgado, V.F., Taneco-Hernández, M.A., Baleanu, D., Escobar-Jiménez, R.F., Al Quarashi, M.M.: Analytical solutions of the electrical RLC circuit via Liouville–Caputo operators with local and non-local kernels. Entropy **18**, 1–22 (2016)
23. Abro, K.A., Memon, A.A., Uqaili, M.A.: A comparative mathematical analysis of RL and RC electrical circuits via Atangana–Baleanu and Caputo–Fabrizio fractional derivatives. Eur. Phys. J. Plus **133**(3), 1–9 (2018)
24. Koca, I., Atangana, A.: Solutions of Cattaneo-Hristov model of elastic heat diffusion with Caputo–Fabrizio and Atangana–Baleanu fractional derivatives. Therm. Sci. **21**(6), 2299–2305 (2017)
25. Abro, K.A., Hussain, M., Baig, M.M.: An analytic study of molybdenum disulfide nanofluids using the modern approach of Atangana–Baleanu fractional derivatives. Eur. Phys. J. Plus **132**(10), 1–12 (2017)
26. Ludwin, C.: Blood alcohol content. Undergrad. Math. Model. **3**(2), 1–8 (2011)
27. Almeida, R., Bastos, N.R., Monteiro, M.T.T.: Modeling some real phenomena by fractional differential equations. Math. Methods Appl. Sci. **39**(16), 4846–4855 (2016)
28. Nazir, A., Ahmed, N., Khan, U., Mohyud-Din, S.T.: Analytical approach to study a mathematical model of $CD4+T-$cells. Int. J. Biomath. **11**(04), 1–18 (2018)
29. Atangana, A., Baleanu, D.: New fractional derivatives with nonlocal and nonsingular kernel: theory and application to heat transfer model. Therm. Sci. **20**(2), 763–769 (2016)
30. Abro, K.A., Chandio, A.D., Irfan, A.A., Khan, I.: Dual thermal analysis of magnetohydrodynamic flow of nanofluids via modern approaches of Caputo–Fabrizio and Atangana–Baleanu fractional derivatives embedded in porous medium. J. Therm. Anal. Calorim. **1**, 1–11 (2018)
31. Kashif, A.A., Mohammad, M.R., Khan, I., Irfan, A.A., Asifa, T.: Analysis of Stokes' second problem for nanofluids using modern fractional derivatives. J. Nanofluids **7**, 738–747 (2018)
32. Caputo, M., Fabrizio, M.: A new definition of fractional derivative without singular kernel. Progr. Fract. Differ. Appl. **1**(2), 73–85 (2015)
33. Abro, K.A., Khan, I.: Analysis of heat and mass transfer in MHD flow of generalized casson fluid in a porous space via non-integer order derivative without singular kernel. Chin. J. Phys. **55**(4), 1583–1595 (2017)
34. Khan, A., Abro, K.A., Tassaddiq, A., Khan, I.: Atangana–Baleanu and Caputo Fabrizio analysis of fractional derivatives for heat and mass transfer of second grade fluids over a vertical plate: a comparative study. Entropy **19**(8), 1–12 (2017)
35. Al-Mdallal, Q., Abro, K.A., Khan, I.: Analytical solutions of fractional Walter's B fluid with applications. Complexity **1**, 1–12 (2018)
36. Laghari, M.H., Abro, K.A., Shaikh, A.A.: Helical flows of fractional viscoelastic fluid in a circular pipe. Int. J. Adv. Appl. Sci. **4**(10), 97–105 (2017)

Parameter Estimation of Fractional Gompertz Model Using Cuckoo Search Algorithm

J. E. Solís-Pérez, J. F. Gómez-Aguilar, R. F. Escobar-Jiménez, L. Torres and V. H. Olivares-Peregrino

Abstract In this chapter, a meta-heuristic optimization algorithm, called cuckoo search algorithm is applied to determine the optimal parameters of the fractional Gompertz model via Liouville–Caputo and Atangana–Baleanu–Caputo fractional derivatives. The numerical solutions of the proposed models were obtained using the Adams method. The proposed methodology is tested on epidemiological examples. In the interval considered, the fractional models had the best fit for the epidemiological data considered. The effectiveness of the methodology is shown by a comparison with the classical models. A comparison between the fractional models and the classical models was carried out to show the effectiveness of our methodology.

Keywords Fractional calculus · Liouville–Caputo fractional derivative · Atangana–Baleanu fractional derivative · Cuckoo search algorithm · Gompertz model

1 Introduction

Benjamin Gompertz introduced in 1825 the Gompertz model [1], the model is based on a sigmoid function and it can be applicable to the growth of animals and plants, as well as to model the number or volume of bacteria and cancer cells [2–6]. Horiuchi in [7], developed mathematical expressions to model the human mortality. In this

J. E. Solís-Pérez · R. F. Escobar-Jiménez · V. H. Olivares-Peregrino
Tecnológico Nacional de México, Centro Nacional de Investigación y Desarrollo Tecnológico, Interna del Internado, Palmira, 62490 Cuernavaca, Morelos, México

J. F. Gómez-Aguilar (✉)
CONACYT-Tecnológico Nacional de México, Centro Nacional de Investigación y Desarrollo Tecnológico, Interna del Internado, Palmira, 62490 Cuernavaca, Morelos, México
e-mail: jgomez@cenidet.edu.mx

L. Torres
CONACYT-Instituto de Ingeniería, Universidad Nacional Autónoma de México, México City, México

J. F. Gómez et al. (eds.), *Fractional Derivatives with Mittag-Leffler Kernel*, Studies in Systems, Decision and Control 194,
https://doi.org/10.1007/978-3-030-11662-0_6

work, six different mortality models were presented, in the presence or absence of the Makeham term. In [8], the model was used for describing the emergence patterns of the growing weed based on the temperature and soil moisture. Moummou in [8] proposed a model that integrates the stochastic growth of a tumor cell population and the effect of two therapies on the cell growth. One of them had an internal immunological effect, while the other was controllable from outside of the cell population. Other applications of the Gompertz model are given in [9, 10].

Fractional Calculus (FC) represents complex physical phenomena accurate and efficiently. In nature, many physical phenomena have "intrinsic" fractional order descriptions, so FC is a necessary mathematical tool to explain them. FC modeling has shown to be useful to describe and explain certain types of physical problems and processes where memory effects have to be considered. These models carry information from its different previous states and can be a useful way to include memory in a dynamic process (the fractional derivative order can be interpreted as an index of the memory [11]). Particularly, biology is a rich source for mathematical ideas [12–14]. There exist various definitions of fractional derivatives and integrals, these definitions includes, Riemann-Liouville, Liouville–Caputo, Grünwald-Letnikov, Marchaud, Weyl, Riesz, Feller, Caputo-Fabrizio, Atangana–Baleanu, among others [15–17]. This diversity of definitions is due to the lack of a consistent geometric and physical interpretation of the fractional derivative and the fact that fractional derivatives take different kernel representations.

Cuckoo Search (CS) algorithm is a relatively recent modern optimization method, this algorithm has a vast number of applications [18, 19]. It was developed by Yang and Deb [20], as a bio-inspired technique that mimics the particular brood parasitism behavior of certain species of cuckoos in nature. CS algorithm can be described in few words as a mutation-based swarm algorithm with Lévy-Flights.

The main aim of this article is to obtain a parameter estimation of fractional Gompertz model using Cuckoo search algorithm. The Liouville–Caputo and Atangana–Baleanu fractional derivatives were considered. We included some epidemiological examples to compared with the ordinary Gompertz model. We considered fractional orders $\gamma \in (0; 1]$ and considering the CS algorithm, we investigated which fractional order would help the fractional derivative to fit better to the experimental data.

2 Basic Definitions

The Liouville–Caputo fractional order derivative (C) for ($\gamma > 0$) is defined as follows [15]

$$ {}_{0}^{C}\mathscr{D}_t^{\gamma}\{f(t)\} = \frac{1}{\Gamma(n-\gamma)}\int_0^t \frac{d^n}{dt^n}f(\theta)(t-\theta)^{n-\gamma-1}d\theta, \qquad n-1 < \gamma \leq n, \tag{1} $$

where $f^{(n)}$ is the derivative of integer nth order of $f(t)$, $n = 1, 2, \ldots \in N$.

Now, we consider the following equation

$$ {}_0^C\mathscr{D}_t^\gamma f(t) = g(t, f(t)), \quad f^k(0) = f_0^k, \quad k = 0, 1, \ldots, n-1. \tag{2}$$

Considering that the Eq. (2) has a unique solution defined in the interval $t \in [0, T]$ and the above equation satisfies the following Volterra integral equation

$$ f(t) = \sum_{k=0}^{n-1} f_0^{(k)} \frac{t^k}{k!} + \frac{1}{\Gamma(\gamma)} \int_0^t (t-u)^{\gamma-1} g(u, f(u)) du, \qquad t < T, \tag{3}$$

where $\gamma > 0$ and ${}_0^C\mathscr{D}_t^\gamma$ is the Liouville–Caputo fractional derivative given by Eq. (1). The solution scheme used to solve Eq. (2) is known as predictor-corrector Adams–Bashforth–Moulton method [21, 22]. The iterative solution is given by

$$ \begin{aligned} f_{k+1}^P &= \sum_{j=0}^{n-1} \frac{t_{k+1}^j}{j!} f_0^{(j)} + \frac{1}{\Gamma(\gamma)} \sum_{j=0}^{k} b_{j,k+1} g(t_j, f_j), \\ f_{k+1} &= \sum_{j=0}^{n-1} \frac{t_{k+1}^j}{j!} f_0^{(j)} + \frac{1}{\Gamma(\gamma)} \left(\sum_{j=0}^{k} a_{j,k+1} g(t_j, f_j) + a_{k+1,k+1} g(t_{k+1}, f_{k+1}^P) \right), \end{aligned} \tag{4}$$

where

$$ a_{j,k+1} = \frac{h^\gamma}{\gamma(\gamma+1)} \cdot \begin{cases} (k^{\gamma+1} - (k-\gamma)(k+1)^\gamma) & j = 0, \\ ((k-j+2)^{\gamma+1} + (k-j)^{\gamma+1} - 2(k-j+1)^{\gamma+1}) & 1 \le j \le k, \\ 1 & j = k+1, \end{cases}$$

$$ b_{j,k+1} = \frac{h^\gamma}{\gamma}((k+1-j)^\gamma - (k-j)^\gamma), \qquad j = 0, 1, 2, \ldots, k.$$

The fractional derivative of type Atangana–Baleanu in the Liouville–Caputo sense is defined as follows [17]

$$ {}_0^{ABC}\mathscr{D}_t^\gamma \{f(t)\} = \frac{B(\gamma)}{n-\gamma} \int_0^t \frac{d^n}{dt^n} f(\theta) E_\gamma\Big[-\gamma \frac{(t-\theta)^\gamma}{n-\gamma} \Big] d\theta, \tag{5}$$

where, $B(\gamma)$ is a normalization function, $B(0) = B(1) = 1$. This fractional operator uses the Mittag-Leffler law as nonsingular and nonlocal kernel.

The numerical approximation of Atangana–Baleanu fractional integral using the Adams-Moulton rule is given by [23]

$$ {}^{AB}_{0}\mathscr{I}_t^{\gamma}[f(t_{n+1})] = \frac{1-\gamma}{B(\gamma)}\Big[\frac{f(t_{n+1})-f(t_n)}{2}\Big] + \frac{\gamma}{\Gamma(\gamma)}\sum_{k=0}^{\infty}\Big[\frac{f(t_{k+1})-f(t_k)}{2}\Big]b_k^{\gamma}, \quad (6) $$

where, $b_k^{\gamma} = (k+1)^{1-\gamma} - (k)^{1-\gamma}$.

Gompertz Model

The Gompertz differential equation is given by

$$ \frac{dN(t)}{dt} = K \cdot N(t) \cdot \ln\left(\frac{A}{N(t)}\right), \quad (7) $$

where $N(t)$ represents the size of the population at time t, K and A the intrinsic rate of growth and the maximum value of growth respectively.

Fractional Gompertz model is obtained from classical equation by replacing the first order time derivative by a fractional derivative of order γ

$$ {}_0\mathscr{D}_t^{\gamma} N(t) = K \cdot N(t) \cdot \ln\left(\frac{A}{N(t)}\right), \quad (8) $$

where the fractional operator ${}_0\mathscr{D}_t^{\gamma}$ is of type Liouville–Caputo or Atangana–Baleanu–Caputo.

- **First case**. In the Liouville–Caputo sense, we have

$$ N(t) = \sum_{i=0}^{n-1} N(0)^{(i)}\frac{t^i}{i!} + \frac{1}{\Gamma(\alpha)}\int_0^t (t-k)^{\alpha-1}\left(K \cdot N(k) \cdot \ln\left(\frac{A}{N(k)}\right)\right)dk, \ t < T. \quad (9) $$

The numerical approximation of Eq. (9) is obtained using the algorithm given by Eq. (4).

- **Second case**. We use the numerical scheme developed in [23] for the fractional derivative based on the Mittag-Leffler kernel

$$ {}^{AB}_{0}I_t^{\alpha}[f(t_{l+1})] = \frac{1-\alpha}{B(\alpha)}\left[\frac{f(t_{l+1}-f(t_l)}{2}\right] + \frac{\alpha}{\Gamma(\alpha)}\sum_{z=0}^{\infty}\left[\frac{f(t_{z+1})-f(t_z)}{2}\right]b_z^{\alpha}, \quad (10) $$

where

$$ b_z^{\alpha} = (z+1)^{1-\alpha} - (z)^{1-\alpha}, \quad (11) $$

and the model (8) is represented by

$$N_{(l+1)}(t) - N_{(l)}(t) = N^{l}_{(1)}(t) + \left\{ \frac{1-\alpha}{B(\alpha)} \left[K \left(\frac{N_{(l+1)}(t) \cdot \ln\left(\frac{A}{N_{(l+1)}(t)}\right) - N_{(l)}(t) \cdot \ln\left(\frac{A}{N_{(l)}(t)}\right)}{2} \right) \right] \right\} + \frac{\alpha}{B(\alpha)} \sum_{z=0}^{\infty} b_z^{\alpha} \cdot \left[K \left(\frac{N_{(z+1)}(t) \cdot \ln\left(\frac{A}{N_{(z+1)}(t)}\right) - N_{(z)}(t) \cdot \ln\left(\frac{A}{N_{(z)}(t)}\right)}{2} \right) \right]. \tag{12}$$

The parameters of the Gompertz model can be obtained by: least squares, linear and nonlinear regression, maximum likelihood estimation, partial derivatives and region of exact confidence [24, 25]. In this chapter, we consider the CS algorithm to find the optimal parameters.

3 Cuckoo Search

Cuckoo Search is a metaheuristic optimization algorithm based on the Nature and developed in 2009 by Xin-She Yang and Suash Deb [20]. It is based on the parasitism of some Cuckoo species and involves Lévy flights to offer a more efficient algorithm compared to Genetic Algorithms and Particle Swarm Optimization.

The three main rules governing this algorithm are

- A cuckoo deposits an egg in nests of other bird species.
- The best nests containing high-quality eggs pass to the next generation.
- There is a fixed number of nests. If a host bird discovers that the eggs in the nest are not their own, they can be discarded from the nest, or abandoned, and build another nest nearby.

Based on these three rules, the basic elements of the CS algorithm can be summarized in Algorithm 1.

Algorithm 1 Cuckoo Search via Lévy Flights

```
initialization of n host nests
while stop criterion do
    choose a cuckoo egg by Lévy flights and evaluate your fitness (F_i = min {RMSE})
    an egg from another nest randomly choose and calculate your fitness (F_j = min {RMSE})
    if F_i > F_j then
        replace the jth egg for the ith egg
        a fraction (p_a) of the worst nests is demolished and replaced by new
        good nests are preserved (better solutions)
    end if
end while
```

The general equation of the CS algorithm is given by

$$X_{g+1;i} = X_{g;i} + \alpha \otimes \text{Lévy}(\beta), \tag{13}$$

where g and i denote the generation number ($g = 1, 2, 3, \ldots, \text{MaxGen}$) and the pattern number ($i = 1, 2, \ldots, n$), respectively. In addition $\alpha > 0$ is the scaled step size that could be related to the scales of the problem of interest and $\otimes$ means input multiplications. The jth attributes of the ith pattern are initiated by

$$X_{g=0;j,i} = \text{rand} \cdot (Ub_i - Lb_i) + Lb_i, \tag{14}$$

where Ub_i and Lb_i are the upper and lower limits of the jth attributes respectively.

The step ϕ used by Yang and Deb for the Lévy flight is given by

$$\phi = \left(\frac{\Gamma(1+\beta) \cdot \sin(\pi \cdot \beta/2)}{\Gamma(((1+\beta)/2) \cdot \beta \cdot 2^{(\beta-1)/2})} \right)^{1/\beta}, \tag{15}$$

and the step size required by the jth attributes can be calculated by

$$s_j = 0.01 \cdot \left(\frac{u_j}{v_j} \right)^{1/\beta} \cdot (v - x_{best}), \tag{16}$$

where $u = \phi \cdot \text{rand}n[D]$ and $v = \text{rand}n[D]$. The donor pattern v is randomly adjusted by

$$v = v + s_j \cdot \text{rand}n[D]. \tag{17}$$

The CS algorithm will evaluate the fitting of the random pattern. If the best solution is found, the x_{best} pattern is updated. The non-viable solutions are reviewed by the cross operator given by the following equation

$$v_i = \begin{cases} x_i + \text{rand} \cdot (x_{r1} - x_{r2}), & \text{rand}_i > p_0 \\ x_i, & \text{other}, \end{cases} \tag{18}$$

where p_0 is the probability of mutation ($p_0 = 0.25$).

Accuracy Prediction Criterion

In this work, the following indices are used to evaluate the accuracy of the proposed model, e.g., the mean absolute percentage error (MAPE) and root mean square error (RMSE).

If A_t is the tth value of the dataset, P_t is the estimated value of A_t and n the amount of data in the set, so the mean absolute percentage error (MAPE) is defined by

$$MAPE = \frac{\sum_{t=1}^{n} |(A_t - P_t)/A_t|}{n} \times 100, \% \tag{19}$$

Table 1 MAPE prediction levels proposed by [26]

MAPE (%)	Level of prediction
<10	Highly accurate
10 – 20	Good
20 – 50	Reasonable
>50	Imprecise

and the *RMSE* (function to be minimized by CS)

$$RMSE = \sqrt{\frac{\sum_{t=1}^{n}(A_t - P_t)^2}{n}}, \tag{20}$$

$$\underset{\gamma,K,A}{\text{minimize}} \quad \{RMSE\},$$

where a small value of MAPE (see Table 1) or RMSE implies greater prediction accuracy.

4 Tests

Example 1 People living with human immunodeficiency virus (HIV) in the world. The estimated population living with HIV in the world was obtained from [27] which includes, men and women of all ages. The data obtained was normalized to implement the CS algorithm and for obtaining the parameters of the particular Gompertz models in the sense of Liouville–Caputo (9) and Atangana–Baleanu–Caputo (10). In the Table 2 we show the parameters obtained by CS algorithm, as well as, the evaluation indices for each obtained models, the result of prediction is shown in Fig. 1.

Example 2 Zika virus in Latin America.
The number of confirmed cases with Zika virus in Latin America was obtained from [28]. The CS algorithm was implemented to obtain the parameters of the models given by Eqs. (9) and (10), its result is shown in Table 2 and Fig. 2.

Example 3 Breast cancer in France.
In the locality of France, tests were carried out to detect breast and cervix cancer, specifically, cases of malignant neoplasm of breast. Statistical data of the implemented program were obtained from [29]. The particular models given by Eqs. (9) and (10) were obtained with CS and the results are shown in Table 2 and Fig. 3.

Table 2 Estimated parameter values for each model through CS

Example	K	A	γ	Model	Example	K	A	γ	Model
Case 1	0.1719	0.2489	1	Classical	*Case 4*	0.4698	1.5423	1	Classical
	0.1657	0.2412	0.9981	Caputo		0.3305	2.8711	0.6306	Caputo
	0.1657	0.2412	0.9998	Atangana		0.3430	3.1758	0.7220	Atangana
Case 2	0.5302	0.8510	1	Classical	*Case 5*	0.0878	0.3746	1	Classical
	0.5760	1.8189	0.3201	Caputo		0.1231	2.4152	0.4374	Caputo
	0.7250	1.7263	0.5060	Atangana		0.1838	0.9823	0.6150	Atangana
Case 3	0.7514	0.3332	1	Classical	*Case 6*	0.1369	0.2555	1	Classical
	0.6912	0.3282	0.9874	Caputo		0.0894	0.2111	0.9973	Caputo
	0.6270	0.3295	0.9987	Atangana		0.0768	0.2223	0.9986	Atangana

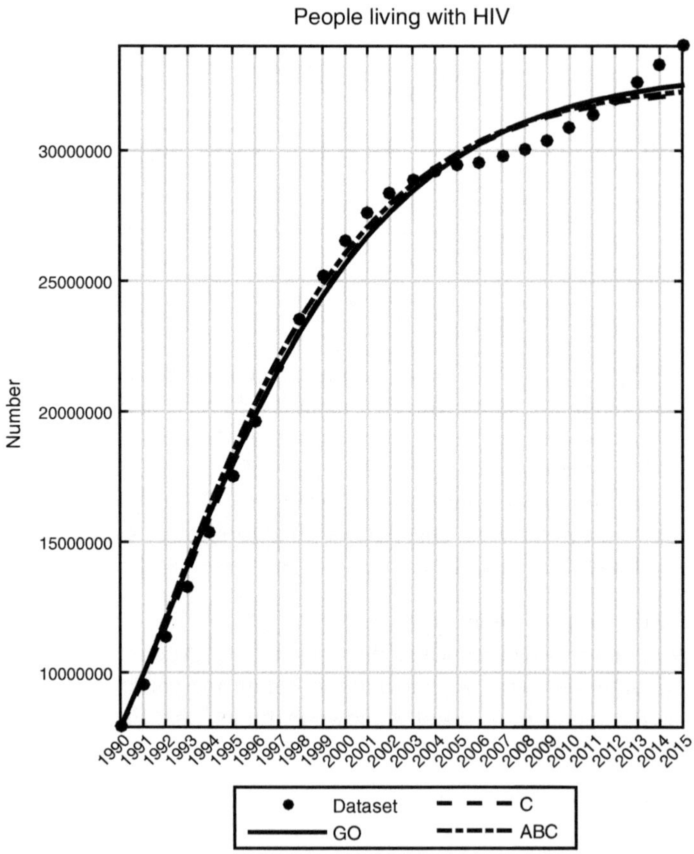

Fig. 1 People living with HIV in the world. • Dataset,▬▬ Classical model,- - - Fractional model with Liouville–Caputo derivative,- . . . Fractional model with Atangana–Baleanu–Caputo derivative

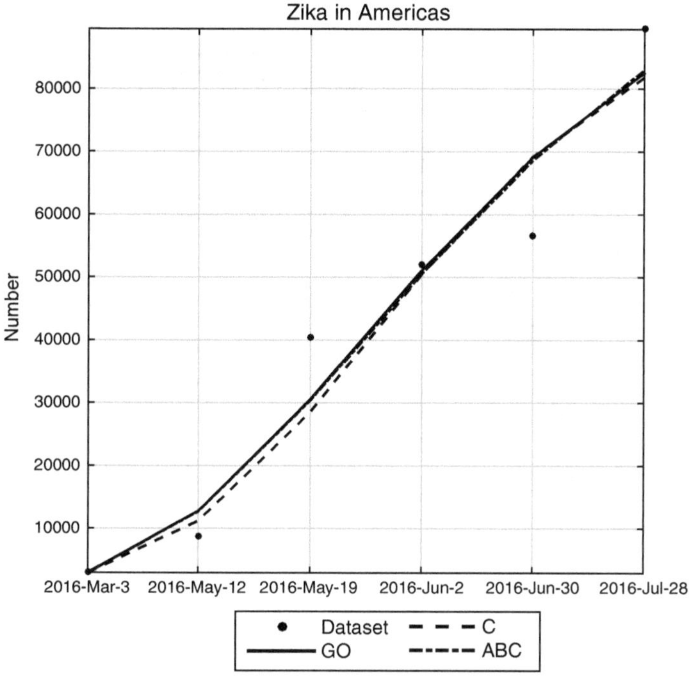

Fig. 2 Confirmed cases of Zika virus in Latin America. • Dataset, —— Classical model, - - - Fractional model with Liouville–Caputo derivative, -.... Fractional model with Atangana–Baleanu–Caputo derivative

Example 4 Capability to predict cancer in mice.
Bolton [14] proposes a prediction model in the Caputo sense to predict the growth of malignant tumors in mice. In this paper, he proposes the least squares optimization algorithm to find the model parameters and to minimize the RMSE (20), for this case we implement the models given in Eqs. (9) and (10) and the CS algorithm. The results are presented in Table 2 and Fig. 4.

Example 5 Patients with kidney transplantation due to end-stage renal disease.
For this example, we consider statistical data of patients with kidney transplantation in Spain [30]. The data are provided in absolute numbers and in population-standardized rates (per 100 000 inhabitants). To predict the cases number of kidney transplantation in the next years are used the models given in Eqs. (9) and (10) and the CS algorithm. The results are presented in Table 2 and Fig. 5.

Example 6 Rate of detection of tuberculosis cases in Mexico.
The World Bank for Nutrition in Health and Population Statistics provides statistics compiled from various international sources, where the main topic of interest to apply the models (9) and (10) is: infectious diseases, specifically tuberculosis. This statistical information was compiled and obtained from [31], where the parameters

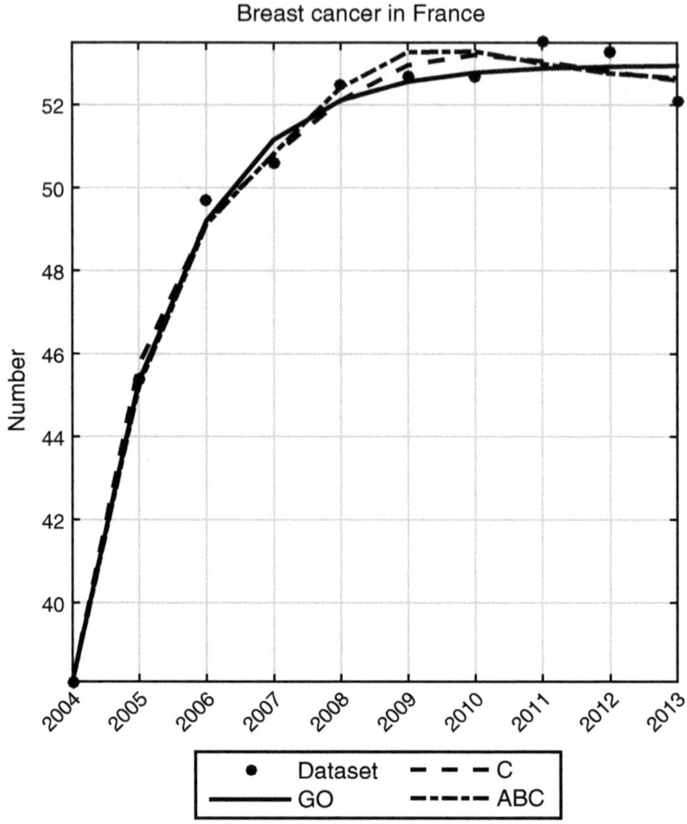

Fig. 3 Confirmed cases of breast cancer in France. • Dataset,—— Classical model,- - - Fractional model with Liouville–Caputo derivative,.... Fractional model with Atangana–Baleanu–Caputo derivative

obtained by CS are presented in Table 2 and the result of the prediction models is shown in Fig. 6.

The parameters showed in the Table 2 represented the better fit to a particular set of epidemiological dates analyzed. The fractional orders were obtained in the range $\gamma \in (0; 1]$.

Table 3 summarizes the prediction criteria used in the different examples for each fractional model. This information is easy to observe in Fig. 7. The chosen models have a small value for RMSE as well as a MAPE value less than 10%.

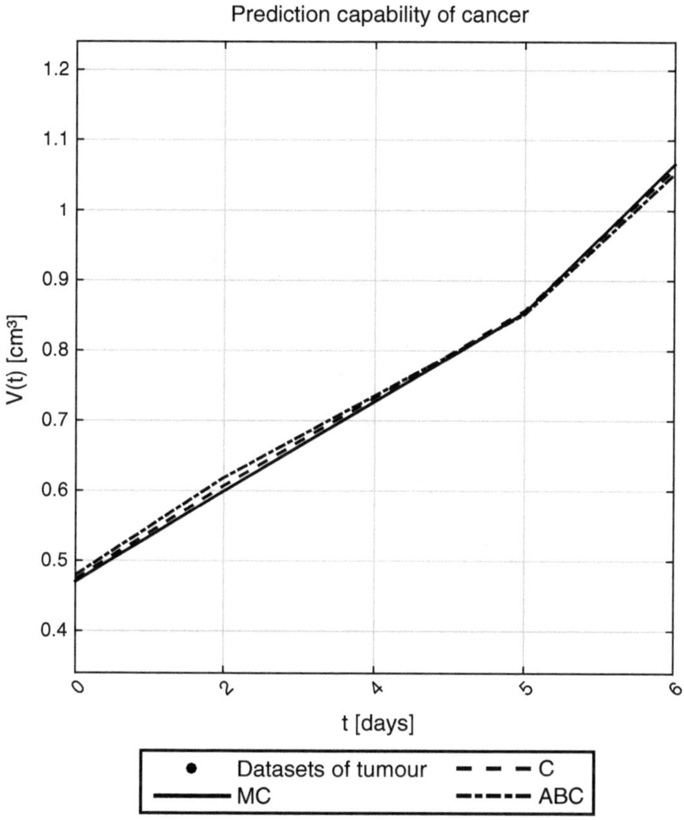

Fig. 4 Prediction of cancer in mice. • Dataset,—— Classical model,- - - Fractional model with Liouville–Caputo derivative,-.... Fractional model with Atangana–Baleanu–Caputo derivative

Table 3 Prediction criterion for all fractional Gompertz models

	Error				Error		
Example	RMSE	MAPE	Model	Example	RMSE	MAPE	Model
Case 1	0.005308	**2.6035**	Classical	*Case 4*	0.019401	**1.8252**	Classical
	0.005244	**2.2512**	Caputo		0.014949	**1.4691**	Caputo
	0.005576	**2.7218**	Atangana		**0.008053**	**0.7451**	Atangana
Case 2	0.058251	17.2277	Classical	*Case 5*	0.010379	**7.5878**	Classical
	0.062618	15.3974	Caputo		**0.008778**	11.9823	Caputo
	0.056949	17.0980	Atangana		0.008932	13.6197	Atangana
Case 3	0.002804	**0.6903**	Classical	*Case 6*	0.017245	**8.7268**	Classical
	0.002506	**0.7068**	Caputo		0.015797	**7.0239**	Caputo
	0.002772	**0.7390**	Atangana		**0.015609**	**6.8655**	Atangana

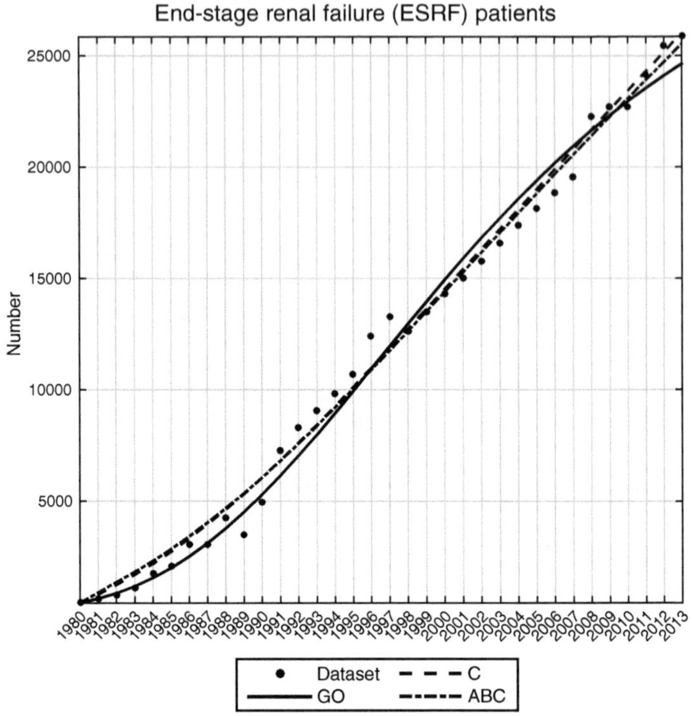

Fig. 5 Number of patients with kidney transplantation due to end-stage renal disease. • Dataset,▬▬ Classical model,- - - Fractional model with Liouville–Caputo derivative,-.-.- Fractional model with Atangana–Baleanu–Caputo derivative

5 Conclusions

In this chapter, the Cuckoo search algorithm was proposed to solve the Gompertz model considering fractional derivatives of Liouville–Caputo and Atangana–Baleanu–Caputo types. Also, they were developed six epidemiological examples using Cuckoo Search algorithm to show the effectiveness of this methodology. Each optimisation procedure was repeated 50 times for statistical inferences. For future work, this methodology can be applied considering other reparameterizations of the Gompertz model, for example, four-parameter Gompertz, the Zwietering modification, the Zweifel and Lasker reparameterization, and the Gompertz-Laird or the Unified Gompertz. For these models, it is more difficult to finding and interpret their parameters. Furthermore, this optimization strategy can be extended in a similar way to the population-based algorithms, such as PSO and genetic algorithms to study multi-objective constrained optimization problems.

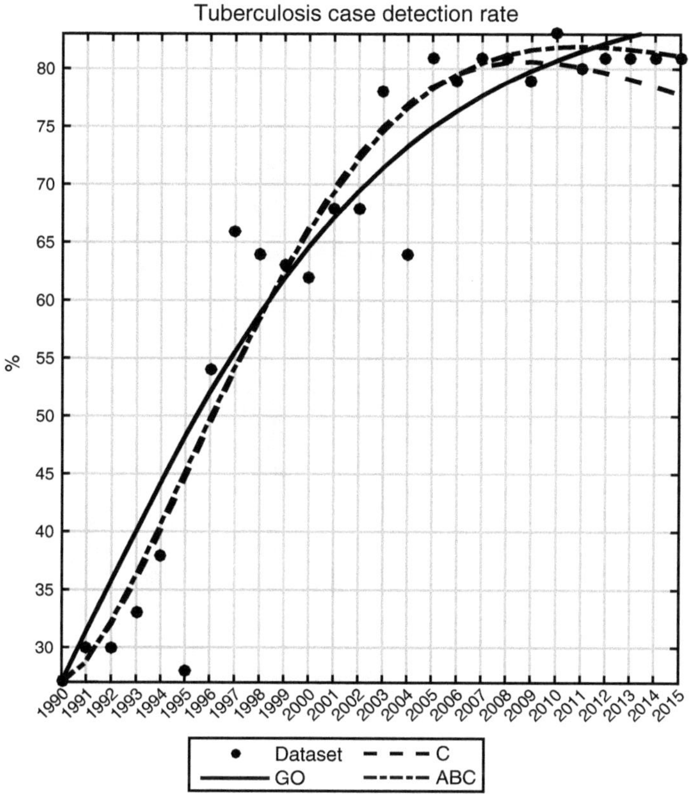

Fig. 6 Percentage of people with Tuberculosis in Mexico. • Dataset,—— Classical model,- - - Fractional model with Liouville–Caputo derivative,-.... Fractional model with Atangana–Baleanu–Caputo derivative

In this chapter, was showed that the fractional Gompertz model via Liouville–Caputo and Atangana–Baleanu–Caputo fit better the dataset than the classical Gompertz model. The fractional order derivatives with non-singular kernel can describe two different waiting times distributions, this is ideal waiting time distribution as such is observed in biological systems. The crossover behavior of these operators is due to their capacity of not obeying the classical index-law imposed in fractional calculus.

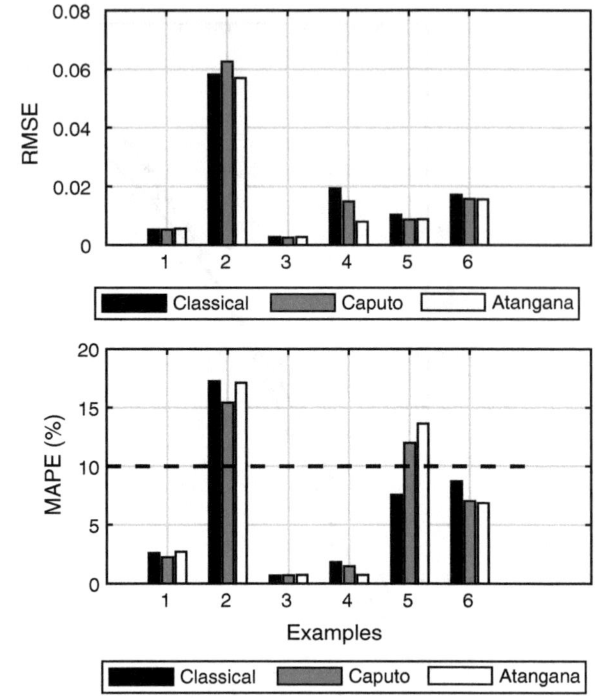

Fig. 7 RMSE and MAPE for the six models fitting

Acknowledgements Jesús Emmanuel Solís Pérez acknowledges the support provided by CONACyT through the assignment doctoral fellowship. José Francisco Gómez Aguilar acknowledges the support provided by CONACyT: Cátedras CONACyT para jóvenes investigadores 2014. José Francisco Gómez Aguilar and Ricardo Fabricio Escobar Jiménez acknowledges the support provided by SNI-CONACyT.

References

1. Gompertz, B.: On the nature of the function expressive of the law of human mortality, and on a new mode of determining the value of life contingencies. Philos. Trans. R. Soc. Lond. B Biol. Sci. **182**, 513–585 (1825)
2. Chatterjee, T., Chatterjee, B.K., Majumdar, D., Chakrabarti, P.: Antibacterial effect of silver nanoparticles and the modeling of bacterial growth kinetics using a modified Gompertz model. Biochim. Biophys. ca (BBA)-Gen. Subj. **1850**(2), 299–306 (2015)
3. Budiyono, I., Sumardiono, S.: Kinetic model of biogas yield production from vinasse at various initial pH: comparison between modified Gompertz model and first order kinetic model. Res. J. Appl. Sci. Eng. Technol. **7**(13), 2798–2805 (2014)
4. Tjorve, K.M., Tjorve, E.: The use of Gompertz models in growth analyses, and new Gompertz-model approach: an addition to the Unified-Richards family. PloS one **12**(6), 1–9 (2017)
5. Liu, H., Chen, N., Feng, C., Tong, S., Li, R.: Impact of electro-stimulation on denitrifying bacterial growth and analysis of bacterial growth kinetics using a modified Gompertz model in a bio-electrochemical denitrification reactor. Bioresour. Technol. **232**, 344–353 (2017)

6. Costa, B.A., Lemos, J.M.: Drug administration design for cancer Gompertz model based on the Lyapunov method. In: Controlo 2016, vol. 1, pp. 131–141. Springer International Publishing, Berlin (2017)
7. Horiuchi, S., Ouellette, N., Cheung, S.L.K., Robine, J.M.: Modal age at death: lifespan indicator in the era of longevity extension. Vienna Yearb. Popul. Res. **1**, 37–69 (2013)
8. Izquierdo, F., Prats, C., López, D.: The use of the Gompertz model in its differential form for weed emergence modelling. In: XV Congreso de la Sociedad Española de Malherbología, SEMh, 2015, vol. 1, pp. 367–373 (2015)
9. Ryan, C.A., Billington, S.L., Criddle, C.S.: Assessment of models for anaerobic biodegradation of a model bioplastic: poly (hydroxybutyrate-co-hydroxyvalerate). Bioresour. Technol. **227**, 205–213 (2017)
10. Mohammadi Farrokhran, E., Mahmoodi, M., Mohammad, K., Rahimi, A., Majlesi, F., Parsaeian, M.: Study of factors affecting first birth interval using modified Gompertz cure model in west Azarbaijan province, Iran. Iran. J. Epidemiol. **9**(1), 41–51 (2013)
11. Du, M., Wang, Z., Hu, H.: Measuring memory with the order of fractional derivative. Sci. Rep. **3**(1), 1–3 (2013)
12. Magin, R.L.: Fractional calculus models of complex dynamics in biological tissues. Comput. Math. Appl. **59**(5), 1586–1593 (2010)
13. Atici, F.M., Şengul, S.: Modeling with fractional difference equations. J. Math. Anal. Appl. **369**(1), 1–9 (2010)
14. Bolton, L., Cloot, A.H., Schoombie, S.W., Slabbert, J.P.: A proposed fractional-order Gompertz model and its application to tumour growth data. Math. Med. Biol. J. IMA **32**(2), 187–207 (2014)
15. Atangana, A., Secer, A.: A note on fractional order derivatives and table of fractional derivatives of some special functions. Abstr. Appl. Anal. **2013**, 1–15 (2013)
16. Caputo, M., Fabrizio, M.: A new definition of fractional derivative without singular kernel. Prog. Fract. Differ. Appl. **1**(2), 73–85 (2015)
17. Atangana, A., Baleanu, D.: New fractional derivatives with nonlocal and non-singular kernel. Theory and application to heat transfer model. Therm. Sci. **20**(2), 763–769 (2016)
18. Yang, X.S., Deb, S.: Cuckoo search: recent advances and applications. Neural Comput. Appl. **24**(1), 169–174 (2014)
19. Abdel-Basset, M., Hessin, A.N., Abdel-Fatah, L.: A comprehensive study of cuckoo-inspired algorithms. Neural Comput. Appl. **1**, 1–17 (2016)
20. Yang, X.S., Deb, S.: Engineering optimisation by cuckoo search. Int. J. Math. Model. Numer. Optim. **1**(4), 330–343 (2010)
21. Changpin, L., Chunxing, T.: On the fractional Adams method. Comput. Math. Appl. **58**(8), 1573–1588 (2009)
22. Diethelm, K., Ford, N.J., Freed, A.D.: Detailed error analysis for a fractional Adams method. Numer. Algorithms **36**(1), 31–52 (2004)
23. Alkahtani, B.S.T.: Chua's circuit model with Atangana–Baleanu derivative with fractional order. Chaos Solitons Fractals **89**, 1–5 (2016)
24. Wu, J.W., Hung, W.L., Tsai, C.H.: Estimation of parameters of the Gompertz distribution using the least squares method. Appl. Math. Comput. **158**(1), 133–147 (2004)
25. Fekedulegn, D., Mac Siurtain, M.P., Colbert, J.J.: Parameter estimation of nonlinear growth models in forestry. Silva Fenn. **33**(4), 327–336 (1999)
26. Lewis, C.D.: Industrial and Business Forecasting Methods: A Practical Guide to Exponential Smoothing and Curve Fitting. Butterworth-Heinemann, Oxford (1982)
27. https://knoema.com/UNAIDSS2016/united-nations-aids-statistics-2016
28. https://knoema.com/WZVO2016jul/world-zika-virus-epidemic-2015-16-monthly-update
29. https://knoema.com/hlth-ps-scre/breast-cancer-and-cervical-cancer-screenings
30. https://knoema.com/hlth-co-ren/end-stage-renal-failure-esrf-patients
31. https://knoema.com/WBHNPStats2016May/health-nutrition-and-population-statistics-world-bank

Existence and Uniqueness Results for a Novel Complex Chaotic Fractional Order System

Ilknur Koca and A. Atangana

Abstract The Atangana–Baleanu fractional differential and integral operators have been used in this chapter to describe the crossover behavior of a chaotic complex system. The existing model was extended and modified by replacing the conventional time local operator by the fractional differential operator with non-local and non-singular kernel. We established the conditions under which the existence of a uniquely exact solution can be found. A newly established numerical scheme was used to solve the modified model and numerical solutions are displayed for different values of fractional order.

Keywords Fractional calculus · Atangana–Baleanu fractional derivative · Chaotic complex system

1 Introduction

Complex systems have attracted attention of all my kind due to their occurrence in our daily life. They are omnipresent in the field of chaos, solitons, fractal, epidemiology and other fields where complexities are observed such as groundwater and biological models portraying the interaction among pieces. The description of these complex natural occurrences can be achieved using the mathematical tools known as derivative, and they can be classified in two big classes, the first class is the local differential operator that uses the rate of chance to express the variation of a moving object or change taking place in time and space [1, 2]. This first class was greatly used in the classical mechanic, where there is no sign of complexity like heterogeneity,

I. Koca (✉)
Department of Mathematics, Faculty of Sciences, Mehmet Akif Ersoy University, 15100 Burdur, Turkey
e-mail: ikoca@mehmetakif.edu.tr

A. Atangana
Faculty of Natural and Agricultural Sciences, Institute of Groundwater Studies, University of Free State, Bloemfontein 9301, South Africa

J. F. Gómez et al. (eds.), *Fractional Derivatives with Mittag-Leffler Kernel*, Studies in Systems, Decision and Control 194,
https://doi.org/10.1007/978-3-030-11662-0_7

self-similarities, and a crossover in mean-square displacement. The connective evolution equation of this differential operator obeys the so-called semi-group principle thus, it cannot replicate the well-known Non-Markovian processes. The second class was well-developed in the last years and is known as nonlocal differential operators. This last class was born as result of discussion between Sir L' Hopital and Leibniz, where the nth derivative could be $\frac{1}{2}$. derivative. Then this was developed later on by Riemann and Liouville, modified by Michele Caputo by transforming the derivative of a convolution of a given function and the power law decay function to a convolution of derivative of a derivative of first and the power law decay function [3–6]. The last class witness a split in the last decade, as the power law Riemann–Liouville and Caputo derivatives have posed some problems, as it is imposing a kind of singularity to those models with no singularities. To solve this problem, a new class was suggested where the power law kernel was replaced by exponential decay and the generalized Mittag-Leffler function [7–16]. An analysis done by three senior Brazilian researchers suggested that the two last suggested kernels have added a real plus in the field as they are able to describe a crossover behavior that is observed in many field of science, technology and engineering. With the new weapons brought by the new class of non-local operators one can describe materials or moving object taking place in different scales as they possesses a mean-square displacement with crossover from normal to sub-diffusion and confined diffusion. In this chapter, we aim to apply the Mittag-Leffler kernel derivative to a well-known complex system able to describe chaotic behavior [17].

2 New Fractional Derivative with Non-singular and Non-local Kernel

Let us remind the definitions of the new fractional derivative with non-singular and non-local kernel [18–27].

Definition 1 Let $f \in H^1(a,b),\ b > a,\ \alpha \in [0,1]$ then, the definition of the new fractional derivative (Atangana–Baleanu derivative in Caputo sense) is given as:

$$ {}_{a}^{ABC}D_t^{\alpha} f(t) = \frac{B(\alpha)}{1-\alpha} \int_a^t f^{'}(x) E_{\alpha}\left[-\alpha \frac{(t-x)^{\alpha}}{1-\alpha}\right] dx, \tag{1} $$

where ${}_{a}^{ABC}D_t^{\alpha}$ is fractional operator with Mittag-Leffler kernel in the Caputo sense with order α with respect to t and $B(\alpha) = B(0) = B(1) = 1$ is a normalization function [5].

It can be noted that the above definition is helpful to model real world problems. Also it has a great advantage while using the Laplace transform to solve problem with initial condition.

Definition 2 Let $f \in H^1(a,b),\ b > a,\ \alpha \in [0,1]$ and not differentiable then, the definition of the new fractional derivative (Atangana–Baleanu fractional derivative in Riemann–Liouville sense) is given as:

$$ {}_{a}^{ABR}D_t^{\alpha} f(t) = \frac{B(\alpha)}{1-\alpha}\frac{d}{dt}\int_a^t f(x)E_{\alpha}\left[-\alpha\frac{(t-x)^{\alpha}}{1-\alpha}\right]dx. \tag{2} $$

Definition 3 The fractional integral of order α of a new fractional derivative is defined as:

$$ {}_{a}^{AB}I_t^{\alpha} f(t) = \frac{1-\alpha}{B(\alpha)} f(t) + \frac{\alpha}{B(\alpha)\Gamma(\alpha)}\int_a^t f(y)(t-y)^{\alpha-1}dy. \tag{3} $$

When α is zero, initial function is obtained and when α is 1, the ordinary integral is obtained.

Theorem 1 *The following time fractional ordinary differential equation*

$$ {}_{0}^{ABC}D_t^{\alpha} f(t) = u(t), \tag{4} $$

has a unique solution with taking the inverse Laplace transform and using the convolution theorem below [4]:

$$ f(t) = \frac{1-\alpha}{B(\alpha)} u(t) + \frac{\alpha}{B(\alpha)\Gamma(\alpha)}\int_a^t u(y)(t-y)^{\alpha-1}dy. \tag{5} $$

3 Picard's Existence and Uniqueness Theorem for Atangana–Baleanu Fractional Complex System in Caputo Sense

In this section, we will present the following existence and uniqueness theorems for Atangana–Baleanu fractional complex system in Caputo sense via Picard's theorem. The theorem considered here is very easy to understand and has same the idea with classical theorems known in the case of first order system of equations. Atangana–Baleanu fractional complex system in Caputo sense is given below:

$$\begin{aligned}
{}_{0}^{ABC}D_{t}^{\alpha}y_{1}(t) &= ay_{3}(t)\,y_{5}(t),\\
{}_{0}^{ABC}D_{t}^{\alpha}y_{2}(t) &= ay_{4}(t)\,y_{5}(t),\\
{}_{0}^{ABC}D_{t}^{\alpha}y_{3}(t) &= b(y_{1}(t)-y_{3}(t)),\\
{}_{0}^{ABC}D_{t}^{\alpha}y_{4}(t) &= b(y_{2}(t)-y_{4}(t)),\\
{}_{0}^{ABC}D_{t}^{\alpha}y_{5}(t) &= 1-y_{1}^{2}(t)-y_{2}^{2}(t),
\end{aligned} \tag{6}$$

and initial conditions

$$\begin{aligned}
&y_{1}(t_{0})=y_{1,0},\; y_{2}(t_{0})=y_{2,0},\; y_{3}(t_{0})=y_{3,0},\\
&y_{4}(t_{0})=y_{4,0},\; y_{5}(t_{0})=y_{5,0}.
\end{aligned} \tag{7}$$

Let us consider the right side of the system with a new expression as below:

$$\begin{aligned}
C_{1}(t,y_{1}(t)) &= ay_{3}(t)\,y_{5}(t),\\
C_{2}(t,y_{2}(t)) &= ay_{4}(t)\,y_{5}(t),\\
C_{3}(t,y_{3}(t)) &= b(y_{1}(t)-y_{3}(t)),\\
C_{4}(t,y_{4}(t)) &= b(y_{2}(t)-y_{4}(t)),\\
C_{5}(t,y_{5}(t)) &= 1-y_{1}^{2}(t)-y_{2}^{2}(t).
\end{aligned} \tag{8}$$

Then applying the Volterra type integral on the above complex fractional system, the following integral system is written:

$$\begin{cases}
y_{1}(t)=y_{1,0}(0)+\frac{1-\alpha}{B(\alpha)}C_{1}(t,y_{1}(t))+\frac{\alpha}{B(\alpha)\Gamma(\alpha)}\int\limits_{0}^{t}C_{1}(\tau,y_{1}(\tau))(t-\tau)^{\alpha-1}d\tau,\\
y_{2}(t)=y_{2,0}(0)+\frac{1-\alpha}{B(\alpha)}C_{2}(t,y_{2}(t))+\frac{\alpha}{B(\alpha)\Gamma(\alpha)}\int\limits_{0}^{t}C_{2}(\tau,y_{2}(\tau))(t-\tau)^{\alpha-1}d\tau,\\
y_{3}(t)=y_{3,0}(0)+\frac{1-\alpha}{B(\alpha)}C_{3}(t,y_{3}(t))+\frac{\alpha}{B(\alpha)\Gamma(\alpha)}\int\limits_{0}^{t}C_{3}(\tau,y_{3}(\tau))(t-\tau)^{\alpha-1}d\tau,\\
y_{4}(t)=y_{4,0}(0)+\frac{1-\alpha}{B(\alpha)}C_{4}(t,y_{4}(t))+\frac{\alpha}{B(\alpha)\Gamma(\alpha)}\int\limits_{0}^{t}C_{4}(\tau,y_{4}(\tau))(t-\tau)^{\alpha-1}d\tau,\\
y_{5}(t)=y_{5,0}(0)+\frac{1-\alpha}{B(\alpha)}C_{5}(t,y_{5}(t))+\frac{\alpha}{B(\alpha)\Gamma(\alpha)}\int\limits_{0}^{t}C_{5}(\tau,y_{5}(\tau))(t-\tau)^{\alpha-1}d\tau,
\end{cases} \tag{9}$$

with initial conditions

$$\begin{aligned}
&y_{1,0}(0)=0,\; y_{2,0}(0)=0,\; y_{3,0}(0)=0,\\
&y_{4,0}(0)=0,\; y_{5,0}(0)=0.
\end{aligned} \tag{10}$$

Theorem *The kernels of system $C_i(t, y_i(t))$, for $i = 1, 2, 3, \ldots 5$, satisfy the Lipschitz condition and contraction if the following inequality holds:*

$$\begin{cases} 0 \le L_1 < 1, \\ 0 \le L_2 < 1, \\ 0 \le L_3 < 1, \\ 0 \le L_4 < 1, \\ 0 \le L_5 < 1. \end{cases} \tag{11}$$

Proof First we start the kernel $C_1(t, y_1(t)) = ay_3(t)\, y_5(t)$. Let $y_1(t)$ and $x_1(t)$ be two functions, so we have the following:

$$|C_1(t, y_1(t)) - C_1(t, x_1(t))| \le 0. |y_1(t) - x_1(t)|, \tag{12}$$

taking as $L_1 = 0$, the Lipschitz condition and contraction are satisfied for $C_1(t, y_1(t))$. It is easy to see that other kernels are also satisfy Lipschitz condition for $0 \le L_i < 1$, for $i = 2, \ldots, 5$.

Now we can give the existence of solution and uniqueness theorems for system under Lipschitz condition with respect to y_i and continuity condition with respect to t.

Theorem (Picard's existence theorem for system) *Let B be a domain in R^2 and $C_i : B \to R$, for $i = 1, 2, \ldots, 5$ be a real functions of system satisfying the following conditions:*

(1) C_i are continuous on B, for $i = 1, 2, \ldots, 5$.
(2) $C_i(t, y_i(t))$, for $i = 1, 2, \ldots, 5$ are Lipschitz continuous with respect to y_i on D with Lipschitz constants of $L_i > 0$.

Let $(t_0, y_{i,0})$ are an interior point on B and $k > 0$, $m_i > 0$ be constants such that the rectangle

$$R = \left\{(t, y_i) : |t - t_0| \le k, \left|y_i - y_{i,0}\right| \le m_i, \textit{ for } i = 1, 2, \ldots, 5\right\} \subset B. \tag{13}$$

If we take

$$c_i = \max_{(t,y_i)\in R} C_i(t, y_i(t)) \textit{ and } h = \min\left(k, \frac{m_i}{c_i}\right), \tag{14}$$

then the initial value problem has a unique solution of y_i, for $i = 1, 2, \ldots, 5$ on the interval $|t - t_0| \le h$.

Remark Since R is a closed rectangle in B, $C_i(t, y_i(t))$ for $i = 1, 2, \ldots, 5$ are satisfy all properties in R.

$$\begin{cases} \text{If } k < \frac{m_i}{c_i} \text{ then } h = k \Longrightarrow R_1 = R, \\ \text{If } \frac{m_i}{c_i} < k \text{ then } h = \frac{m_i}{c_i} \Longrightarrow R_1 \subset R. \end{cases} \tag{15}$$

Here for $i = 1, 2, \ldots, 5$,

$$R = \left\{(t, y_i) : |t - t_0| \le k, \left|y_i - y_{i,0}\right| \le m_i\right\}, \qquad (16)$$
$$R_1 = \left\{(t, y_i) : |t - t_0| \le h, \left|y_i - y_{i,0}\right| \le m_i\right\}.$$

We prove the theorem by successive approximation of the Picard's iterants $y_{i,n}(t)$ for $i = 1, 2, \ldots, 5$, on $|t - t_0| \le h$ and are defined by

$$\begin{aligned}
y_{1,n}(t) &= y_{1,0}(0) + \tfrac{1-\alpha}{B(\alpha)} C_1(t, y_{1,n-1}(t)) + \tfrac{\alpha}{B(\alpha)\Gamma(\alpha)} \int_0^t C_1(\tau, y_{1,n-1}(\tau))(t-\tau)^{\alpha-1} d\tau, \\
y_{2,n}(t) &= y_{2,0}(0) + \tfrac{1-\alpha}{B(\alpha)} C_2(t, y_{2,n-1}(t)) + \tfrac{\alpha}{B(\alpha)\Gamma(\alpha)} \int_0^t C_2(\tau, y_{2,n-1}(\tau))(t-\tau)^{\alpha-1} d\tau, \\
y_{3,n}(t) &= y_{3,0}(0) + \tfrac{1-\alpha}{B(\alpha)} C_3(t, y_{3,n-1}(t)) + \tfrac{\alpha}{B(\alpha)\Gamma(\alpha)} \int_0^t C_3(\tau, y_{3,n-1}(\tau))(t-\tau)^{\alpha-1} d\tau, \\
y_{4,n}(t) &= y_{4,0}(0) + \tfrac{1-\alpha}{B(\alpha)} C_4(t, y_{4,n-1}(t)) + \tfrac{\alpha}{B(\alpha)\Gamma(\alpha)} \int_0^t C_4(\tau, y_{4,n-1}(\tau))(t-\tau)^{\alpha-1} d\tau, \\
y_{5,n}(t) &= y_{5,0}(0) + \tfrac{1-\alpha}{B(\alpha)} C_5(t, y_{5,n-1}(t)) + \tfrac{\alpha}{B(\alpha)\Gamma(\alpha)} \int_0^t C_5(\tau, y_{5,n-1}(\tau))(t-\tau)^{\alpha-1} d\tau.
\end{aligned} \qquad (17)$$

Now we divide the proof into 4 parts.

Part 1: In this part we will show some properties of the equations $\left\{y_{i,n}(t)\right\}$ for $i = 1, 2, \ldots, 5$. Let us give step by step of what we will obtain in part 1.

(*i*) The functions $\left\{y_{i,n}(t)\right\}$ for $i = 1, 2, \ldots, 5$ defined above are well defined.
(*ii*) $y_{i,n}(t)$' s for $i = 1, 2, \ldots, 5$ have continuous derivatives.
(*iii*) $\left|y_{i,n}(t) - y_{i,0}(0)\right| \le \left(\frac{1-\alpha}{B(\alpha)} + \frac{t^\alpha}{B(\alpha)\Gamma(\alpha)}\right) c_i$ for $i = 1, 2, \ldots, 5$ on $[t_0, t_0 + h]$.
(*iv*) $C_i(t, y_{i,n}(t))$, for $i = 1, 2, \ldots, 5$ are well defined.

Proof of Part 1: We prove this part by mathematical induction. Assume that $y_{i,n-1}(t)$ exists, has continuous derivative on $[t_0, t_0 + h]$ and it satisfies

$$\left|y_{i,n-1}(t) - y_{i,0}(0)\right| \le m_i, \text{ for } i = 1, 2, \ldots, 5 \text{ on } t \in [t_0, t_0 + h]. \qquad (18)$$

Here

$$m_i = \left(\frac{1-\alpha}{B(\alpha)} + \frac{t^\alpha}{B(\alpha)\Gamma(\alpha)}\right) c_i. \qquad (19)$$

This implies $\left(t, y_{i,n-1}(t)\right) \in R_1$. Also we have $C_i(t, y_{i,n-1}(t))$ are defined and continuous on $[t_0, t_0 + h]$. Further $\left|C_i(t, y_{i,n-1}(t))\right| \le c_i$ on $[t_0, t_0 + h]$. Let us consider absolute value on both sides of equation

$$\begin{aligned}
\left|y_{i,n}(t) - y_{i,0}(0)\right| &= \frac{1-\alpha}{B(\alpha)} \left|C_i(t, y_{i,n-1}(t))\right| \\
&\quad + \frac{\alpha}{B(\alpha)\Gamma(\alpha)} \int_0^t \left|C_i(\tau, y_{i,n-1}(\tau))\right| (t-\tau)^{\alpha-1} d\tau,
\end{aligned} \qquad (20)$$

with using triangle inequality we have

$$\left|y_{i,n}(t) - y_{i,0}(0)\right| \le \frac{1-\alpha}{B(\alpha)} c_i + \frac{\alpha}{B(\alpha)\Gamma(\alpha)} \int_0^t c_i (t-\tau)^{\alpha-1} d\tau \le \left(\frac{1-\alpha}{B(\alpha)} + \frac{t^\alpha}{B(\alpha)\Gamma(\alpha)}\right) c_i, \tag{21}$$

then

$$\left|y_{i,n}(t) - y_{i,0}(0)\right| \le m_i. \tag{22}$$

So $(t, y_{i,n}(t))$ lies in the rectangle R_1 and hence $C_i(t, y_{i,n}(t))$ is defined and continuous on $[t_0, t_0 + h]$.

When $n = 1$,

$$y_{i,1}(t) = y_{i,0}(0) + \frac{1-\alpha}{B(\alpha)} C_i(t, y_{i,0}(t)) + \frac{\alpha}{B(\alpha)\Gamma(\alpha)} \int_0^t C_i(\tau, y_{i,0}(\tau))(t-\tau)^{\alpha-1} d\tau. \tag{23}$$

Obviously, $y_{i,1}(t)$ is defined, has continuous derivative on $[t_0, t_0 + h]$. Also

$$\left|y_{i,1}(t) - y_{i,0}(0)\right| \le \frac{1-\alpha}{B(\alpha)} \left|C_i(t, y_{i,0}(t))\right| + \frac{\alpha}{B(\alpha)\Gamma(\alpha)} \int_0^t \left|C_i(\tau, y_{i,0}(\tau))\right| (t-\tau)^{\alpha-1} d\tau, \tag{24}$$

$$\left|y_{i,1}(t) - y_{i,0}(0)\right| \le \left(\frac{1-\alpha}{B(\alpha)} + \frac{t^\alpha}{B(\alpha)\Gamma(\alpha)}\right) c_i \le m_i.$$

So $(t, y_{i,1}(t))$ lies in the rectangle R_1 and hence $C_i(t, y_{i,1}(t))$ is continuous on $[t_0, t_0 + h]$. Properties are true for $n = 1$. Thus, by the method of mathematical induction $\{y_{i,n}(t)\}$ sequence functions defined in integral system are possessing all desired properties in $[t_0, t_0 + h]$. Hence part 1 of the proof is completed.

Part 2: The functions $\{y_{i,n}(t)\}$, for $i = 1, 2, \ldots, 5$ satisfy the following inequality as below:

$$\left|y_{i,n}(t) - y_{i,n-1}(t)\right| \le \left(\frac{1-\alpha}{B(\alpha)} + \frac{t^\alpha}{B(\alpha)\Gamma(\alpha)}\right)^n L_i^{n-1} c_i \text{ on } [t_0, t_0 + h]. \tag{25}$$

Proof of Part 2: We prove this part also by mathematical induction. Assume that

$$\left|y_{i,n-1}(t)-y_{i,n-2}(t)\right| \leq \left(\frac{1-\alpha}{B(\alpha)}+\frac{t^{\alpha}}{B(\alpha)\Gamma(\alpha)}\right)^{n-1} L_i^{n-2} c_i t \in [x_0, x_0+h]. \quad (26)$$

Then

$$\begin{aligned}
&\left|y_{i,n}(t)-y_{i,n-1}(t)\right| \quad (27)\\
&\leq \frac{1-\alpha}{B(\alpha)}\left|C_i(t, y_{i,n-1}(t))-C_i(t, y_{i,n-2}(t))\right|\\
&+\frac{\alpha}{B(\alpha)\Gamma(\alpha)}\int_0^t \left|C_i(\tau, y_{i,n-1}(\tau))-C_i(\tau, y_{i,n-2}(\tau))\right|(t-\tau)^{\alpha-1}d\tau.
\end{aligned}$$

From part 1, we have

$$\left|y_{i,n}(t)-y_{i,0}(0)\right| \leq \left(\frac{1-\alpha}{B(\alpha)}+\frac{t^{\alpha}}{B(\alpha)\Gamma(\alpha)}\right)c_i, \text{ for } \forall n \in R_1. \quad (28)$$

Hence $(t, y_{i,n-1}(t))$, $(t, y_{i,n-2}(t))$ also belong to in R_1. From Lipschitz continuity of all C_i for $i = 1, 2, \ldots 5$, we have

$$\begin{aligned}
\left|y_{i,n}(t)-y_{i,n-1}(t)\right| \leq{}& \frac{1-\alpha}{B(\alpha)}L_i\left|y_{i,n-1}(t)-y_{i,n-2}(t)\right| \quad (29)\\
&+\frac{\alpha L_i}{B(\alpha)\Gamma(\alpha)}\int_0^t \left|y_{i,n-1}(\tau)-y_{i,n-2}(\tau)\right|(t-\tau)^{\alpha-1}d\tau.
\end{aligned}$$

From assumption,

$$\begin{aligned}
\left|y_{i,n}(t)-y_{i,n-1}(t)\right| \leq{}& \frac{1-\alpha}{B(\alpha)}\left(\frac{1-\alpha}{B(\alpha)}+\frac{t^{\alpha}}{B(\alpha)\Gamma(\alpha)}\right)^{n-1} L_i^{n-1}c_i\\
&+\frac{t^{\alpha}}{B(\alpha)\Gamma(\alpha)}\left(\frac{1-\alpha}{B(\alpha)}+\frac{t^{\alpha}}{B(\alpha)\Gamma(\alpha)}\right)^{n-1} L_i^{n-1}c_i\\
\leq{}& \left(\frac{1-\alpha}{B(\alpha)}+\frac{t^{\alpha}}{B(\alpha)\Gamma(\alpha)}\right)^{n} L_i^{n-1}c_i, \left|t-t_0\right| \leq h.
\end{aligned}$$

The inequality is true for n. Let take $n = 1$,

$$\left|y_{i,1}(t)-y_{i,0}(t)\right| \leq \left(\frac{1-\alpha}{B(\alpha)}+\frac{t^{\alpha}}{B(\alpha)\Gamma(\alpha)}\right)c_i. \quad (30)$$

By mathematical induction the inequality is true for all n.

Part 3: While $n \to \infty$, $\{y_{i,n}(t)\}$, for $i = 1, 2, \ldots 5$, converges uniformly to a continuous function y_i on $[t_0, t_0 + h]$.

Proof of Part 3: From proof of part 2, we got inequality as below:

$$\left|y_{i,n}(t) - y_{i,n-1}(t)\right| \le \left(\frac{1-\alpha}{B(\alpha)} + \frac{t^{\alpha}}{B(\alpha)\Gamma(\alpha)}\right)^{n} L_i^{n-1} c_i \text{ on } [t_0, t_0 + h]. \tag{31}$$

Let consider right side of equality as

$$\sum_{n=1}^{\infty}\left(\frac{1-\alpha}{B(\alpha)} + \frac{t^{\alpha}}{B(\alpha)\Gamma(\alpha)}\right)^{n} L_i^{n-1} c_i. \tag{32}$$

It is clear that this series converges ıf

$$\left(\frac{1-\alpha}{B(\alpha)} + \frac{t^{\alpha}}{B(\alpha)\Gamma(\alpha)}\right) L_i < 1 \text{ for } i = 1, 2, \ldots 5. \tag{33}$$

Now consider left side of equality as

$$\sum_{n=1}^{\infty}\left|y_{i,n}(t) - y_{i,n-1}(t)\right|. \tag{34}$$

Since

$$\sum_{n=1}^{\infty}\left(\frac{1-\alpha}{B(\alpha)} + \frac{t^{\alpha}}{B(\alpha)\Gamma(\alpha)}\right)^{n} L_i^{n-1} c_i, \tag{35}$$

converges then by Weierstrass M-test

$$\sum_{n=1}^{\infty}\left|y_{i,n}(t) - y_{i,n-1}(t)\right|, \tag{36}$$

converges on $[t_0, t_0 + h]$.

If it is converges what is the limit of y_i? Let us try to answer this question below: Consider the sequence of partial sum of the above series with $S_n(t)$.

$$S_n(t) = y_{i,0}(t) + \sum_{k=1}^{n}\left|y_{i,k}(t) - y_{i,k-1}(t)\right| = y_{i,n}(t). \tag{37}$$

Here $\{S_n(t)\} = \{y_{i,n}(t)\}$ converges uniformly to a limit function y_i on $[t_0, t_0 + h]$. The sequence of functions $\{y_{i,n}(t)\}$ defined by the Picard's iterative scheme converges uniformly to y_i on $[t_0, t_0 + h]$. From part 1, each $y_{i,n}(t)$ is continuous on $[t_0, t_0 + h]$ and hence the limit function y_i itself is continuous on $[t_0, t_0 + h]$.

Conclusion of part 3: $\{y_{i,n}(t)\} \to \{y_i\}$ for $i = 1, 2, \ldots 5$, on $[t_0, t_0 + h]$ and $y_i \in C[t_0, t_0 + h]$.

Part 4: The limit function y_i for $i = 1, 2, \ldots 5$, satisfies the complex fractional order system on the interval $[t_0, t_0 + h]$.

Proof of Part 4: Since each $\{y_{i,n}(t)\}$ for $i = 1, 2, \ldots 5$, satisfies

$$\left|y_{i,n}(t) - y_{i,0}(t)\right| \leq \left(\frac{1-\alpha}{B(\alpha)} + \frac{t^{\alpha}}{B(\alpha)\Gamma(\alpha)}\right) c_{i,} \tag{38}$$

on the interval $[t_0, t_0 + h]$. So we get

$$\left|y_i(t) - y_{i,0}(t)\right| \leq \left(\frac{1-\alpha}{B(\alpha)} + \frac{t^{\alpha}}{B(\alpha)\Gamma(\alpha)}\right) c_{i,} \tag{39}$$

on the interval $[t_0, t_0 + h]$ for $i = 1, 2, \ldots 5$. We also have $y_{i,n}(t) \to y_i(t)$ uniformly converges. We will prove that $C_i(t, y_{i,n}(t)) \to C_i(t, y_i(t))$ uniformly on $[t_0, t_0 + h]$ for $i = 1, 2, \ldots 5$. If we find that by using Lipschitz argument

$$\left|C_i(t, y_{i,n}(t)) \to C_i(t, y_i(t))\right| \leq L_i \left|y_{i,n}(t) - y_i(t)\right|. \tag{40}$$

Let us give the uniform convergence of $\{y_{i,n}(t)\}$ as below:

For $\forall\varepsilon > 0$, $\exists N(\varepsilon) > 0$ such that

$$\left|y_{i,n}(t) - y_i(t)\right| < \frac{\varepsilon}{L_i}, \tag{41}$$

for $\forall n > N(\varepsilon)$. So for $\forall n > N(\varepsilon)$

$$\begin{aligned}\left|C_i(t, y_{i,n}(t)) \to C_i(t, y_i(t))\right| &\leq L_i.\frac{\varepsilon}{L_i} \\ &\leq \varepsilon.\end{aligned} \tag{42}$$

This shows $C_i(t, y_{i,n}(t)) \to C_i(t, y_i(t))$ uniformly on $[t_0, t_0 + h]$ for $i = 1, 2, \ldots 5$. Since $C_i(t, y_{i,n}(t)$ is continuous for each n on $[t_0, t_0 + h]$.

So, therefore

$$y_i(t) = \lim_{n\to\infty} \left(\begin{array}{c} y_{i,n}(t) = y_{i,0}(0) + \frac{1-\alpha}{B(\alpha)} C_i(t, y_{i,n}(t)) \\ + \frac{\alpha}{B(\alpha)\Gamma(\alpha)} \int\limits_0^t C_i(\tau, y_{i,n}(\tau))(t-\tau)^{\alpha-1} d\tau \end{array} \right), \tag{43}$$

$$y_i(t) = \lim_{n\to\infty} y_{i,n}(t) = y_{i,0}(0) + \lim_{n\to\infty} \frac{1-\alpha}{B(\alpha)} C_i(t, y_{i,n}(t))$$

$$+ \frac{\alpha}{B(\alpha)\Gamma(\alpha)} \int\limits_0^t \lim_{n\to\infty} C_i(\tau, y_{i,n}(\tau))(t-\tau)^{\alpha-1} d\tau,$$

$$y_i(t) = y_{i,0}(0) + \frac{1-\alpha}{B(\alpha)} C_i(t, y_i(t)) + \frac{\alpha}{B(\alpha)\Gamma(\alpha)} \int_0^t C_i(\tau, y_i(\tau))(t-\tau)^{\alpha-1} d\tau, \quad (44)$$

where $y_i(t)$ is limit function of $\{y_{i,n}(t)\}$. From the basic lemma, the function $y_{i,n}(t)$ satisfies the initial value problem. This proves the existence of a solution of complex fractional Atangana–Baleanu system in Caputo sense.

Uniqueness of Solution

In this part, we will show the uniqueness of the solutions of the system. Assume that we have other solution of complex fractional system as $x_i(t)$ for $i = 1, 2, 3, \ldots 5$. Then consider two different integral equations as below:

$$y_i(t) = y_{i,0}(0) + \frac{1-\alpha}{B(\alpha)} C_i(t, y_i(t)) + \frac{\alpha}{B(\alpha)\Gamma(\alpha)} \int_0^t C_i(\tau, y_i(\tau))(t-\tau)^{\alpha-1} d\tau, \quad (45)$$

and

$$x_i(t) = x_{i,0}(0) + \frac{1-\alpha}{B(\alpha)} C_i(t, x_i(t)) + \frac{\alpha}{B(\alpha)\Gamma(\alpha)} \int_0^t C_i(\tau, x_i(\tau))(t-\tau)^{\alpha-1} d\tau, \quad (46)$$

for $|t - t_0| \leq h$. Then we have

$$y_i(t) - x_i(t) = \frac{1-\alpha}{B(\alpha)} \left(C_i(t, y_i(t)) - C_i(t, x_i(t))\right) + \frac{\alpha}{B(\alpha)\Gamma(\alpha)} \int_0^t \left(C_i(\tau, y_i(\tau)) - C_i(\tau, x_i(\tau))\right)(t-\tau)^{\alpha-1} d\tau. \quad (47)$$

Let us put absolute value on both side of above equality and consider Lipschitz condition, we have the following:

$$|y_i(t) - x_i(t)| = \frac{1-\alpha}{B(\alpha)} \left|\left(C_i(t, y_i(t)) - C_i(t, x_i(t))\right)\right| + \frac{\alpha}{B(\alpha)\Gamma(\alpha)} \int_0^t \left|\left(C_i(\tau, y_i(\tau)) - C_i(\tau, x_i(\tau))\right)\right|(t-\tau)^{\alpha-1} d\tau, \quad (48)$$

$$\begin{aligned}|y_i(t) - x_i(t)| \leq & \frac{1-\alpha}{B(\alpha)} L_i |y_i(t) - x_i(t)| \\ & + \frac{t^\alpha}{B(\alpha)\Gamma(\alpha)} L_i |y_i(t) - x_i(t)| .\end{aligned} \tag{49}$$

Then this gives,

$$|y_i(t) - x_i(t)| \left(1 - \left(\frac{1-\alpha}{B(\alpha)} + \frac{t^\alpha}{B(\alpha)\Gamma(\alpha)}\right) L_i\right) \leq 0. \tag{50}$$

It is verified with

$$|y_i(t) - x_i(t)| = 0 \Rightarrow y_i(t) = x_i(t), \text{ for } i = 1, 2, 3, \ldots 5. \tag{51}$$

So we have that equation has a unique solution.

4 Numerical Scheme

Recently, Toufik and Atangana have developed a novel numerical scheme to solve some special problems of fractional derivative with non-local and non-singular kernel [28]. In their paper, it can be easily seen that their method not only converges quickly to the exact solutions but also is highly accurate. To explain their method, let us consider the following non-linear fractional ordinary equation:

$$\begin{cases} {}_0^{ABC}D_t^\alpha x(t) = f(t, x(t)), \\ \quad x(0) = x_0. \end{cases} \tag{52}$$

This initial value problem is equivalent to fractional integral as below:

$$x(t) - x(0) = \frac{1-\alpha}{B(\alpha)} f(t, x(t)) + \frac{\alpha}{B(\alpha)\Gamma(\alpha)} \int_0^t f(y, x(y))(t-y)^{\alpha-1} dy. \tag{53}$$

At a given point $t = t_{n+1}$, $n = 0, 1, 2, \ldots$ the above integral equation is written as

$$x(t_{n+1}) - x(0) = \frac{1-\alpha}{B(\alpha)} f(t_n, x(t_n)) + \frac{\alpha}{B(\alpha)\Gamma(\alpha)} \int_0^{t_{n+1}} f(y, x(y))(t_{n+1}-y)^{\alpha-1} dy. \tag{54}$$

If we consider $f(y, x(y))$ via two-step Lagrange polynomial interpolation, the following expression will be obtained in the interval $[t_k, t_{k+1}]$.

$$\begin{aligned} p_k(y) &= f(y, x(y)) \\ &= \frac{y-t_{k-1}}{t_k-t_{k-1}} f(t_k, x(t_k)) - \frac{y-t_k}{t_k-t_{k-1}} f(t_{k-1}, x(t_{k-1})) \\ &= \frac{f(t_k, x(t_k))}{h}(y-t_{k-1}) - \frac{f(t_{k-1}, x(t_{k-1}))}{h}(y-t_k) \\ &\simeq \frac{f(t_k, x_k)}{h}(y-t_{k-1}) - \frac{f(t_{k-1}, x_{k-1})}{h}(y-t_k)\,. \end{aligned} \tag{55}$$

If we put the above expression in where $f(y, x(y))$, then we have

$$\begin{aligned} x_{n+1} &= x_0 + \frac{1-\alpha}{B(\alpha)} f(t_n, x(t_n)) \\ &+ \frac{\alpha}{B(\alpha)\Gamma(\alpha)} \sum_{k=0}^{n} \left(\begin{array}{c} \frac{f(t_k, x_k)}{h} \int_{t_k}^{t_{k+1}} (y-t_{k-1})(t_{n+1}-y)^{\alpha-1} dy \\ -\frac{f(t_{k-1}, x_{k-1})}{h} \int_{t_k}^{t_{k+1}} (y-t_k)(t_{n+1}-y)^{\alpha-1} dy \end{array} \right). \end{aligned} \tag{56}$$

After calculating the integral expression in the above sum, we have the following equality,

$$\begin{aligned} x_{n+1} &= x_0 + \frac{1-\alpha}{B(\alpha)} f(t_n, x(t_n)) \\ &+ \frac{\alpha}{B(\alpha)} \sum_{k=0}^{n} \left(\begin{array}{c} \frac{h^\alpha f(t_k, x_k)}{\Gamma(\alpha+2)} \left(\begin{array}{c} (n+1-k)^\alpha (n-k+2+\alpha) \\ -(n-k)^\alpha (n-k+2+2\alpha) \end{array} \right) \\ -\frac{h^\alpha f(t_{k-1}, x_{k-1})}{\Gamma(\alpha+2)} \left(\begin{array}{c} (n+1-k)^{\alpha+1} \\ -(n-k)^\alpha (n-k+1+\alpha) \end{array} \right) \end{array} \right) \\ &+ R_n^\alpha, \end{aligned} \tag{57}$$

where R_n^α is the remainder term that is given by

$$\begin{aligned} R_n^\alpha = \frac{\alpha}{B(\alpha)\Gamma(\alpha)} \sum_{k=0}^{n} \int_{t_k}^{t_{k-1}} & \frac{(y-t_k)(y-t_{k-1})}{2!} \\ & \cdot \frac{\partial^2}{\partial y^2} \left[f(y, x(y)) \right]_{y=\varepsilon_y} (t_{n+1}-y)^{\alpha-1} dy. \end{aligned} \tag{58}$$

The upper boundary of the error has been provided in their paper [4].

4.1 Numerical Scheme for a Complex Fractional Order System

Let us consider the complex fractional order system (6). We saw that by applying on both sides, the Atangana–Baleanu fractional integral model can be written with C_i, $i = 1, 2, 3, 4$ kernels as below:

$$\begin{cases} y_1(t) = y_{1,0}(0) + \frac{1-\alpha}{B(\alpha)} C_1(t, y_1(t)) + \frac{\alpha}{B(\alpha)\Gamma(\alpha)} \int\limits_0^t C_1(\tau, y_1(\tau))(t-\tau)^{\alpha-1} d\tau, \\ y_2(t) = y_{2,0}(0) + \frac{1-\alpha}{B(\alpha)} C_2(t, y_2(t)) + \frac{\alpha}{B(\alpha)\Gamma(\alpha)} \int\limits_0^t C_2(\tau, y_2(\tau))(t-\tau)^{\alpha-1} d\tau, \\ y_3(t) = y_{3,0}(0) + \frac{1-\alpha}{B(\alpha)} C_3(t, y_3(t)) + \frac{\alpha}{B(\alpha)\Gamma(\alpha)} \int\limits_0^t C_3(\tau, y_3(\tau))(t-\tau)^{\alpha-1} d\tau, \\ y_4(t) = y_{4,0}(0) + \frac{1-\alpha}{B(\alpha)} C_4(t, y_4(t)) + \frac{\alpha}{B(\alpha)\Gamma(\alpha)} \int\limits_0^t C_4(\tau, y_4(\tau))(t-\tau)^{\alpha-1} d\tau, \\ y_5(t) = y_{5,0}(0) + \frac{1-\alpha}{B(\alpha)} C_5(t, y_5(t)) + \frac{\alpha}{B(\alpha)\Gamma(\alpha)} \int\limits_0^t C_5(\tau, y_5(\tau))(t-\tau)^{\alpha-1} d\tau, \end{cases} \tag{59}$$

with initial conditions

$$\begin{array}{c} y_{1,0}(0) = 0,\ y_{2,0}(0) = 0,\ y_{3,0}(0) = 0, \\ y_{4,0}(0) = 0,\ y_{5,0}(0) = 0. \end{array} \tag{60}$$

Now we can apply new numerical scheme for the system above at a given point $t = t_{n+1}$.

$$\begin{aligned} y_{1,n+1} &= y_{1,0} + \frac{1-\alpha}{B(\alpha)} C_1(t_n, y_1(t_n)) \\ &+ \frac{\alpha}{B(\alpha)} \sum_{k=0}^{n} \left(\begin{array}{c} \frac{h^\alpha C_1(t_k, y_{1,k})}{\Gamma(\alpha+2)} \left(\begin{array}{c} (n+1-k)^\alpha (n-k+2+\alpha) \\ -(n-k)^\alpha (n-k+2+2\alpha) \end{array} \right) \\ -\frac{h^\alpha C_1(t_{k-1}, y_{1,k-1})}{\Gamma(\alpha+2)} \left(\begin{array}{c} (n+1-k)^{\alpha+1} \\ -(n-k)^\alpha (n-k+1+\alpha) \end{array} \right) \end{array} \right) \\ &+ {}^1R_n^\alpha, \end{aligned} \tag{61}$$

$$\begin{aligned} y_{2,n+1} &= y_{2,0} + \frac{1-\alpha}{B(\alpha)} C_2(t_n, y_2(t_n)) \\ &+ \frac{\alpha}{B(\alpha)} \sum_{k=0}^{n} \left(\begin{array}{c} \frac{h^\alpha C_2(t_k, y_{2,k})}{\Gamma(\alpha+2)} \left(\begin{array}{c} (n+1-k)^\alpha (n-k+2+\alpha) \\ -(n-k)^\alpha (n-k+2+2\alpha) \end{array} \right) \\ -\frac{h^\alpha C_2(t_{k-1}, y_{2,k-1})}{\Gamma(\alpha+2)} \left(\begin{array}{c} (n+1-k)^{\alpha+1} \\ -(n-k)^\alpha (n-k+1+\alpha) \end{array} \right) \end{array} \right) \\ &+ {}^2R_n^\alpha, \end{aligned} \tag{62}$$

$$y_{3,n+1} = y_{3,0} + \frac{1-\alpha}{B(\alpha)} C_3(t_n, y_3(t_n)) \tag{63}$$
$$+ \frac{\alpha}{B(\alpha)} \sum_{k=0}^{n} \left(\begin{array}{c} \frac{h^\alpha C_3(t_k, y_{3,k})}{\Gamma(\alpha+2)} \left(\begin{array}{c} (n+1-k)^\alpha (n-k+2+\alpha) \\ -(n-k)^\alpha (n-k+2+2\alpha) \end{array} \right) \\ -\frac{h^\alpha C_3(t_{k-1}, y_{3,k-1})}{\Gamma(\alpha+2)} \left(\begin{array}{c} (n+1-k)^{\alpha+1} \\ -(n-k)^\alpha (n-k+1+\alpha) \end{array} \right) \end{array} \right)$$
$$+ {}^3R_n^\alpha,$$

$$y_{4,n+1} = y_{4,0} + \frac{1-\alpha}{B(\alpha)} C_4(t_n, y_4(t_n)) \tag{64}$$
$$+ \frac{\alpha}{B(\alpha)} \sum_{k=0}^{n} \left(\begin{array}{c} \frac{h^\alpha C_4(t_k, y_{4,k})}{\Gamma(\alpha+2)} \left(\begin{array}{c} (n+1-k)^\alpha (n-k+2+\alpha) \\ -(n-k)^\alpha (n-k+2+2\alpha) \end{array} \right) \\ -\frac{h^\alpha C_4(t_{k-1}, y_{4,k-1})}{\Gamma(\alpha+2)} \left(\begin{array}{c} (n+1-k)^{\alpha+1} \\ -(n-k)^\alpha (n-k+1+\alpha) \end{array} \right) \end{array} \right)$$
$$+ {}^4R_n^\alpha,$$

$$y_{5,n+1} = y_{5,0} + \frac{1-\alpha}{B(\alpha)} C_5(t_n, y_5(t_n)) \tag{65}$$
$$+ \frac{\alpha}{B(\alpha)} \sum_{k=0}^{n} \left(\begin{array}{c} \frac{h^\alpha C_5(t_k, y_{5,k})}{\Gamma(\alpha+2)} \left(\begin{array}{c} (n+1-k)^\alpha (n-k+2+\alpha) \\ -(n-k)^\alpha (n-k+2+2\alpha) \end{array} \right) \\ -\frac{h^\alpha C_5(t_{k-1}, y_{5,k-1})}{\Gamma(\alpha+2)} \left(\begin{array}{c} (n+1-k)^{\alpha+1} \\ -(n-k)^\alpha (n-k+1+\alpha) \end{array} \right) \end{array} \right)$$
$$+ {}^5R_n^\alpha,$$

where ${}^iR_n^\alpha$, $i = 1, 2, 3, 4, 5$ are remainder terms given as below:

$$\begin{cases} {}^1R_n^\alpha = \frac{\alpha}{B(\alpha)\Gamma(\alpha)} \sum\limits_{k=0}^{n} \int\limits_{t_k}^{t_{k-1}} \frac{(\tau-t_k)(\tau-t_{k-1})}{2!} \frac{\partial^2}{\partial\tau^2} \left[C_1(\tau, y_1(\tau))\right]_{\tau=\varepsilon_\tau} (t_{n+1}-\tau)^{\alpha-1} d\tau, \\ {}^2R_n^\alpha = \frac{\alpha}{B(\alpha)\Gamma(\alpha)} \sum\limits_{k=0}^{n} \int\limits_{t_k}^{t_{k-1}} \frac{(\tau-t_k)(\tau-t_{k-1})}{2!} \frac{\partial^2}{\partial\tau^2} \left[C_2(\tau, y_2(\tau))\right]_{\tau=\varepsilon_\tau} (t_{n+1}-\tau)^{\alpha-1} d\tau, \\ {}^3R_n^\alpha = \frac{\alpha}{B(\alpha)\Gamma(\alpha)} \sum\limits_{k=0}^{n} \int\limits_{t_k}^{t_{k-1}} \frac{(\tau-t_k)(\tau-t_{k-1})}{2!} \frac{\partial^2}{\partial\tau^2} \left[C_3(\tau, y_3(\tau))\right]_{\tau=\varepsilon_\tau} (t_{n+1}-\tau)^{\alpha-1} d\tau, \\ {}^4R_n^\alpha = \frac{\alpha}{B(\alpha)\Gamma(\alpha)} \sum\limits_{k=0}^{n} \int\limits_{t_k}^{t_{k-1}} \frac{(\tau-t_k)(\tau-t_{k-1})}{2!} \frac{\partial^2}{\partial\tau^2} \left[C_4(\tau, y_4(\tau))\right]_{\tau=\varepsilon_\tau} (t_{n+1}-\tau)^{\alpha-1} d\tau, \\ {}^5R_n^\alpha = \frac{\alpha}{B(\alpha)\Gamma(\alpha)} \sum\limits_{k=0}^{n} \int\limits_{t_k}^{t_{k-1}} \frac{(\tau-t_k)(\tau-t_{k-1})}{2!} \frac{\partial^2}{\partial\tau^2} \left[C_5(\tau, y_5(\tau))\right]_{\tau=\varepsilon_\tau} (t_{n+1}-\tau)^{\alpha-1} d\tau. \end{cases} \tag{66}$$

Using the numerical scheme of (60)–(65) we obtain the following numerical simulations. We give these simulations in Figs. 1, 2, 3 and 4 for different values of alpha.

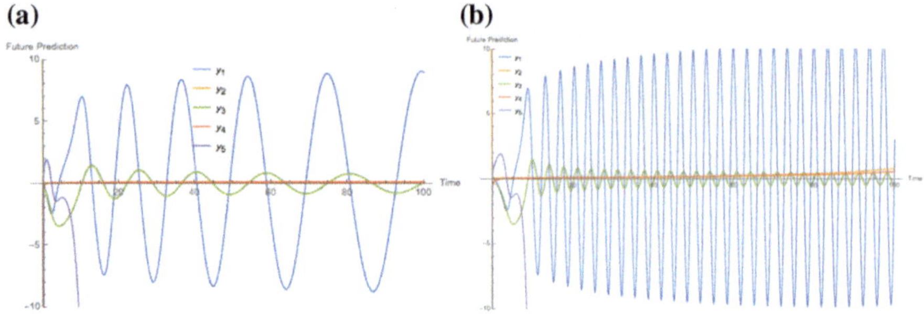

Fig. 1 Numerical solution for $\alpha = 0.15$. and numerical solution for $\alpha = 0.45$, respectively

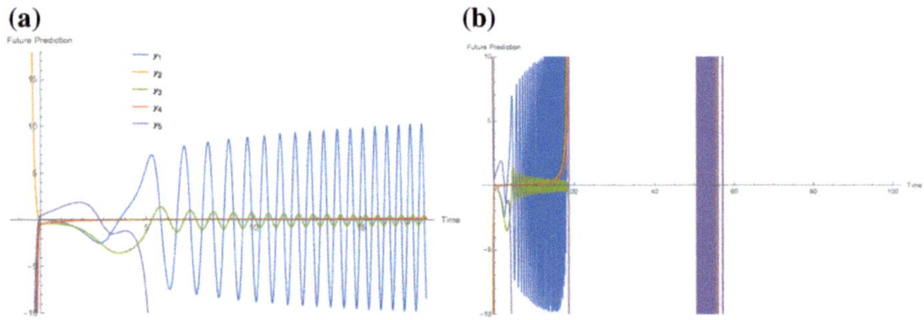

Fig. 2 Numerical solution for $\alpha = 0.85$. and numerical solution for $\alpha = 1$, respectively

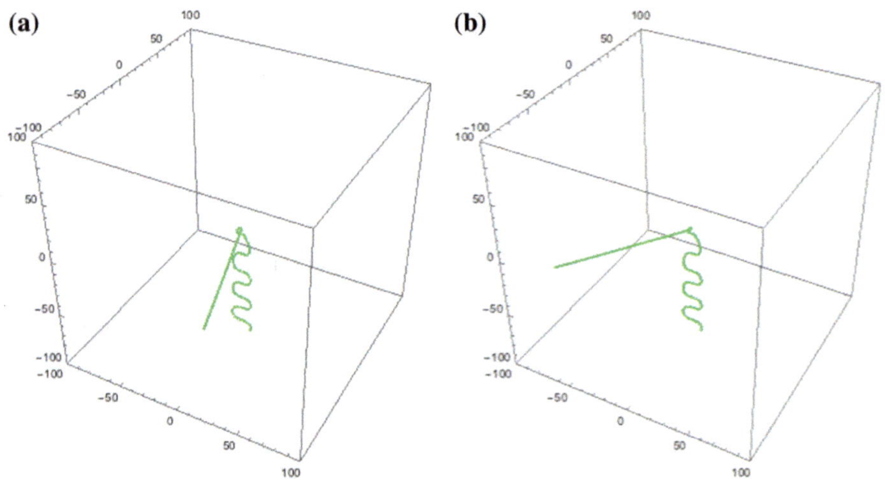

Fig. 3 Chaotic attractor in y_1, y_2, y_3 for $\alpha = 0.45$ and Chaotic attractor in y_1, y_2, y_3 for $\alpha = 0.85$, respectively

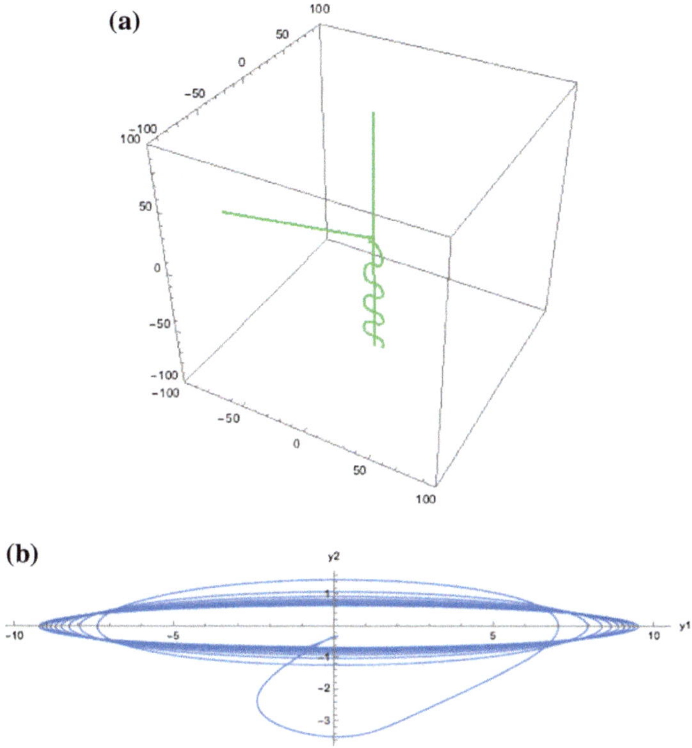

Fig. 4 Chaotic attractor in y_1, y_2, y_3 for $\alpha = 1$ and Chaotic attractor in y_1, y_2 for $\alpha = 1$, respectively

5 Conclusion

To include into mathematical formulation the Markovian and non-Markovian processes to a complex system describing chaotic behavior, we replaced the time derivative based on the concept of rate of change with that with nonlocal and non-singular kernel. A detailed analysis of existence using the Picard's method and the connection of the Banach space with contraction operator to establish the uniqueness of the exact solution was developed. Very recently, a new numerical method was suggested, combining the fundamental theorem of fractional calculus and the Lagrange interpolation formulation. The method was found to be efficient than the well-known Adams–Bashforth as the method is fast, accurate and friendly user. We use this new numerical scheme to solve numerically the modified model and present the numerical simulations.

References

1. Ross, B.A.: Brief history and exposition of the fundamental theory of fractional calculus. Fractional Calculus and Its Applications. Lecture Notes in Mathematics, vol. 457, pp. 1–36 (1975)
2. Debnath, L.: A brief historical introduction to fractional calculus. Int. J. Math. Educ. Sci. Technol. **35**, 487–501 (2004)
3. Caputo, M.: Linear model of dissipation whose Q is almost frequency independent-II. Geophys. J. R. Astron. Soc. Can. **13**, 529–539 (1967)
4. Benson, D., Wheatcraft, S., Meerschaert, M.: Application of a fractional advection-dispersion equation. Water Resour. Res. **36**, 1403–1412 (2000)
5. Caputo, M., Fabrizio, M.: A new definition of fractional derivative without singular kernel. Prog. Fract. Differ. Appl. **1**(2), 73–85 (2015)
6. Losada, J., Nieto, J.J.: Properties of a new fractional derivative without singular kernel. Prog. Fract. Differ. Appl. **1**(2), 87–92 (2015)
7. Alkahtani, B.S.T., Koca, I., Atangana, A.: Analysis of a new model of H1N1 spread: model obtained via Mittag-Leffler function. Adv. Mech. Eng. **9**(8), 1–7 (2017)
8. Gómez-Aguilar, J.F.: Fundamental solutions to electrical circuits of non-integer order via fractional derivatives with and without singular kernels. Eur. Phys. J. Plus **133**(5), 1–25 (2018)
9. Alkahtani, B.S.T., Atangana, A., Koca, I.: A new nonlinear triadic model of predator-prey based on derivative with non-local and non-singular kernel. Adv. Mech. Eng. **8**(11), 1–9 (2016)
10. Gómez-Aguilar, J.F.: Analytical and Numerical solutions of a nonlinear alcoholism model via variable-order fractional differential equations. Phys. A: Stat. Mech. Its Appl. **494**, 52–75 (2018)
11. Morales-Delgado, V.F., Taneco-Hernández, M.A., Gómez-Aguilar, J.F.: On the solutions of fractional order of evolution equations. Eur. Phys. J. Plus **132**(1), 1–17 (2017)
12. Saad, K.M., Gómez-Aguilar, J.F.: Analysis of reaction diffusion system via a new fractional derivative with non-singular kernel. Phys. A: Stat. Mech. Its Appl. **509**, 703–716 (2018)
13. Gómez-Aguilar, J.F., Escobar-Jiménez, R.F., López-López, M.G., Alvarado-Martínez, V.M.: Atangana-Baleanu fractional derivative applied to electromagnetic waves in dielectric media. J. Electromagn. Waves Appl. **30**(15), 1937–1952 (2016)
14. Gómez-Aguilar, J.F., Dumitru, B.: Fractional transmission line with losses. Zeitschrift für Naturforschung A **69**(10–11), 539–546 (2014)
15. Gómez-Aguilar, J.F., Torres, L., Yépez-Martínez, H., Baleanu, D., Reyes, J.M., Sosa, I.O.: Fractional Liénard type model of a pipeline within the fractional derivative without singular kernel. Adv. Differ. Equ. **2016**(1), 1–17 (2016)
16. Atangana, A., Koca, I.: Model of thin viscous fluid sheet flow within the scope of fractional calculus: fractional derivative with and no singular kernel. Fundamenta Informaticae **151**(1–4), 145–159 (2017)
17. Vishal, K., Agrawal, S.K.: On the dynamics, existence of chaos, control and synchronization of a novel complex chaotic system. Chin. J. Phys. **55**(2), 519–532 (2017)
18. Atangana, A., Baleanu, D.: New fractional derivatives with nonlocal and non-singular kernel. Theory and application to heat transfer model. Therm. Sci. **20**(2), 763–769 (2016)
19. Atangana, A., Gómez-Aguilar, J.F.: Decolonisation of fractional calculus rules: breaking commutativity and associativity to capture more natural phenomena. Eur. Phys. J. Plus **133**, 1–22 (2018)
20. Gómez-Aguilar, J.F., Atangana, A., Morales-Delgado, J.F.: Electrical circuits RC, LC, and RL described by Atangana-Baleanu fractional derivatives. Int. J. Circ. Theor. Appl. **1**, 1–22 (2017)
21. Coronel-Escamilla, A., Gómez-Aguilar, J.F., Torres, L., Escobar-Jiménez, R.F., Valtierra-Rodríguez, M.: Synchronization of chaotic systems involving fractional operators of Liouville-Caputo type with variable-order. Phys. A: Stat. Mech. Its Appl. **487**, 1–21 (2017)
22. Coronel-Escamilla, A., Gómez-Aguilar, J.F., López-López, M.G., Alvarado-Martínez, V.M., Guerrero-Ramírez, G.V.: Triple pendulum model involving fractional derivatives with different kernels. Chaos Solitons Fractals **91**, 248–261 (2016)

23. Gómez-Aguilar, J.F., López-López, M.G., Alvarado-Martínez, V.M., Baleanu, D., Khan, H.: Chaos in a cancer model via fractional derivatives with exponential decay and Mittag-Leffler law. Entropy **19**(12), 1–18 (2017)
24. Gómez-Aguilar, J.F.: Behavior characteristics of a cap-resistor, memcapacitor, and a memristor from the response obtained of RC and RL electrical circuits described by fractional differential equations. Turk. J. Electr. Eng. Comput. Sci. **24**(3), 1–16 (2016)
25. Coronel-Escamilla, A., Gómez-Aguilar, J.F., Baleanu, D., Córdova-Fraga, T., Escobar-Jiménez, R.F., Olivares-Peregrino, V.H., Qurashi, M.M.A.: Bateman-Feshbach tikochinsky and Caldirola-Kanai oscillators with new fractional differentiation. Entropy **19**(2), 1–21 (2017)
26. Morales-Delgado, V.F., Gómez-Aguilar, J.F., Kumar, S., Taneco-Hernández, M.A.: Analytical solutions of the Keller-Segel chemotaxis model involving fractional operators without singular kernel. Eur. Phys. J. Plus **133**(5), 1–19 (2018)
27. Atangana, A., Koca, I.: Chaos in a simple nonlinear system with Atangana-Baleanu derivatives with fractional order. Chaos Solitons Fractals **89**, 447–454 (2016)
28. Toufik, M., Atangana, A.: New numerical approximation of fractional derivative with non-local and non-singular kernel: application to chaotic models. Eur. Phys. J. Plus **132**, 1–14 (2017)

On the Chaotic Pole of Attraction with Nonlocal and Nonsingular Operators in Neurobiology

Emile F. Doungmo Goufo, Abdon Atangana and Melusi Khumalo

Abstract Until the neurologists J.L. Hindmarsh and R.M. Rose improved the Hodgkin–Huxley model to provide a better understanding on the diversity of neural response, features like pole of attraction unfolding complex bifurcation for the membrane potential was still a mystery. This work explores the possible existence of chaotic poles of attraction in the dynamics of Hindmarsh–Rose neurons with external current input. Combining with fractional differentiation, the model is generalized with introduction of an additional parameter, the non-integer order of the derivative σ and solved numerically thanks to the Haar Wavelets. Numerical simulations of the membrane potential dynamic show that in the standard case the control parameter is $\sigma = 1$, the nerve cell's behavior seems irregular with a pole of attraction generating a limit cycle. This irregularity accentuates as σ decreases ($\sigma = 0.8$ and $\sigma = 0.5$) with the pole of attraction becoming chaotic.

Keywords Fractional calculus · Atangana–Baleanu fractional derivative · Hindmarsh Rose neuron

1 Introduction to the Model

A considerable progress on improving the understanding of nerve cells' functions has been successfully achieved during the recent past decades. Living beings' nervous system is made of various type of cells with the nerve cells seen as the most remarkable ones called neurons. A typical neuron, as depicted in Fig. 1, comprises

E. F. Doungmo Goufo (✉) · M. Khumalo
Department of Mathematical Sciences, University of South Africa, Florida 0003, South Africa
e-mail: dgoufef@unisa.ac.za

A. Atangana
Institute for Groundwater Studies, University of the Free State, Bloemfontein 9300, South Africa

J. F. Gómez et al. (eds.), *Fractional Derivatives with Mittag-Leffler Kernel*,
Studies in Systems, Decision and Control 194,
https://doi.org/10.1007/978-3-030-11662-0_8

an axon that extends the cell body to the terminal branches and dendrites. This cell internal connection is very significant, especially in the transmission of information both internally to the said cell and externally to other nerve cells. Transmission of an electrical or chemical signal from a neuron to another is made possible via a structure called synapses.

Literature on neurophysiology is well documented thanks to works like [1–6]. However, it was also essential to mathematically study the neuronal activities in order to assess the spiking and bursting behavior of nerve cell's components like the membrane potential observed in some experiments involving a neuron. In that momentum, pioneers like Hodgkin and Huxley [7, 8], two renowned neurophysiologists, started by developing an empirical kinetic description of ionic mechanisms in a nerve cell. This yielded the Hodgkin–Huxley mathematical model, a set of non-linear differential equations based on conductance and that describes the initiation and propagation of action potentials in nerve cells. This model was eventually improved by two other scientists, Hindmarsh and Rose [9] who simplified the model of Hodgkin and Huxley and proposed the Hindmarsh–Rose (HR) neuron model of Fitzhugh–Nagumo type where they substituted some variables by constants and established the relations between those various unknowns. Hindmarsh and Rose proposed a system of two (2D) and three (3D) non-linear first order differential equations that also study the spiking and bursting dynamic of the neuron membrane potential while taking into account the dynamics of ions across the membrane via the ion channels.

Let us recall that the main reason why the HR neuron model was established is to be able to reproduce bursting behavior of nerve cells. On the other hand, recently during experimentation on single rat neocortical pyramidal nerve cells [10], it was observed that the firing rate of adaptation multiple time-scales is consistent even for generalized models with fractional-order differentiation, and that HR neuron models expressed with both integer and non-integer derivative order can successfully exhibit bursting dynamics as well. Therefore, our prime objective is this work is to investigate the dynamics of the 2D and 3D-generalized HR neuron model where we have considered an additional parameter on top of other well known parameters of the HR neuron model, including the external current I^{ext}.

One main goal of this chapter is to study the dynamic of the HR neuron that is usually expressed by a system of 3D nonlinear differential equations and expressed in its general form by

$$\begin{cases} {}^{ABC}D_t^{\sigma} x(t) = I^e + y - ax^3 + bx^2 - z, \\ {}^{ABC}D_t^{\sigma} y(t) = c - dx^2 - y, \\ {}^{ABC}D_t^{\sigma} z(t) = \eta_1[\eta_2(x - x_r) - z], \end{cases} \tag{1}$$

where I^e represents the the external current input while the system state $x = x(t)$ plays the role of the membrane potential, $y = y(t)$ plays the role of a recovery variable linked to the fast current of Na^+ or K^+ ions, and $z = z(t)$ represents the adaptation current related to the slow current of Ca^{2+} ion. Moreover, the parameters $a,\ b,\ c,\ d,\ \eta_1,\ \eta_2,\ x_r$ are real numbers. The term ${}^{ABC}D_t^{\sigma}$ represents the Atangana–Baleanu fractional derivative of order σ, with $0, \sigma \leq 1$ that will be comprehensively

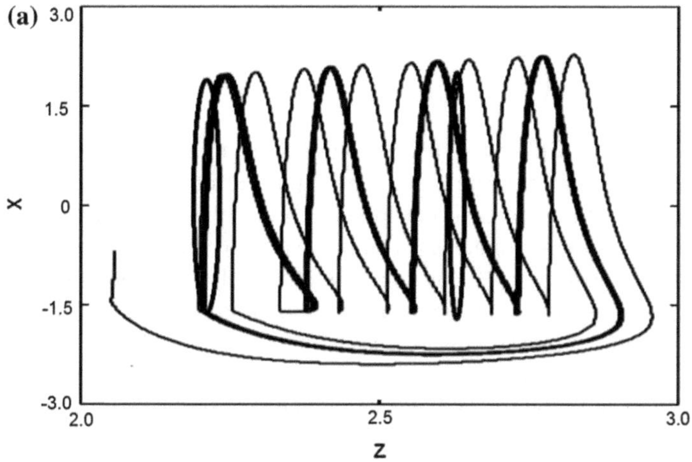

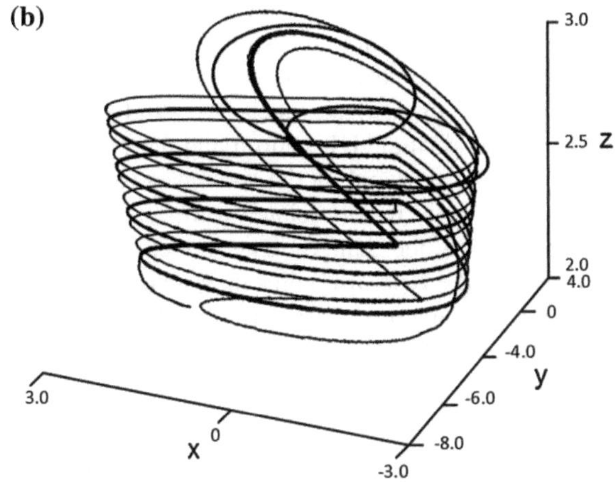

Fig. 1 Representation of the HR neuron dynamic (15)–(16) for $\alpha = \gamma = 1,\ \delta = -1$ and the control parameters at $a = c = 1,\ \ b = 3.1,\ \ d = 4,\ \eta_1 = 65 \times 10^{-3},\ \ \eta_2 = 5,\ x_r = -3/2,$ and, $I^e = 2.6$, with $\sigma = 1$. The HR neuron behavior seems regular and simply periodic

defined in the next section. Knowing that

$$ {}^{ABC}D_t^1 u(t) \sim \frac{du(t)}{dt}, \tag{2} $$

then, for $\sigma = 1$ the model (1) obviously reduces to the standard well-know Hindmarsh Rose neurons with external current input given by the 3D non-linear differential equations

$$\begin{cases} \frac{dx(t)}{dt} = I^e + y - ax^3 + bx^2 - z, \\ \frac{dy(t)}{dt} = c - dx^2 - y, \\ \frac{dz(t)}{dt} = \eta_1[\eta_2(x - x_r) - z]. \end{cases} \tag{3}$$

Each parameter in the system has a biological meaning and plays a specific role. For example, I^e incarnates the membrane input current for nerve cells while the switch between bursting process and spiking is controlled by the parameter b which also regulates the spiking frequency. The slow variable symbolized by z in model (1) has a speed regulated by η_1 (It controls how efficiently slow routes exchange ions). Parameter η_2 is most concerned about adaptation for the system. This means one unit of η_2 ($\eta_2 = 1$) symbolizes spiking process without accommodation and subthreshold adaptation, while an increasing value of η_2 ($\eta_2 = 4$ for instance) provides a strong accommodation and subthreshold overshoot, or even oscillations. Lastly the variable x_r stands for the resting potential of the whole system. Recall that the model (3) has been comprehensively analysed in numerous works [11–16].

For instance, the authors in [11] successfully study a plethora of chaotic phenomena in the HR neuron model using various computational techniques including the bifurcation parameter continuation and spike-quantification. In the work [14] was presented an adaptive neural network based sliding mode control for unidirectional synchronization of HR neurons in a master-slave configuration. The authors first established the dynamics of single HR neuron and thereafter formulated the problem. The fractional-order HR neuronal model was investigated in [13], in this paper the authors proved some useful results related to different chaotic and periodic firing modes as the fractional order changes. The model (1) can appear to be tough and hard to analyze due to the type of nonlinearity it shows as well as the external factor. Hence, to seize its dynamic, we first use the Haar wavelet numerical method to approximate it in its general form and provide simulations for some particular cases.

2 Some Definitions on the Non-integer Order Derivatives

The concept of derivative with fractional order has been used intensively these last decades to enhance, generalize and investigate non-linear mathematical models that describe real life phenomena [17–39]. This was possible thanks to various definitions of fractional derivatives that have never stopped growing. The version used in the analysis of this paper is the Atangana–Baleanu fractional derivative born alongside with the following fractional integral of order σ

$$I^\sigma u(t) = \frac{\sigma}{W(\sigma)\Gamma(\sigma)} \int_0^t (t-\tau)^{\sigma-1} u(\tau) d\tau + \frac{1-\sigma}{W(\sigma)} u(t), \tag{4}$$

that is based on Euler transform when applied to analytic function and Cauchy's formula for calculating iterated integrals. Hence, the Atangana–Baleanu fractional derivative of order σ is defined for any $t > 0$ as

$$ {}^{ABC}D_t^{\sigma} u(t) = \frac{W(\sigma)}{(1-\sigma)} \int_0^t \dot{u}(\tau)\, E_{\sigma}\left[-\frac{\sigma(t-\tau)^{\sigma}}{1-\sigma}\right] d\tau, \tag{5} $$

with the derivative order $\sigma \in [0, 1]$ and where $W(\sigma)$ defines a normalization function such that

$$ W(0) = W(1) = 1, \tag{6} $$

where the function g that is from the first order Sobolev space

$$ H^1(a, e) = \{g : g, \frac{dg}{dt} \in L^2(a, e)\}. \tag{7} $$

Before continuing let us at least mention some other versions of fractional derivative that piles up in the related literature. The list includes the oldest version that is the Riemann–Liouville derivative and more recently, the Caputo–Fabrizio derivative, the new Riemann–Liouville fractional derivative, the one and two-parameter derivatives with non-local and non-singular kernel. For more details on old and recent developments in fractional calculus, please refer to the works [17–20, 45–48] and also to the references therein.

3 Generalities on the Haar Wavelets Method for Non-linear Differential Equations

The Haar wavelet function $\mathbf{h}(t)$ defined on the set of real numbers $\mathbb{R}$ is given by [40–42]

$$ \mathbf{h}(t) = \begin{cases} 1, & \text{for } 0 \le t < 1/2; \\ -1, & \text{for } 1/2 < t \le 1; \\ 0, & \text{elsewhere.} \end{cases} \tag{8} $$

We can define the family

$$ \mathbf{H}_i(t) = \begin{cases} 2^{\frac{j}{2}}\mathbf{h}(2^j t - l), & \text{for } i = 1, 2, \ldots; \\ 1, & \text{for } i = 0, \end{cases} \tag{9} $$

for each $i = 0, 1, 2, 3, \ldots$ with $t \in [0, 1)$, and knowing that each $i = 0, 1, 2, \ldots$ can be expressed into the form $i = 2^j + l$ with $j = 0, 1, 2, \ldots$ and $l = 0, 1, 2, \ldots, 2^j - 1$. Hence, this yields the following result: $\{\mathbf{H}_i(t)\}_{i=0}^{\infty}$ forms a complete orthonormal system in the Banach space of square-integrable function $L^2[0, 1)$. Furthermore, by taking $\mathbf{s}$ in the Banach space of continuous functions $C[0, 1)$ then, the series $\sum_{i=0}^{\infty}\langle \mathbf{s}, \mathbf{H}_i\rangle \mathbf{H}_i$ converges uniformly to $\mathbf{s}$ with $\langle \mathbf{s}, \mathbf{H}_i\rangle = \int_0^{\infty} \mathbf{s}(t)\mathbf{H}_i(t)dt$. The same function $\mathbf{s}$ can be expanded to become

$$\mathbf{s}(t) = \sum_{i=0}^{\infty} A_i \mathbf{H}_i(t),$$

where $A_i = \langle \mathbf{s}, \mathbf{H}_i \rangle$. For reasons of practicability, the approximated solution reads as

$$\mathbf{s}(t) \approx \mathbf{s}_l(t) = \sum_{i=0}^{l-1} A_i \mathbf{H}_i(t),$$

where $l \in \{2^j : \; j = 0, 1, 2, \ldots\}$.

Let $e \in \mathbb{N}$, we can now use the Haar function translation on $[0, e)$ to provide the following definition:

$$\mathbf{H}_{r,i}(t) = \mathbf{H}_i(t - r + 1) \quad r = 1, 2, \ldots, e \quad \text{and} \quad i = 0, 1, 2, \ldots, \tag{10}$$

where $\mathbf{H}_i$ is given by (9). Obviously all the properties that are satisfied by $\mathbf{H}_i$ also hold for $\mathbf{H}_{r,i}$. Namely, the family $\{\mathbf{H}_{r,i}(t)\}_{i=0}^{\infty}$, $(r = 1, 2, \ldots, e)$ also forms a complete orthonormal system in the Banach space of square-integrable function $L^2[0, e)$. Hence it is also possible to exploit the following Haar orthonormal basis functions

$$A_{r,i} = \langle \mathbf{s}, \mathbf{H}_{r,i} \rangle = \int_0^{\infty} \mathbf{s}(t) \mathbf{H}_{r,i}(t) dt,$$

to expand the solution $\mathbf{s} \in L^2[0, e)$ as the series

$$\mathbf{s}(t) = \sum_{r=1}^{e} \sum_{i=0}^{\infty} A_{r,i} \mathbf{H}_{r,i}(t). \tag{11}$$

In the same way and for reasons of simplicity, the approximated solution is given by

$$\mathbf{s}(t) \approx \mathbf{s}_l(t) = \sum_{r=1}^{e} \sum_{i=0}^{l-1} A_{r,i} \mathbf{H}_{r,i}(t), \tag{12}$$

where $l \in \{2^j : \; j = 0, 1, 2, \ldots\}$. Solution (12) can be expressed into the compact form

$$\mathbf{s}(t) \approx \mathbf{s}_l(t) = (\varphi_{el \times 1})^T \mathbf{h}_{el \times 1}, \tag{13}$$

with T symbol of the transpose vector so that $(\varphi_{el \times 1})^T$ is the vector given by

$$(\varphi_{el \times 1})^T = \left(A_{1,0}, \ldots, A_{1,l-1}, A_{2,0}, \ldots, A_{2,l-1}, \ldots, A_{e,0}, \ldots, A_{e,l-1}\right),$$

and

$$\mathbf{h}_{el \times 1} = \left(\mathbf{H}_{1,0}, \ldots, \mathbf{H}_{1,l-1}, \mathbf{H}_{2,0}, \ldots, \mathbf{H}_{2,l-1}, \ldots, \mathbf{H}_{e,0}, \ldots, \mathbf{H}_{e,l-1}\right)^T.$$

4 Haar Wavelets Numerical Method for the System (1)

In this section, a comprehensive solvability of the model (1) is conducted. Let us first define the system states variable

$$\mathbf{s}(t) = (x(t), y(t), z(t))^T,$$

and

$$W_0(x, y, z) = \mathbf{s}(0) = (x(0), y(0), z(0))^T.$$

Setting

$$x(0) = \alpha, \quad y(0) = \delta \text{ and } z(0) = \gamma,$$

and the matrix operator

$$\begin{aligned} \mathbb{M}(\mathbf{s}(t), t) &= \mathbb{M}(x(t), y(t), z(t), t), \\ &= (\mathbb{M}_1(\mathbf{s}(t), t), \ \mathbb{M}_2(\mathbf{s}(t), t), \ \mathbb{M}_3(\mathbf{s}(t), t))^T, \\ &= (\mathbb{M}_1(x(t), y(t), z(t), t), \ \mathbb{M}_2(x(t), y(t), z(t), t), \ \mathbb{M}_3(x(t), y(t), z(t), t))^T, \end{aligned} \tag{14}$$

where

$$\begin{cases} \mathbb{M}_1(\mathbf{s}(t), t) = I^e + x^2(b - ax) + y - z, \\ \mathbb{M}_2(\mathbf{s}(t), t) = c - dx^2 - y, \\ \mathbb{M}_3(\mathbf{s}(t), t) = \eta x - \eta x_r - \eta_1 z. \end{cases}$$

We can now use those notations to express in a compact form the model (1) equivalently written as

$$\begin{cases} {}^{ABC}D_t^{\sigma} x(t) = I^e + x^2(b - ax) + y - z, \\ {}^{ABC}D_t^{\sigma} y(t) = c - dx^2 - y, \\ {}^{ABC}D_t^{\sigma} z(t) = \eta x - \eta x_r - \eta_1 z, \end{cases} \tag{15}$$

with $\eta = \eta_1 \eta_2$ and assumed to satisfy the following initial conditions

$$x(0) = \alpha(x), \ y(0) = \delta(y), \ z(0) = \gamma(z). \tag{16}$$

Hence, the compact form of (15) is given by

$${}^{ABC}D_t^{\sigma} \mathbf{s}(t) = \mathbb{M}(\mathbf{s}(t), t),$$

which is also expressed by

$$\begin{aligned} {}^{ABC}D_t^{\sigma} x(t) &= \mathbb{M}_1(\mathbf{s}(t), t), \\ {}^{ABC}D_t^{\sigma} y(t) &= \mathbb{M}_2(\mathbf{s}(t), t), \\ {}^{ABC}D_t^{\sigma} z(t) &= \mathbb{M}_3(\mathbf{s}(t), t), \end{aligned} \tag{17}$$

and assumed to verify the initial conditions

$$x(0) = \alpha(x),\ y(0) = \delta(y),\ z(0) = \gamma(z). \tag{18}$$

Approximating the Atangana–Baleanu fractional model (17)–(18) using Haar wavelets numerical scheme given in (13) leads to the system

$$\begin{aligned} {}^{ABC}D_t^{\sigma} x(t) = \mathbb{M}_1(\mathbf{s}(t), t) \approx {}^{ABC}D_t^{\sigma} x_l(t) ={}^T \varphi^1_{el\times 1}\mathbf{h}_{el\times 1},\\ {}^{ABC}D_t^{\sigma} y(t) = \mathbb{M}_2(\mathbf{s}(t), t) \approx {}^{ABC}D_t^{\sigma} y_l(t) ={}^T \varphi^2_{el\times 1}\mathbf{h}_{el\times 1},\\ {}^{ABC}D_t^{\sigma} z(t) = \mathbb{M}_3(\mathbf{s}(t), t) \approx {}^{ABC}D_t^{\sigma} z_l(t) ={}^T \varphi^3_{el\times 1}\mathbf{h}_{el\times 1}. \end{aligned} \tag{19}$$

To obtain the solution we apply the Riemann–Liouville fractional integral (4) on both side of each equation of (19) which gives

$$\begin{aligned} x(t) - \alpha \approx {}^{ABC}D_t^{\sigma} x_l(t) ={}^T \varphi^1_{el\times 1} F^{\sigma}_{el\times el}\mathbf{h}_{el\times 1},\\ y(t) - \delta \approx {}^{ABC}D_t^{\sigma} y_l(t) ={}^T \varphi^2_{el\times 1} F^{\sigma}_{el\times el}\mathbf{h}_{el\times 1},\\ z(t) - \gamma \approx {}^{ABC}D_t^{\sigma} z_l(t) ={}^T \varphi^3_{el\times 1} F^{\sigma}_{el\times el}\mathbf{h}_{el\times 1}, \end{aligned} \tag{20}$$

also expressed by

$$\begin{aligned} x(t) \approx x_l(t) ={}^T \varphi^1_{el\times 1} F^{\sigma}_{el\times el}\mathbf{h}_{el\times 1} + \alpha,\\ y(t) \approx y_l(t) ={}^T \varphi^2_{el\times 1} F^{\sigma}_{el\times el}\mathbf{h}_{el\times 1} + \delta,\\ z(t) \approx z_l(t) ={}^T \varphi^3_{el\times 1} F^{\sigma}_{el\times el}\mathbf{h}_{el\times 1} + \gamma, \end{aligned} \tag{21}$$

where $F^{\sigma}_{el\times el}$ defines the haar wavelets fractional operational matrix as developed in the works [40, 41]. The next step is to exploit the Galerkin's method based on collocation points, to be able to solve the model (15)–(16). For that, we substitute the approximated systems (19) and (21) into (15) and the three residual errors $err_1,\ err_2,\ err_3$ caused by such a substitution are obviously strongly dependent on Haar orthonormal basis functions $A_{r,i}$. They all read as

$$\begin{aligned} &err_1\left(\kappa^1, \kappa^2, \kappa^3, t\right)\\ &={}^T \varphi^1_{el\times 1}\mathbf{h}_{el\times 1} - \mathbb{M}_1\left({}^T\varphi^1_{el\times 1} F^{\sigma}_{el\times el}\mathbf{h}_{el\times 1}, {}^T \varphi^2_{el\times 1} F^{\sigma}_{el\times el}\mathbf{h}_{el\times 1}, {}^T \varphi^3_{el\times 1} F^{\sigma}_{el\times el}\mathbf{h}_{el\times 1}, t\right),\\ &err_2\left(\kappa^1, \kappa^2, \kappa^3, t\right)\\ &={}^T \varphi^2_{el\times 1}\mathbf{h}_{el\times 1} - \mathbb{M}_2\left({}^T\varphi^1_{el\times 1} F^{\sigma}_{el\times el}\mathbf{h}_{el\times 1}, {}^T \varphi^2_{el\times 1} F^{\sigma}_{el\times el}\mathbf{h}_{el\times 1}, {}^T \varphi^3_{el\times 1} F^{\sigma}_{el\times el}\mathbf{h}_{el\times 1}, t\right),\\ &err_3\left(\kappa^1, \kappa^2, \kappa^3, t\right)\\ &={}^T \varphi^3_{el\times 1}\mathbf{h}_{el\times 1} - \mathbb{M}_3\left({}^T\varphi^1_{el\times 1} F^{\sigma}_{el\times el}\mathbf{h}_{el\times 1}, {}^T \varphi^2_{el\times 1} F^{\sigma}_{el\times el}\mathbf{h}_{el\times 1}, {}^T \varphi^3_{el\times 1} F^{\sigma}_{el\times el}\mathbf{h}_{el\times 1}, t\right), \end{aligned} \tag{22}$$

where

$$\kappa^1 = A^1_{1,0}, \ldots, A^1_{1,l-1}, \ldots, A^1_{e,0}, \ldots, A^1_{e,l-1},$$

$$\kappa^2 = A^2_{1,0}, \ldots, A^2_{1,l-1}, \ldots, A^2_{e,0}, \ldots, A^2_{e,l-1},$$

$$\kappa^3 = A^3_{1,0}, \ldots, A^3_{1,l-1}, \ldots, A^3_{e,0}, \ldots, A^3_{e,l-1},$$

and $A^i_{\cdot,\cdot}$ represents the ith components of $(\varphi^i_{\cdot\times\cdot})^T$.

Assuming that

$$err_1\left(\kappa^1, \kappa^2, \kappa^3, t_{r,i}\right) = 0,$$

$$err_2\left(\kappa^1, \kappa^2, \kappa^3, t_{r,i}\right) = 0,$$

$$err_3\left(\kappa^1, \kappa^2, \kappa^3, t_{r,i}\right) = 0,$$

where

$$t_{r,i} = \frac{2i-1}{2l} + r - 1, \qquad r = 1, 2, \ldots, e; \qquad i = 1, 2, \ldots, l,$$

represent a $e \times l$ number of collocation points, this leaves us with a system of $3e \times l$ equations, with $3e \times l$ unknowns defined by

$$A^1_{1,0}, \ldots, A^1_{1,l-1}, \ldots, A^1_{e,0}, \ldots, A^1_{e,l-1},$$

$$A^2_{1,0}, \ldots, A^2_{1,l-1}, \ldots, A^2_{e,0}, \ldots, A^2_{e,l-1},$$

$$A^3_{1,0}, \ldots, A^3_{1,l-1}, \ldots, A^3_{e,0}, \ldots, A^3_{e,l-1}.$$

Therefore, we easily obtain these unknowns and after substituting them into (21), we obtain the desired approximation for the solution

$$\mathbf{s}(t) \approx (x_l(t),\ y_l(t),\ z_l(t))^T.$$

5 Error Analysis and Convergence for the Numerical Approximation by Haar Wavelets Method

Before visualizing the dynamic of solutions to our HR neuron model (15)–(16), especially the dynamic of the membrane potential x, we have to address the convergence of the Haar wavelets method used here above. We proceed by error analysis method and aim to obtain here exact error bounds that were used in the mathematical process. With the assumption that $\mathbf{s} \in L^2[0, e)$, we can also consider $x \in L^2[0, e)$, $y \in L^2[0, e)$ and $z \in L^2[0, e)$ and define

$$\|\mathbf{s}\|_2 = \left(\|x\|^2_{L^2} + \|y\|^2_{L^2} + \|z\|^2_{L^2}\right)^{1/2}, \tag{23}$$

which is evidently a norm with

$$\|x\|_{L^2} = \left(\int_0^e |x(t)|^2 dt\right)^{1/2}, \quad \|y\|_{L^2} = \left(\int_0^e |y(t)|^2 dt\right)^{1/2}, \quad \|z\|_{L^2} = \left(\int_0^e |z(t)|^2 dt\right)^{1/2}.$$

This norm (23) is essential here since its makes easy the use of some properties related to the Sobolev space H^1 that is first considered in our analysis. Hence, we can now give and prove that the following Theorem showing the desired convergence result, particularly for functions x, y and z belonging to the Sobolev space $H^1[0, e)$.

Theorem 1 *Let $0 \le \sigma \le 1$ and assume that $x \in H^1[0, e)$, $y \in H^1[0, e)$ and $z \in H^1[0, e)$. Assuming that the approximation ${}^{ABC}D_t^\sigma \mathbf{s}_l(t) \simeq {}^{ABC}D_t^\sigma \mathbf{s}(t)$ holds and was obtained thanks to Haar wavelet approximation schemes defined above then, the resulting exact upper bound is given by:*

$$\|{}^{ABC}D_t^\sigma \mathbf{s}(t) - {}^{ABC}D_t^\sigma \mathbf{s}_l(t)\|_2 \le \frac{\beth\beth_\sigma^{-1}\sqrt{3}}{W(\sigma)\Gamma(1-\sigma)}, \tag{24}$$

where $\beth$ is a real non negative number and $\beth_\sigma^{-1} = \frac{2^{\frac{3}{2}-\sigma}(1-\sigma)}{e}\sqrt{\left(\frac{(1-l^{(1-\sigma)})}{2^{2\sigma-1}-1} + \frac{(1-l^{2(1-\sigma)})}{2^{2\sigma-1}-2}\right)}$.

Proof Making use of (12) and (13), we may consider, as done in (21), that the Atangana–Baleanu derivative ${}^{ABC}D_t^\sigma \mathbf{s}_l(t)$ of fractional order σ represents an approximation ${}^{ABC}D_t^\sigma \mathbf{s}(t)$ given by

$${}^{ABC}D_t^\sigma \mathbf{s}(t) \approx {}^{ABC}D_t^\sigma \mathbf{s}_l(t) = \sum_{r=1}^{e}\sum_{i=0}^{l-1} A_{r,i}\mathbf{H}_{r,i}(t),$$

also written as

$$\begin{aligned}\left({}^{ABC}D_t^\sigma x_l(t), {}^{ABC}D_t^\sigma y_l(t), {}^{ABC}D_t^\sigma y_l(t)\right)^T &= {}^{ABC}D_t^\sigma \mathbf{s}_l(t),\\ &= \sum_{r=1}^{e}\sum_{i=0}^{l-1} A_{r,i}\mathbf{H}_{r,i}(t),\\ &= \left(\sum_{r=1}^{e}\sum_{i=0}^{l-1} A_{r,i}^1\mathbf{H}_{r,i}(t), \sum_{r=1}^{e}\sum_{i=0}^{l-1} A_{r,i}^2\mathbf{H}_{r,i}(t),\right.\\ &\quad \left.\sum_{r=1}^{e}\sum_{i=0}^{l-1} A_{r,i}^3\mathbf{H}_{r,i}(t)\right)^T,\end{aligned}$$

where $l \in \{2^j : j = 0, 1, 2, \ldots\}$ and $A_{r,i} = \langle {}^{ABC}D_t^\sigma \mathbf{s}_l, \mathbf{H}_{r,i}\rangle_e = \int_0^e {}^{ABC}D_t^\sigma \mathbf{s}_l(t)\mathbf{H}_{r,i}(t)dt$,

$$A_{r,i}^1 = \langle {}^{ABC}D_t^\sigma x_l, \mathbf{H}_{r,i}\rangle_e = \int_0^e {}^{ABC}D_t^\sigma x_l(t)\mathbf{H}_{r,i}(t)dt,$$

$$A_{r,i}^2 = \langle {}^{ABC}D_t^\sigma y_l, \mathbf{H}_{r,i} \rangle_e = \int_0^e {}^{ABC}D_t^\sigma y_l(t) \mathbf{H}_{r,i}(t) dt, \tag{25}$$

$$A_{r,i}^3 = \langle {}^{ABC}D_t^\sigma z_l, \mathbf{H}_{r,i} \rangle_e = \int_0^e {}^{ABC}D_t^\sigma z_l(t) \mathbf{H}_{r,i}(t) dt.$$

Then,

$$\begin{aligned} {}^{ABC}D_t^\sigma \mathbf{s}(t) - {}^{ABC}D_t^\sigma \mathbf{s}_l(t) &= \sum_{r=1}^{e} \sum_{i=l}^{\infty} A_{r,i} \mathbf{H}_{r,i}(t), \\ &= \sum_{r=1}^{e} \sum_{i=2^j}^{\infty} A_{r,i} \mathbf{H}_{r,i}(t) \quad j = 0, 1, 2, \ldots, \\ &= \left(\sum_{r=1}^{e} \sum_{i=2^j}^{\infty} A_{r,i}^1 \mathbf{H}_{r,i}(t), \sum_{r=1}^{e} \sum_{i=2^j}^{\infty} A_{r,i}^2 \mathbf{H}_{r,i}(t), \sum_{r=1}^{e} \sum_{i=2^j}^{\infty} A_{r,i}^3 \mathbf{H}_{r,i}(t) \right)^T, \end{aligned} \tag{26}$$

with $j = 0, 1, 2, \ldots$. From (23) and exploiting the haar wavelet expression (26) we have

$$\begin{aligned} &\| {}^{ABC}D_t^\sigma \mathbf{s}(t) - {}^{ABC}D_t^\sigma \mathbf{s}_l(t) \|_2 \\ &= \sqrt{\left(\| {}^{ABC}D_t^\sigma x - {}^{ABC}D_t^\sigma x_l \|_{L2}^2 + \| {}^{ABC}D_t^\sigma y - {}^{ABC}D_t^\sigma y_l \|_{L2}^2 + \| {}^{ABC}D_t^\sigma z - {}^{ABC}D_t^\sigma z_l \|_{L2}^2 \right)}, \\ &= \sqrt{\left(\int_0^e | {}^{ABC}D_t^\sigma x(t) - {}^{ABC}D_t^\sigma x_l(t) |^2 dt + \int_0^e | {}^{ABC}D_t^\sigma y(t) - {}^{ABC}D_t^\sigma y_l(t) |^2 dt + \int_0^e | {}^{ABC}D_t^\sigma z(t) - {}^{ABC}D_t^\sigma z_l(t) |^2 dt \right)}, \\ &= \sqrt{\left(\int_0^e \left| \sum_{r=1}^{e} \sum_{i=l}^{\infty} A_{r,i}^1 \mathbf{H}_{r,i}(t) \right|^2 dt + \int_0^e \left| \sum_{r=1}^{e} \sum_{i=l}^{\infty} A_{r,i}^2 \mathbf{H}_{r,i}(t) \right|^2 dt + \int_0^e \left| \sum_{r=1}^{e} \sum_{i=l}^{\infty} A_{r,i}^3 \mathbf{H}_{r,i}(t) \right|^2 dt \right)}. \end{aligned} \tag{27}$$

Recall as we mentioned in Sect. 3 that the sequence $\{\mathbf{H}_i(t)\}_{i=0}^{\infty}$ forms a complete orthonormal system on $L^2[0, e)$, and this means that the following equality holds: $\int_0^e \mathbf{h}_{el}(t)^T \mathbf{h}_{el}(t) dt = \mathbb{J}_{el}$ (identity matrix). Hence exploiting that property together with Fubini–Tonelli theorem for positive functions [43, 44] and knowing that l can be written as the form of powers of 2 ($l \in \{2^j : \; j = 0, 1, 2, \ldots\}$) all lead to

$$\begin{aligned} &\| {}^{ABC}D_t^\sigma \mathbf{s}(t) - {}^{ABC}D_t^\sigma \mathbf{s}_l(t) \|_2 \\ &\le \sqrt{\left(\sum_{r=1}^{e} \sum_{j=0}^{\infty} \sum_{i=2^j}^{2^{j+1}} \int_0^e \left| A_{r,i}^1 \mathbf{H}_{r,i}(t) \right|^2 dt + \sum_{r=1}^{e} \sum_{j=0}^{\infty} \sum_{i=2^j}^{2^{j+1}} \int_0^e \left| A_{r,i}^2 \mathbf{H}_{r,i}(t) \right|^2 dt + \sum_{r=1}^{e} \sum_{j=0}^{\infty} \sum_{i=2^j}^{2^{j+1}} \int_0^e \left| A_{r,i}^3 \mathbf{H}_{r,i}(t) \right|^2 dt \right)}, \\ &\le \sqrt{\left(\sum_{r=1}^{e} \sum_{j=0}^{\infty} \sum_{i=2^j}^{2^{j+1}} \int_0^e \left| A_{r,i}^1 \right|^2 dt + \sum_{r=1}^{e} \sum_{j=0}^{\infty} \sum_{i=2^j}^{2^{j+1}} \int_0^e \left| A_{r,i}^2 \right|^2 dt + \sum_{r=1}^{e} \sum_{j=0}^{\infty} \sum_{i=2^j}^{2^{j+1}} \int_0^e \left| A_{r,i}^3 \right|^2 dt \right)}, \end{aligned} \tag{28}$$

where the coefficients $A_{r,i}^q, \quad q = 1, 2, 3$ are defined by (25). The calculation of each $A_{r,i}^q$ is now possible at this stage thanks to (25) and the definitions (9) and (10) of $\mathbf{H}_{r,i}$. Therefore we obtain

$$A_{r,i}^1 = 2^{j/2}\left(\int_{\frac{l}{2^j}-1+r}^{\frac{l+\frac{1}{2}}{2^j}-1+r} {}^{ABC}D_t^{\sigma}x(t)dt - \int_{\frac{l+\frac{1}{2}}{2^j}-1+r}^{\frac{l+1}{2^j}-1+r} {}^{ABC}D_t^{\sigma}x(t)dt\right),$$

$$A_{r,i}^2 = 2^{j/2}\left(\int_{\frac{l}{2^j}-1+r}^{\frac{l+\frac{1}{2}}{2^j}-1+r} {}^{ABC}D_t^{\sigma}y(t)dt - \int_{\frac{l+\frac{1}{2}}{2^j}-1+r}^{\frac{l+1}{2^j}-1+r} {}^{ABC}D_t^{\sigma}y(t)dt\right), \tag{29}$$

$$A_{r,i}^3 = 2^{j/2}\left(\int_{\frac{l}{2^j}-1+r}^{\frac{l+\frac{1}{2}}{2^j}-1+r} {}^{ABC}D_t^{\sigma}z(t)dt - \int_{\frac{l+\frac{1}{2}}{2^j}-1+r}^{\frac{l+1}{2^j}-1+r} {}^{ABC}D_t^{\sigma}z(t)dt\right).$$

We now make use of the Mean value theorem for definite integrals, to state the existence of two times $T_x \in \left(\frac{l}{2^j} - 1 + r,\ \frac{l+\frac{1}{2}}{2^j} - 1 + r\right)$ and $\widetilde{T}_x \in \left(\frac{l+\frac{1}{2}}{2^j} - 1 + r,\ \frac{l+1}{2^j} - 1 + r\right)$ so that

$$\begin{aligned} A_{r,i}^1 &= (\sqrt{2})^j\left(\frac{1}{2^{j+1}}{}^{ABC}D_t^{\sigma}x(T_x)dt - \frac{1}{2^{j+1}}{}^{ABC}D_t^{\sigma}x(\widetilde{T}_x)dt\right), \\ &= 2^{-\left(\frac{j}{2}+1\right)}\left({}^{ABC}D_t^{\sigma}x(T_x)dt - {}^{ABC}D_t^{\sigma}x(\widetilde{T}_x)dt\right). \end{aligned} \tag{30}$$

The definition (5) of Atangana–Baleanu derivative substituted into the later equation implies that

$$\begin{aligned} \left|A_{r,i}^1\right| &= 2^{-\left(\frac{j}{2}+1\right)}\left|{}^{ABC}D_t^{\sigma}x(T_x)dt - {}^{ABC}D_t^{\sigma}x(\widetilde{T}_x)dt\right|, \\ &=2^{-\left(\frac{j}{2}+1\right)}\frac{1}{W(\sigma)\Gamma(1-\sigma)}\left|\int_0^{T_x}(T_x-\xi)^{-\sigma}\frac{dx(\xi)}{d\xi}d\xi - \int_0^{\widetilde{T}_x}\left(\widetilde{T}_x-\xi\right)^{-\sigma}\frac{dx(\xi)}{d\xi}d\xi\right|. \end{aligned}$$

Under the assumption that $x \in H^1[0, e)$ hence, there exists a non-negative constant $\beth_x$ such that $\|\dot{x}(\xi)\| \leq \beth_x$ for all $\xi \in (0, T_x)$ and $\xi \in (0, \widetilde{T}_x)$. This yields

$$\left|A_{r,i}^1\right| \leq \beth_x 2^{-\left(\frac{j}{2}+1\right)}\frac{1}{W(\sigma)\Gamma(1-\sigma)}\left|\int_0^{T_x}(T_x-\xi)^{-\sigma}d\xi - \int_0^{\widetilde{T}_x}\left(\widetilde{T}_x-\xi\right)^{-\sigma}d\xi\right|.$$

Integrating and simplifying finally leave us with

$$\begin{aligned} \left|A_{r,i}^1\right| &\leq \frac{\beth_x 2^{-\left(\frac{j}{2}+1\right)}}{(1-\sigma)\,W(\sigma)\Gamma(1-\sigma)}\left|{T_x}^{(1-\sigma)} - (\widetilde{T}_x)^{(1-\sigma)}\right| \\ &\leq \frac{\beth_x 2^{-\left(\frac{j}{2}+1\right)}}{(1-\sigma)\,W(\sigma)\Gamma(1-\sigma)}2^{j(1-\sigma)}, \end{aligned} \tag{31}$$

where the facts that $0 \leq \sigma \leq 1$, $T_x \in \left(\frac{l}{2^j} - 1 + r,\ \frac{l+\frac{1}{2}}{2^j} - 1 + r\right)$ and $\widetilde{T}_x \in \left(\frac{l+\frac{1}{2}}{2^j} - 1 + r,\ \frac{l+1}{2^j} - 1 + r\right)$ have been used.

We can proceed in the same manner as above, for the variables y and z to easily show that there also exist two non-negative constants $\daleth_y$ and $\daleth_z$ such that

$$\left|A^2_{r,i}\right| \leq \frac{\daleth_y 2^{-\left(\frac{j}{2}+1\right)}}{(1-\sigma)\, W(\sigma)\Gamma(1-\sigma)} 2^{j(1-\sigma)}, \tag{32}$$

and

$$\left|A^3_{r,i}\right| \leq \frac{\daleth_z 2^{-\left(\frac{j}{2}+1\right)}}{(1-\sigma)\, W(\sigma)\Gamma(1-\sigma)} 2^{j(1-\sigma)}. \tag{33}$$

We now choose $\daleth$ to be $\daleth = \max(\daleth_x, \daleth_y, \daleth_z)$. Substituting (31)–(33) into (28) leaves us with

$$\begin{aligned}
\|{}^{ABC}D_t^\sigma \mathbf{s}(t) - {}^{ABC}D_t^\sigma \mathbf{s}_l(t)\|_2 &\leq \sqrt{\left(\frac{3e\daleth^2}{4(W(\sigma)\Gamma(1-\sigma))^2(1-\sigma)^2}\sum_{r=1}^{e}\sum_{j=0}^{\infty}\sum_{i=2^j}^{2^{j+1}}\frac{2^{2j(1-\sigma)}}{2^j}\right)}, \\
&\leq \sqrt{\left(\frac{3e^2\daleth^2}{4(W(\sigma)\Gamma(1-\sigma))^2(1-\sigma)^2}\left(\frac{2^{2\sigma}l^{(1-\sigma)} - 2^{2\sigma}}{2-2^{2\sigma}} + \frac{2^{2\sigma}l^{2(1-\sigma)} - 2^{2\sigma}}{4-2^{2\sigma}}\right)\right)}, \\
&\leq \frac{e\daleth}{2(W(\sigma)\Gamma(1-\sigma))(1-\sigma)}\sqrt{\left(3\frac{2^{2\sigma}\left(l^{(1-\sigma)}-1\right)}{2-2^{2\sigma}} + 3\frac{2^{2\sigma}\left(l^{2(1-\sigma)}-1\right)}{4-2^{2\sigma}}\right)}, \\
&\leq \frac{\daleth}{W(\sigma)\Gamma(1-\sigma)} \cdot \frac{2^{\frac{3}{2}-\sigma}e\sqrt{3}}{(1-\sigma)}\sqrt{\left(\frac{\left(1-l^{(1-\sigma)}\right)}{2^{2\sigma-1}-1} + \frac{\left(1-l^{2(1-\sigma)}\right)}{2^{2\sigma-1}-2}\right)}.
\end{aligned} \tag{34}$$

and the proof of the theorem is complete.

We may have the cases where all the conditions of the aforementioned Theorem 1 are not satisfied, especially those related to the variable functions x, y and z. For instance we might have $x \notin H^1[0,e)$, $y \notin H^1[0,e)$ and $z \notin H^1[0,e)$. In this situation, the conditions $x \in L^2[0,e)$, $y \in L^2[0,e)$ and $z \in L^2[0,e)$ only are not enough to prove the Theorem 1 here above. This is essentially due to the interval $[0, e)$ that is not closed and will have consequence that variable functions x, y and z together with their respective first order derivatives $\frac{dx(t)}{dt}$, $\frac{dy(t)}{dt}$ and $\frac{dz(t)}{dt}$ might not be bounded nor attain their bounds on the real interval $[0, e)$. Thus, the proof of following Corollary has just been done.

Corollary 1 *Let $0 \leq \sigma \leq 1$. Consider that the variable functions $x \in L^2[0,e)$, $y \in L^2[0,e)$, $z \in L^2[0,e)$ and their respective first order derivatives $\frac{dx(t)}{dt}$, $\frac{dy(t)}{dt}$ and $\frac{dz(t)}{dt}$ are continuous and bounded on $[0, e)$. Assuming that the approximation ${}^{ABC}D_t^\sigma \mathbf{s}_l(t) \simeq {}^{ABC}D_t^\sigma \mathbf{s}(t)$ holds and was obtained thanks to Haar wavelet approximation schemes defined above then, the resulting exact upper bound is given by:*

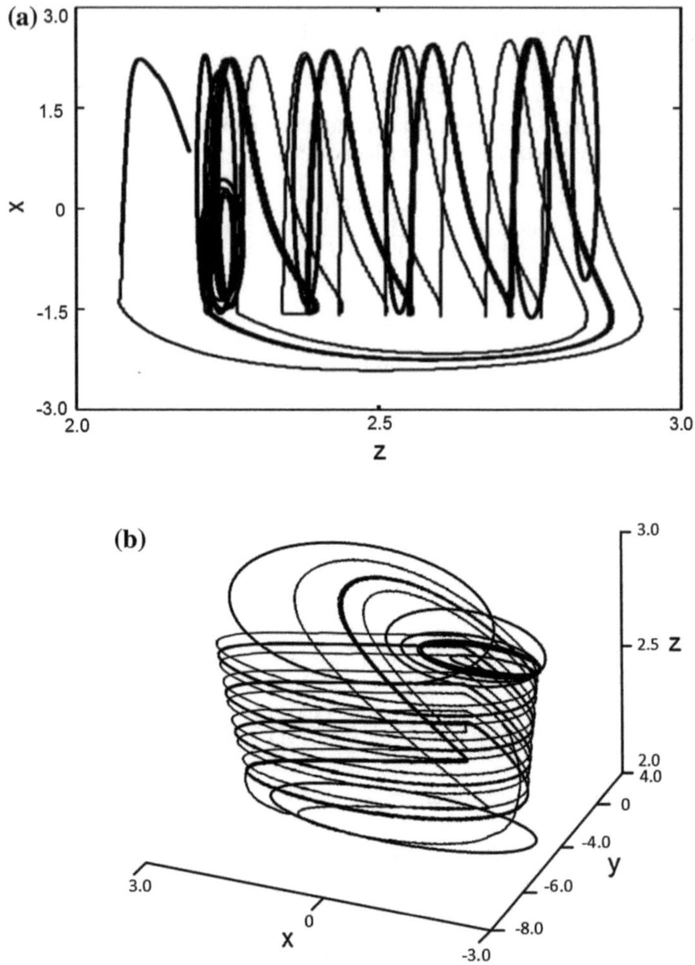

Fig. 2 Representation of the HR neuron dynamic (15)–(16) for $\alpha = \gamma = 1,\ \delta = -1$ and the control parameters at $a = c = 1,\ \ b = 3.1,\ \ d = 4,\ \eta_1 = 65 \times 10^{-3},\ \ \eta_2 = 5,\ x_r = -3/2,$ and, $I^e = 2.6$, with $\sigma = 0.8$. The HR neuron behavior seems irregular with a pole of attraction that generates a limit cycle

$$\|{}^{ABC}D_t^{\sigma}\mathbf{s}(t) - {}^{ABC}D_t^{\sigma}\mathbf{s}_l(t)\|_2 \leq \frac{\daleth\daleth_{\sigma}^{-1}\sqrt{3}}{W(\sigma)\Gamma(1-\sigma)}, \tag{35}$$

where $\daleth$ *is a real non negative number and* $\daleth_{\sigma}^{-1} = \frac{2^{\frac{3}{2}-\sigma}(1-\sigma)}{e}\sqrt{\left(\frac{(1-l^{(1-\sigma)})}{2^{2\sigma-1}-1} + \frac{(1-l^{2(1-\sigma)})}{2^{2\sigma-1}-2}\right)}.$

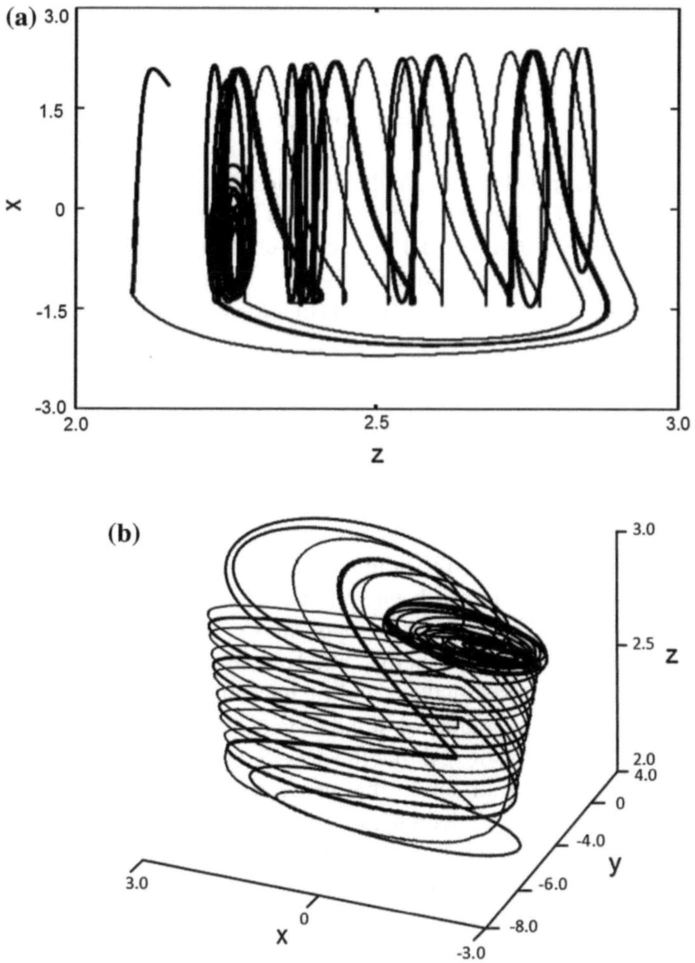

Fig. 3 Representation of the HR neuron dynamic (15)–(16) for $\alpha = \gamma = 1,\ \delta = -1$ and the control parameters at $a = c = 1,\ \ b = 3.1,\ \ d = 4,\ \ \eta_1 = 65 \times 10^{-3},\ \ \eta_2 = 5,\ x_r = -3/2,$ and, $I^e = 2.6$, with $\sigma = 0.5$. The irregularity of HR neuron behavior seems accentuates with the pole of attraction becoming chaotic

6 Application to Particular Cases of HR Neuron with External Current Input

We now wrap up this work by providing some visualization related to the evolution of the variable functions, especially the one symbolizing the membrane potential. Hence we perform some numerical simulations using the Haar wavelet scheme presented above and proven to be convergent. Setting $\alpha = \gamma = 1,\ \delta = -1$ and the the control parameters at

$$a = c = 1,\ \ b = 3.1,\ \ d = 4,\ \ \eta_1 = 65 \times 10^{-3},\ \ \eta_2 = 5,\ \ x_r = -3/2,\ \ \text{and,}\ \ I^e = 2.6,$$

yields the phase portraits in the 2D-plane x and the 3D-space x as depicted in Figs. 1, 2 and 3 for $\sigma = 1$, $\sigma = 0.9$ and $\sigma = 0.85$ respectively. Indeed, as expected in the standard well-known case $\sigma = 1$, the HR neuron is shown to exhibit a dynamic with a behavior that seems regular and simply periodic (Fig. 1a, b). When we change the value of the fractional derivative to $\sigma = 0.8$, the HR neuron is shown to exhibit a dynamic with a behavior that seems irregular and characterized by a pole of attraction that generates a limit cycle (Fig. 2a, b). When σ is decreased further ($\sigma = 0.5$), the irregularity seems to accentuate with the pole of attraction becoming chaotic (Fig. 3a, b).

7 Conclusion

We have explored the possible existence of chaotic poles of attraction in neurobiology, especially in the dynamics of Hindmarsh–Rose neurons with an external current input. The HR neuron model has been generalized with the introduction of an additional parameter in order to give it the opportunity to exhibit that feature that was still not pointed out yet. With the help of the Haar Wavelets numerical scheme, we showed the convergence error analysis and we have managed to solve the non-linear model and provided some graphical simulations with σ as the control parameter. Those graphical simulations reveal that the nerve cell's behavior is characterized by irregularity with a pole of attraction that generates a limit cycle. The irregularity nevertheless accentuates as σ decreases from 1 to 0, leaving us with a pole of attraction that becomes chaotic. This is a great observation showing some features of Atangana–Baleanu derivative applied to neurobiology models like that of JL Hindmarsh and RM Rose that were unknown until now. More analysis with certainly follow.

Acknowledgements The work of EF Doungmo Goufo was partially supported by the grant No: 105932 from the National Research Foundation (NRF) of South Africa.

References

1. Department of Biochemistry and Molecular Biophysics, Jessell, T., Siegelbaum, S., Hudspeth, A.J.: Principles of Neural Science. Kandel, E.R., Schwartz, J.H., Jessell, T.M. (eds.), vol. 4, pp. 1227–1246. McGraw-hill, New York (2000)
2. Logothetis, N.K., Pauls, J., Augath, M., Trinath, T., Oeltermann, A.: Neurophysiological investigation of the basis of the fMRI signal. Nature **412**(6843), 150–157 (2001)
3. Ren, H.P., Bai, C., Baptista, M.S., Grebogi, C.: Weak connections form an infinite number of patterns in the brain. Sci. Rep. **7**, 1–12 (2017)
4. Rizzolatti, G., Craighero, L.: The mirror-neuron system. Annu. Rev. Neurosci. **27**, 169–192 (2004)

5. Thompson, R.F., Spencer, W.A.: Habituation: a model phenomenon for the study of neuronal substrates of behavior. Psychol. Rev. **73**(1), 1–16 (1966)
6. Misiaszek, J.E.: The H-reflex as a tool in neurophysiology: its limitations and uses in understanding nervous system function. Muscle Nerve **28**(2), 144–160 (2003)
7. Hodgkin, A., Huxley, A.: A quantitative description of membrane current and its application to conduction and excitation in nerve. Bull. Math. Biol. **52**(1–2), 25–71 (1990)
8. Hodgkin, A.L., Huxley, A.F.: A quantitative description of membrane current and its application to conduction and excitation in nerve. J. Physiol. **117**(4), 500–544 (1952)
9. Hindmarsh, J., Rose, R.: A model of the nerve impulse using two first-order differential equations. Nature **296**(5853), 162–164 (1982)
10. Yamada, Y., Kashimori, Y.: Neural mechanism of dynamic responses of neurons in inferior temporal cortex in face perception. Cogn. Neurodynamics **7**(1), 23–38 (2013)
11. Barrio, R., Angeles Martínez, M., Serrano, S., Shilnikov, A.: Macro-and micro-chaotic structures in the Hindmarsh-Rose model of bursting neurons. Chaos: Interdiscip. J. Nonlinear Sci. **24**(2), 1–11 (2014)
12. Hindmarsh, J.L., Rose, R.: A model of neuronal bursting using three coupled first order differential equations. Proc. R. Soc. Lond. B: Biol. Sci. **221**(1222), 87–102 (1984)
13. Jun, D., Guang-jun, Z., Yong, X., Hong, Y., Jue, W.: Dynamic behavior analysis of fractional-order Hindmarsh-Rose neuronal model. Cogn. Neurodynamics **8**(2), 167–175 (2014)
14. Che, Y.-Q., Wang, J., Tsang, K.-M., Chan, W.-L.: Unidirectional synchronization for Hindmarsh-Rose neurons via robust adaptive sliding mode control. Nonlinear Anal. R. World Appl. **11**(2), 1096–1104 (2010)
15. Ostojic, S., Brunel, N., Hakim, V.: Synchronization properties of networks of electrically coupled neurons in the presence of noise and heterogeneities. J. Comput. Neurosci. **26**(3), 1–24 (2009)
16. Storace, M., Linaro, D., de Lange, E.: The Hindmarsh–Rose neuron model: bifurcation analysis and piecewise-linear approximations. Chaos: Interdiscip. J. Nonlinear Sci. **18**(3), 1–10 (2008)
17. Caputo, M.: Linear models of dissipation whose Q is almost frequency independent-II. Geophys. J. Int. **13**(5), 529–539 (1967)
18. Doungmo Goufo, E.F., Atangana, A.: Analytical and numerical schemes for a derivative with filtering property and no singular kernel with applications to diffusion. Eur. Phys. J. Plus **131**(8), 1–26 (2016)
19. Doungmo Goufo, E.F.: Chaotic processes using the two-parameter derivative with non-singular and nonlocal kernel: basic theory and applications. Chaos: Interdiscip. J. Nonlinear Sci. **26**(8), 1–21 (2016)
20. Atangana, A., Baleanu, D.: New fractional derivatives with non-local and non-singular kernel. Therm. Sci. **20**(2), 763–769 (2016)
21. Gómez-Aguilar, J.F.: Analytical and numerical solutions of a nonlinear alcoholism model via variable-order fractional differential equations. Phys. A: Stat. Mech. Its Appl. **494**, 52–75 (2018)
22. Atangana, A., Gómez-Aguilar, J.F.: Decolonisation of fractional calculus rules: breaking commutativity and associativity to capture more natural phenomena. Eur. Phys. J. Plus **133**, 1–22 (2018)
23. Atangana, A.: Non validity of index law in fractional calculus: a fractional differential operator with markovian and non-markovian properties. Phys. A: Stat. Mech. Its Appl. **505**, 688–706 (2018)
24. Atangana, A., Nieto, J.: Numerical solution for the model of RLC circuit via the fractional derivative without singular kernel. Adv. Mech. Eng. **7**(10), 1–7 (2015)
25. Morales-Delgado, V.F., Taneco-Hernández, M.A., Gómez-Aguilar, J.F.: On the solutions of fractional order of evolution equations. Eur. Phys. J. Plus **132**(1), 1–17 (2017)
26. Gómez-Aguilar, J.F., Escobar-Jiménez, R.F., López-López, M.G., Alvarado-Martínez, V.M.: Atangana-Baleanu fractional derivative applied to electromagnetic waves in dielectric media. J. Electromagn. Waves Appl. **30**(15), 1937–1952 (2016)

27. Brockmann, D., Hufnagel, L.: Front propagation in reaction-superdiffusion dynamics: taming Lévy flights with fluctuations. Phys. Rev. Lett. **98**(17), 178–301 (2007)
28. Doungmo Goufo, E.F.: Speeding up chaos and limit cycles in evolutionary language and learning processes. Math. Methods Appl. Sci. **40**(8), 3055–3065 (2017)
29. Doungmo Goufo, E.F.: Application of the Caputo-Fabrizio fractional derivative without singular kernel to Korteweg-de Vries-Bergers equation. Math. Model. Anal. **21**(2), 188–198 (2016)
30. Das, S.: Convergence of Riemann-Liouville and caputo derivative definitions for practical solution of fractional order differential equation. Int. J. Appl. Math. Stat. **23**(D11), 64–74 (2011)
31. Doungmo Goufo, E.F.: Solvability of chaotic fractional systems with 3D four-scroll attractors. Chaos Solitons Fractals **104**, 443–451 (2017)
32. Hilfer, R.: Application of Fractional Calculus in Physics. World Scientific, Singapore (1999)
33. Kilbas, A.A., Srivastava, H.M., Trujillo, J.J.: Theory and Applications of Fractional Differential Equations. Elsevier Science Limited (2006)
34. Coronel-Escamilla, A., Torres, F., Gómez-Aguilar, J.F., Escobar-Jiménez, R.F., Guerrero-Ramírez, G.V.: On the trajectory tracking control for an SCARA robot manipulator in a fractional model driven by induction motors with PSO tuning. Multibody Syst. Dyn. **43**(3), 257–277 (2018)
35. Yépez-Martínez, H., Gómez-Aguilar, J.F.: A new modified definition of Caputo-Fabrizio fractional-order derivative and their applications to the multi step homotopy analysis method (MHAM). J. Comput. Appl. Math. **346**, 247–260 (2018)
36. Gómez-Aguilar, J.F., Yépez-Martínez, H., Escobar-Jiménez, R.F., Astorga-Zaragoza, C.M., Reyes-Reyes, J.: Analytical and numerical solutions of electrical circuits described by fractional derivatives. Appl. Math. Model. **40**(21–22), 9079–9094 (2016)
37. Rosales, J., Guía, M., Gómez, F., Aguilar, F., Martínez, J.: Two dimensional fractional projectile motion in a resisting medium. Open Phys. **12**(7), 517–520 (2014)
38. Gómez-Aguilar, J.F., Yépez-Martínez, H., Escobar-Jiménez, R.F., Astorga-Zaragoza, C.M., Morales-Mendoza, L.J., González-Lee, M.: Universal character of the fractional space-time electromagnetic waves in dielectric media. J. Electromagn. Waves Appl. **29**(6), 727–740 (2015)
39. Coronel-Escamilla, A., Gómez-Aguilar, J.F., Alvarado-Méndez, E., Guerrero-Ramírez, G.V., Escobar-Jiménez, R.F.: Fractional dynamics of charged particles in magnetic fields. Int. J. Mod. Phys. C **27**(08), 1–16 (2016)
40. Babolian, E., Shahsavaran, A.: Numerical solution of nonlinear fredholm integral equations of the second kind using haar wavelets. J. Comput. Appl. Math. **225**(1), 87–95 (2009)
41. Chen, Y., Yi, M., Yu, C.: Error analysis for numerical solution of fractional differential equation by Haar wavelets method. J. Comput. Sci. **3**(5), 367–373 (2012)
42. Lepik, Ü., Hein, H.: Haar Wavelets: With Applications. Springer Science & Business Media (2014)
43. Tonelli, L.: Sullintegrazione per parti. Rend. Acc. Naz. Lincei **5**(18), 246–253 (1909)
44. Fubini, G.: Opere scelte II. Cremonese, Roma (1958)
45. Morales-Delgado, V.F., Gómez-Aguilar, J.F., Kumar, S., Taneco-Hernández, M.A.: Analytical solutions of the Keller-Segel chemotaxis model involving fractional operators without singular kernel. Eur. Phys. J. Plus **133**(5), 1–26 (2018)
46. Coronel-Escamilla, A., Gómez-Aguilar, J.F., Baleanu, D., Córdova-Fraga, T., Escobar-Jiménez, R.F., Olivares-Peregrino, V.H., Qurashi, MMAl: Bateman-Feshbach tikochinsky and Caldirola–Kanai oscillators with new fractional differentiation. Entropy **19**(2), 1–21 (2017)
47. Cuahutenango-Barro, B., Taneco-Hernández, M.A., Gómez-Aguilar, J.F.: On the solutions of fractional-time wave equation with memory effect involving operators with regular kernel. Chaos Solitons Fractals **115**, 283–299 (2018)
48. Doungmo Goufo, E.F., Nieto, J.J.: Attractors for fractional differential problems of transition to turbulent flows. J. Comput. Appl. Math. **339**, 329–342 (2017)

Modulating Chaotic Oscillations in Autocatalytic Reaction Networks Using Atangana–Baleanu Operator

Emile F. Doungmo Goufo and A. Atangana

Abstract Many mathematical models describing dynamics that occur in autocatalytic reaction networks have been proved to be chaotic, exhibiting orbits with unpredictable outcomes. Is it always possible to modulate that chaos? We use Haar wavelet numerical method to investigate a fractional system modeling autocatalytic reaction networks, where particular attention is made on biochemical systems of four-component networks. The convergence of the method is detailed through error analysis. Graphical representations reveal that the dynamic of the whole system is characterized by limit-cycles followed by period-doubling bifurcations that culminate with chaos, depending on the change of the total concentration of cofactor. The behavior of the system becomes more unpredictable as the concentration of cofactor increases, but the phenomenon is shown to be regulated by an additional parameter, the order of the fractional derivative γ, which plays an important role in triggering and controlling the appearance of chaos. Moreover, the chaotic behavior observed in the cascade diagram of the pure fractional case is proven to appear earlier, showing that the parameter γ is a valuable tool to regulate the chaos observed in some biochemical systems.

Keywords Fractional calculus · Atangana–Baleanu fractional derivative Autocatalytic reaction networks

E. F. Doungmo Goufo (✉)
Department of Mathematical Sciences, University of South Africa, Florida 0003, South Africa
e-mail: dgoufef@unisa.ac.za

A. Atangana
Institute of Groundwater Studies, Faculty of Natural and Agricultural Sciences, University of Free State, Bloemfontein 9301, South Africa

J. F. Gómez et al. (eds.), *Fractional Derivatives with Mittag-Leffler Kernel*, Studies in Systems, Decision and Control 194,
https://doi.org/10.1007/978-3-030-11662-0_9

1 Introduction

Catalysis in replication networks is defined as a system that comprises certain elements (like molecules), and is characterized by a copying mechanism that autoproduces a number of these elements and has become a significant issue in biology and several domains of natural sciences. As examples we have polynucleotides molecules (RNA or DNA) catalysis, idiotype recognition in the immune response, and dynamical models of Maynard-Smith games in sociobiology [1–3]. Many cyclic patterns of temporal self-organization were observed during the processes of various reactions and biochemical regulation [1–5]. Those observations were carried out for systems satisfying some global properties such as their availability to energy transfer or to mass change, their distance away from equilibrium state (causing the steps of particular reaction to be essentially irreversible), or the nonlinearity of their chemical reaction mechanism. Therefore, this type of systems is dominated by stable and conservative chemical oscillations in cyclic and autocatalytic reaction networks of p elements which may replicate independently according to the kinetic equation:

$$G_k + G_{k+1} \xrightarrow{F_k} 2G_{k+1}, \qquad k = 1, 2, \ldots, p, \tag{1}$$

where G_k is the number of elements of the system with the constrain $G_{k+p} = G_k$, and F_k represents a kinetic constant involved in the replication dynamic symbolized in (1) by the arrow. Note that the kinetic constant F_k contains a fixed energy source that keeps the network away from equilibrium state. Let us denote by x_k the concentration of element G_k. Hence, x_k changes in time t according to the differential equation

$$\frac{dx_k}{dt} = x_k(F_{k-1}x_{k-1} - F_k x_{k+1}), \qquad k = 1, 2, \ldots, p. \tag{2}$$

In this model, the element G_k is lost by interaction with the successive element, but is gained by interaction with the preceding element. Then, expression (2) represents the expansion of reaction rates in terms of concentrations.

1.1 Model's Description

Herein, we consider the system of differential equations (2) with $p = 4$. This yields a model of autocatalytic networks with four components driven from the outside by a constant energy source. We assume that the reaction network contains not only the four elements $G_k, \quad k = 1, 2, 3, 4$, but also considers a cofactor assumed not to be degraded and present in a constant quantity. A proton taken up at one step and released in another step, in a chemical reaction is a concrete example of such a cofactor. The kinetics are expressed as

$$
\begin{aligned}
C + G_1 + G_2 &\xrightarrow{F_1} 2G_2, \\
G_2 + G_3 &\xrightarrow{F_2} 2G_3 + C, \\
G_3 + G_4 &\xrightarrow{F_3} 2G_4, \\
G_4 + G_1 &\xrightarrow{F_4} 2G_1,
\end{aligned} \tag{3}
$$

with C representing the cofactor which interacts with G_1 to produce G_2 and is liberated from G_2 to produce G_3. Hence, the resulting system of differential equations reads as

$$
\begin{cases}
\dot{x}_1(t) = x_1(F_4x_4 - F_1zx_2), \\
\dot{x}_2(t) = x_2(F_1zx_1 - F_2x_3), \\
\dot{x}_3(t) = x_3(F_2x_2 - F_3x_4), \\
\dot{x}_4(t) = x_4(F_3x_3 - F_4x_1),
\end{cases} \tag{4}
$$

where z is a real number denoting the cofactor activity. To better assess the modulation effects of the limit-cycle oscillations susceptible to appear in those autocatalytic reaction networks, we extent the system (4) by considering the following model:

$$
\begin{cases}
{}^{ABC}D_t^{\gamma}\, x_1(t) = x_1(F_4x_4 - F_1zx_2), \\
{}^{ABC}D_t^{\gamma}\, x_2(t) = x_2(F_1zx_1 - F_2x_3), \\
{}^{ABC}D_t^{\gamma}\, x_3(t) = x_3(F_2x_2 - F_3x_4), \\
{}^{ABC}D_t^{\gamma}\, x_4(t) = x_4(F_3x_3 - F_4x_1),
\end{cases} \tag{5}
$$

assumed to verify the initial conditions

$$
x_1(0) = \nu_1, \ x_2(0) = \nu_2, \ x_3(0) = \nu_3, \ \ x_4(0) = \nu_4. \tag{6}
$$

Here ${}^{ABC}D_t^{\gamma}$ is the Atangana–Baleanu derivative of order γ $(0 < \gamma \leq 1)$ as detailed in the follow-up section and $\nu_i, \ \ i = 1, 2, 3, 4$, are real non negative functions. The variables x_k represent the concentrations of elements G_k respectively.

1. Note that the later model reduces to (4) for $\gamma = 1$.
2. Summing the four lines of (5) always gives zero and this is due to the fact that the system is closed to its other elements.
3. The concentration sum $m = z + x_2$, denotes the total concentration of cofactor with x_2 giving the quantity of cofactor linked to element G_2.

We solve the model (5)–(6) by means of numerical simulations, and we evaluate its dynamics using the Haar wavelet method.

1.2 Fractional Derivatives with Non-local Kernel and Variants

The past decades have seen a growing interest in using derivatives with non-integer orders (real or complex) for systems modeling and analysis [6–35]. This is due to observations in which many real life phenomena, including natural sciences, engineering and biology can be described successfully by models using differential equations with fractional orders. Differential equations with fractional derivatives happen to be useful assets to describe and investigate non-linear phenomena, found in various branches of ecology, chemistry, engineering, biology and number of fields in applied sciences. Recent literature in fractional calculus has been enriched by concrete models proven to give a better description of the phenomenon under investigation when using non-integer order derivative. Some of those phenomena are from mathematical epidemiology, porous media, biomedical engineering, signal and image processing, control theory, phase-locked loops, filters motion and nonlocal phenomena [30, 36–39].

The derivatives with fractional order mostly used by researchers are the formulations by Riemann–Liouville, Caputo and Atangana–Baleanu [10, 11, 20, 27, 40] respectively, defined by

$$\begin{aligned} {}^{RL}D_t^{\gamma} x(t) &= \frac{d}{dt} I^{1-\gamma} x(t), \\ &= \frac{1}{W(\gamma)\Gamma(\gamma)} \frac{d}{dt} \int_a^t x(\tau)\,(t-\tau)^{-\gamma} d\tau, \quad 0<\gamma\leq 1, \qquad t>0, \end{aligned} \tag{7}$$

$$\begin{aligned} {}^{C}D_t^{\gamma} x(t) &= I^{1-\gamma} \frac{d}{dt} x(t), \\ &= \frac{1}{W(\gamma)\Gamma(\gamma)} \int_a^t \frac{\frac{dx(\tau)}{d\tau}}{(t-\tau)^{\gamma}} d\tau, \quad 0<\gamma\leq 1, \qquad t>0, \end{aligned} \tag{8}$$

where I^{γ} is the fractional integral of order γ defined as

$$I^{\gamma} x(t) = \frac{1}{\Gamma(\gamma)} \int_a^t \frac{x(\tau)}{(t-\tau)^{1-\gamma}} d\tau. \tag{9}$$

and

$${}^{ABC}D_t^{\gamma} x(t) = \frac{W(\gamma)}{(1-\gamma)} \int_0^t \dot{x}(\tau)\, E_{\gamma}\left[-\frac{\gamma(t-\tau)^{\gamma}}{1-\gamma}\right] d\tau, \tag{10}$$

with the derivative order $\gamma \in [0,1]$ and where $W(\gamma)$ defines a normalization function such that

$$W(0) = W(1) = 1. \tag{11}$$

It is important to recall the antiderivative that comes with Atangana–Baleanu derivative used in this analysis. It is given by the following antiderivative (also called fractional integral of order γ):

$$I^{\gamma}x(t) = \frac{\gamma}{W(\gamma)\Gamma(\gamma)}\int_0^t (t-\tau)^{\gamma-1}x(\tau)d\tau + \frac{1-\gamma}{W(\gamma)}x(t). \tag{12}$$

Here we have $\infty \leq a < t, \;\; b > a$ and $x : (a, b) \longrightarrow \mathbb{R}$ an arbitrary real and locally integrable function, usually taken from the first order Sobolev space

$$H^1(a, b) = \left\{ x : x, \; \frac{dx}{dt} \in L^2(a, b) \right\}.$$

Other definitions of derivatives with fractional order have been proposed and introduced recently into the literature of fractional calculus [8, 9, 38].

We believe that the choice of Atangana–Baleanu derivative in this paper is the best in the sense that this fractional derivative offers better conditions of regularity compared to any other fractional derivative, like for instance, the Riemann–Liouville or Caputo derivative. A simple example is the fact that the Riemann–Liouville derivative of a constant function is different from zero, while the Caputo sense Atangana–Baleanu derivative of a constant is zero but requires some conditions of differentiability. Furthermore, it is important to mention that [7, 27] fractional-order initial states, as required by fractional differential equations using Riemann–Liouville derivative, are very difficult to obtain and sometimes appear to be physically non-realizable. Liouville and Riemann, in their respective initialization, chose the lower limit to be $-\infty$, but they were in fact addressing problems related to the same type of initialization [27]. However, Atangana–Baleanu [20, 27, 40] highlighted that to verify the composition of the fractional differintegrals, the integrated function together with all its integer-order derivatives must be equal to zero. Hence, by using Atangana–Baleanu fractional derivative we have the mathematical advantage that the derivative of a constant is zero and the model requires only integer order initial conditions (6). However, we keep in mind that there are still, in the use of this derivative, many insufficiencies in the physical reality of the initialization effects, especially when applied to fractional differential equations. Indeed, a constant initialization of the past lacks generality, as shown in the well-known Laplace transform for differintegrals based on that assumption. This insurgency can also be shown in solutions of fractional differential equations with histories assumed to be the set of initializing constants representing the values of fractional differintegrals at the beginning time [27, 30].

1.3 Haar Wavelet Method for System of Differential Equations

We call Haar wavelet the real function expressed by [41–43]

$$\mathbb{H}(t) = \begin{cases} 1, & 0 \leq t < 1/2; \\ -1, & 1/2 \leq t < 1; \\ 0, & \text{elsewhere}, \end{cases}$$

that is defined on the real set $\mathbb{R}$. Knowing that if we consider any $i = 0, 1, 2, \ldots$, it can always take the form $i = 2^j + k$ with $j = 0, 1, 2, \ldots$, and $k = 0, 1, 2, \ldots, 2^j - 1$, we can define for each $i = 0, 1, 2, 3, \ldots$, the family

$$h_i(t) = \begin{cases} 2^{\frac{j}{2}} \mathbb{H}(2^j t - k), & \text{for } i = 1, 2, \ldots, \\ 1, & \text{for } i = 0, \end{cases} \tag{13}$$

for $t \in [0, 1)$. Therefore, the resulting family $\{h_i(t)\}_{i=0}^{\infty}$ has been proven to form a complete orthonormal system in the space of square-integrable function $L^2[0, 1)$ [42, 43]. Considering the continuous function $l \in C[0, 1)$, the series $\sum_{i=0}^{\infty} \langle l, h_i \rangle h_i$ is uniformly convergent to l with $\langle l, h_i \rangle = \int_0^{\infty} l(t) h_i(t) dt$. It is therefore possible to decompose l in such a way that it becomes

$$l(t) = \sum_{i=0}^{\infty} a_i h_i(t),$$

where $a_i = \langle l, h_i \rangle$. Consequently, we can consider the approximated solution given by

$$l(t) \approx l_k(t) = \sum_{i=0}^{k-1} a_i h_i(t),$$

where $k \in \{2^j : \; j = 0, 1, 2, \ldots\}$.

Taking $n \in \mathbb{N}$, we can use the Haar function's translation on $[0, n)$ to define the function

$$h_{m,i}(t) = h_i(t - m + 1) \quad m = 1, 2, \ldots, n \quad \text{and } i = 0, 1, 2, \ldots, \tag{14}$$

where h_i is defined in (13). Note that the same properties as those of h_i hold for $h_{m,i}$. Hence, the family $\{h_{m,i}(t)\}_{i=0}^{\infty}$, $m = 1, 2, \ldots, n$, forms a complete orthonormal system in the space of square-integrable function $L^2[0, 1)$. Let us now exploit the following Haar orthonormal basis functions

$$a_{m,i} = \langle l, h_{m,i} \rangle = \int_0^{\infty} l(t) h_{m,i}(t) dt,$$

to expand the solution $l \in L^2[0, n)$ as the series

$$l(t) = \sum_{m=1}^{n} \sum_{i=0}^{\infty} a_{m,i} h_{m,i}(t). \tag{15}$$

Similarly, this brings us to consider the approximated solution

$$w(t) \approx w_k(t) = \sum_{m=1}^{n} \sum_{i=0}^{k-1} a_{m,i} h_{m,i}(t), \tag{16}$$

with $k \in \{2^j : \; j = 0, 1, 2, \ldots\}$.

Note that (16) can take the following compact form that will be significant in the coming analysis.

$$w(t) \approx w_k(t) =^T \mathbf{A}_{nk\times 1} \mathbb{H}_{nk\times 1}, \tag{17}$$

with ${}^T\mathbf{A}_{nk\times 1}$ meaning the transpose vector of

$$\mathbf{A}_{nk\times 1} = \begin{pmatrix} a_{1,0} \\ \vdots \\ a_{1,k-1} \\ a_{2,0} \\ \vdots \\ a_{2,k-1} \\ \vdots \\ a_{n,0} \\ \vdots \\ a_{n,k-1} \end{pmatrix} \quad \text{and} \quad \mathbb{H}_{nk\times 1} = \begin{pmatrix} h_{1,0} \\ \vdots \\ h_{1,k-1} \\ h_{2,0} \\ \vdots \\ h_{2,k-1} \\ \vdots \\ h_{n,0} \\ \vdots \\ h_{n,k-1} \end{pmatrix}.$$

2 Solvability of the Model by Means of Haar Wavelets

We exploit in this section the Haar wavelets numerical method to solve the model (5)

$$\begin{cases} {}^{ABC}D_t^{\gamma} x_1(t) = x_1(F_4 x_4 - F_1 z x_2), \\ {}^{ABC}D_t^{\gamma} x_2(t) = x_2(F_1 z x_1 - F_2 x_3), \\ {}^{ABC}D_t^{\gamma} x_3(t) = x_3(F_2 x_2 - F_3 x_4), \\ {}^{ABC}D_t^{\gamma} x_4(t) = x_4(F_3 x_3 - F_4 x_1), \end{cases} \tag{18}$$

that verifies the initial conditions

$$x_1(0) = \nu_1,\ x_2(0) = \nu_2,\ x_3(0) = \nu_3,\ \ x_4(0) = \nu_4. \tag{19}$$

To put the model (18)–(19) into the compact form, we can define the system state vectors and matrices by

$$w(t) = \begin{pmatrix} x_1(t) \\ x_2(t) \\ x_3(t) \\ x_4(t) \end{pmatrix} \quad \text{and} \quad f_0(x_1, x_2, x_3, x_4) = w(0) = \begin{pmatrix} x_1(0) \\ x_2(0) \\ x_3(0) \\ x_4(0) \end{pmatrix} = \begin{pmatrix} \nu_1 \\ \nu_2 \\ \nu_3 \\ \nu_4 \end{pmatrix}.$$

$$\mathscr{S}(w(t), t) = \mathscr{S}(x_1(t), x_2(t), x_3(t), x_4(t), t) = \begin{pmatrix} \mathscr{S}_1(w(t), t) \\ \mathscr{S}_2(w(t), t) \\ \mathscr{S}_3(w(t), t) \\ \mathscr{S}_4(w(t), t) \end{pmatrix} = \begin{pmatrix} \mathscr{S}_1(x_1(t), x_2(t), x_3(t), x_4(t), t) \\ \mathscr{S}_2(x_1(t), x_2(t), x_3(t), x_4(t), t) \\ \mathscr{S}_3(x_1(t), x_2(t), x_3(t), x_4(t), t) \\ \mathscr{S}_4(x_1(t), x_2(t), x_3(t), x_4(t), t) \end{pmatrix},$$

where

$$\begin{cases} \mathscr{S}_1(w(t), t) = x_1(F_4x_4 - F_1zx_2), \\ \mathscr{S}_2(w(t), t) = x_2(F_1zx_1 - F_2x_3), \\ \mathscr{S}_3(w(t), t) = x_3(F_2x_2 - F_3x_4), \\ \mathscr{S}_4(w(t), t) = x_4(F_3x_3 - F_4x_1). \end{cases}$$

Therefore, (18) takes the form

$${}^{ABC}D_t^{\gamma} w(t) = \mathscr{S}(w(t), t),$$

which is equivalent to

$$\begin{aligned} {}^{ABC}D_t^{\gamma} x_1(t) &= \mathscr{S}_1(w(t), t), \\ {}^{ABC}D_t^{\gamma} x_2(t) &= \mathscr{S}_2(w(t), t), \\ {}^{ABC}D_t^{\gamma} x_3(t) &= \mathscr{S}_3(w(t), t), \\ {}^{ABC}D_t^{\gamma} x_4(t) &= \mathscr{S}_4(w(t), t), \end{aligned} \tag{20}$$

and verifies the initial conditions

$$x_1(0) = \nu_1(x_1),\ x_2(0) = \nu_2(x_2),\ x_3(0) = \nu_3(x_3),\ x_4(0) = \nu_4(x_4).$$

Exploiting the Haar wavelets numerical scheme in (17) leads us to approximate the system with the Atangana–Baleanu fractional derivative (20) to get

$$\begin{aligned}
{}^{ABC}D_t^{\gamma} x_1(t) &= \mathscr{S}_1(w(t), t) \approx^{ABC} D_t^{\gamma} x_1^k(t) =^T \mathbf{A}^1_{nk\times 1}\mathbb{H}_{nk\times 1},\\
{}^{ABC}D_t^{\gamma} x_2(t) &= \mathscr{S}_2(w(t), t) \approx^{ABC} D_t^{\gamma} x_2^k(t) =^T \mathbf{A}^2_{nk\times 1}\mathbb{H}_{nk\times 1},\\
{}^{ABC}D_t^{\gamma} x_3(t) &= \mathscr{S}_3(w(t), t) \approx^{ABC} D_t^{\gamma} x_3^k(t) =^T \mathbf{A}^3_{nk\times 1}\mathbb{H}_{nk\times 1},\\
{}^{ABC}D_t^{\gamma} x_4(t) &= \mathscr{S}_4(w(t), t) \approx^{ABC} D_t^{\gamma} x_4^k(t) =^T \mathbf{A}^4_{nk\times 1}\mathbb{H}_{nk\times 1}.
\end{aligned} \tag{21}$$

We now apply the Atangana–Baleanu operator (12) on both sides of (21) to obtain

$$\begin{aligned}
x_1(t) - \nu_1 &\approx^{ABC} D_t^{\gamma} x_1^k(t) =^T \mathbf{A}^1_{nk\times 1} Q^{\gamma}_{nk\times nk}\mathbb{H}_{nk\times 1},\\
x_2(t) - \nu_2 &\approx^{ABC} D_t^{\gamma} x_2^k(t) =^T \mathbf{A}^2_{nk\times 1} Q^{\gamma}_{nk\times nk}\mathbb{H}_{nk\times 1},\\
x_3(t) - \nu_3 &\approx^{ABC} D_t^{\gamma} x_3^k(t) =^T \mathbf{A}^3_{nk\times 1} Q^{\gamma}_{nk\times nk}\mathbb{H}_{nk\times 1},\\
x_4(t) - \nu_4 &\approx^{ABC} D_t^{\gamma} x_4^k(t) =^T \mathbf{A}^4_{nk\times 1} Q^{\gamma}_{nk\times nk}\mathbb{H}_{nk\times 1},
\end{aligned} \tag{22}$$

which is equivalent to

$$\begin{aligned}
x_1(t) &\approx x_1^k(t) =^T \mathbf{A}^1_{nk\times 1} Q^{\gamma}_{nk\times nk}\mathbb{H}_{nk\times 1} + \nu_1,\\
x_2(t) &\approx x_2^k(t) =^T \mathbf{A}^2_{nk\times 1} Q^{\gamma}_{nk\times nk}\mathbb{H}_{nk\times 1} + \nu_2,\\
x_3(t) &\approx x_3^k(t) =^T \mathbf{A}^3_{nk\times 1} Q^{\gamma}_{nk\times nk}\mathbb{H}_{nk\times 1} + \nu_3,\\
x_4(t) &\approx x_4^k(t) =^T \mathbf{A}^4_{nk\times 1} Q^{\gamma}_{nk\times nk}\mathbb{H}_{nk\times 1} + \nu_4,
\end{aligned} \tag{23}$$

with $Q^{\gamma}_{nk\times nk}$ representing the Haar wavelets operational matrix [41, 42]. Solving the system (18)–(19) through the Galerkin's method on collocation points and substituting the approximations (21) and (23) into (18) leads to the residual errors expressed by

$$\begin{aligned}
&\varepsilon_1\left(\eta^1, \eta^2, \eta^3, \eta^4, t\right) =^T \mathbf{A}^1_{nk\times 1}\mathbb{H}_{nk\times 1}\\
&- \mathscr{S}_1\left({}^T\mathbf{A}^1_{nk\times 1} Q^{\gamma}_{nk\times nk}\mathbb{H}_{nk\times 1}, {}^T\mathbf{A}^2_{nk\times 1} Q^{\gamma}_{nk\times nk}\mathbb{H}_{nk\times 1}, {}^T\mathbf{A}^3_{nk\times 1} Q^{\gamma}_{nk\times nk}\mathbb{H}_{nk\times 1}, {}^T\mathbf{A}^4_{nk\times 1} Q^{\gamma}_{nk\times nk}\mathbb{H}_{nk\times 1}, t\right),\\
&\varepsilon_2\left(\eta^1, \eta^2, \eta^3, \eta^4, t\right) =^T \mathbf{A}^2_{nk\times 1}\mathbb{H}_{nk\times 1}\\
&- \mathscr{S}_2\left({}^T\mathbf{A}^1_{nk\times 1} Q^{\gamma}_{nk\times nk}\mathbb{H}_{nk\times 1}, {}^T\mathbf{A}^2_{nk\times 1} Q^{\gamma}_{nk\times nk}\mathbb{H}_{nk\times 1}, {}^T\mathbf{A}^3_{nk\times 1} Q^{\gamma}_{nk\times nk}\mathbb{H}_{nk\times 1}, {}^T\mathbf{A}^4_{nk\times 1} Q^{\gamma}_{nk\times nk}\mathbb{H}_{nk\times 1}, t\right),\\
&\varepsilon_3\left(\eta^1, \eta^2, \eta^3, \eta^4, t\right) =^T \mathbf{A}^3_{nk\times 1}\mathbb{H}_{nk\times 1}\\
&- \mathscr{S}_3\left({}^T\mathbf{A}^1_{nk\times 1} Q^{\gamma}_{nk\times nk}\mathbb{H}_{nk\times 1}, {}^T\mathbf{A}^2_{nk\times 1} Q^{\gamma}_{nk\times nk}\mathbb{H}_{nk\times 1}, {}^T\mathbf{A}^3_{nk\times 1} Q^{\gamma}_{nk\times nk}\mathbb{H}_{nk\times 1}, {}^T\mathbf{A}^4_{nk\times 1} Q^{\gamma}_{nk\times nk}\mathbb{H}_{nk\times 1}, t\right),\\
&\varepsilon_4\left(\eta^1, \eta^2, \eta^3, \eta^4, t\right) =^T \mathbf{A}^4_{nk\times 1}\mathbb{H}_{nk\times 1}\\
&- \mathscr{S}_4\left({}^T\mathbf{A}^1_{nk\times 1} Q^{\gamma}_{nk\times nk}\mathbb{H}_{nk\times 1}, {}^T\mathbf{A}^2_{nk\times 1} Q^{\gamma}_{nk\times nk}\mathbb{H}_{nk\times 1}, {}^T\mathbf{A}^3_{nk\times 1} Q^{\gamma}_{nk\times nk}\mathbb{H}_{nk\times 1}, {}^T\mathbf{A}^4_{nk\times 1} Q^{\gamma}_{nk\times nk}\mathbb{H}_{nk\times 1}, t\right),
\end{aligned} \tag{24}$$

with

$$\eta^1 = a^1_{1,0}, \ldots, a^1_{1,k-1}, \ldots, a^1_{n,0}, \ldots, a^1_{n,k-1},$$

$$\eta^2 = a^2_{1,0}, \ldots, a^2_{1,k-1}, \ldots, a^2_{n,0}, \ldots, a^2_{n,k-1},$$

$$\eta^3 = a^3_{1,0}, \ldots, a^3_{1,k-1}, \ldots, a^3_{n,0}, \ldots, a^3_{n,k-1},$$

$$\eta^4 = a^4_{1,0}, \ldots, a^4_{1,k-1}, \ldots, a^4_{n,0}, \ldots, a^4_{n,k-1},$$

so that $a^i_{\cdot,\cdot}$ represent the components of ${}^T\mathbf{A}^i_{\cdot\times\cdot}$.

The following assumptions

$$\varepsilon_1\left(\eta^1, \eta^2, \eta^3, \eta^4, t_{m,i}\right) = 0,$$

$$\varepsilon_2\left(\eta^1, \eta^2, \eta^3, \eta^4, t_{m,i}\right) = 0,$$

$$\varepsilon_3\left(\eta^1, \eta^2, \eta^3, \eta^4, t_{m,i}\right) = 0,$$

$$\varepsilon_4\left(\eta^1, \eta^2, \eta^3, \eta^4, t_{m,i}\right) = 0,$$

lead to the system of $3nk$ equations, with $4nk$ unknowns given by

$$a^1_{1,0}, \ldots, a^1_{1,k-1}, \ldots, a^1_{n,0}, \ldots, a^1_{n,k-1},$$

$$a^2_{1,0}, \ldots, a^2_{1,k-1}, \ldots, a^2_{n,0}, \ldots, a^2_{n,k-1},$$

$$a^3_{1,0}, \ldots, a^3_{1,k-1}, \ldots, a^3_{n,0}, \ldots, a^3_{n,k-1},$$

$$a^4_{1,0}, \ldots, a^4_{1,k-1}, \ldots, a^4_{n,0}, \ldots, a^4_{n,k-1},$$

where

$$t_{m,i} = \frac{2i-1}{2k} + m - 1, \qquad m = 1, 2, \ldots, n, \qquad i = 1, 2, \ldots, k,$$

denote a nk number of collocation points. Hence, solving for those unknowns and substituting into (23) yields

$$w(t) \approx \begin{pmatrix} x^k_1(t) \\ x^k_2(t) \\ x^k_3(t) \\ x^k_4(t) \end{pmatrix},$$

which is the desired solution.

3 Error Analysis and Convergence

Our goal is to present in this section the exact error bounds used in the above Haar wavelet method to analyse the model (18)–(19). Because $w \in L^2[0, n)$, we take $x \in L^2[0, n)$, $y \in L^2[0, n)$ and $z \in L^2[0, n)$ and define the norm

$$\|w\|_2 = \left(\|x_1\|_{L^2}^2 + \|x_2\|_{L^2}^2 + \|x_3\|_{L^2}^2 + \|x_4\|_{L^2}^2\right)^{1/2}, \tag{25}$$

where

$$\|x_1\|_{L^2} = \left(\int_0^n |x_1(t)|^2 dt\right)^{1/2}, \quad \|x_2\|_{L^2} = \left(\int_0^n |x_2(t)|^2 dt\right)^{1/2},$$

$$\|x_3\|_{L^2} = \left(\int_0^n |x_3(t)|^2 dt\right)^{1/2}, \quad \|x_4\|_{L^2} = \left(\int_0^n |x_4(t)|^2 dt\right)^{1/2}.$$

From expressions (16), (17) and (23), the fractional derivative ${}^{ABC}D_t^{\gamma} r_K(t)$ is considered as an approximation of ${}^{ABC}D_t^{\gamma} w(t)$ and given as

$${}^{ABC}D_t^{\gamma} w(t) \approx {}^{ABC}D_t^{\gamma} w^k(t) = \sum_{m=1}^{n} \sum_{i=0}^{k-1} a_{m,i} h_{m,i}(t).$$

This equation is equivalent to

$$\begin{pmatrix} {}^{ABC}D_t^{\gamma} x_1^k(t) \\ {}^{ABC}D_t^{\gamma} x_2^k(t) \\ {}^{ABC}D_t^{\gamma} x_2^k(t) \\ {}^{ABC}D_t^{\gamma} x_3^k(t) \end{pmatrix} = {}^{ABC}D_t^{\gamma} w^k(t) = \sum_{m=1}^{n} \sum_{i=0}^{k-1} a_{m,i} h_{m,i}(t) = \begin{pmatrix} \sum_{m=1}^{n} \sum_{i=0}^{k-1} a_{m,i}^1 h_{m,i}(t) \\ \sum_{m=1}^{n} \sum_{i=0}^{k-1} a_{m,i}^2 h_{m,i}(t) \\ \sum_{m=1}^{n} \sum_{i=0}^{k-1} a_{m,i}^3 h_{m,i}(t) \\ \sum_{m=1}^{n} \sum_{i=0}^{k-1} a_{m,i}^4 h_{m,i}(t) \end{pmatrix},$$

with $k \in \{2^j : \ j = 0, 1, 2, \ldots\}$ and $a_{m,i} = \langle {}^{ABC}D_t^{\gamma} w^k, h_{m,i}\rangle_n = \int_0^n {}^{ABC}D_t^{\gamma} w^k(t)$ $h_{m,i}(t)dt$,

$$\begin{aligned} a_{m,i}^1 &= \langle {}^{ABC}D_t^{\gamma} x_1^k, h_{m,i}\rangle_n = \int_0^n {}^{ABC}D_t^{\gamma} x_1^k(t) h_{m,i}(t) dt, \\ a_{m,i}^2 &= \langle {}^{ABC}D_t^{\gamma} x_2^k, h_{m,i}\rangle_n = \int_0^n {}^{ABC}D_t^{\gamma} x_2^k(t) h_{m,i}(t) dt, \\ a_{m,i}^3 &= \langle {}^{ABC}D_t^{\gamma} x_3^k, h_{m,i}\rangle_n = \int_0^n {}^{ABC}D_t^{\gamma} x_3^k(t) h_{m,i}(t) dt, \\ a_{m,i}^4 &= \langle {}^{ABC}D_t^{\gamma} x_4^k, h_{m,i}\rangle_n = \int_0^n {}^{ABC}D_t^{\gamma} x_4^k(t) h_{m,i}(t) dt. \end{aligned} \tag{26}$$

Therefore,

$$
\begin{aligned}
{}^{ABC}D_t^{\gamma} w(t) - {}^{ABC}D_t^{\gamma} w^k(t) &= \sum_{m=1}^{n}\sum_{i=k}^{\infty} a_{m,i}h_{m,i}(t) \\
&= \sum_{m=1}^{n}\sum_{i=2^j}^{\infty} a_{m,i}h_{m,i}(t) \quad j = 0, 1, 2, \ldots \\
&= \begin{pmatrix} \sum_{m=1}^{n}\sum_{i=2^j}^{\infty} a^1_{m,i}h_{m,i}(t) \\ \sum_{m=1}^{n}\sum_{i=2^j}^{\infty} a^2_{m,i}h_{m,i}(t) \\ \sum_{m=1}^{n}\sum_{i=2^j}^{\infty} a^3_{m,i}h_{m,i}(t) \\ \sum_{m=1}^{n}\sum_{i=2^j}^{\infty} a^4_{m,i}h_{m,i}(t) \end{pmatrix} \qquad j = 0, 1, 2, \ldots
\end{aligned} \tag{27}
$$

From the norm (25), it is now possible to formulate and prove the following proposition on the convergence of the numerical method. Note that functions x_1, x_2, x_3 and x_4 are first taken in the Sobolev space $H^1[0, n)$.

Proposition 1 *Let $0 \le \gamma \le 1$ and choose $x_1, x_2, x_3 \in H^1[0, n)$. The following exact upper bound holds*

$$
\|{}^{ABC}D_t^{\gamma} w(t) - {}^{ABC}D_t^{\gamma} w^k(t)\|_2 \le \frac{\mathcal{G}}{\mathcal{G}_\gamma W(\gamma)\Gamma(\gamma)}, \tag{28}
$$

where ${}^{ABC}D_t^{\gamma} w(t)$ are approximated by the functions ${}^{ABC}D_t^{\gamma} w^k(t)$ through Haar wavelet method, $\mathcal{G} \in \mathbb{R}^+$ and $\mathcal{G}_\gamma = \frac{2(1-\gamma)}{2^\gamma n}\left(\frac{(3-3k^{(1-\gamma)})}{2^{2\gamma}-2} + \frac{(3-3k^{(2-2\gamma)})}{2^{2\gamma}-4}\right)^{-1/2}$.

Proof Considering (25) and (27) gives

$$
\begin{aligned}
&\|{}^{ABC}D_t^{\gamma} w(t) - {}^{ABC}D_t^{\gamma} w^k(t)\|_2, \\
&= \left(\|{}^{ABC}D_t^{\gamma} x - {}^{ABC}D_t^{\gamma} x_1^k\|^2_{L^2} + \|{}^{ABC}D_t^{\gamma} y - {}^{ABC}D_t^{\gamma} x_2^k\|^2_{L^2} + \|{}^{ABC}D_t^{\gamma} z - {}^{ABC}D_t^{\gamma} x_3^k\|^2_{L^2} - {}^{ABC}D_t^{\gamma} x_4^k\|^2_{L^2}\right)^{1/2}, \\
&= \left(\int_0^n |{}^{ABC}D_t^{\gamma} x_1(t) - {}^{ABC}D_t^{\gamma} x_1^k(t)|^2 dt + \int_0^n |{}^{ABC}D_t^{\gamma} x_2(t)\right. \\
&\left. - {}^{ABC}D_t^{\gamma} x_2^k(t)|^2 dt + \int_0^n |{}^{ABC}D_t^{\gamma} x_3(t) - {}^{ABC}D_t^{\gamma} x_3^k(t)|^2 dt + \int_0^n |{}^{ABC}D_t^{\gamma} x_4(t) - {}^{ABC}D_t^{\gamma} x_4^k(t)|^2 dt\right)^{1/2}, \\
&= \left(\int_0^n \left|\sum_{m=1}^{n}\sum_{i=k}^{\infty} a^1_{m,i}h_{m,i}(t)\right|^2 dt + \int_0^n \left|\sum_{m=1}^{n}\sum_{i=k}^{\infty} a^2_{m,i}h_{m,i}(t)\right|^2 dt\right. \\
&\left. + \int_0^n \left|\sum_{m=1}^{n}\sum_{i=k}^{\infty} a^3_{m,i}h_{m,i}(t)\right|^2 dt + \int_0^n \left|\sum_{m=1}^{n}\sum_{i=k}^{\infty} a^4_{m,i}h_{m,i}(t)\right|^2 dt\right)^{1/2}.
\end{aligned}
$$

Since $\{h_i(t)\}_{i=0}^{\infty}$ forms a complete orthonormal system on $[0, n)$, (meaning $\int_0^n \mathbb{H}_{nk}(t)^T \mathbb{H}_{nk}(t)dt = \mathbb{I}_{nk}$ (where $\mathbb{I}_{nk}$ is the identity matrix), we can make use of Fubini-Tonelli theorem [44, 45] to write

$$
\begin{aligned}
&\|{}^{ABC}D_t^{\gamma} w(t) - {}^{ABC}D_t^{\gamma} w^k(t)\|_2 \\
&\le \left(\sum_{m=1}^{n}\sum_{j=0}^{\infty}\sum_{i=2^j}^{2^{j+1}} \int_0^n \left|a_{m,i}^1 h_{m,i}(t)\right|^2 dt + \sum_{m=1}^{n}\sum_{j=0}^{\infty}\sum_{i=2^j}^{2^{j+1}} \int_0^n \left|a_{m,i}^2 h_{m,i}(t)\right|^2 dt \right. \\
&\left. + \sum_{m=1}^{n}\sum_{j=0}^{\infty}\sum_{i=2^j}^{2^{j+1}} \int_0^n \left|a_{m,i}^3 h_{m,i}(t)\right|^2 dt + \sum_{m=1}^{n}\sum_{j=0}^{\infty}\sum_{i=2^j}^{2^{j+1}} \int_0^n \left|a_{m,i}^4 h_{m,i}(t)\right|^2 dt \right)^{1/2} \qquad (29) \\
&\le \left(\sum_{m=1}^{n}\sum_{j=0}^{\infty}\sum_{i=2^j}^{2^{j+1}} \int_0^n \left|a_{m,i}^1\right|^2 dt + \sum_{m=1}^{n}\sum_{j=0}^{\infty}\sum_{i=2^j}^{2^{j+1}} \int_0^n \left|a_{m,i}^2\right|^2 dt \right. \\
&\left. + \sum_{m=1}^{n}\sum_{j=0}^{\infty}\sum_{i=2^j}^{2^{j+1}} \int_0^n \left|a_{m,i}^3\right|^2 dt + \sum_{m=1}^{n}\sum_{j=0}^{\infty}\sum_{i=2^j}^{2^{j+1}} \int_0^n \left|a_{m,i}^4\right|^2 dt \right)^{1/2}.
\end{aligned}
$$

Here $a_{m,i}^r$, $r = 1, 2, 3, 4$, are defined by (26) and k belongs to $\{2^j : \; j = 0, 1, 2, \ldots\})$. We can now compute each $a_{m,i}^r$ using (26) together with definitions (13) and (14) of $h_{m,i}$. Hence,

$$
\begin{aligned}
a_{m,i}^1 &= (\sqrt{2})^j \left[\int_{\frac{k}{2^j}-1+m}^{\frac{k+\frac{1}{2}}{2^j}-1+m} {}^{ABC}D_t^{\gamma} x_1(t)dt - \int_{\frac{k+\frac{1}{2}}{2^j}-1+m}^{\frac{k+1}{2^j}-1+m} {}^{ABC}D_t^{\gamma} x_1(t)dt \right], \\
a_{m,i}^2 &= (\sqrt{2})^j \left[\int_{\frac{k}{2^j}-1+m}^{\frac{k+\frac{1}{2}}{2^j}-1+m} {}^{ABC}D_t^{\gamma} x_2(t)dt - \int_{\frac{k+\frac{1}{2}}{2^j}-1+m}^{\frac{k+1}{2^j}-1+m} {}^{ABC}D_t^{\gamma} x_2(t)dt \right], \\
a_{m,i}^3 &= (\sqrt{2})^j \left[\int_{\frac{k}{2^j}-1+m}^{\frac{k+\frac{1}{2}}{2^j}-1+m} {}^{ABC}D_t^{\gamma} x_3(t)dt - \int_{\frac{k+\frac{1}{2}}{2^j}-1+m}^{\frac{k+1}{2^j}-1+m} {}^{ABC}D_t^{\gamma} x_3(t)dt \right], \qquad (30) \\
a_{m,i}^4 &= (\sqrt{2})^j \left[\int_{\frac{k}{2^j}-1+m}^{\frac{k+\frac{1}{2}}{2^j}-1+m} {}^{ABC}D_t^{\gamma} x_4(t)dt - \int_{\frac{k+\frac{1}{2}}{2^j}-1+m}^{\frac{k+1}{2^j}-1+m} {}^{ABC}D_t^{\gamma} x_4(t)dt \right].
\end{aligned}
$$

According to the mean value theorem for definite integrals, there exist $\tau_{x_1} \in \left(\frac{k}{2^j} - 1 + m,\ \frac{k+\frac{1}{2}}{2^j} - 1 + m\right)$ and $\widetilde{\tau}_{x_1} \in \left(\frac{k+\frac{1}{2}}{2^j} - 1 + m,\ \frac{k+1}{2^j} - 1 + m\right)$ such that

$$a_{m,i}^{1} = (\sqrt{2})^{j}\left(\frac{1}{2^{j+1}}\,{}^{ABC}D_t^{\gamma}x_1(\tau_{x_1})dt - \frac{1}{2^{j+1}}\,{}^{ABC}D_t^{\gamma}x_1(\widetilde{\tau}_{x_1})dt\right)$$
$$= 2^{-\left(\frac{j}{2}+1\right)}\left({}^{ABC}D_t^{\gamma}x_1(\tau_{x_1})dt - {}^{ABC}D_t^{\gamma}x_1(\widetilde{\tau}_{x_1})dt\right). \tag{31}$$

Using the definition of Atangana–Baleanu derivative leads to

$$\left|a_{m,i}^{1}\right| = 2^{-\left(\frac{j}{2}+1\right)}\left|{}^{ABC}D_t^{\gamma}x_1(\tau_{x_1})dt - {}^{ABC}D_t^{\gamma}x_1(\widetilde{\tau}_{x_1})dt\right|$$
$$=2^{-\left(\frac{j}{2}+1\right)}\frac{1}{W(\gamma)\Gamma(\gamma)}\left|\int_0^{\tau_{x_1}}(\tau_{x_1}-\xi)^{-\gamma}\frac{dx_1(\xi)}{d\xi}d\xi - \int_0^{\widetilde{\tau}_{x_1}}(\widetilde{\tau}_{x_1}-\xi)^{-\gamma}\frac{dx_1(\xi)}{d\xi}d\xi\right|.$$

Because $x_1 \in H^1[0,n)$, there exists $\mathscr{G}_{x_1} \in \mathbb{R}^+$ so that $\|\dot{x}(\xi)\| \leq \mathscr{G}_{x_1}$ for $\xi \in (0, \tau_{x_1})$ and $\xi \in (0, \widetilde{\tau}_{x_1})$. Therefore,

$$|a_{m,i}^{1}| \leq \mathscr{G}_{x_1} 2^{-\left(\frac{j}{2}+1\right)}\frac{1}{W(\gamma)\Gamma(\gamma)}\left|\int_0^{\tau_{x_1}}(\tau_{x_1}-\xi)^{-\gamma}d\xi - \int_0^{\widetilde{\tau}_{x_1}}(\widetilde{\tau}_{x_1}-\xi)^{-\gamma}d\xi\right|.$$

Keeping in mind that $0 \leq \gamma \leq 1$, $\tau_{x_1} \in \left(\frac{k}{2^j} - 1 + m,\ \frac{k+\frac{1}{2}}{2^j} - 1 + m\right)$ and $\widetilde{\tau}_{x_1} \in \left(\frac{k+\frac{1}{2}}{2^j} - 1 + m,\ \frac{k+1}{2^j} - 1 + m\right)$, we can integrate and simplify the previous expression and we have

$$|a_{m,i}^{1}| \leq \frac{\mathscr{G}_{x_1}2^{-\left(\frac{j}{2}+1\right)}}{(1-\gamma)\,W(\gamma)\Gamma(\gamma)}\left|{\tau_{x_1}}^{(1-\gamma)} - (\widetilde{\tau}_{x_1})^{(1-\gamma)}\right| \leq \frac{\mathscr{G}_{x_1}2^{-\left(\frac{j}{2}+1\right)}}{(1-\gamma)\,W(\gamma)\Gamma(\gamma)}2^{j(1-\gamma)}. \tag{32}$$

A similar reasoning leads to constants $\mathscr{G}_{x_2}$, $\mathscr{G}_{x_3}$, $\mathscr{G}_{x_4} \in \mathbb{R}^+$ such that

$$|a_{m,i}^{2}| \leq \frac{\mathscr{G}_{x_2}2^{-\left(\frac{j}{2}+1\right)}}{(1-\gamma)\,W(\gamma)\Gamma(\gamma)}2^{j(1-\gamma)}, \tag{33}$$

$$|a_{m,i}^{3}| \leq \frac{\mathscr{G}_{x_3}2^{-\left(\frac{j}{2}+1\right)}}{(1-\gamma)\,W(\gamma)\Gamma(\gamma)}2^{j(1-\gamma)}, \tag{34}$$

$$|a_{m,i}^{4}| \leq \frac{\mathscr{G}_{x_4}2^{-\left(\frac{j}{2}+1\right)}}{(1-\gamma)\,W(\gamma)\Gamma(\gamma)}2^{j(1-\gamma)}. \tag{35}$$

Now we set $\mathscr{G} = \max(\mathscr{G}_{x_1}, \mathscr{G}_{x_2}, \mathscr{G}_{x_3}, \mathscr{G}_{x_4})$ and substituting (32), (33) and (35) into (29) yields

$$
\begin{aligned}
\|{}^{ABC}D_t^{\gamma}w(t) - {}^{ABC}D_t^{\gamma}w^k(t)\|_2 &\le \left(\frac{3n\mathcal{G}^2}{4(W(\gamma)\Gamma(\gamma))^2(1-\gamma)^2} \sum_{m=1}^{n}\sum_{j=0}^{\infty}\sum_{i=2j}^{2j+1} \frac{2^{2j(1-\gamma)}}{2^j} \right)^{1/2}, \\
&\le \left(\frac{3n^2\mathcal{G}^2}{4(W(\gamma)\Gamma(\gamma))^2(1-\gamma)^2} \left(\frac{2^{2\gamma} - 2^{2\gamma}k^{(1-\gamma)}}{2^{2\gamma}-2} + \frac{2^{2\gamma} - 2^{2\gamma}k^{(2-2\gamma)}}{2^{2\gamma}-4} \right)\right)^{1/2}, \\
&\le \frac{n\mathcal{G}}{2(W(\gamma)\Gamma(\gamma))(1-\gamma)} \left(3\frac{2^{2\gamma}\left(1-k^{(1-\gamma)}\right)}{2^{2\gamma}-2} + 3\frac{2^{2\gamma}\left(1-k^{(2-2\gamma)}\right)}{2^{2\gamma}-4} \right)^{1/2}.
\end{aligned} \tag{36}
$$

and the proof is concluded.

If the functions x_1, x_2, x_3 and x_4 are not in $H^1[0, n)$ but are in $L^2[0, n)$, then we need additional statement for the Corollary 3.1, because $[0, n)$ is not a closed interval. That is, the functions x_1, x_2, x_3, x_4 and their derivatives of order one may not be bounded and may not attain their bounds on $[0, n)$. Hence we have proved the following

Corollary 3.1 *Let* $0 \le \gamma \le 1$, $x_1 \in L^2[0, n)$, $x_2 \in L^2[0, n)$, $x_3 \in L^2[0, n)$ *and* $x_4 \in L^2[0, n)$ *and let us assume that* $\dot{x_1}(t)$, $\dot{x_2}(t)$, $\dot{x_3}(t)$ *and* $\dot{x_4}(t)$ *are continuous and bounded on* $[0, n)$. *The following exact upper bound holds*

$$
\|{}^{ABC}D_t^{\gamma}w(t) - {}^{ABC}D_t^{\gamma}w^k(t)\|_2 \le \frac{\mathcal{G}}{\mathcal{G}_\gamma W(\gamma)\Gamma(\gamma)}, \tag{37}
$$

where ${}^{ABC}D_t^{\gamma}w(t)$ *are approximated by the functions* ${}^{ABC}D_t^{\gamma}w^k(t)$ *through Haar wavelet method,* $\mathcal{G} \in \mathbb{R}^+$ *and* $\mathcal{G}_\gamma = \frac{2(1-\gamma)}{2^\gamma n}\left(\frac{(3-3k^{(1-\gamma)})}{2^{2\gamma}-2} + \frac{(3-3k^{(2-2\gamma)})}{2^{2\gamma}-4}\right)^{-1/2}$.

4 Applications to Biochemical Systems of Four-Component Networks: Period-Doubling Bifurcations, Chaos and Interpretations

The model that was solved in the previous section is typical in biochemical systems, where many networks elements can be seen as various allosteric states of a polyfunctional macromolecule [2, 4, 5]. Recall that some biochemical systems involve structures, functions and interactions of biological macromolecules, such as proteins, nucleic acids, carbohydrates and lipids, which provide the structure of cells and performing several of the functions associated with life. Then, autocatalysis in the process itself necessitates contact (direct or indirect) between elements and therefore, implies straight collision of the macromolecular states. Simple examples are found in biological membranes where proteins, nucleic acids, carbohydrates and lipids are sometimes closely packed. The movement of those biological macromolecules is regulated and affected by the activity of a cofactor when it is involved in the concerned autocatalytic reaction network (see Figs. 1 and 2).

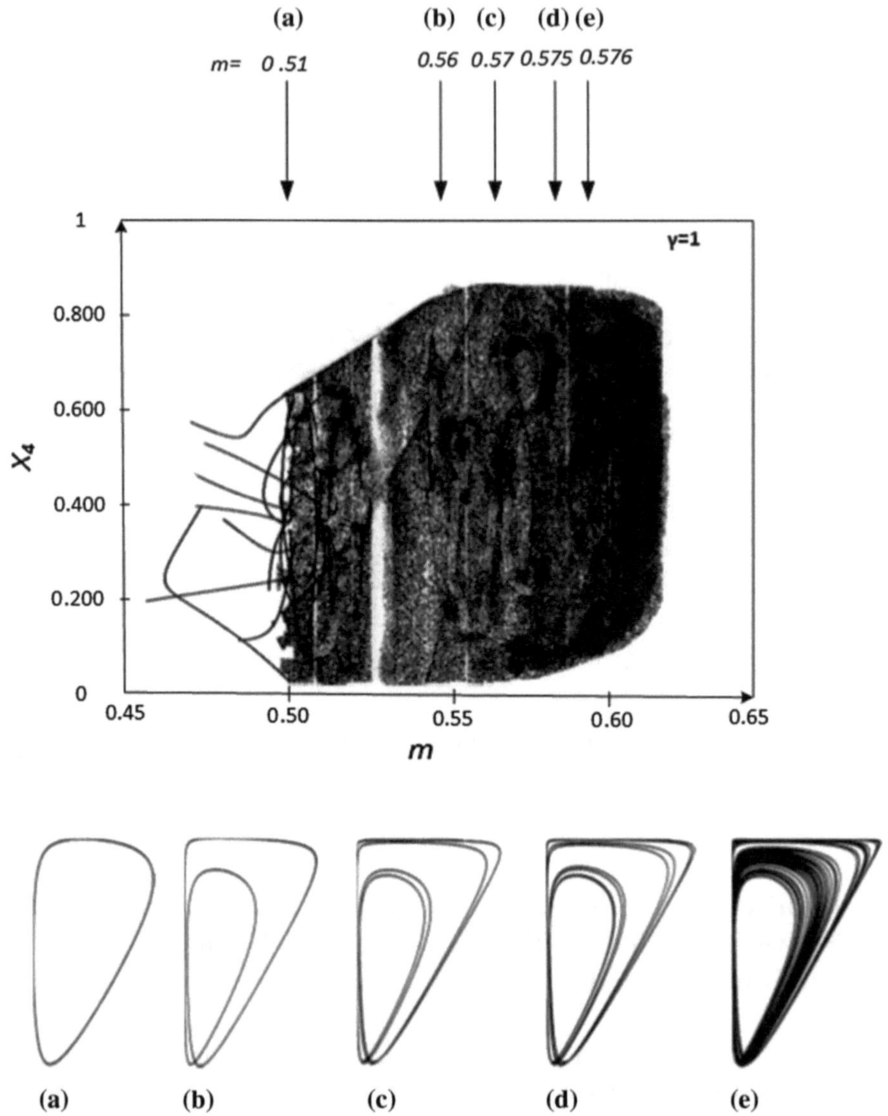

Fig. 1 Representation of cascade diagram for the model (18)–(19) in the standard case $\gamma = 1$ with $\nu_1 = 0.95$, $\nu_2 = \nu_3 = 0.01$, $\nu_4 = 0.02$, and the constants of the kinetic equations $F_1 = 12\ \text{s}^{-1}$, $F_2 = F_3 = F_4 = 1\ \text{s}^{-1}$. It shows limit-cycles followed by period-doubling bifurcations that culminate with chaos, (**a**–**e**), versus the variation of the total concentration of cofactor m. In **a**, the orbit for $m = 0.51$ has only two extremes that start to be doubled in **b** and it continues the dynamics by increasing irregularities

The simulations of the system (18) assumed to be subject to initial conditions (19) show a complex behavior in the dynamics of the network that seems to loose its conservative property. Those simulations are performed in the standard case ($\gamma = 1$)

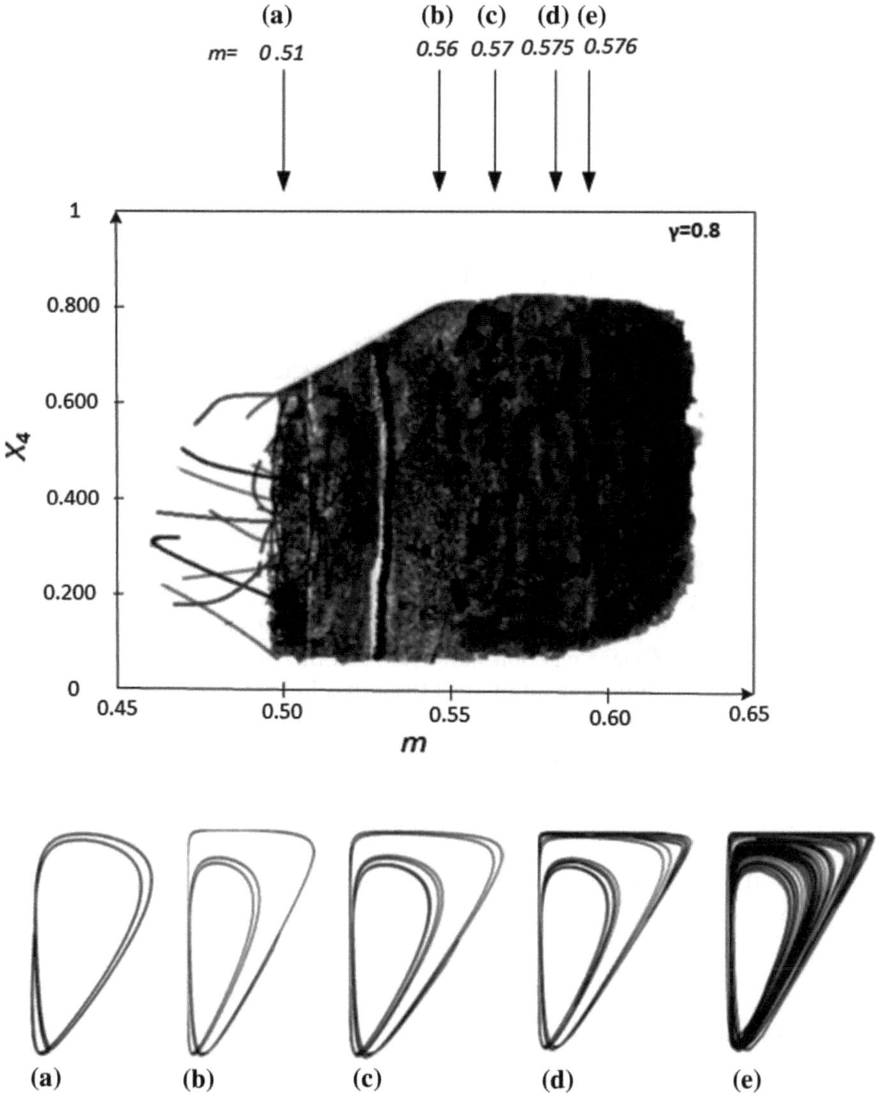

Fig. 2 Representation of cascade diagram for the model (18)–(19) in the pure fractional case $\gamma = 0.8$ with $\nu_1 = 0.95,\ \nu_2 = \nu_3 = 0.01,\ \nu_4 = 0.02$, and the constants of the kinetic equations $F_1 = 12\ \mathrm{s}^{-1},\ \ F_2 = F_3 = F_4 = 1\ \mathrm{s}^{-1}$. It shows again limit-cycles followed by period-doubling bifurcations that culminate with chaos, (**a–e**), versus the variation of the total concentration of cofactor m. However in **a**, in comparison to Fig. 1, the orbit for $m = 0.51$ does not have only two extremes, but already the double of that. Irregular behaviors are shown to appear earlier in this case

and in the total fractional case ($\gamma = 0.8$ and 0.6). All the Figs. 1, 2 and 3 exhibit different cascade diagrams that are characterized by limit-cycles followed by period-doubling bifurcations caused, in one hand by the presence of fuzzy steady states, and the other side by the variation of the concentration sum $m = z + x_2$ denoting the total concentration of cofactor. Note that x_2 represents the quantity of cofactor linked to element G_2. To obtain these results, the following control values were adopted: $\nu_1 = 0.95,\ \nu_2 = \nu_3 = 0.01,\ \nu_4 = 0.02$, and the constants of the kinetic equations $F_1 = 12\ \mathrm{s}^{-1},\quad F_2 = F_3 = F_4 = 1.00\ \mathrm{s}^{-1}$. In Fig. 1, the horizontal axis represents some values taken by the total concentration of cofactor m as it varies. For each value of m corresponds an orbit within the cascade diagram that can easily be traced. Some of those orbits are represented in Fig. 1a–e for $m = 0.51,\ 0.56,\ 0.57,\ 0.575,\ 0.576$ respectively. Those values of m also hold for Fig. 2a–e (and Fig. 3a–e) respectively. The resulting orbits are shown to be characterized by behaviors with complexity proportional to the total concentration of cofactor m. The level of complexity of the dynamics increases with m and ends up by becoming chaotic (Figs. 1e, 2e and 3e).

Another parameter comes into play here to make the difference between Figs. 1 and 2 (or Fig. 3). It is γ with chosen values $\gamma = 1.00$ for Fig. 1 and $\gamma = 0.80$ for Fig. 2 (or $\gamma = 0.60$ for Fig. 3). Indeed, the same value of m generates a more unpredictable behavior within orbits in Fig. 2 (or Fig. 3) compared to those in Fig. 1. For instance, limit-cycle in Fig. 1a has only two extremes while it is doubled for the corresponding case in Fig. 2a (or Fig. 3a). Hence, chaos is shown to appear earlier in the cascade diagram of Fig. 2 (or Fig. 3). This points out the very important role of the parameter γ in controlling the dynamics that emerge in biochemical systems of four-component reaction networks. Another illustration is depicted by Fig. 4 that shows the limit cycle of model (18)–(19) plotted for the parameter values $\gamma = 1,\ 0.95,\ 0.90,\ 0.85,\ 0.8$ with $\nu_1 = 0.95,\ \nu_2 = \nu_3 = 0.01,\ \nu_4 = 0.02$, and the constants of the kinetic equations $F_1 = 12\,\mathrm{s}^{-1},\ \ F_2 = F_3 = F_4 = 1.00\,\mathrm{s}^{-1}$. It shows again limit-cycle oscillations with chaotic behavior becoming more irregular as γ decreases. In Fig. 5, the same scenario and Fig. 4 occurs for $F_1 = 18\ \mathrm{s}^{-1},\quad F_2 = F_3 = F_4 = 2.5\ \mathrm{s}^{-1}$ and $\gamma = 1,\ 0.90,\ 0.80,\ 0.65,\ 0.50$ with $\nu_1 = 1.05,\ \nu_2 = \nu_3 = 0.01,\ \nu_4 = 0.02$. Here we observe limit-cycles with a chaotic jump and trajectories growing as γ decreases from 1 to 0.5.

5 Concluding Remarks

This paper, by means of Haar wavelets, solved a system modeling autocatalytic reaction networks, with a special focus on biochemical systems of four-component network. After showing the convergence of the method through error analysis, several numerical experiments revealed that the dynamics of the system is characterized by limit-cycles followed by period-doubling bifurcations that culminate with chaos according to the variation of the concentration of cofactor m. The level of complexity

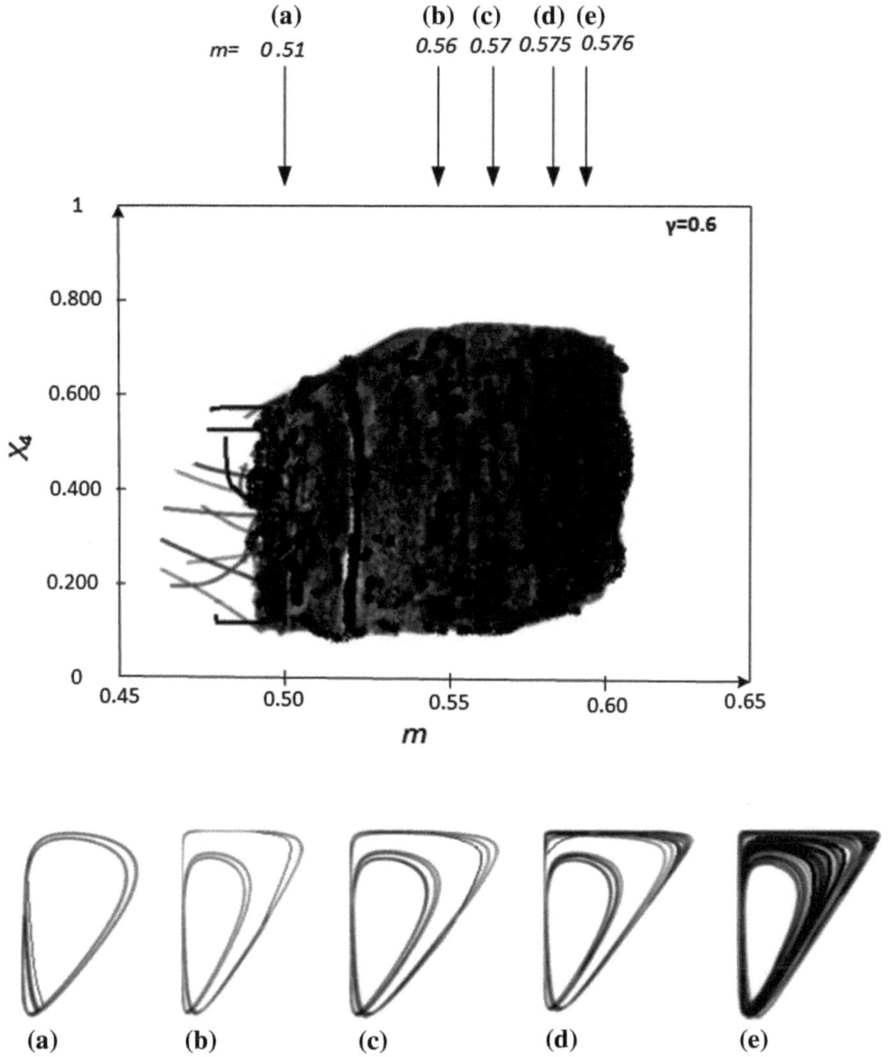

Fig. 3 Representation of cascade diagram for the model (18)–(19) in the pure fractional case $\gamma = 0.6$ with $\nu_1 = 0.95$, $\nu_2 = \nu_3 = 0.01$, $\nu_4 = 0.02$, and the constants of the kinetic equations $F_1 = 12\ \text{s}^{-1}$, $F_2 = F_3 = F_4 = 1\ \text{s}^{-1}$. It shows a dynamic closely similar to the one in Fig. 2

was found to be proportional to m. The parameter γ, that represents the derivative order of the model, also plays an important role in triggering and controlling the appearance of chaos. Moreover, the chaos is shown to appear earlier in the cascade diagram of the fractional case. Hence, fractional chaotical models generalizing the integer-order models in function of activity and total concentration, the parameter γ is a valuable and promising tool to regulate this chaos.

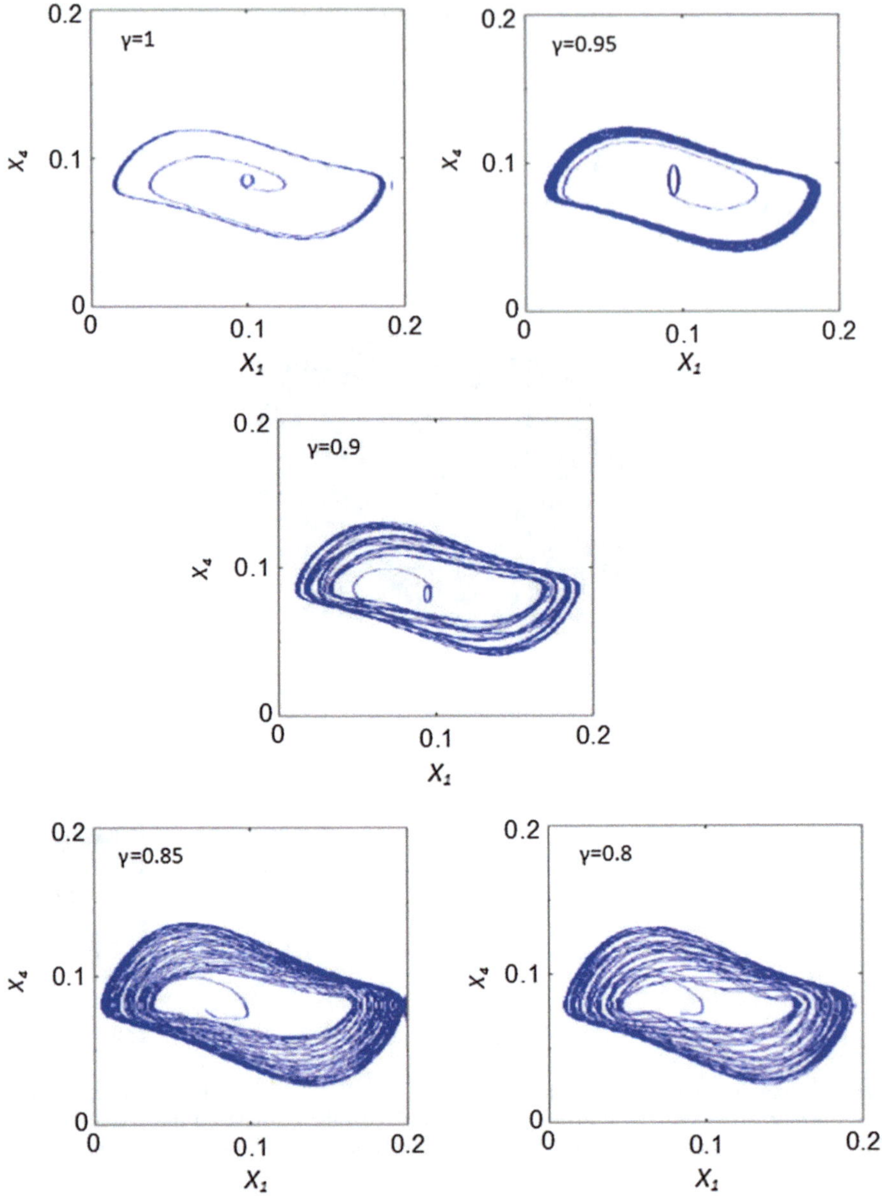

Fig. 4 The plot representing the projection of the limit cycle of model (18)–(19) for some values of parameter γ with $\nu_1 = 0.95,\ \nu_2 = \nu_3 = 0.01,\ \nu_4 = 0.02$, and the constants of the kinetic equations $F_1 = 12\,\mathrm{s}^{-1},\ F_2 = F_3 = F_4 = 1\,\mathrm{s}^{-1}$. It shows again limit-cycles with chaotic trajectories growing as γ decreases from 1 to 0.8

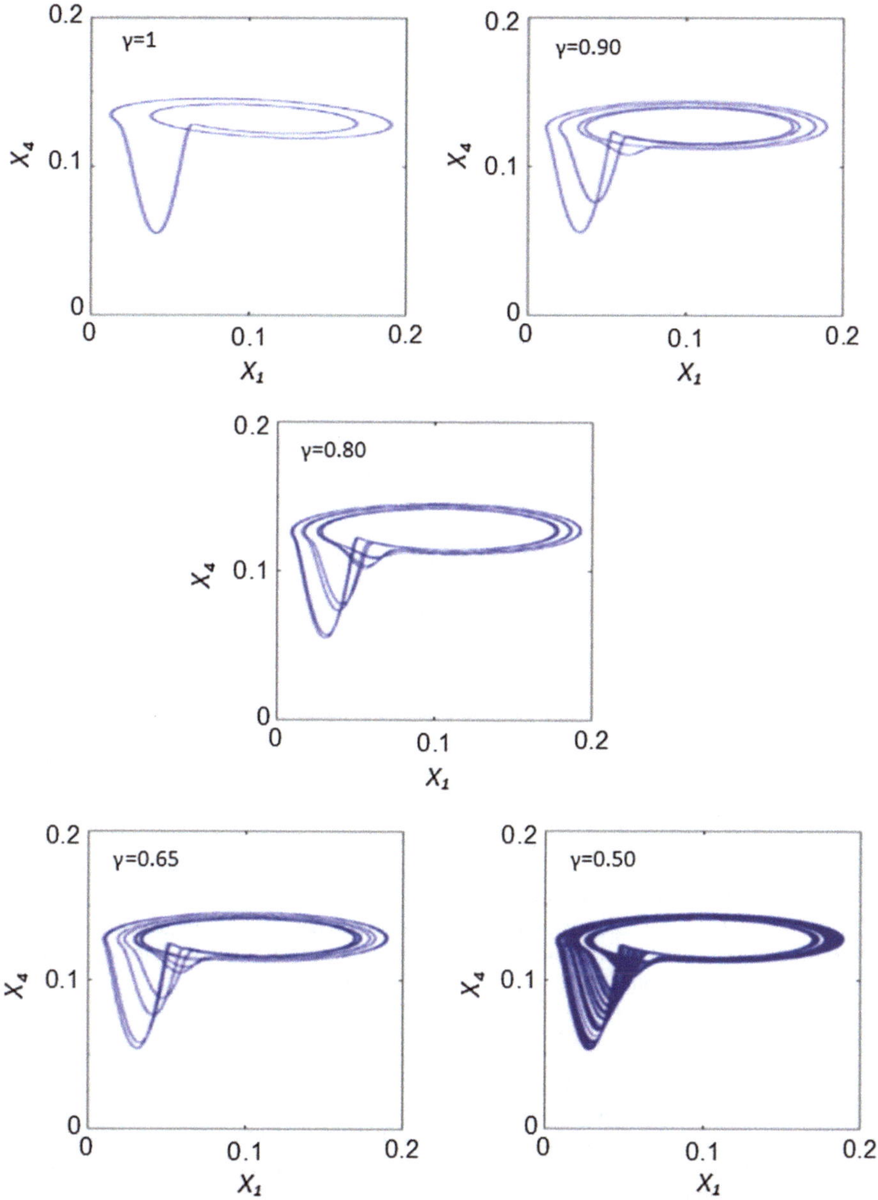

Fig. 5 The plot representing the projection of the limit cycle of model (18)–(19) for some values of parameter γ with $\nu_1 = 1.05,\ \nu_2 = \nu_3 = 0.01,\ \nu_4 = 0.02$, and the constants of the kinetic equations $F_1 = 18\ s^{-1},\quad F_2 = F_3 = F_4 = 2.5\ s^{-1}$. It shows again limit-cycles with a chaotic jump and trajectories growing as γ decreases from 1 to 0.5

Acknowledgements The work of EF Doungmo Goufo was partially supported by the grant No: 105932 from the National Research Foundation (NRF) of South Africa and a grant from the Simons Foundation. EF. Doungmo Goufo would also like to thank his post-doctoral fellows for their valuable comments.

References

1. Stadler, P.F., Schuster, P.: Mutation in autocatalytic reaction networks. J. Math. Biol. **30**(6), 597–631 (1992)
2. Gerdts, C.J., Sharoyan, D.E., Ismagilov, R.F.: A synthetic reaction network: chemical amplification using nonequilibrium autocatalytic reactions coupled in time. J. Am. Chem. Soc. **126**(20), 6327–6331 (2004)
3. Blackmond, D.G.: An examination of the role of autocatalytic cycles in the chemistry of proposed primordial reactions. Angewandte Chemie **121**(2), 392–396 (2009)
4. Di Cera, E., Phillipson, P.E., Wyman, J.: Limit-cycle oscillations and chaos in reaction networks subject to conservation of mass. Proc. Natl. Acad. Sci. **86**(1), 142–146 (1989)
5. Filisetti, A., Graudenzi, A., Serra, R., Villani, M., Füchslin, R.M., Packard, N., Kauffman, S.A., Poli, I.: A stochastic model of autocatalytic reaction networks. Theory Biosci. **131**(2), 85–93 (2012)
6. Bazhlekova, E.G.: Subordination principle for fractional evolution equations. Fract. Calc. Appl. Anal. **3**(3), 213–230 (2000)
7. Caputo, M.: Linear models of dissipation whose Q is almost frequency independent-II. Geophys. J. Int. **13**(5), 529–539 (1967)
8. Caputo, M., Fabrizio, M.: A new definition of fractional derivative without singular kernel. Progr. Fract. Differ. Appl. **1**(2), 1–13 (2015)
9. Losada, J., Nieto, J.J.: Properties of a new fractional derivative without singular kernel. Progr. Fract. Differ. Appl. **1**(2), 87–92 (2015)
10. Atangana, A., Koca, I.: Chaos in a simple nonlinear system with Atangana-Baleanu derivatives with fractional order. Chaos Solitons & Fractals **89**, 447–454 (2016)
11. Atangana, A., Baleanu, D.: New fractional derivatives with non-local and non-singular kernel. Therm. Sci. **20**(2), 763–769 (2016)
12. Gómez-Aguilar, J.F., Yépez-Martínez, H., Escobar-Jiménez, R.F., Astorga-Zaragoza, C.M., Reyes-Reyes, J.: Analytical and numerical solutions of electrical circuits described by fractional derivatives. Appl. Math. Modell. **40**(21–22), 9079–9094 (2016)
13. Gómez-Aguilar, J.F., Dumitru, B.: Fractional transmission line with losses. Zeitschrift für Naturforschung A **69**(10–11), 539–546 (2014)
14. Gómez-Aguilar, J.F., Torres, L., Yépez-Martínez, H., Baleanu, D., Reyes, J.M., Sosa, I.O.: Fractional Liénard type model of a pipeline within the fractional derivative without singular kernel. Adv. Differ. Equ. **2016**(1), 1–17 (2016)
15. Yépez-Martínez, H., Gómez-Aguilar, J.F., Sosa, I.O., Reyes, J.M., Torres-Jiménez, J.: The Feng's first integral method applied to the nonlinear mKdV space-time fractional partial differential equation. Rev. Mex. Fís **62**(4), 310–316 (2016)
16. Atangana, A., Gómez-Aguilar, J.F.: A new derivative with normal distribution kernel: theory, methods and applications. Phys. A: Stat. Mech. Appl. **476**, 1–14 (2017)
17. Gómez-Aguilar, J.F.: Analytical and numerical solutions of a nonlinear alcoholism model via variable-order fractional differential equations. Phys. A: Stat. Mech. Appl. **494**, 52–75 (2018)
18. Gómez-Aguilar, J.F.: Novel analytical solutions of the fractional Drude model. Optik **168**, 728–740 (2018)
19. Saad, K., Gómez, F.: Coupled reaction-diffusion waves in a chemical system via fractional derivatives in Liouville-Caputo sense. Rev. Mex. Fís **64**(5), 539–547 (2018)

20. Atangana, A., Gómez-Aguilar, J.F.: Decolonisation of fractional calculus rules: breaking commutativity and associativity to capture more natural phenomena. Eur. Phys. J. Plus **133**, 1–22 (2018)
21. Gómez-Aguilar JF. Irving-Mullineux oscillator via fractional derivatives with Mittag-Leffler kernel. Chaos Soliton. Fract. 2017; (95) 35: 1-7
22. Gómez-Aguilar, J.F., Escobar-Jiménez, R.F., López-López, M.G., Alvarado-Martínez, V.M.: Atangana-Baleanu fractional derivative applied to electromagnetic waves in dielectric media. J. Electromag. Waves Appl. **30**(15), 1937–1952 (2016)
23. Saad, K.M., Gómez-Aguilar, J.F.: Analysis of reaction diffusion system via a new fractional derivative with non-singular kernel. Phys. A: Stat. Mech. Appl. **509**, 703–716 (2018)
24. Ghanbari, B., Gómez-Aguilar, J.F.: Modeling the dynamics of nutrient-phytoplankton-zooplankton system with variable-order fractional derivatives. Chaos, Solitons & Fractals **116**, 114–120 (2018)
25. Coronel-Escamilla, A., Gómez-Aguilar, J.F., Torres, L., Escobar-Jiménez, R.F.: A numerical solution for a variable-order reaction-diffusion model by using fractional derivatives with non-local and non-singular kernel. Phys. A: Stat. Mech. Appl. **491**, 406–424 (2018)
26. Coronel-Escamilla, A., Gómez-Aguilar, J.F., Torres, L., Escobar-Jiménez, R.F., Valtierra-Rodríguez, M.: Synchronization of chaotic systems involving fractional operators of Liouville-Caputo type with variable-order. Phys. A: Stat. Mech. Appl. **487**, 1–21 (2017)
27. Doungmo Goufo, E.F., Atangana, A.: Analytical and numerical schemes for a derivative with filtering property and no singular kernel with applications to diffusion. Eur. Phys. J. Plus **131**(8), 1–26 (2016)
28. Doungmo Goufo EF. Chaotic processes using the two-parameter derivative with non-singular and nonlocal kernel: Basic theory and applications, Chaos: An Interdisciplinary Journal of Nonlinear Science, 2016; 26(8), 1-21
29. Miller, K.S., Ross, B.: An Introduction to the Fractional Calculus and Fractional Differential Equations. Wiley, New York (1993)
30. Kilbas, A.A., Srivastava, H.M., Trujillo, J.J.: Theory and Applications of Fractional Differential Equations. Elsevier Science Limited, Amsterdam (2006)
31. Gómez-Aguilar, J.F., Atangana, A.: New insight in fractional differentiation: power, exponential decay and Mittag-Leffler laws and applications. Eur. Phys. J. Plus **132**(1), 1–13 (2017)
32. Oldham, K.B., Spanier, J.: The Fractional Calculus: Theory and Applications of Differentiation and Integration to Arbitrary order. Academic Press Inc, Cambridge (1974)
33. Podlubny, I.: The Laplace transform method for Linear Differential Equations of the Fractional order (1997). arXiv.org/pdf/func-an/9710005
34. Morales-Delgado, V.F., Taneco-Hernández, M.A., Gómez-Aguilar, J.F.: On the solutions of fractional order of evolution equations. Eur. Phys. J. Plus **132**(1), 1–17 (2017)
35. Prüss, J.: Evolutionary Integral Equations and Applications. Birkhäuser, Basel (2013)
36. Brockmann, D., Hufnagel, L.: Front propagation in reaction-superdiffusion dynamics: taming Lévy flights with fluctuations. Phys. Rev. Lett. **98**(17), 178–301 (2007)
37. Doungmo Goufo, E.F.: Stability and convergence analysis of a variable-order replicator-mutator process in a moving medium. J. Theor. Biol. **403**, 178–187 (2016)
38. Doungmo Goufo, E.F.: Application of the Caputo-Fabrizio fractional derivative without singular kernel to Korteweg-de Vries-Bergers equation. Math. Modell. Anal. **21**(2), 188–198 (2016)
39. Pooseh, S., Rodrigues, H.S., Torres, D.F.: Fractional derivatives in dengue epidemics. AIP Conf. Proc. **1389**(1), 739–742 (2011)
40. Atangana, A.: Non validity of index law in fractional calculus: a fractional differential operator with markovian and non-markovian properties. Phys. A: Stat. Mech. Appl. **505**, 688–706 (2018)
41. Babolian, E., Shahsavaran, A.: Numerical solution of nonlinear fredholm integral equations of the second kind using haar wavelets. J. Comput. Appl. Math. **225**(1), 87–95 (2009)
42. Chen, Y., Yi, M., Yu, C.: Error analysis for numerical solution of fractional differential equation by Haar wavelets method. J. Comput. Sci. **3**(5), 367–373 (2012)
43. Lepik Ü, Hein H. Haar Wavelets: With Applications. Springer Science & Business Media, Berlin (2014)

44. Tonelli, L.: Sullintegrazione per parti. Rend. Acc. Naz. Lincei **5**(18), 246–253 (1909)
45. Fubini, G.: Opere scelte II. Cremonese, Roma (1958)

Development and Elaboration of a Compound Structure of Chaotic Attractors with Atangana–Baleanu Operator

Emile F. Doungmo Goufo

Abstract After the finding of the compound structure for standard chaotic attractors, the main concern was related to how to regulate such a fascinating dynamic. Hence, the question about the existence of a compound structure for chaotic attractors generated by fractional systems was raised. In this work, we investigate the existence of compound structure of a chaotic attractor generated from a Atangana–Baleanu fractional system where two cases are studied: the integer case and the fractional one. The model is first solved numerically thanks to the Haar Wavelets scheme whose convergence is proved via error analysis. Numerical simulations are performed and clearly reveal the existence of the desired compound structure in both cases and characterized by the generation of a left-attractor seen as the reflection of a right attractor through the mirror operation. Moreover, those two simple attractors can always be combined together to form the resulting chaotic attractor. The mechanism of forming those simple attractors is shown and leads to a bounded partial attractor. Furthermore, that same mechanism appears to be strongly dependent on two parameters, the model parameter u and the Atangana–Baleanu derivative with order α, important in controlling the systems. It is observed that, in the fractional case ($\alpha = 0.9$), the period-doubling bifurcations start at a higher value of u compared to the integer case ($\alpha = 1$).

Keywords Fractional calculus · Atangana–Baleanu fractional derivative · Chaotic attractors

E. F. Doungmo Goufo (✉)
Department of Mathematical Sciences, University of South Africa, Florida 0003, South Africa
e-mail: dgoufef@unisa.ac.za

J. F. Gómez et al. (eds.), *Fractional Derivatives with Mittag-Leffler Kernel*,
Studies in Systems, Decision and Control 194,
https://doi.org/10.1007/978-3-030-11662-0_10

1 Introduction

Most of chaotic attractors that govern three-dimensional systems were shown for the first time in the years 1960s. They include the chaotic Rössler system [1–3] and the Lorenz system [2–4] respectively given by

$$\begin{cases} \frac{d}{dt}x(t) = -y - z, \\ \frac{d}{dt}y(t) = x + \sigma y, \\ \frac{d}{dt}z(t) = \rho x + z(x - \delta), \end{cases} \tag{1}$$

and

$$\begin{cases} \frac{d}{dt}x(t) = \sigma(y - x), \\ \frac{d}{dt}y(t) = x(\rho - z) - y, \\ \frac{d}{dt}z(t) = xy - \delta z, \end{cases} \tag{2}$$

with $t > 0$, where $x = x(t),\ y = y(t),\ z = z(t)$ are the system state and $\sigma,\ \rho,\ \delta$ some real parameters to the systems. However, more recently some authors [5] managed to develop and propose an equivalent chaotic (Chen's) attractor which, nevertheless is not similar to the previous ones (1) and (2). The authors rather considered that new chaotic attractor as the dual to Lorenz attractor was generated by the system (2). The main reason for that is due to the observations that Lorenz system and the system that generates Chen's attractor meet similar conditions related to the constant matrix for the linear part of their systems [6, 7]. Consequently, an analysis linking Lorenz attractor and Chen attractor was successfully proposed by the same authors. In the same momentum, the Lorenz system was modified in [8] and shown to exhibit a compound structure. The same analysis was done for Chen's system in [9]. Therefore, existence of a compound nature of both Lorenz and Chen attractors is no longer a mystery and their chaotic character was successfully investigated. On the other hand, models of types (1), (2) or Chen's model were comprehensively analyzed using the fractional calculus, and their chaotic features were pointed out successfully [2, 10–13]. For example in [12], the Atangana–Baleanu fractional derivative was used together with Haar wavelet numerical scheme to investigate a chaotic four-wing attractors generated by a three-dimensional (3D) system similar to the types (1), (2) or Chen's model. A similar analysis was performed in [2] making use of the two-parameter derivative with non-local and non-singular kernel. The question that predominates the present circumstance is about the existence of a compound structure for fractional systems of types (1), (2) or Chen's model when these are expressed by means of fractional derivative like Atangana–Baleanu's. Moreover, if such structure exists, we would like to know how it is created, how it evolves and how can it be controlled. This paper aims to address those concerns and for that, we consider the following system:

$$\begin{cases} {}^{ABC}D_t^{\alpha}x(t) = a(x-y), \\ {}^{ABC}D_t^{\alpha}y(t) = -xz + cz + u, \\ {}^{ABC}D_t^{\alpha}z(t) = xz - bz, \end{cases} \tag{3}$$

where $a,\ b,\ c \in \mathbb{R},\ x = x(t),\ y = y(t),\ z = z(t)$ the system state and the term ${}^{ABC}D_t^{\alpha}$ representing the Atangana–Baleanu fractional derivative [14–29] defined as

$${}^{ABC}D_t^{\alpha}u(t) = \frac{W(\alpha)}{(1-\alpha)}\int_0^t \dot{u}(\tau)\,E_{\alpha} - \frac{\alpha(t-\tau)^{\alpha}}{1-\alpha}d\tau, \tag{4}$$

with the derivative order $\alpha \in [0, 1]$ and where $W(\gamma)$ defines a normalization function such that

$$W(0) = W(1) = 1. \tag{5}$$

It is important to recall the antiderivative that comes with Atangana–Baleanu derivative and used in this analysis. It is given by the following antiderivative (also called fractional integral of order α):

$$I^{\alpha}u(t) = \frac{\alpha}{W(\alpha)\Gamma(\alpha)}\int_0^t (t-\tau)^{\alpha-1}u(\tau)d\tau + \frac{1-\alpha}{W(\alpha)}u(t). \tag{6}$$

Because of the complex non-linear features of ${}^{ABC}D_t^{\alpha}$, model (3) is not always straightforward to solve. That is why we are going to address the solvability of the model by means of numerical approximations, namely the method of Haar wavelets [10–12, 30].

1.1 A Particular Case for $\alpha = 1$

Recall that when the Atangana–Baleanu derivative order reduces to $\alpha = 1$, the system (3) takes the particular form

$$\begin{cases} \frac{d}{dt}x(t) = a(x-y), \\ \frac{d}{dt}y(t) = -xz + cz + u, \\ \frac{d}{dt}z(t) = xz - bz. \end{cases} \tag{7}$$

This model with $u = 0$ is believed to be a kind of connection between the Lorenz and Chens attractors [19]. It is characterized by a chaotic attractor when the parameters a, b and c satisfy $a = 35,\ b = 3,\ c = 22$ as depicted in Fig. 1c. Now what happens for α taken randomly in [0, 1]! We answer this question by means of Haar wavelet method.

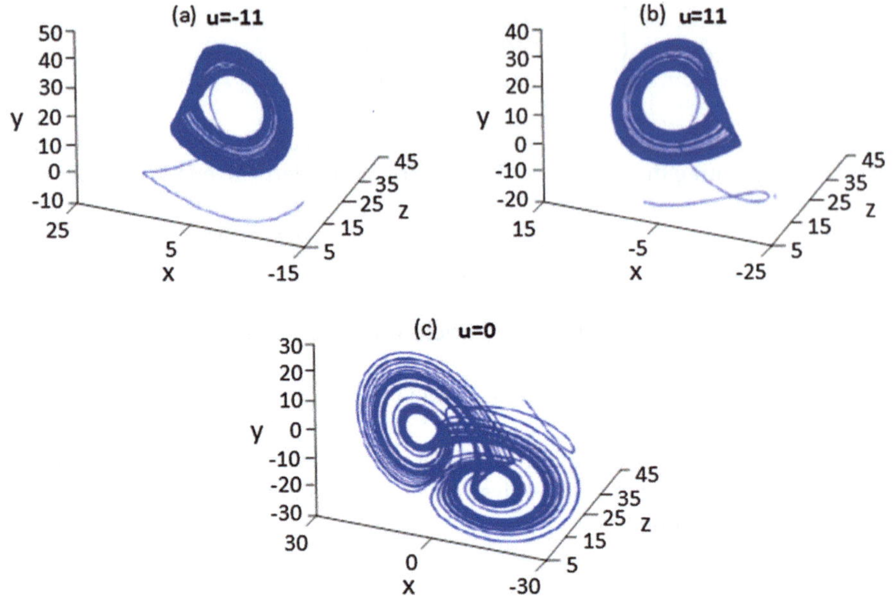

Fig. 1 Chaotic dynamic given by the model (3), with initial conditions $x(0) = 1,\ y(0) = 1,\ z(0) = 12$ and control parameters $a = 35,\ b = 3,\ c = 22,\ u = -11,\ u = 11$ and $u = 0$. The compound structure dynamic manifests itself with the chaotic attractor in **c** and its left-attractor in **a** and right-attractor in **b**

2 A Note on Haar wavelets

The following function that takes its values on the real set $\mathbb{R}$ and that reads as

$$\mathbb{H}(t) = \begin{cases} 1, & \text{for } t \in [0,\ 1/2); \\ -1, & \text{for } t \in [1/2,\ 1); \\ 0, & \text{elsewhere.} \end{cases} \tag{8}$$

is called the Haar wavelet [10, 11, 30]. Recall that each $q = 0, 1, 2, \ldots,$ can always take the form $q = 2^k + i$ with $k = 0, 1, 2, \ldots,$ and $i = 0, 1, 2, \ldots, 2^k - 1$. Considering for each $q = 0, 1, 2, \ldots$ the family

$$\mathbf{h}_q(t) = \begin{cases} 2^{\frac{k}{2}}\mathbb{H}(2^k t - i), & \text{for } q = 1, 2, \ldots, \\ 1, & \text{for } q = 0, \end{cases} \tag{9}$$

for $t \in [0, 1)$, then, the system $\{\mathbf{h}_{q(t)}\}_{q=0}^{\infty} = 0$ is well defined to form a complete orthonormal system in $L^2[0, 1)$ (see [11, 30]). Assuming $g \in C[0, 1)$, therefore, the series $\sum_{q=0}^{\infty} \langle g,\ \mathbf{h}_q \rangle \mathbf{h}_q$ is uniformly convergent to g with $\langle g,\ \mathbf{h}_q \rangle = \int_0^{\infty} g(t)\mathbf{h}_q(t)dt$ Thus, we are able to rewrite g as

$$g(t) = \sum_{q=0}^{\infty} \varphi_q \mathbf{h}_q(t),$$

with $\varphi_q = \langle g, \mathbf{h}_q \rangle$. We obviously get the approximated solution

$$g(t) \approx g_i(t) = \sum_{q=0}^{i-1} \varphi_q \mathbf{h}_q(t),$$

where $i \in \{2^k : \ k = 0, 1, 2, \ldots\}$.

With $m \in \mathbb{N}$, let us use the interval $[0, m)$, as the translation of the Haar function. It allows us to set

$$\mathbf{h}_{p,q}(t) = \mathbf{h}_q(t - p + 1) \quad p = 1, 2, \ldots, m \quad \text{and } q = 0, 1, 2, \ldots, \tag{10}$$

where the family $\mathbf{h}_q$ is given by (9) and which verify the same properties as $\mathbf{h}_{p,q}$ does. Thus, $\{\mathbf{h}_{p,q}(t)\}_{q=0}^{\infty}$, $p = 1, 2, \ldots, m$, also forms complete orthonormal system in $L^2[0, 1)$. This generates the basis functions defined as

$$\varphi_{p,q} = \langle g, \mathbf{h}_{p,q} \rangle = \int_0^{\infty} g(t)\mathbf{h}_{p,q}(t)dt,$$

and also called the Haar basis functions. Note that the family of Haar basis functions forms an orthonormal system. Hence, the solution $g \in L^2[0, m)$ can be written as the series

$$g(t) = \sum_{p=1}^{m} \sum_{q=0}^{\infty} \varphi_{p,q} \mathbf{h}_{p,q}(t), \tag{11}$$

where we have used Haar basis functions. Similarly, the following approximation can be performed:

$$g(t) \approx g_i(t) = \sum_{p=1}^{m} \sum_{q=0}^{i-1} \varphi_{p,q} \mathbf{h}_{p,q}(t), \tag{12}$$

with $i \in \{2^k : \ k = 0, 1, 2, \ldots\}$.

For the sake of notation simplicity in our analysis, the model (12) can take the compact form when we set

$$g(t) \approx g_i(t) = \mathbf{Q}_{mi\times 1}{}^T \mathbb{H}_{mi\times 1}, \tag{13}$$

with ${}^T\mathbf{Q}_{mi\times 1}$ the vector

$$\mathbf{Q}_{mi\times 1} = \left(\varphi_{1,0}, \ldots, \varphi_{1,i-1}, \varphi_{2,0}, \ldots, \varphi_{2,i-1}, \ldots, \varphi_{m,0}, \ldots, \varphi_{m,i-1}\right),$$

and ${}^{T}\mathbb{H}_{mi\times 1}$ the transpose vector of

$$\mathbb{H}_{mi\times 1} = \left(\mathbf{h}_{1,0}, \ldots, \mathbf{h}_{1,i-1}, \mathbf{h}_{2,0}, \ldots, \mathbf{h}_{2,i-1}, \ldots, \mathbf{h}_{m,0}\mathbf{h}_{m,i-1}\right).$$

3 Solvability of the Model (3)

In this section, the model (3) is solved numerically by means of Haar wavelets. Thus, we follow the same steps as [12]. We provide in the following lines some important steps of the solvability. The model reads as

$$\begin{cases} {}^{ABC}D_t^{\alpha} x(t) = a(x-y), \\ {}^{ABC}D_t^{\alpha} y(t) = -xz + cz + u, \\ {}^{ABC}D_t^{\alpha} z(t) = xz - bz, \end{cases} \tag{14}$$

and for the reasons of applicabilities, it is provided with to the initial conditions

$$X(0) = u(x),\ y(0) = v(y),\ z(0) = w(z). \tag{15}$$

System (14)–(15) can take the a compact form when we proceed as follows: Let g be the system state vector, then,

$$g(t) = (x(t),\ y(t),\ z(t)) \text{ and } g_0(x,y,z) = g(0) = (x(0), y(0), z(0)) = (u, v, w).$$

Let us set the matrix

$$\mathscr{P}(g(t),t) = \mathscr{P}(x(t),\ y(t),\ z(t),t) = (\mathscr{P}_1(g(t),t),\ \mathscr{P}_2(g(t),t),\ \mathscr{P}_3(g(t),t)),$$

$$(\mathscr{P}_1(x(t),\ y(t),\ z(t),t),\ \mathscr{P}_2(x(t),\ y(t),\ z(t),t),\ \mathscr{P}_3(x(t),\ y(t),\ z(t),t)),$$

where

$$(\mathscr{P}_1(g(t),t) = a(x-y),\ (\mathscr{P}_2(g(t),t) = -xz + cz + u,\ (\mathscr{P}_3(g(t),t) = xz - bz.$$

Thus, (14) takes the form

$${}^{ABC}D_t^{\alpha} g(t) = \mathscr{P}(g(t),t),$$

equivalently,

$$\begin{cases} {}^{ABC}D_t^{\alpha} x(t) = \mathscr{P}_1(g(t),t), \\ {}^{ABC}D_t^{\alpha} y(t) = \mathscr{P}_2(g(t),t), \\ {}^{ABC}D_t^{\alpha} z(t) = \mathscr{P}_3(g(t),t), \end{cases} \tag{16}$$

still provided with to the initial conditions $X(0) = u(x),\ y(0) = v(y),\ z(0) = w(z)$. Considering the haar wavelet scheme (13) used to approximate the Atangana–Baleanu fractional derivative as shown in (16), then,

$$\begin{aligned} {}^{ABC}D_t^\alpha x(t) &= \mathscr{P}_1(g(t), t) \approx^{ABC} D_t^\alpha x_i(t) = \mathbf{Q}^1_{mi\times 1}{}^T \mathbb{H}_{mi\times 1}, \\ {}^{ABC}D_t^\alpha y(t) &= \mathscr{P}_2(g(t), t) \approx^{ABC} D_t^\alpha y_i(t) = \mathbf{Q}^2_{mi\times 1}{}^T \mathbb{H}_{mi\times 1}, \\ {}^{ABC}D_t^\alpha z(t) &= \mathscr{P}_3(g(t), t) \approx^{ABC} D_t^\alpha z_i(t) = \mathbf{Q}^3_{mi\times 1}{}^T \mathbb{H}_{mi\times 1}. \end{aligned} \tag{17}$$

Applying the Atangana–Baleanu fractional integral (6) on both side of (17), we get

$$\begin{aligned} x(t) - u &\approx^{ABC} D_t^\alpha x_i(t) = \mathbf{Q}^1_{mi\times 1} J^\alpha_{mi\times mi}{}^T \mathbb{H}_{mi\times 1}, \\ y(t) - v &\approx^{ABC} D_t^\alpha y_i(t) = \mathbf{Q}^2_{mi\times 1} J^\alpha_{mi\times mi}{}^T \mathbb{H}_{mi\times 1}, \\ z(t) - w &\approx^{ABC} D_t^\alpha z_i(t) = \mathbf{Q}^3_{mi\times 1} J^\alpha_{mi\times mi}{}^T \mathbb{H}_{mi\times 1}, \end{aligned} \tag{18}$$

equivalently

$$\begin{aligned} x(t) \approx x_i(t) &= \mathbf{Q}^1_{mi\times 1} J^\alpha_{mi\times mi}{}^T \mathbb{H}_{mi\times 1} + u, \\ y(t) \approx y_i(t) &= \mathbf{Q}^2_{mi\times 1} J^\alpha_{mi\times mi}{}^T \mathbb{H}_{mi\times 1} + v, \\ z(t) \approx z_i(t) &= \mathbf{Q}^3_{mi\times 1} J^\alpha_{mi\times mi}{}^T \mathbb{H}_{mi\times 1} + w, \end{aligned} \tag{19}$$

Here the operator $J^\alpha_{mi\times mi}$ is the Haar wavelet fractional matrix. For more details about the Haar wavelets fractional operational matrix, the reader can consult the articles [10, 11] and other references therein. By means of Galerkin's techniques related to collocation points, the system (14)–(15) is solved by substituting (17) and (19) into (14). This substitution causes the following residual errors:

$$\begin{aligned} &\zeta_1\left(\nu^1, \nu^2, \nu^3, t\right) \\ &= \mathbf{Q}^1_{mi\times 1}{}^T \mathbb{H}_{mi\times 1} - \mathscr{P}_1\left(\mathbf{Q}^1_{mi\times 1} J^\alpha_{mi\times mi}{}^T \mathbb{H}_{mi\times 1}, \mathbf{Q}^2_{mi\times 1} J^\alpha_{mi\times mi}{}^T \mathbb{H}_{mi\times 1}, \mathbf{Q}^3_{mi\times 1} J^\alpha_{mi\times mi}{}^T \mathbb{H}_{mi\times 1}, t\right); \\ &\zeta_2\left(\nu^1, \nu^2, \nu^3, t\right) \\ &= \mathbf{Q}^2_{mi\times 1}{}^T \mathbb{H}_{mi\times 1} - \mathscr{P}_2\left(\mathbf{Q}^1_{mi\times 1} J^\alpha_{mi\times mi}{}^T \mathbb{H}_{mi\times 1}, \mathbf{Q}^2_{mi\times 1} J^\alpha_{mi\times mi}{}^T \mathbb{H}_{mi\times 1}, \mathbf{Q}^3_{mi\times 1} J^\alpha_{mi\times mi}{}^T \mathbb{H}_{mi\times 1}, t\right); \\ &\zeta_3\left(\nu^1, \nu^2, \nu^3, t\right) \\ &= \mathbf{Q}^3_{mi\times 1}{}^T \mathbb{H}_{mi\times 1} - \mathscr{P}_3\left(\mathbf{Q}^1_{mi\times 1} J^\alpha_{mi\times mi}{}^T \mathbb{H}_{mi\times 1}, \mathbf{Q}^2_{mi\times 1} J^\alpha_{mi\times mi}{}^T \mathbb{H}_{mi\times 1}, \mathbf{Q}^3_{mi\times 1} J^\alpha_{mi\times mi}{}^T \mathbb{H}_{mi\times 1}, t\right), \end{aligned} \tag{20}$$

where

$$\begin{aligned} \nu^1 &= \varphi^1_{1,0}, \ldots, \varphi^1_{1,i-1}, \ldots, \varphi^1_{m,0}, \ldots, \varphi^1_{m,i-1}; \\ \nu^2 &= \varphi^2_{1,0}, \ldots, \varphi^2_{1,i-1}, \ldots, \varphi^2_{m,0}, \ldots, \varphi^2_{m,i-1}; \\ \nu^3 &= \varphi^3_{1,0}, \ldots, \varphi^3_{1,i-1}, \ldots, \varphi^3_{m,0}, \ldots, \varphi^3_{m,i-1}, \end{aligned}$$

and where $\varphi^q_{\cdot,\cdot}$ are the components of $\mathbf{Q}^q_{\cdot\times\cdot}$.

When it is assumed that

$$\zeta_1\left(\nu^1, \nu^2, \nu^3, t_{p,q}\right) = 0;$$

$$\zeta_2\left(\nu^1, \nu^2, \nu^3, t_{p,q}\right) = 0;$$

$$\zeta_3\left(\nu^1, \nu^2, \nu^3, t_{p,q}\right) = 0,$$

then, we are left with a system of $3mi$ equations, with $3mi$ unknowns given by

$$\varphi^1_{1,0}, \ldots, \varphi^1_{1,i-1}, \ldots, \varphi^1_{m,0}, \ldots, \varphi^1_{m,i-1};$$

$$\varphi^2_{1,0}, \ldots, \varphi^2_{1,i-1}, \ldots, \varphi^2_{m,0}, \ldots, \varphi^2_{m,i-1};$$

$$\varphi^3_{1,0}, \ldots, \varphi^3_{1,i-1}, \ldots, \varphi^3_{m,0}, \ldots, \varphi^3_{m,i-1}.$$

Here the quantities

$$t_{p,q} = \frac{2q-1}{2i} + p - 1, \qquad p = 1, 2, \ldots, m; \qquad q = 1, 2, \ldots, i$$

give a $3mi$ number of collocation points. Finally, we easily get those unknowns and substitute them into (19) to have the desired approximated solution reading as

$$g(t) \approx \begin{pmatrix} x_i(t) \\ y_i(t) \\ z_i(t) \end{pmatrix}.$$

4 Convergence Results

To check the convergence of that numerical method, we can proceed by means of error analysis, which gives us the resulting exact error bounds used from applying the method for solving (14)–(15). Thus, because $g \in L^2[0, m)$, we take $x \in L^2[0, m)$, $y \in L^2[0, m)$ and $z \in L^2[0, m)$ and set

$$\|g\|_2 = \left(\|x\|^2_{L^2} + \|y\|^2_{L^2} + \|z\|^2_{L^2}\right)^{1/2}, \tag{21}$$

where

$$\|x\|_{L^2} = \left(\int_0^m |x(t)|^2 dt\right)^{1/2}, \quad \|y\|_{L^2} = \left(\int_0^m |y(t)|^2 dt\right)^{1/2}, \quad \|z\|_{L^2} = \left(\int_0^m |z(t)|^2 dt\right)^{1/2}.$$

Obviously $\|g\|_2$ defines a norm. From (12) and (13) we consider that, like (19), the Atangana–Baleanu derivative ${}^{ABC}D_t^{\alpha} g_i(t)$ approximates ${}^{ABC}D_t^{\alpha} g(t)$ so that

$$^{ABC}D_t^{\alpha} g(t) \approx {}^{ABC}D_t^{\alpha} g_i(t) = \sum_{p=1}^{m} \sum_{q=0}^{i-1} \varphi_{p,q} \mathbf{h}_{p,q}(t).$$

Where,

$$\begin{pmatrix} ^{ABC}D_t^{\alpha} x_i(t) \\ ^{ABC}D_t^{\alpha} y_i(t) \\ ^{ABC}D_t^{\alpha} y_i(t) \end{pmatrix} = {}^{ABC}D_t^{\alpha} g_i(t) = \sum_{p=1}^{m} \sum_{q=0}^{i-1} \varphi_{p,q} \mathbf{h}_{p,q}(t) = \begin{pmatrix} \sum_{p=1}^{m} \sum_{q=0}^{i-1} \varphi_{p,q}^{1} \mathbf{h}_{p,q}(t) \\ \sum_{p=1}^{m} \sum_{q=0}^{i-1} \varphi_{p,q}^{2} \mathbf{h}_{p,q}(t) \\ \sum_{p=1}^{m} \sum_{q=0}^{i-1} \varphi_{p,q}^{3} \mathbf{h}_{p,q}(t) \end{pmatrix},$$

where $i \in \{2^k : \ k = 0, 1, 2, \ldots\}$ and $\varphi_{p,q} = \langle {}^{ABC}D_t^{\alpha} g_i, h_{p,q} \rangle_m = \int_0^m {}^{ABC}D_t^{\alpha} g_i(t)$ $h_{p,q}(t)dt$,

$$\begin{aligned} \varphi_{p,q}^{1} &= \langle {}^{ABC}D_t^{\alpha} x_i, h_{p,q} \rangle_m = \int_0^m {}^{ABC}D_t^{\alpha} x_i(t) h_{p,q}(t)dt, \\ \varphi_{p,q}^{2} &= \langle {}^{ABC}D_t^{\alpha} y_i, h_{p,q} \rangle_m = \int_0^m {}^{ABC}D_t^{\alpha} y_i(t) h_{p,q}(t)dt, \qquad (22) \\ \varphi_{p,q}^{3} &= \langle {}^{ABC}D_t^{\alpha} z_i, h_{p,q} \rangle_m = \int_0^m {}^{ABC}D_t^{\alpha} z_i(t) h_{p,q}(t)dt, \end{aligned}$$

Hence,

$$\begin{aligned} ^{ABC}D_t^{\alpha} g(t) - {}^{ABC}D_t^{\alpha} g_i(t) &= \sum_{p=1}^{m} \sum_{q=i}^{\infty} \varphi_{p,q} \mathbf{h}_{p,q}(t), \\ &= \sum_{p=1}^{m} \sum_{q=2^k}^{\infty} \varphi_{p,q} \mathbf{h}_{p,q}(t), \quad k = 0, 1, 2, \ldots \\ &= \begin{pmatrix} \sum_{p=1}^{m} \sum_{q=2^k}^{\infty} \varphi_{p,q}^{1} \mathbf{h}_{p,q}(t) \\ \sum_{p=1}^{m} \sum_{q=2^k}^{\infty} \varphi_{p,q}^{2} \mathbf{h}_{p,q}(t) \\ \sum_{p=1}^{m} \sum_{q=2^k}^{\infty} \varphi_{p,q}^{3} \mathbf{h}_{p,q}(t) \end{pmatrix} \quad k = 0, 1, 2, \ldots \end{aligned} \qquad (23)$$

At this stage, we exploit the norm (21) to state the following convergence theorem, valid especially for the state functions $x, \ y$ and z taken from the Sobolev space $H^1[0, m)$.

Theorem 1 *Considering* $0 \le \alpha \le 1$ *and assuming* $x \in H^1[0, m), \ y \in H^1[0, m)$ *and* $z \in H^1[0, m)$. *If the Atangana–Baleanu fractional operator* ${}^{ABC}D_t^{\alpha} g_i(t)$ *approximate* ${}^{ABC}D_t^{\alpha} g(t)$ *via the Haar wavelet technique detailed here above, then the resulting exact upper bound is given by:*

$$\|^{ABC} D_t^\alpha g(t) -^{ABC} D_t^\alpha g_i(t)\|_2 \leq \frac{AA_\alpha^{-1}}{\Gamma(1-\alpha)}, \tag{24}$$

where A is a real positive number and $A_\alpha = \frac{(1-\alpha)}{2^{\alpha-1}m}\left(\frac{3-3i^{(1-\alpha)}}{2^{2\alpha}-2} + \frac{3-3i^{(2-2\alpha)}}{2^{2\alpha}-4}\right)^{-1/2}$.

Proof The proof follows from [11, Theorem 6.1], where we apply the exact same steps.

For the functions x, y and z that do not belong to $H^1[0.m)$, then the only conditions $x \in L^2[0.m)$, $y \in L^2[0.m)$ and $z \in L^2[0.m)$ are no longer enough to state the previous theorem. The reason is that the interval $[0.m)$ is not closed and the state functions x, y, z and their first order derivatives may not be bounded nor attain their bounds on $[0.m)$. Hence the previous theorem, in this case adapts as follow [12, Corollary 6.1]

Corollary 4.1 *Considering* $0 \leq \alpha \leq 1$, $x \in L^2[0, m)$, $y \in L^2[0, m)$, $z \in L^2[0, m)$ *and assuming that* x', y' *and* z' *are continuous and bounded on* $[0, m)$. *If the Atangana–Baleanu fractional operator* ${}^{ABC}D_t^\alpha g_i(t)$ *approximates* ${}^{ABC}D_t^\alpha g(t)$ *via the Haar wavelet technique detailed here above, then the resulting exact upper bound is given by:*

$$\|^{ABC} D_t^\alpha g(t) -^{ABC} D_t^\alpha g_i(t)\|_2 \leq \frac{AA_\alpha^{-1}}{\Gamma(1-\alpha)}, \tag{25}$$

where A is a real positive number and $A_\alpha = \frac{(1-\alpha)}{2^{\alpha-1}m}\left(\frac{3-3i^{(1-\alpha)}}{2^{2\alpha}-2} + \frac{3-3i^{(2-2\alpha)}}{2^{2\alpha}-4}\right)^{-1/2}$.

5 Simulations and Discussion on Mechanism of Forming the “Fractional” Attractors

With the peace of mind that we can neglect the error made by applying the Haar wavelets scheme as described here above, we are now studying the mechanism that leads to the forming of the chaotic attractor observed in the dynamic of the system (3). We proceed by numerical simulations in order to better access the resulting compound structure and point out the topology that sustains such a compound structure. As expected, Figs. 1 and 3 show chaotic dynamics and prove the existence of a compound structure of chaotic attractors for both the standard integer case (where $\alpha = 1$) and fractional case (where $\alpha = 0.9$). Indeed, we observe in both cases that it possible to obtain the compound structure of the resulting chaotic attractor simply by combining together two simple attractors: The one on the left (Fig. 1a or 3a) and the one on the right (Fig. 1b or 3 b). In that structure, the left-attractor is considered as the reflection of the right-attractor through the mirror operation. This result is clearly depicted by the pairs (Fig. 1a, b) and (Fig. 3a, b) which, as u continues to vary, will end-up with a full chaotic attractor as shown in Figs. 1c and 3c respectively.

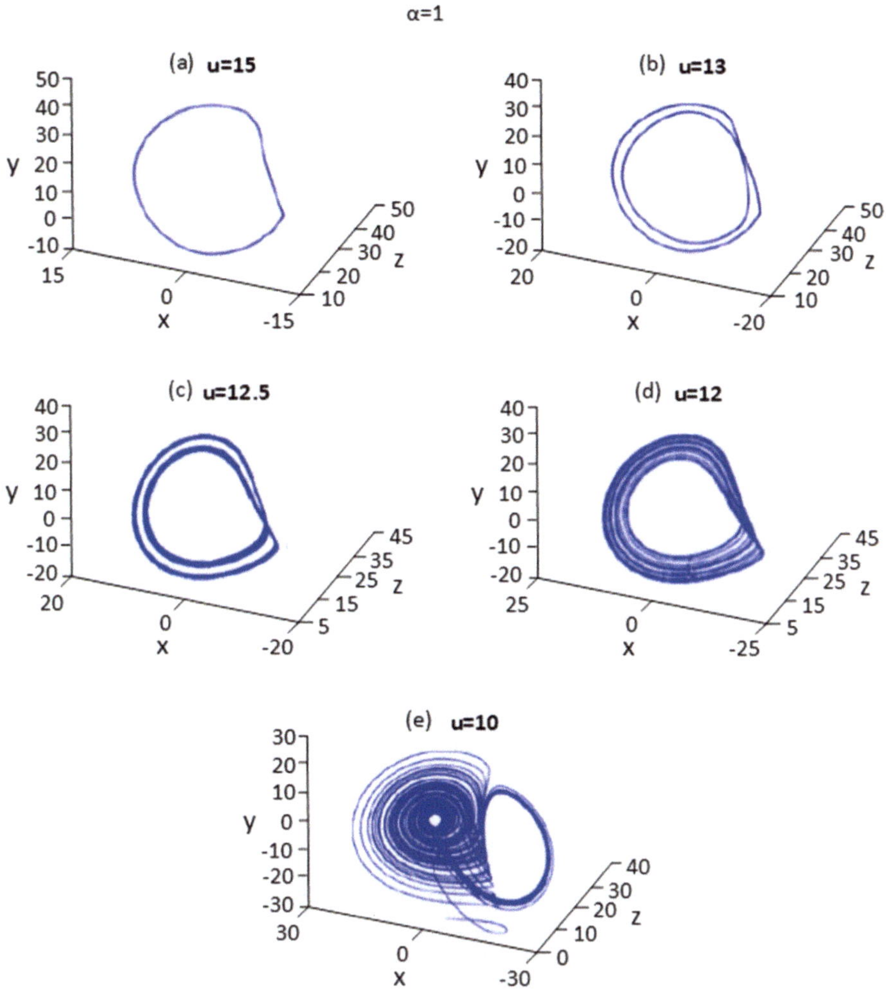

Fig. 2 Phase portraits of the model (3), figures showing the mechanism of forming the simple attractor and the partial attractor. The initial conditions used are $x(0) = 1,\ y(0) = 1,\ z(0) = 12$ and the control parameters are $a = 35,\ b = 3,\ c = 22,\ u = 15,\ u = 13,\ u = 12.5,\ u = 12$ and $u = 10$

Figures 2 and 4 show mechanism of Forming those attractors with the standard integer case of $\alpha = 1$ and the fractional case of $\alpha = 0.9$. The mechanism in the case $\alpha = 1$ starts by a limit cycle when $u = 15$ (Fig. 2a), followed by period-doubling bifurcations when u decreases a bit as $u = 13,\ 12.5,\ 12$ (Fig. 2a–c respectively). The mechanism pursues its dynamics as u continues to decrease by the creation of a right-attractor for $u = 11$ (Fig. 1b) and the creation of a partial attractor for $u = 10$. Moreover, this partial attractor is bounded as clearly depicted in Fig. 2e. When the parameter u hits 0, we observe a complete attractor as depicted in Fig. 1c.

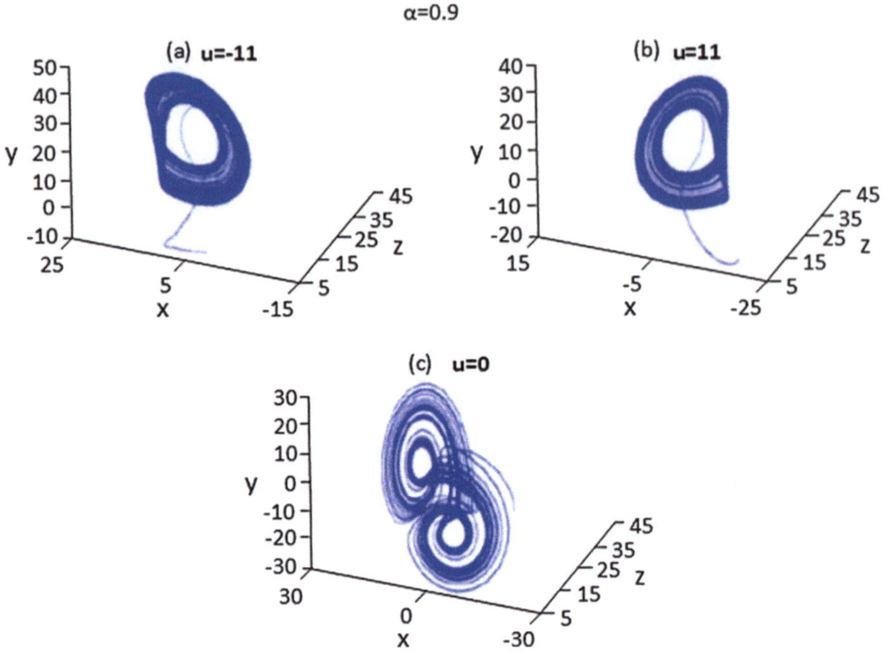

Fig. 3 The fractional chaotic dynamic given by the model (3), with initial conditions $x(0) = 1,\ y(0) = 1,\ z(0) = 12$ and control parameters $a = 35,\ b = 3,\ c = 22,\ u = -11,\ u = 11$ and $u = 0$. It reveals existence of a compound structure of the chaotic attractor in **c** with the left-attractor in **a** and right-attractor in **b**. Those two simple attractors can be combined together to generate the desired compound structure

The same type of dynamics is globally observed for the fractional case where $\alpha = 0.9$. The mechanism of forming the 'fractional' attractors is revealed to be similar to the one of the standard integer case. Such mechanism is depicted by Figs. 3 and 4. We used the word "globally" because there is a slight difference here and it is due to the fact that the period-doubling bifurcations start at a higher value of u ($u = 15$) compared to $u = 13$ of the integer case $\alpha = 1$. Hence, the parameter α when combined with a decreasing u, appears to be an early trigger of chaotic behavior for systems of type (3). Note that we have chosen some discrete values of both parameters u and α to sustain our results here. The dynamics the the system (3) varies continuously as u and α change. Therefore, the role played by both parameters u and α is very important in regulating and controlling the system. The complete dynamics are summarized by Table 1 (for $\alpha = 1$) and Table 2 (for $\alpha = 0.9$). They correspond to the results in [31], proven for the integer case, but nevertheless, support our statement by displaying various and different dynamics in both cases. The intervals of dynamic applicability for u is given by those two tables.

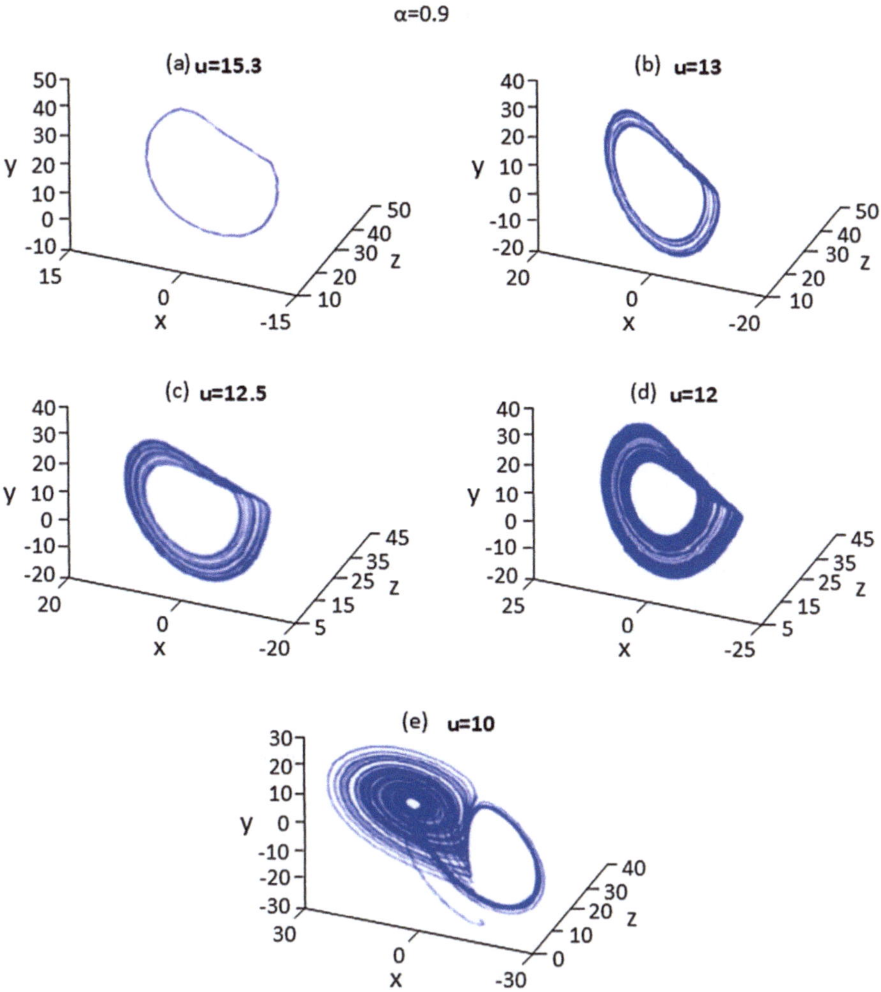

Fig. 4 Phase portraits of the model (3), and showing the mechanism of forming the "fractional" simple attractor and the partial attractor. The initial conditions used are $x(0) = 1,\ y(0) = 1,\ z(0) = 12$ and the control parameters are $a = 35,\ b = 3,\ c = 22,\ u = 15.3,\ u = 13,\ u = 12, 5,\ u = 12$ and $u = 10$. Here, the period-doubling bifurcations start at $u = 15$, higher value compared to that of Fig. 2. Therefore the parameter α, when combined with a decreasing u, appears to be a early trigger of chaotic behavior for our system

6 Concluding Remarks

We have investigated the existence of compound structure of a chaotic attractor issued from a Atangana–Baleanu fractional dynamical system. We focussed on studying two particular cases: the integer case where the Atangana–Baleanu derivative order was set at $\alpha = 1$) and the fractional one with $\alpha = 0.9$. The model has first been solved

Table 1 Types of dynamics due to the controller parameters u and α.

$\alpha = 1$	
Intervals applicability for u	Resulting dynamics for system (3)
$(-\infty, -48] \cup [48, +\infty)$	Convergence to a point
$[-47, -14.8] \cup [14.8, 47]$	Existence of a limit cycle (Fig. 2a)
$[-14.7, -11.5) \cup (11.5, 14.7]$	Formation of period-doubling bifurcations (Fig. 2b–d)
$[-11.5, -10.9] \cup [10.9, 11.5]$	Formation of left- or right-attractor (Fig. 1a or b)
$[-10.8, -3] \cup [3, 10.8]$	Existence of a partial attractor, shown to be bounded (Fig. 2e)
$[-2; 2]$	Existence of a full attractor (Fig. 1c)

Table 2 Types of dynamics due to the controller parameters u and α.

$\alpha = 0.9$	
Intervals applicability for u	Resulting dynamics for system (3)
$(-\infty, -48] \cup [48, +\infty)$	Convergence to a point
$[-47, -15.3] \cup [15.3, 47]$	Existence of a limit cycle (Fig. 4a)
$[-15.2, -11.8) \cup (11.8, 15.2]$	Formation of period-doubling bifurcations (Fig. 4b–d)
$[-11.7, -10.9] \cup [10.9, 11.7]$	Formation of left- or right-attractor (Fig. 3a or b)
$[-10.8, -3] \cup [3, 10.8]$	Existence of a partial attractor, shown to be bounded (Fig. 4e)
$[-2; 2]$	Existence of a full attractor (Fig. 3c)

numerically by means of Haar Wavelets method. Its convergence was proved via error analysis before performing some numerical simulations. Those simulations clearly revealed the existence of the compound structure in both cases and characterized by the generation of a left attractor seen as the reflection of a right-attractor through the mirror operation. We have observed in both cases the possibility of obtaining the compound structure of the resulting chaotic attractor by combining together the two simple attractors (left- and right-attractor). The mechanism of forming those two simple attractors has been given in detail and shown to be strongly dependent on the parameter α, as well as the model parameter u. This gives to those parameters the important status of system's controller or regulator. It has also been observed that, in the fractional case ($\alpha = 0.9$), the period-doubling bifurcations start at a higher value of u compared to the integer case ($\alpha = 1$). Therefore the parameter α, when combined with a decreasing model parameter u, appears to be an early trigger of a chaotic behavior for certain type of dynamical systems like the one analyzed in this paper. This is a great observation that may be substantial in future related works.

References

1. Rössler, O.E.: An equation for continuous chaos. Phys. Lett. A **57**(5), 397–398 (1976)
2. Doungmo Goufo, E.F. Chaotic processes using the two-parameter derivative with non-singular and nonlocal kernel: basic theory and applications. Chaos: Interdiscip. J. Nonlinear Sci. **26**(8), 1–21 (2016)
3. Wang, Z., Sun, Y., Van Wyk, B.J., Qi, G., Van Wyk, M.A.: A 3-D four-wing attractor and its analysis. Braz. J. Phys. **39**(3), 547–553 (2009)
4. Lorenz, E.N.: Deterministic nonperiodic flow. J. Atmos. Sci. **20**(2), 130–141 (1963)
5. Chen, G., Ueta, T.: Yet another chaotic attractor. Int. J. Bifurc. Chaos **9**(07), 1465–1466 (1999)
6. Vanecek, A., Celikovsky, S.C.: Control Systems: from Linear Analysis to Synthesis of Chaos. Prentice Hall International (UK) Ltd, Prentice (1996)
7. Lü, J., Chen, G.: A new chaotic attractor coined. Int. J. Bifurc. Chaos **12**(03), 659–661 (2002)
8. Özoguz, S., Elwakil, A.S., Kennedy, M.: Experimental verification of the butterfly attractor in a modified lorenz system. Int. J. Bifurc. Chaos **12**(07), 1627–1632 (2002)
9. Lü, J., Zhou, T., Chen, G., Zhang, S.: The compound structure of chen's attractor. Int. J. Bifurc. Chaos **12**(04), 855–858 (2002)
10. Babolian, E., Shahsavaran, A.: Numerical solution of nonlinear fredholm integral equations of the second kind using haar wavelets. J. Comput. Appl. Math. **225**(1), 87–95 (2009)
11. Chen, Y., Yi, M., Yu, C.: Error analysis for numerical solution of fractional differential equation by Haar wavelets method. J. Comput. Sci. **3**(5), 367–373 (2012)
12. Doungmo Goufo, E.F.: Solvability of chaotic fractional systems with 3D four-scroll attractors. Chaos Solitons & Fractals **104**, 443–451 (2017)
13. Doungmo Goufo, E.F., Nieto, J.J.: Attractors for fractional differential problems of transition to turbulent flows. J. Comput. Appl. Math. **339**, 329–342 (2017)
14. Atangana, A., Koca, I.: Chaos in a simple nonlinear system with Atangana-Baleanu derivatives with fractional order. Chaos Solitons & Fractals **89**, 447–454 (2016)
15. Atangana, A., Baleanu, D.: New fractional derivatives with non-local and non-singular kernel. Therm. Sci. **20**(2), 763–769 (2016)
16. Gómez-Aguilar, J.F.: Analytical and Numerical solutions of a nonlinear alcoholism model via variable-order fractional differential equations. Phys. A: Stat. Mech. Appl. **494**, 52–75 (2018)
17. Atangana, A., Gómez-Aguilar, J.F.: Decolonisation of fractional calculus rules: breaking commutativity and associativity to capture more natural phenomena. Eur. Phys. J. Plus **133**, 1–22 (2018)
18. Morales-Delgado, V.F., Taneco-Hernández, M.A., Gómez-Aguilar, J.F.: On the solutions of fractional order of evolution equations. Eur. Phys. J. Plus **132**(1), 1–17 (2017)
19. Gómez-Aguilar, J.F., Escobar-Jiménez, R.F., López-López, M.G., Alvarado-Martínez, V.M.: Atangana-Baleanu fractional derivative applied to electromagnetic waves in dielectric media. J. Electromag. Waves Appl. **30**(15), 1937–1952 (2016)
20. Ghanbari, B., Gómez-Aguilar, J.F.: Modeling the dynamics of nutrient-phytoplankton-zooplankton system with variable-order fractional derivatives. Chaos, Solitons & Fractals **116**, 114–120 (2018)
21. Gómez-Aguilar, J.F., Dumitru, B.: Fractional transmission line with losses. Zeitschrift für Naturforschung A **69**(10–11), 539–546 (2014)
22. Gómez-Aguilar, J.F., Torres, L., Yépez-Martínez, H., Baleanu, D., Reyes, J.M., Sosa, I.O.: Fractional Liénard type model of a pipeline within the fractional derivative without singular kernel. Adv. Differ. Equ. **2016**(1), 1–17 (2016)
23. Gómez-Aguilar, J.F.: Novel analytical solutions of the fractional Drude model. Optik **168**, 728–740 (2018)
24. Yépez-Martínez, H., Gómez-Aguilar, J.F., Sosa, I.O., Reyes, J.M., Torres-Jiménez, J.: The Feng's first integral method applied to the nonlinear mKdV space-time fractional partial differential equation. Rev. Mex. Fís **62**(4), 310–316 (2016)

25. Cuahutenango-Barro, B., Taneco-Hernández, M.A., Gómez-Aguilar, J.F.: On the solutions of fractional-time wave equation with memory effect involving operators with regular kernel. Chaos, Solitons & Fractals **115**, 283–299 (2018)
26. Atangana, A., Gómez-Aguilar, J.F.: Fractional derivatives with no-index law property: Application to chaos and statistics. Chaos, Solitons & Fractals **114**, 516–535 (2018)
27. Saad, K.M., Gómez-Aguilar, J.F.: Analysis of reaction diffusion system via a new fractional derivative with non-singular kernel. Phys. A: Stat. Mech. Appl. **509**, 703–716 (2018)
28. Atangana, A.: Non validity of index law in fractional calculus: a fractional differential operator with markovian and non-markovian properties. Phys. A: Stat. Mech. Appl. **505**, 688–706 (2018)
29. Doungmo Goufo, E.F., Atangana, A.: Analytical and numerical schemes for a derivative with filtering property and no singular kernel with applications to diffusion. Eur. Phys. J. Plus **131**(8), 1–26 (2016)
30. Lepik, Ü., Hein, H.: Haar Wavelets: With Applications. Springer Science & Business Media, Berlin (2014)
31. Lü, J., Chen, G., Zhang, S.: Dynamical analysis of a new chaotic attractor. Int. J. Bifurc. Chaos **12**(05), 1001–1015 (2002)

On the Atangana–Baleanu Derivative and Its Relation to the Fading Memory Concept: The Diffusion Equation Formulation

Jordan Hristov

Abstract The constructions of physically adequate forms of the diffusion equation with implementation of the Atangana–Baleanu derivative with Mittag-Leffler exponential kernel have been discussed. The specific form of the corresponding Atangana–Baleanu integral relates it directly to the fading memory concept, following the Boltzmann linear superposition principle with the standard Riemann-Liouville integral as the time-fading term. This approach relates the new fractional operators with non-singular kernel to the classical Riemann-Liouville integral. Using the concept of the fading memory and the specific form of the Atangana–Baleanu integral three forms of the diffusion equation have been investigated. The adequate definition of the flux to gradient relationship has been the main focus of the study resulting in two physically adequate formulations of the diffusion equation. The direct (formalistic) fractionalization of the classical diffusion equation results in physically inadequate relationships.

Keywords Fractional calculus · Atangana–Baleanu fractional derivative · Diffusion equation

1 Introduction

1.1 Fractional Diffusion Equation with Non-singular Kernel

This study addresses the construction of the fractional diffusion equation when the time-fractional derivative is with non-singular kernel based on the Mittag-Leffler function [2, 5, 12, 14, 16–19, 43]. The main idea relates the associated fractional

J. Hristov (✉)
Department of Chemical Engineering, University of Chemical Technology and Metallurgy, 8 Kliment Ohridsky, Blvd., 1756 Sofia, Bulgaria

J. F. Gómez et al. (eds.), *Fractional Derivatives with Mittag-Leffler Kernel*,
Studies in Systems, Decision and Control 194,
https://doi.org/10.1007/978-3-030-11662-0_11

integral to the concept of fading memory, modelling the flux relaxation in simple materials [10, 11, 21, 22, 39–41]. The study demonstrates that starting from a simple rate equation defined irrespectively of the type of the fractional derivative and the associated fractional integral; it is possible to obtain formulations of the diffusion equation. The development of the idea, expressed above, uses the new results of Baleanu and Fernandez [3, 5, 14] and the classic formulations of heat diffusion in materials with memory [11, 21]. Recently, Tateishi et al. [46] investigated the effects of different fractional kernels $K(t)$ in the right hand of the diffusion equation formulated as

$$\frac{\partial f}{\partial t} = D_0 F_t^{\alpha}\left(\frac{\partial^2 f(x,t)}{\partial x^2}\right), \quad F_t^{\alpha} = \frac{\partial}{\partial t}\int_0^t f(x,t)K(t-\tau)d\tau. \tag{1}$$

This diffusion equation focuses on the formulation of the functional in the right-hand side relating the flux to the space derivative. The latter is especially investigated in the present study when the time derivative is the Atangana–Baleanu time-fractional derivative [2, 5, 14] and the flux is related to the fading memory concept.

1.2 Problem Formulation

Consider the following general *constitutive formulation* of the continuity Eq. (2)

$$\frac{\partial^{\alpha} f(x,t)}{\partial t^{\alpha}} = -\nabla j(x,t), \tag{2}$$

that should result in a diffusion equation generally formulated as (3)

$$\frac{\partial^{\alpha} f(x,t)}{\partial t^{\alpha}} = \Phi\left[\frac{\partial^2}{\partial x^2} f(x,t)\right]. \tag{3}$$

The right-hand side in (3) is a function of the gradient $\partial f/\partial x$, while $\partial^{\alpha} f/\partial t^{\alpha}$ is any time-fractional derivative of causal function $f(x,t)$. In general, the functional Φ should be dependent on the gradient $\partial f/\partial x$, that is, we consider a simple medium, in sense of Gurtin [11, 21, 38]. Precisely, in simple materials the flux $j(x,t)$ is related to the gradient, that is $j(x,t) = \Phi'(\nabla f)$ so that $\nabla \bullet j(x,t) = \Phi(\Delta \bullet f)$ [45]. The functional Φ is generally related to a memory integral with a kernel $R(t)$

$$j(x,t) \equiv -D(0)\int_0^t R(t-\tau)\frac{\partial f(x,\tau)}{\partial t}d\tau, \tag{4}$$

following the Boltzmann [6] and Volterra [47] concepts (see the sequel).

In the case of instantaneous flux propagation (infinite speed), when $\alpha = 1$ (local time derivative), we have the kernel $R(t) = \delta_D(t)$ (Dirac delta) so that

$\int_0^\infty R(t)dt = \int_0^\infty \delta_D(t)dt = 1$ and consequently the Fourier (Fick) law $j(x,t) = -D(0)\nabla \bullet f(x,t)$ is recovered. Then, with the continuity equation (2) we get the classical diffusion equation.

In this context, in the framework of the classical fractional calculus, if the Continuous Time Random Walk (CTRW) mechanism describes the diffusion process [23, 33, 36, 37] with memory kernel $R(t) \propto t^{-\alpha}/\Gamma(\alpha)$, $\alpha \in (0,1)$, the derivative $\partial^\alpha f/\partial t^\alpha$ is either Riemann–Liouville or Liouville-Caputo with singular kernels, namely [42]

$$ {}_{RL}D_t^\alpha = \frac{1}{\Gamma(1-\alpha)} \int_0^t f(s) \frac{1}{(t-s)^\alpha} ds, \quad \alpha \in (0,1), \quad t > 0, \tag{5}$$

$$ {}_{C}D_t^\alpha = \frac{1}{\Gamma(1-\alpha)} \int_0^t \frac{df(s)}{dt} \frac{1}{(t-s)^\alpha} ds, \tag{6}$$

Then the flux $j(x,t)$ is related to its history through the Riemann–Liouville fractional integral [42]

$$ {}^{RL}I^\alpha(\nabla \bullet f) = \frac{1}{\Gamma(\alpha)} \int_0^t \frac{\partial f(x,t)}{\partial x} (t-\tau)^{-\alpha} d\tau, \quad \alpha \in (0,1), \quad t > 0. \tag{7}$$

As a result, the diffusion equation (2) takes the form [36, 42]

$$ \frac{\partial^\alpha f(x,t)}{\partial t^\alpha} = D_0 \frac{\partial^2 f(x,t)}{\partial x^2}, \quad \alpha \in (0,1), \quad t > 0. \tag{8}$$

Now, the problem of interest is the formulation of the diffusion equation when the time-fractional derivative has the non-singular Mittag-Leffler function E_α as a memory kernel, that is $\partial^\alpha f/\partial t^\alpha$ is the Atangana–Baleanu (AB) time-fractional derivative [2, 5] in two basic definitions, precisely ABR and ABC are defined in the sequel.

2 Necessary Background

2.1 Derivatives with Non-singular Mittag-Leffler Kernels: Definitions

One of the modern trends in fractional calculus is the development of fractional operators with non-singular kernels. Here we focus the attention on Atangana–Baleanu derivatives (AB derivatives) in two basic definitions

Riemann–Liouville sense (ABR derivative)

$$ {}^{ABR}D^{\alpha}_{a+}f(t) = \frac{B(\alpha)}{1-\alpha}\frac{d}{dt}\int_0^t f(z)E_{\alpha}\left[\frac{-\alpha}{1-\alpha}(t-z)^{\alpha}\right]dz, \tag{9} $$

with $0 < \alpha < 1, a < t < b$ and $f(x,t) \in l^1[a,b]$

Caputo sense (ABC derivative)

$$ {}^{ABC}D^{\alpha}_{a+}f(t) = \frac{B(\alpha)}{1-\alpha}\int_0^t \frac{df(z)}{dt}E_{\alpha}\left[\frac{-\alpha}{1-\alpha}(t-z)^{\alpha}\right]dz, \tag{10} $$

with $0 < \alpha < 1$, $a < t < b$ and $f(x,t)$ is a differentiable function on $[a,b]$ such that $df(x,t)/dt \in l^1[a,b]$.

The normalization function $B(\alpha)$ can be any function satisfying the conditions $B(0) = B(1) = 1$ [2] (see also [5] for details). In these definitions E_{α} is one-parameter Mittag-Leffler function [42]

$$ E_{\alpha} = \sum_0^{\infty}\frac{z^k}{\Gamma(\alpha k+1}. \tag{11} $$

The Mittag-Leffler function is an entire function of z and the series (11) are converging locally uniformly in the whole complex plane.

2.1.1 AB Derivatives: Laplace Transforms and Relationships

The Laplace transform of ABR derivative is [2]

$$ L\left\{{}_0^{ABR}D_t^{\alpha}\left[f(t)\right]\right\}(p) = \frac{B(\alpha)}{1-\alpha}\frac{p^{\alpha}}{p^{\alpha}+\frac{\alpha}{1-\alpha}}L\left\{f(t)\right\}(p). \tag{12} $$

Similarly for the ABC derivative

$$ L\left\{{}_0^{ABR}D_t^{\alpha}\left[f(t)\right]\right\}(p) = \frac{B(\alpha)}{1-\alpha}\frac{p^{\alpha}}{p^{\alpha}+\frac{\alpha}{1-\alpha}}\left[L\left\{f(t)\right\}(p) - p^{\alpha-1}f(0)\right]. \tag{13} $$

Consequently, the relation between ABR and ABC is [2]

$$ {}_0^{ABC}D_t^{\alpha}\left[f(t)\right] = {}_0^{ABR}D_t^{\alpha}\left[f(t)\right] - \frac{B(\alpha)}{1-\alpha}f(0)E_{\alpha}\left(-\frac{\alpha}{1-\alpha}t^{\alpha}\right). \tag{14} $$

Hence, with zero initial conditions both derivatives are identical, a property already known from the classical Riemann-Liouville and Caputo-Liouville derivatives.

2.1.2 ABR Derivative and the Related Fractional Integral

As it was demonstrated by Atangana and Baleanu [2] the following fractional differential equation

$$ {}_{0}^{ABR}D_{t}^{\alpha}\left[f(t)\right] = u(t), \tag{15}$$

has a unique solution

$$f(t) = \frac{1-\alpha}{B(\alpha)}u(t) + \frac{\alpha}{B(\alpha)}\frac{1}{\Gamma(\alpha)}\int_{0}^{t} u(\tau)(t-\tau)^{\alpha-1}d\tau, \tag{16}$$

where for $\alpha = 0$ we recover the initial function, while for $\alpha = 1$ we get the ordinary Riemann integral.

The ABR fractional derivative can be expressed as [5]

$$ {}_{0}^{ABR}D_{t}^{\alpha}\left[f(t)\right] = \frac{B(\alpha)}{1-\alpha}\sum_{k=0}^{\infty}\left(\frac{-\alpha}{1-\alpha}\right)^{k}\frac{d}{dt}\left[{}^{RL}I_{a+}^{\alpha k+1}f(t)\right]. \tag{17}$$

The AB fractional integral operator ${}^{AB}I_{a+}^{\alpha}$ follows directly from the solution (16) and can be precisely defined as [5]

$$ {}^{AB}I_{a+}^{\alpha}f(t) = \frac{1-\alpha}{B(\alpha)}f(t) + \frac{\alpha}{B(\alpha)}\,{}^{RL}I_{a+}^{\alpha}f(t), \tag{18}$$

where ${}^{RL}I_{a+}^{\alpha k+1}f(t)$ is the Riemann-Liouville fractional integral [42]. The relation (18) can be easily developed by applying the Laplace transform to equation (15) as it was demonstrated by Baleanu and Fernandez [5]. Further, we have the following left and right inverse properties [5]

$$ {}^{AB}I_{a+}^{\alpha}\left[{}^{ABR}D_{a+}^{\alpha}f(t)\right] = f(t), \tag{19}$$

$$ {}^{ABR}D_{a+}^{\alpha}\left[{}^{AB}I_{a+}^{\alpha}f(t)\right] = f(t), \tag{20}$$

and the commutative properties for $\beta \in (0, 1)$

$$ {}^{ABR}D_{a+}^{\alpha}\left[{}^{ABR}D_{a+}^{\beta}f(t)\right] = {}^{ABR}D_{a+}^{\beta}\left[{}^{ABR}D_{a+}^{\alpha}f(t)\right], \tag{21}$$

$$ {}^{AB}I_{a+}^{\alpha}\left[{}^{AB}I_{a+}^{\beta}f(t)\right] = {}^{AB}I_{a+}^{\beta}\left[{}^{AB}I_{a+}^{\alpha}f(t)\right], \tag{22}$$

$$ {}^{ABR}D_{a+}^{\alpha}\left[{}^{AB}I_{a+}^{\alpha}f(t)\right] = {}^{AB}I_{a+}^{\alpha}\left[{}^{ABR}D_{a+}^{\alpha}f(t)\right]. \tag{23}$$

For the sake of the simplicity, assuming hereafter $B(\alpha) = 1$, we get from (18)

$$ {}^{AB}I^{\alpha}_{a+}u(t) = (1-\alpha)u(x,t) + \alpha\, {}^{RL}I^{\alpha}_{a+}u(x,t) = f(t), \tag{24} $$

or equivalently

$$ {}^{AB}I^{\alpha}_{a+}u(t) = m(\alpha)u(x,t) + \lambda(\alpha)\, {}^{RL}I^{\alpha}_{a+}u(x,t). \tag{25} $$

The construction of Eqs. (24) and (25) are the same as that of the Boltzmann linear superposition functional [6] expressing the fading memory concept [21, 40] with a time-dependent memory (influence) function) $R(t)$, namely

$$ \varphi(x,t) = m\left[v_x(x,t)\right] + \lambda \int_0^t R(t,z)v_x(z)dz. \tag{26} $$

The memory integral (the 2nd terms in (24) or (25) is the standard Riemann-Liouville fractional integral ${}^{RL}I^{\alpha}_{a+}u(x,t)$. In (26) $v_x(z) = \nabla v(z)$ and $\nabla \bullet \nabla = \Delta$ is the Laplacian. The coefficients m and λ are weighting functions (transport coefficients) depending on the character of the modelled diffusion process (heat or mass). The basic idea of the fading memory concept is explained in the next point.

2.2 *Fading Memory Concept in Case of Diffusion of Heat (Mass)*

The fading memory concept relating the flux to its gradient, for simple materials [21, 38, 45], is modelled by the following integro-differential equation

$$ j(x,t) = -D_0\nabla C(x,t) - D' \int_{-\infty}^{t} R(t-\tau)\nabla C(x,\tau)d\tau, \tag{27} $$

as a manifestation of the Boltzmann linear superposition functional (see (26) expressing the flux history [6, 21, 38] through the function of influence (memory kernel) $R(t,z)$. In (27) D_0 and D' are transport coefficients (diffusivities).

The appropriate history value problem for(27) is related to the following integral [38]

$$ d(t) = -\int_{-\infty}^{t} R(t-\tau)\nabla C(x,\tau)d\tau, \tag{28} $$

allowing to give a function $C(x,t)$ on $-\infty t < 0$.

From (27) and (28) it follows that

$$ \nabla \bullet j(x,t) = -D_0\Delta C(x,t) - D' \int_0^t R(t-\tau)\nabla C(x,\tau)d\tau + \nabla \bullet d(t). \tag{29} $$

Since $C(x,t)$ is a causal function (vanishing for $t < 0$) and considered only for $0 < t < \infty$, then we accept $d(t) = \nabla \bullet d(t) = 0$ and therefore (29) can be rewritten as

$$\nabla \bullet j(x,t) = -D_0 \Delta C(x,t) - D' \int_0^t R(t-\tau)\nabla C(x,\tau)d\tau. \tag{30}$$

3 Diffusion Equation in Terms of ABR Derivative

3.1 Conjectures

Now, let us turn on to the basic formulation of the rate equation (15) expressed through either ABR or ABC derivative [2, 5] allowing the time fractional derivative to be related to the flux expressed as a function of the gradient $\partial f/\partial x$. The following model development is based on two constitutive conjectures.

3.1.1 Conjecture 1

Assume that in (15) the function in the right-hand side is defined as $u(x,t) = (-\partial f/\partial x)$, without loss of the generality of this equation. Consequently the AB fractional integral operator ${}^{AB}I^{\alpha}_{a+}u(t)$ (see (24) as a construction) can be expressed as

$${}^{AB}I^{\alpha}_{a+}\left[u(x,t)\right] = -\left\{m(\alpha)\frac{\partial f(x,t)}{\partial x} + \lambda(\alpha)\,{}^{RL}I^{\alpha}_{a+}\left[\frac{\partial f(x,t)}{\partial x}\right]\right\}. \tag{31}$$

Now, the weighting functions $m(\alpha) = (1-\alpha)$ and $\lambda(\alpha) = \alpha$ depend on the degree of fractionality of the modelled diffusion process. For $\alpha = 1$ we get $m(\alpha) = 0$ and $\lambda(\alpha) = 1$. Actually, the expressions (30) and (31) coincide with the Coleman-Noll definitions [11, 21, 38] (see the relation (27) and the others related to it) about the flux of diffusant in simple materials (where the flux is proportional to the gradient $j(x,t) \equiv -\partial f/\partial x$ [21, 45], following the Boltzmann superposition principle (26).

3.1.2 Conjecture 2

The flux relaxation follows the fading memory concept expressed by (30) or (26), that is

$$j_a(x,t) = {}^{AB}I^{\alpha}_{a+}\left[-D_0\frac{\partial f}{\partial x}\right] = -D_0\left\{m(\alpha)\frac{\partial f}{\partial x} + \lambda(\alpha)\,{}^{RL}I^{\alpha}_{a+}\left[\frac{\partial f}{\partial x}\right]\right\}, \tag{32}$$

where D_0 is the transport coefficients (the diffusivity) for $\alpha = 1$ (instantaneous flux with an infinite speed).

3.2 To the Diffusion Equation: Model 1

Hence, we may rewrite (15) as a continuity equation in terms of ABR derivative, equivalent to (1), namely

$$\frac{\partial^\alpha f(x,t)}{\partial t^\alpha} = -\frac{d}{dx} j_a(x,t). \tag{33}$$

Differentiation with respect to x in (32) yields

$$-\frac{d}{dx} j_a(x,t) = D_0 \left\{ m(\alpha) \frac{\partial^2 f(x,t)}{\partial x^2} + \lambda(\alpha)^{RL} I^\alpha_{a+} \left[\frac{\partial^2 f(x,t)}{\partial x^2} \right] \right\}. \tag{34}$$

Therefore, using the above conjecture, the diffusion equation takes the form

$$\frac{\partial^\alpha f(x,t)}{\partial t^\alpha} = D_0 \left[{}^{AB} I^\alpha_{a+} \frac{\partial^2 f(x,t)}{\partial x^2} \right], \tag{35}$$

or

$${}^{ABR}D^\alpha_{a+} f(x,t) = D_0 \left[{}^{AB} I^\alpha_{a+} \frac{\partial^2 f(x,t)}{\partial x^2} \right]. \tag{36}$$

Explicitly, setting the lower terminal the ABR derivative and the Riemann-Liouville integral as $a = 0$, we get

$${}^{ABR}D^\alpha_{a+} f(x,t) = D_0 \left\{ m(\alpha) \frac{\partial^2 f(x,t)}{\partial x^2} + \lambda(\alpha)^{RL} I^\alpha_{a+} \left[\frac{\partial^2 f(x,t)}{\partial x^2} \right] \right\}. \tag{37}$$

As commented above, for $\alpha = 1$ we have $m(\alpha) = 0$ and $\lambda(\alpha) = 1$, and the memory integral in (37) becomes $\left[{}^{AB} I^\alpha_t \left(\partial^2 f(x,t)/\partial x^2\right)\right]$. Hence, for $\alpha = 1$ from (33) we recover the Fourier (Fick) law $j_{\alpha=1} = -D_0(\partial f/\partial x)$ and equation (37) reduces to the classical diffusion equation. Alternatively, for the sake of clarity of this statement, we may express (37) as (with $m(\alpha) = 1 - \alpha$ and $\lambda(\alpha) = \alpha$)

$${}^{ABR}D^\alpha_{a+} f(x,t) = D_0 \left\{ \frac{\partial^2 f(x,t)}{\partial x^2} + \left[\alpha I^\alpha_t \left(\frac{\partial^2 f(x,t)}{\partial x^2} \right) - \frac{\partial^2 f(x,t)}{\partial x^2} \right] \right\}. \tag{38}$$

For $\alpha \to 1$ the second term (in the squared brackets) tends to zero and we recover the classical diffusion equation. Moreover, we can see that when the power-law kernel controls the memory integral (see Eq. (7)) then $m(\alpha) \to 0$ and $\lambda(\alpha) \to 1$. However, this is a formal comparison because the singular power-law kernel and the non-singular one based on the Mittag-Leffler function *describe different relaxation processes* and *model different physical phenomena*. Moreover, the formulations (33) and (34) differ from the directly fractionalized diffusion model (discussed further in this chapter)

$$\frac{\partial^\alpha f(x,t)}{\partial t^\alpha} = D_0 \frac{\partial^2 f(x,t)}{\partial x^2}, \quad -\infty < x < +\infty, \quad t > 0, \tag{39}$$

with conditions $f(x, 0) = \delta_D(x)$, $f(\pm, t) = 0$, and $\frac{\partial}{\partial x} f(\pm\infty, t) = 0$.

3.2.1 Front Propagation, Finite Speed Concept and Approximate Integral Solution

The solution of the classical diffusion equation has infinite speed of propagation due to missing damping flux relationship to the gradient, in accordance with the Fourier (Fick) law. However, the incorporation of damping (memory) integral in the flux to gradient relationship (irrespective to the type of memory kernel used) leads to the concept of finite speed of propagation. In the light of these thoughts the diffusion zone is defined as $0 \leq x \leq \delta$ with $f(x, t) > 0$ and a virgin zone $\delta \leq x \leq \infty$ where $f(x, t) = 0$. This assumption means that at $x = \delta$ we have the boundary conditions $f(\delta(t)) = 0$ and $\partial f(\delta)/\partial x = 0$, and consequently $j_\delta(\delta) = 0$. To estimate how the front behaves when the fractionality of the process varies we suggest that the solution may be approximated by a parabolic profile with unspecified exponent [24, 26, 28, 32]

$$f_a = f_s \left(1 - \frac{x}{\delta}\right)^n. \tag{40}$$

The solution about $\delta(t)$ will be performed by the double-integration method (DIM) successfully used to solve approximately subdiffusion [24, 28] and superdiffusion [29, 30] problems. For the sake of simplicity we consider the Dirichlet problem that means $f_s = f(0, t) = 1$.

The application of the double integration method to the diffusion equation (37) yields

$$\int_0^\delta \int_x^\delta \left[{}_0^{ABR} D_t^\alpha f(x,t)\right] dx dx =$$
$$= \int_0^\delta \int_x^\delta D_0 \left\{ \frac{\partial^2 f(x,t)}{\partial x^2} + \left[\alpha I_t^\alpha \left(\frac{\partial^2 f(x,t)}{\partial x^2} \right) - \frac{\partial^2 f(x,t)}{\partial x^2} \right] \right\} dx dx. \tag{41}$$

The integration in the right-hand side of (41) results in

$$\int_0^\delta \int_x^\delta \left[{}_0^{ABR} D_t^\alpha f(x,t)\right] dx dx = D_0 \left\{ m(\alpha) f(0,t) + \lambda(\alpha) {}_0^{RL} I_t^\alpha \left[f(0,t) \right] \right\}. \tag{42}$$

The integral relation (42), from a physical point of view, is a mass (energy) balance over the disturbed zone $0 \leq x \leq \delta$, that is the accumulated mass (energy) in the diffusion layer (the integral in the left side of (41) is controlled by the flux of mass (energy) at the boundary (the right-hand side of (42)).

3.2.2 Front Propagation

Now, replacing $f(x,t)$ in (42) by the assumed profile (40) and performing the double integration we get

$$ {}_{0}^{ABR}D_t^{\alpha}\left[\frac{\delta^2}{(n+1)(n+2)}\right] = D_0 m(\alpha) + \lambda(\alpha){}_0^{RL}I_t^{\alpha}[1]. \tag{43}$$

Simply, denoting $N=(n+1)(n+2)$ we get

$$ {}_{0}^{ABR}D_t^{\alpha}[\delta^2] = D_0 N\left[(1-\alpha)+\alpha\frac{t^{\alpha}}{\Gamma(\alpha+1)}\right]. \tag{44}$$

Equation (44) should have an initial condition $\delta(0)=0$ since there is no diffusant penetration at $t=0$. Moreover, the solution of (44) with respect to δ^2 follows directly from the definition of ${}^{AB}I_0^{\alpha}f(t)$. Hence, we have

$$\delta^2 = D_0 N\left[(1-\alpha)+\alpha\frac{t^{\alpha}}{\Gamma(\alpha+1)}\right] + {}_0^{RL}I_t^{\alpha}\left\{D_0 N\left[(1-\alpha)+\alpha\frac{t^{\alpha}}{\Gamma(\alpha+1}\right]\right\}. \tag{45}$$

Consequently we get

$$\delta^2 = D_0 N\left[(1-\alpha)+\alpha\frac{t^{\alpha}}{\Gamma(\alpha+1)}\right] + D_0 N\left[(1-\alpha)+\alpha\frac{\Gamma(\alpha+1)}{\Gamma(2\alpha+1}t^{2\alpha}\right], \tag{46}$$

$$\delta^2 = D_0 N\left\{(1-\alpha)+\frac{t^{\alpha}}{\Gamma(\alpha+1)}+\alpha\frac{\Gamma(\alpha+1)}{\Gamma(2\alpha+1)}t^{2\alpha}\right\}. \tag{47}$$

Using the relationship $y\Gamma(y)=\Gamma(y+1)$ we get two equivalent forms (the differences are in the last term)

$$\delta^2 = D_0 N\left\{(1-\alpha)+\frac{t^{\alpha}}{\Gamma(\alpha+1)}+\alpha^2\frac{\Gamma(\alpha)}{\Gamma(2\alpha+1)}t^{2\alpha}\right\}, \tag{48}$$

$$\delta^2 = D_0 N\left\{(1-\alpha)+\frac{t^{\alpha}}{\Gamma(\alpha+1)}+\frac{\alpha}{2}\frac{\Gamma(\alpha)}{\Gamma(2\alpha)}t^{2\alpha}\right\}. \tag{49}$$

From (49), for instance, we get

$$\delta^2 = D_0 N t^{2\alpha}\left\{\frac{(1-\alpha)}{t^{2\alpha}}+\frac{1}{\Gamma(\alpha+1)}\frac{1}{t^{\alpha}}+\frac{\alpha}{2}\frac{\Gamma(\alpha)}{\Gamma(2\alpha)}\right\}. \tag{50}$$

Then, *for long times* we have $\delta^2 \to \left[D_0 N\frac{\alpha}{2}\frac{\Gamma(\alpha)}{\Gamma(2\alpha)}\right]t^{2\alpha}$ and therefore $\delta \propto t^{\alpha}$, that is, this corresponds to a subdiffusive transport.

3.2.3 Asymptotic Behaviour of the Front

Let us now estimate the asymptotic behaviour of δ^2 for $\alpha \to 0$ and $\alpha \to 1$.

For $\alpha \to 0$ we have $(1-\alpha) \to 1$, $\Gamma(\alpha+1) \to 1$, $\Gamma(2\alpha+1) \to 1$ and α^2 could be neglected. Then, the approximation is

$$\delta^2_{\alpha\to 0} \equiv D_0 N \left\{1 + \frac{t^\alpha}{\Gamma(\alpha+1)}\right\}, \quad \delta^2_{\alpha\to 0} \equiv D_0 N t^\alpha \left\{\frac{1}{\Gamma(\alpha+1)} + \frac{1}{t^\alpha}\right\}. \tag{51}$$

Hence for *small* α (strong fractionality) and *long times* the front exhibits *power-law* (subdiffusion) *behaviour* and $\delta_{\alpha\to 0}(t\to\infty) \equiv \sqrt{D_0 N t^\alpha}$ as in the cases solved in [24, 28, 32]. Actually, the first version of the estimation (51) can be presented as

$$\delta^2_{\alpha\to 0} \equiv D_0 N \left\{\frac{(t^\alpha)^0}{\Gamma\alpha\cdot 0+1} + \frac{(t^\alpha)^1}{\Gamma(\alpha\cdot 1+1)}\right\} \equiv D_0 N \sum_0^1 \frac{(t^\alpha)^k}{\Gamma(\alpha k+1)}. \tag{52}$$

Hence, for $\alpha \to 0$ the front propagation is modeled by the first two terms of $E_\alpha(t^\alpha)$. Recall, that for $\alpha \to 0$ the relaxation is extremely slow and it could be assumed that it practically does not happen. Moreover, for long times and $\alpha \to 0$ we get from the second form of (51) that $\delta^2_{\alpha\to 0} \equiv t^\alpha$, i.e. *subdiffusive behaviour is modelled* as in the case of the Riemann-Liouville derivative. The latter statement is in agreement with the results of Atangana [1] and Atangana and Gomez [4] where it was clearly proved that *for long times* the AB derivative behaves as the Riemann-Liouville derivative and thus *the propagator of the solution* (in the present case this is $\delta(t)$) *should exhibit a subdiffusive behaviour.*

For $\alpha \to 1$ we have $(1-\alpha) \to 0$, $\Gamma(\alpha+1) \to 1$, $\Gamma(2\alpha+1) \to 2$ and $\alpha^2 \to 1$. Then, we have

$$\delta^2_{\alpha\to 1} \equiv D_0 N \left\{t^\alpha + \frac{\alpha^2}{2} t^{2\alpha}\right\} \equiv D_0 N t^{2\alpha} \left\{\frac{\alpha^2}{2} + \frac{1}{t^\alpha}\right\}. \tag{53}$$

The approximation (53) indicates that *for long times the movement is subdiffusive.*

$$\delta_{\alpha\to 0} \equiv \sqrt{D_0 N t^\alpha}. \tag{54}$$

For $t = 0$ we have $\delta^2 = 0$ which means that there is no diffusant penetration into the medium, which is a physically correct result. At this moment it is worthnoting that the concept of the finite penetration depth comes from the approximate solutions of integer-order parabolic equations [20, 26] as *ad hoc* correction of the unphysical infinite flux speed. Besides, *for short times* and $\alpha \to 1$ we get $\delta_{\alpha\to 1} \equiv \sqrt{D_0 N t^\alpha}$, a solution known from subdiffusive cases when the fractional derivative of Riemann-Liouville type [24, 28] was used. At this point we have to recall that for small times the Mittag-Leffler function approaches the behaviour of the stretched Kohlrausch exponential [1]. Moreover, as it was discussed in [31] for short times the Kohlrausch relaxation function exhibits the so-called algebraic decay $(t/\tau)^{-\alpha}$ thus matching the behaviour of the singular power-law kernel of the Riemann-Liouville derivative. Hence, despite the different approach using the behaviour of $\delta(t)$ to estimate the type of the transport process modelled by the constituted diffusion equation (36) the results are consistent with those of Atangana [1, 4] developed on the basis of a formal diffusion equation of type (8) and (39) (see also (61) next) where the time-fractional derivative is ABR. The similarities are based on the fact that the asymmetric behaviour of the solution strongly depends on the asymmetric behaviour of the fractional operator rather than on the diffusion term on the right side of the model equation related to the gradient.

In the context of the above expressions about δ^2 we may rearrange (49) as

$$\delta^2 = D_0 N \left\{ \sum_0^1 \frac{(t^\alpha)^k}{\Gamma(\alpha+1)} - \left[-\alpha + (1-\alpha)\frac{t^\alpha}{\Gamma(\alpha+1)} \right] \right\}. \tag{55}$$

3.2.4 The First Passage Time

The first passage time distribution (FPT) distribution [13, 34, 48] is defined as $F(t) = df_{fpt}/dt$ where $f_{fpt} = \int_0^L f(x,t)dx$. With the concept of the finite penetration depth we have the upper terminal $L = \delta$ and consequently with the assumed profile (40) we have

$$f_{fpt} = \int_0^\delta \left(1 - \frac{x}{\delta}\right)^n dx = \frac{\delta(x,t)}{n+1}. \tag{56}$$

Then,

$$F_\delta = \frac{d}{dt}\left[\frac{\delta(x,t)}{n+1}\right]. \tag{57}$$

Moreover, the mean FPT (MFPT) is defined as [13, 34, 48]

$$T_{FPT} = \int_0^\infty f_{fpt} dt = \int_0^\infty \left[\int_0^\delta \left(1 - \frac{x}{\delta}\right)^n dx \right] dt. \tag{58}$$

Taking into account the results (48) and (49) about δ^2 and the asymptotic estimations, it is not possible to get a simple expression about δ allowing easy integration in (62). Because of that, we will use the squared value of FPT termed here as Squared First Passage Time distribution (SFPT) $F^2(t) = (df_{fpt}/dt)^2$ which with (61) results in

$$F_\delta^2 = \frac{d}{dt}\left[\frac{\delta^2(t)}{(n+1)^2}\right]. \tag{59}$$

Consequently,

$$(T_{FPT})^2 = \left(\int_0^\infty f_{fpt} dt\right)^2 = \int_0^\infty \left[\int_0^\delta \left(1 - \frac{x}{\delta}\right)^n dx\right]^2 dt =$$
$$= \int_0^\infty \frac{D_0 N}{(n+1)} \left\{(1-\alpha) + \frac{t^\alpha}{\Gamma(\alpha+1)} + \frac{\alpha}{2} \frac{\Gamma(\alpha)}{\Gamma(2\alpha)} t^{2\alpha}\right\} dt. \tag{60}$$

It is obvious that the integration in (60) (for any $0 < \alpha < 1$) yields $(T_{FPT})^2 \rightarrow \infty$. Therefore, this result is consistent with the results in [13, 34, 48] that MFPT for subdiffusive transport is infinite. To be precise, this statement is correct for any $\alpha < 1/2$. The result (60) shows that it should be also valid for $0.5 < \alpha < 1$ because if $\delta^2 \propto t^{2\alpha}$ as it follows from the second term of (48) we may estimate that $\delta \propto t^\alpha$. Then, repeating the estimation of T_{FPT} we have $T_{FPT} \equiv \int_0^\infty \delta dt \equiv \int_0^\infty t^\alpha dt \rightarrow \infty$.

3.3 To the Diffusion Equation: Model 2

Approach 2 constitutes that the rate equation is

$$\frac{\partial f(x,t)}{\partial t} = -\frac{d}{dx} j_a(x,t), \tag{61}$$

where $j_a(x,t)$ follows the fading memory concept expressed by Eq. (32). This approach leads

$$\frac{\partial f(x,t)}{\partial t} = -\frac{d}{dx}\left\{ {}^{AB}I_{a+}^{\alpha}\left[-D_0 \frac{\partial f(x,t)}{\partial x}\right]\right\}, \tag{62}$$

$$\frac{\partial f(x,t)}{\partial t} = D_0 \frac{d}{dx}\left\{m(\alpha)\frac{\partial f(x,t)}{\partial x} + \lambda(\alpha)^{RL}\int_{a+}^{\alpha}\left[\frac{\partial f(x,t)}{\partial x}\right]\right\}, \tag{63}$$

or equivalently

$$\frac{\partial f(x,t)}{\partial t} = D_0 \left\{m(\alpha)\frac{\partial^2 f(x,t)}{\partial x^2} + \lambda(\alpha)^{RL}\int_{a+}^{\alpha}\left[\frac{\partial^2 f(x,t)}{\partial x^2}\right]\right\}. \tag{64}$$

From (3) and (4) and $\alpha = 1$ we recover the classical diffusion equation. Thus, the compact form of the diffusion equation in terms of the AB integral only is

$$\frac{\partial f(x,t)}{\partial t} = D_0^{AB} I_{a+}^{\alpha}\left[\frac{\partial^2 f(x,t)}{\partial x^2}\right]. \tag{65}$$

Equation (65) seems pretty but let test it with the integral-balance method in a manner already used with Model 1. The double integration of (65) yields

$$\int_0^\delta \int_x^\delta \frac{\partial f(x,t)}{\partial t} dx dx = D_0 \int_0^\delta \int_x^\delta \left\{ {}^{AB}I^\alpha_{a+} \left[\frac{\partial^2 f(x,t)}{\partial x^2} \right] \right\} dx dx. \qquad (66)$$

With the assumed parabolic profile (40) the differential equation about δ is

$$\frac{1}{(n+1)(n+2)} \frac{d\delta^2}{dt} = D_0 \left\{ {}^{AB}I^\alpha_{a+} \left[f_a(0,t) \right] \right\}. \qquad (67)$$

For the Dirichlet problem we have $f_a(0,t) = 1$ and the equation about penetration depth with the initial condition $\delta(0) = 0$ is

$$\frac{d\delta^2}{dt} = D_0 N \left[m(\alpha) + \lambda(\alpha) \frac{t^\alpha}{\Gamma(\alpha+1)} \right] = D_0 N t^\alpha \left[\frac{m(\alpha)}{t^\alpha} + \frac{\lambda(\alpha)}{\Gamma(\alpha+1)} \right]. \qquad (68)$$

For $\alpha \to 1$ we have $m(\alpha) \to 0$ and $\lambda(\alpha) \to 1$, and $\Gamma(\alpha+1) \to 1$. Then, *for long times* we get from (68) that *the front moves in a subdiffusive manner*, namely

$$\frac{d\delta^2}{dt}_{(\alpha\to 1)} \equiv D_0 t^\alpha \Rightarrow \delta_{(\alpha\to 1, t\to\infty)} \equiv \sqrt{D_0 t^\alpha}. \qquad (69)$$

Also, *for small* $\alpha(\alpha \to 0)$ and *small times* we have

$$\frac{d\delta^2}{dt}_{(\alpha\to 1, \alpha\to 0)} \equiv D_0 t^\alpha \Rightarrow \delta_{(\alpha\to 0, t\to 0)} \equiv \sqrt{D_0 t^\alpha}. \qquad (70)$$

Hence, for these cases *the behaviour is subdiffusive*. Further, from (68), the solution about δ is

$$\delta = \sqrt{(D_0 t) + \frac{\lambda(\alpha)}{m(\alpha)} \frac{t^{\alpha+1}}{(\alpha+1)\Gamma(1+\alpha)}} \sqrt{m(\alpha)} \sqrt{N}. \qquad (71)$$

Then the scaled penetration depth is

$$\frac{\delta}{\sqrt{D_0 t}} = \sqrt{1 + \frac{1}{(D_0 t)} \frac{\lambda(\alpha)}{m(\alpha)} \frac{t^\alpha}{(\alpha+1)\Gamma(1+\alpha)}} \sqrt{m(\alpha)} \sqrt{N}. \qquad (72)$$

Again, for small $\alpha(\alpha \to 0)$ we get $\frac{\delta}{\sqrt{D_0 t}} = \sqrt{N}$ which is the well-known result of the integral-balance solution of the linear diffusion equation. Further, for $\alpha \to 1$ the ratio $\lambda(\alpha)/m(\alpha) \to \infty$ and its weight increases, while $(\alpha+1)\Gamma(\alpha+1) = \Gamma(\alpha+2) \to \Gamma(3) \to 2$ and $t^{\alpha+1} \to t^2$. However, with $0 < \alpha < 1$ we have always $\alpha + 1 < 2$ and therefore for the intermediate values of α at any times *the front moves with a behaviour which is a crossover of normal diffusion* (the first term) and *subdiffusion* (second term) because $(\alpha+1)/2 < 1$.

3.4 To the Diffusion Equation: Formal Approach and Model 3

Now we turn on the problem of the *formalistic fractionalization* (denoted by the subscript *FF*) which mechanistically leads to equation (39) with construction different from the results (37) or (38) developed in this work. The result presented as (39), in fact, is a conservative law with a non-locality represented by both the ABR derivative and the Riemann-Liouville integral. If the diffusion equation is constituted (postulated) as (39) then we have a formal **Model 3**

$$ {}_0^{ABR}D_t^{\alpha} f(x,t) = D_0 \frac{\partial^2 f(x,t)}{\partial x^2}. \tag{73}$$

For $\alpha = 1$ Eq. (73) formally recovers the classical diffusion equation, but when $0 < \alpha < 1$ the corresponding integral is (see (16))

$$ f(x,t) = \frac{1-\alpha}{B(\alpha)}\left[\frac{\partial^2 f(x,t)}{\partial x^2}\right] + \frac{\alpha}{B(\alpha)}\frac{1}{\Gamma(\alpha)}\int_0^t \left[\frac{\partial^2 f(x,t)}{\partial x^2}\right](t-\tau)^{\alpha-1}d\tau. \tag{74}$$

Then, the flux $j_{FF}(t) \equiv -\frac{\partial f}{\partial x}$ and its space derivative can be defined as two unphysical relationships (see (75) and (76)) with high order space derivatives

$$ \begin{aligned} j_{FF}(t) \equiv \frac{\partial f}{\partial x} \equiv D_0 \frac{1-\alpha}{B(\alpha)}\left[\frac{\partial^3 f(x,t)}{\partial x^3}\right] + \\ + D_0 \frac{\alpha}{B(\alpha)}\frac{1}{\Gamma(\alpha)}\int_0^t \left[\frac{\partial^3 f(x,t)}{\partial x^3}\right](t-\tau)^{\alpha-1}d\tau, \end{aligned} \tag{75}$$

and with $\frac{dj_{FF}(t)}{dx} \equiv -\frac{\partial^2 f}{\partial x^2}$

$$ \begin{aligned} \frac{dj_{FF}(t)}{dx} \equiv \frac{\partial^2 f}{\partial x^2} \equiv D_0 \frac{1-\alpha}{B(\alpha)}\left[\frac{\partial^4 f(x,t)}{\partial x^4}\right] + \\ + D_0 \frac{\alpha}{B(\alpha)}\frac{1}{\Gamma(\alpha)}\int_0^t \left[\frac{\partial^4 f(x,t)}{\partial x^4}\right](t-\tau)^{\alpha-1}d\tau. \end{aligned} \tag{76}$$

It is hard to explain why from (76) when $\alpha \to 1$ in the left-hand side we have $\frac{\partial^2 f}{\partial x^2}$, while in the right-hand side it appears $\frac{\partial^4 f}{\partial x^4}$, since no physical meaning could be found to support this. In contrast, for $\alpha \to 1$ equation (34) (Model 1) and (65) (Model 2) simply lead to $\frac{dj_a(t)}{dx} = \frac{\partial^2 f}{\partial x^2}$ in accordance with the classical diffusion concept.

Besides, applying the right inverse property (20) to the diffusion equation (36) one gets

$$ {}^{ABR}D_{a+}^{\alpha}\left[{}^{ABR}D_{a+}^{\alpha}f(x,t)\right] = {}^{ABR}D_{t}^{\alpha}\left[D_0 I_{a+}^{\alpha}\frac{\partial^2 f(x,t)}{\partial x^2}\right]. \tag{77} $$

This operation leads to

$$ {}^{ABR}D_{a+}^{\alpha}\left[{}^{ABR}D_{a+}^{\alpha}f(x,t)\right] = D_0\frac{\partial^2 f(x,t)}{\partial x^2}. \tag{78} $$

This result resembles formally the formalistic equation (61) and mimics the classical subdiffusion equation with a power-law memory kernel *but we have to remember that with the ABR derivative* (see (21))

$$ {}^{ABR}D_{a+}^{\alpha}\left[{}^{ABR}D_{a+}^{\alpha}f(x,t)\right] \neq {}^{ABR}D_{a+}^{\alpha+\alpha}f(x,t). \tag{79} $$

Hence, the formal construction of the diffusion equation in terms of AB derivatives, without a physical background about the flux relaxation concept is unreasonable.

4 Discussion

The appearance of fractional derivatives with non-singular kernels [2, 4, 5, 8, 9, 25, 27] provoked a strong flux of articles (see [31] and the references therein) and counted publications based on the pillars of the classical fractional calculus using singular kernels as well. However, since the very begging of this new branch in fractional calculus, precisely in the seminal articles of Caputo and Fabrizio [8, 9], it was especially stated that these new derivatives are especially oriented to models of dissipative phenomena which cannot be adequately described by the classical fractional derivatives (see [31] about the discussion on the Caputo-Fabrizio derivative). Despite this special notion in [8, 9], the formalism in applying the rules of classical fractional calculus and the limited physical knowledge of mathematical society using fractional models resulted in complicated situations with unusual academic conflicts and wrong critical publications.

In this chapter we especially related the results of the Atangana–Baleanu derivative, more precisely the form of the construction of its related integral in order to draw the red line connecting the new and the old fractional operators. The fading memory concept of Boltzmann [6] is about 100 years older that the famous Caputo fractional derivative (1967) now considered as a standard fractional operator with a singular kernel. Since the choice of the memory (influential) function in the memory integral strongly depends on the physical process modelled, there are no restrictions in selection of the memory kernel function if it obeys the only condition to disappear at long times. As it is commonly discussed, it is quite important to see the asymptotic behaviours of the fractional operator for small and long times [1, 4, 46].

Here we have to stress the attention on the forgotten physics in eye of the unusual conflict (battle) between the developers of new derivatives and the defenders of the classics in the fractional calculus. Precisely, as it is clearly demonstrated in the book of Findlay et al. [15], for instance, most of the applications of the integrals with memory are related it should be et al. to non-linear phenomena where the influence function is not a power-law. But, as it was especially considered in [15] only when the response of the materials is considered for very short times, then the power-law-approximation is applicable, and consequently it is possible to apply the tools of the classical fractional calculus. Recall that, the Caputo derivative was conceived in 1967 [7] for description of linear short time elastic responses of deformed solids. It was consequently applied to the field of linear viscoelasticity [35] where the Riemann-Liouville derivative was already applied to describe viscoelastic effects [44].

Hence, all thoughts are following the natural way of evolution, that is when the phenomena cannot be adequately described by the existing linear tools, then it is reasonable to look for and invent if possible, new adequately working fractional operators. Certainly, it is natural that features known from the linear operators cannot be encountered in the non-linear operators such as the semi-group properties and the index law [1, 4]. In this context, it is worth to note that, the Atangana–Baleanu fractional operator, especially the fractional integral, provides a logical relationship to the classical Riemann-Liouville integral [5] which is a basic construction of the classical fractional calculus.

As it is demonstrated in [4], the asymptotic behaviours of either Atangana–Baleanu or Caputo-Fabrizio fractional operators match the power-law behaviour. The models constructed in this chapter relate the old idea of fading memory to the new fractional derivatives with non-singular kernels. The approximate solutions of Model 1 and Model 2 developed here reveal similar asymptotic behaviours for small and long times and for small and large values of the fractional parameter as well.

Moreover, the new fractional operators have stronger and complex memories allowing capturing behaviours combining simultaneously (crossover) classical diffusion and anomalous behaviour [4, 46]. This point is commented in [31], especially for the Caputo-Fabrizio derivative.

To recapitulate, the quest to model more complex and non-linear phenomena invokes new operators, and they cannot be treated with the tools of the known linear technology of the classical fractional calculus. This is natural and a good performance of the evolution of knowledge. We believe this chapter took a step ahead relating the new fractional operators and the forgotten physics that should be known in the new era of fractional calculus.

References

1. Atangana, A.: Non validity of index law in fractional calculus: a fractional differential operator with Markovian and non-Markovian properties. Phys. A **505**, 688–706 (2018)
2. Atangana, A., Baleanu, D.: New fractional derivatives with non-local and non-singular kernel: theory and application to Heat transfer model. Therm. Sci. **20**, 763–769 (2016)

3. Atangana, A., Gómez-Aguilar, J.F.: Fractional derivatives with no-index law property: application to chaos and statistics. Chaos Solitons Fractals **114**, 516–535 (2018)
4. Atangana, A., Gómez-Aguilar, J.F.: Decolonisation of fractional calculus rules: breaking commutativity and associativity to capture more natural phenomena. Eur. Phys. J. Plus **133**, 1–21 (2018)
5. Baleanu, D., Fernandez, A.: On some new properties of fractional derivatives with Mittag-Leffler kernel. Commun. Nonlinear Sci. Numer. Simul. **59**, 444–462 (2018)
6. Boltzmann, L.: Zur Theorie der Elastischen Nachwirkung. Sitzungsber. Akad. Wiss. Wien. Mathem.- Naturwiss. **70**, 275–300 (1874)
7. Caputo, M.: Linear model of dissipation whose Q is almost frequency independent-II. Geophys. J. R. Astron. Soc. **13**, 529–539 (1969)
8. Caputo, M., Fabrizio, M.: A new definition of fractional derivative without singular kernel. Progr. Fract. Differ. Appl. **1**, 73–85 (2015)
9. Caputo, M., Fabrizio, M.: Applications of new time and spatial fractional derivatives with exponential kernels. Progr. Fract. Differ. Appl. **2**, 1–11 (2016)
10. Coleman, B., Gurtin, M.E.: Equipresence and constitutive equations for rigid heat conductors. Z. Angew. Math. Phys. **18**, 188–208 (1967)
11. Coleman, B., Noll, W.: Foundations of linear viscoelasticity. Rev. Mod. Phys. **33**, 239–249 (1961)
12. Coronel-Escamilla, A., Gómez-Aguilar, J.F., Baleanu, D., Córdova-Fraga, T., Escobar-Jiménez, R.F., Olivares-Peregrino, V.H., Qurashi, M.M.A.: Bateman-Feshbach tikochinsky and Caldirola-Kanai oscillators with new fractional differentiation. Entropy **19**(2), 1–21 (2017)
13. Fa, K.S., Lenzi, E.K.: Anomalous diffusion, solutions, and the first passage time: Influence of diffusion coefficient. Phys. Rev. E **71**, 1–8 (2005)
14. Fernandez, A., Baleanu, D.: The mean value theorem and Taylor's theorem for fractional derivatives with Mittag-Leffler kernel. Adv. Diff. Eqs. **1**, 1–18 (2018)
15. Fidley, W.N., Lai, J.S., Onaran, K.: Creep and Relaxation of Nonlinear Viscoelastic Materials. North-Hollad, Amsterdam (1976)
16. Gómez-Aguilar, J.F.: Analytical and numerical solutions of a nonlinear alcoholism model via variable-order fractional differential equations. Phys. A: Stat. Mech. Appl. **494**, 52–75 (2018)
17. Gómez-Aguilar, J.F., Atangana, A.: New insight in fractional differentiation: power, exponential decay and Mittag-Leffler laws and applications. Eur. Phys. J. Plus **132**(1), 1–13 (2017)
18. Gómez-Aguilar, J.F., Escobar-Jiménez, R.F., López-López, M.G., Alvarado-Martínez, V.M.: Atangana-Baleanu fractional derivative applied to electromagnetic waves in dielectric media. J. Electromagn. Waves Appl. **30**(15), 1937–1952 (2016)
19. Gómez-Aguilar, J.F., Atangana, A., Morales-Delgado, J.F.: Electrical circuits RC, LC, and RL described by Atangana-Baleanu fractional derivatives. Int. J. Circ. Theor. Appl. **1**, 1–22 (2017)
20. Goodman, T.R.: Application of Integral Methods to Transient Nonlinear Heat Transfer. Advances in Heat Transfer, vol. 1, pp. 51–122. Academic Press, San Diego (1964)
21. Gurtin, M.E.: On the thermodynamics of materials with memory. Arch. Rational. Mech. Anal. **28**, 40–50 (1968)
22. Gurtin, M.E., Pipkin, A.C.: A general theory of heat conduction with finite wave speeds. Arch. Rational Mech. Anal. **31**, 113–126 (1968)
23. Havlin, S., Ben, Avraham D.: Diffusion in disordered media. Adv. Phys. **36**, 695–798 (1987)
24. Hristov, J.: Approximate solutions to time-fractional models by integral balance approach. In: Cattani, C., Srivastava, H.M., Yang, X.-J. (eds.) Fractional Dynamics, pp. 78–109. De Gruyter Open, Berlin (2015)
25. Hristov, J.: Transient heat diffusion with a non-singular fading memory: from the Cattaneo constitutive equation with Jeffrey's kernel to the Caputo-Fabrizio time-fractional derivative. Therm. Sci. **20**, 765–770 (2016)
26. Hristov, J.: Integral solutions to transient nonlinear heat (mass) diffusion with a power-law diffusivity: a semi-infinite medium with fixed boundary conditions. Heat Mass Transf. **52**, 635–655 (2016)

27. Hristov, J.: Steady-state heat conduction in a medium with spatial non-singular fading memory: derivation of Caputo-Fabrizio space-fractional derivative with Jeffrey's kernel and analytical solutions. Therm. Sci. **21**, 827–839 (2017)
28. Hristov, J.: Double integral-balance method to the fractional subdiffusion equation: approximate solutions, optimization problems to be resolved and numerical simulations. J. Vib. Control **23**, 2795–2818 (2017)
29. Hristov, J.: Space-fractional diffusion with a potential power-law coefficient: transient approximate solution. Progr. Fract. Differ. Appl. **3**, 119–139 (2017)
30. Hristov, J.: Transient space-fractional diffusion with a power-law superdiffusivity: approximate integral-balance approach. Fundam. Inform. **151**, 371–388 (2017)
31. Hristov, J.: Fractional derivative with non-singular kernels: from the Caputo-Fabrizio definition and beyond: appraising analysis with emphasis on diffusion models. In: Bhalekar, S. (ed.) Frontiers in Fractional Calculus, pp. 269–342. Bentham Science Publishers, Sharjah (2017)
32. Hristov, J.: Integral-balance solution to nonlinear subdiffusion equation. In: Bhalekar, S. (ed.) Frontiers in Fractional Calculus, pp. 71–106. Bentham Science Publishers, Sharjah (2017)
33. Hughes, B.D.: Random Walks and Random Environments, vol. 1. Random Walks. Oxford University Press, New York (1995)
34. Le Vot, F., Abad, E., Yuste, S.B.: Continuous time random walk model for anomalous diffusion in expanding media. Phys. Rev. E 1–11 (2017)
35. Mainardi, F.: Fractional Calculus and Waves in Linear Viscoelasticity. Imperial College Press, London (2010)
36. Metzler, R., Klafter, J.: The random walk's guide to anomalous diffusion: a fractional dynamics approach. Phys. Rep. **339**, 1–77 (2000)
37. Metzler, R., Klafter, J.: The restaurant at the end of the random walk: recent developments in the description of anomalous transport by fractional dynamics. J. Phys. A Math. Gen. **37**, 161–208 (2004)
38. Miller, R.K.: An integrodifferential equation for rigid heat conductors with memory. J. Math. Anal. Appl. **66**, 313–332 (1978)
39. Nachlinger, R.R., Weeler, L.: A uniqueness theorem for rigid heat conductors with memory. Q. Appl. Math. **31**, 267–273 (1973)
40. Nunciato, J.W.: On heat conduction in materials with memory. Q. Appl. Math. **29**, 187–273 (1971)
41. Nunciato, J.W.: On uniqueness in the linear theory of heat conduction with finite wave speds. SIAM J. Appl. Math. **25**, 1–4 (1973)
42. Podlubny, I.: Fractional Differential Equations: An Introduction to Fractional Derivatives, Fractional Differential Equations, to Methods of Their Solution and some of Their Applications. Academic Press, San Diego, Calif, USA (1999)
43. Saad, K.M., Gómez-Aguilar, J.F.: Analysis of reaction diffusion system via a new fractional derivative with non-singular kernel. Phys. A: Stat. Mech. Appl. **509**, 703–716 (2018)
44. Scott-Blair, G.W.: Analytical and integrative aspects of the stress-strain-time problem. J. Sci. Instrum. **21**, 80–84 (1944)
45. Storm, M.L.: Heat conduction in simple metals. J. Appl. Phys. **22**, 940–951 (1951)
46. Tateishi, A.A., Ribeiro, H.V., Lenzi, E.K.: The role of fractional time-derivative operators on anomalous diffusion. Front. Phys. **1**, 1–16 (2017)
47. Volterra, V.: Theory of functional, English edn. Blackie and Son, London (1930)
48. Yuste, S.B., Lindenberg, K.: Comments on first passage time for anomalous diffusion. Phys. Rev. E 1–8 (2004)

Numerical Solutions and Pattern Formation Process in Fractional Diffusion-Like Equations

Kolade M. Owolabi

Abstract Nowadays, a lot of researchers have challenged the use of classical diffusion equation to model real life situations. To circumvent some of the up-roaring challenges, time and space fractional derivatives have been proposed as alternative to model some anomalous diffusion or related processes where a particle plume spreads at inconsistent rate with the classical Brownian motion model. In this work, we shall consider the general diffusion equations with fractional order derivatives which describe the diffusion in complex systems. Fractional diffusion equation is obtained by allowing the exponent order α to vary in the intervals $(0, 1)$ and $(1, 2)$ which correspond to subdiffusion and superdiffusion special cases. For the numerical approximations, we propose to use the newly correct version of the Adams-Bashforth scheme which takes into account the nonlinearity of the kernels such as the Mittag-Leffler law for the Atangana-Baleanu case, the power law for the Riemann-Liouville and Caputo derivatives. The efficiency and accuracy of the numerical schemes based on these operators will be justified by reporting their norm infinity and norm relative errors. The complexity of the dynamics in the equations will be discussed theoretically by examining their local and global stability analysis. Our numerical experiment results are expected to give a new direction into pattern formation process in fractional diffusion-like scenarios.

Keywords Fractional calculus · Atangana-Baleanu fractional derivative · Diffusion equations

---

K. M. Owolabi (✉)
Faculty of Natural and Agricultural Sciences, Institute for Groundwater Studies, University of the Free State, Bloemfontein 9300, South Africa
e-mail: kmowolabi@futa.edu.ng

K. M. Owolabi
Department of Mathematical Sciences, Federal University of Technology, PMB 704, Akure, Ondo State, Nigeria

J. F. Gómez et al. (eds.), *Fractional Derivatives with Mittag-Leffler Kernel*, Studies in Systems, Decision and Control 194,
https://doi.org/10.1007/978-3-030-11662-0_12

1 Introduction

In this paper we consider a general time-fractional diffusion equation of the form

$$\frac{\partial^\alpha u(x,t)}{\partial t^\alpha} = d\frac{\partial^2 u(x,t)}{\partial x^2} + f(x,t), \qquad (x,t) \in [0,L] \times [0,T], \tag{1}$$

where x and t are the respective space and time variables, d denotes an arbitrary positive constant usually called the diffusion coefficient, $f(x,t)$ stands for a sufficiently smooth function or any of the nonlinearities often used in most models ranging from combustion theory to ecology, $0 < \alpha \leq 1$ and $0 < \alpha \leq 2$ are the sub-diffusive and super-diffusive scenarios, subject to the appropriate initial conditions with either the Neumann or Dirichlet boundary conditions. In the present case, $u : [0,L] \times [0,T] \to \mathbb{R}$, also $f : [0,L] \times [0,T] \to \mathbb{R}$. We apply both the Atangana-Baleanu in caputo sense [3, 6] and the Caputo [10] fractional derivative operators are respectively defined as

$$\frac{\partial^\alpha u(t)}{\partial t^\alpha} = {}^{AB}_{a}\mathscr{D}^{\alpha}_{t}[u(t)] = \frac{M(\alpha)}{1-\alpha}\int_a^t u'(\xi)E_\alpha\left[-\alpha\frac{(t-\xi)^\alpha}{1-\alpha}\right]d\xi, \tag{2}$$

and

$$\frac{\partial^\alpha u(t)}{\partial t^\alpha} = {}^{C}_{0}\mathscr{D}^{\alpha}_{t}u(t) = \frac{1}{\Gamma(n-\alpha)}\int_0^t \frac{u^{(n)}\xi}{(t-\xi)^{\alpha+1-n}}d\xi, \quad n-1 < \alpha < n, \ n \in \mathbb{N}, \tag{3}$$

Fractional reaction-diffusion equations represent the extension of classical equations of mathematical biology and physics [3, 5, 6, 8, 12–16, 20, 22–28, 33, 34, 40–44]. In recent decades, the concept of differentiation with fractional order, nowadays is increasingly popular and generates a lot of interest; however, it could be seen as an old concept that has been studied by some researchers within the field of fractional calculus.

Over the years, Laplace transform was applied to solve a range of linear problems. This perhaps motivates Michele Caputo to solve linear equation which describes a real-world problem with the Laplace transform method. In his remarks, two definitions based on power law were suggested. However, both definitions posed serious setback to the derivatives due to singularity problem that occurs at the origin [4]. To circumvent the problem of singularity, Caputo and Fabrizio [10, 11] constructed a new derivative with no singularity by using the exponential decay law. Later, Atangana and Baleanu [3] introduced the concept of the generalized Mittag-Leffler function to derive another derivative with no-singular and non-local kernel.

The time fractional reaction-diffusion-like equations have been used to describe the most important physical phenomena arising in engineering and other physical problems, for example, in fractals and groundwater models, biological systems, control, pattern formation processes, amorphous, comb structures, colloid, glassy and porous materials, and percolation clusters, polymers, random and disordered media, dielectrics and semiconductors, signal processing, geophysical and geological processes (see for instance [1, 2, 4, 19, 29, 30]). The major advantage of fractional derivatives over its classical counterpart lies in the fact that they are suitable and robust for modelling memory and hereditary properties of a lot of materials and processes.

Fractional reaction-diffusion equations are an important class of differential equations. In these years, a lot of theory and numerical techniques for fractional differential equations have been addressed [14, 49]. Most of the analytical methods applied to solve these equations have a lot of limitations in terms of application, and the numerical techniques often applied give rise to round-off errors. Recently, Owolabi and Atangana [44] proposed a new fractional Adams-Bashforth method with the Caputo-Fabrizio derivative for the solution of linear and nonlinear fractional differential equations. In line with this, we propose a new two-step Laplace Adams-Bashforth scheme for the solution of time fractional diffusion-like and differential equations. One of the major advantages of this method is due to its flexibility to switch between ordinary and partial differential equations.

In this work, we use the Mittag-Leffler function in conjunction with the Atangana-Baleanu derivative in the sense of Caputo and the power law via Caputo fractional derivative to model a range of time-fractional reaction-diffusion equations. The Atangana-Baleanu fractional order derivatives are known to have possessed two important roles; because, they serve as filters with fractional regulators and display the role of non-integer derivatives. The nonlocal kernel role paves way for better representation of median with different scales as well as the memory within structure, while the second role is the memory effect that the new derivative brought to mathematical equations.

The remainder part of the work is structured as follows: Some useful preliminaries and definitions are given. A viable numerical method for the approximation of fractional derivative is formulated. We finally present some illustrative examples of time-fractional reaction-diffusion problems that are still of current and recurrent interest.

2 Some Basic Properties of Fractional Calculus

Here we briefly highlight some of the basic properties and definitions of fractional calculus.

The fractional derivative denoted by operator $\mathscr{D}_y^\alpha f(y)$ satisfies the properties: (a) $\mathscr{D}_y^0 f(y) = f(y)$, that is, identity property. (b) $\mathscr{D}_y^\alpha f(y) = f(y)$ is standard derivative if α $\alpha \in \mathbb{N}$. (c) Linearity property if $\mathscr{D}_y^\alpha [af(y) + bg(y)] = {}_a\mathscr{D}_y^\alpha f(y) + {}_b\mathscr{D}_y^\alpha g(y)$. (d) The Leibniz product rule, $\mathscr{D}_y^\alpha [f(y)g(y)] = \sum_{k=0}^{\infty} \binom{\alpha}{k} \mathscr{D}_y^\alpha [f(y)] \mathscr{D}_y^\alpha [g(y)]$.

Here we adopt the notation $\mathscr{L}$ to represent a Laplace transform with variable u and $\mathscr{F}$ to represent a Fourier transform with variable, say ω. The term $\mathscr{D}$ denotes a fractional derivative with respect to time t of order α.

The Fourier transform pairs are given as

$$\hat{u}(\omega) = \int_{-\infty}^{+\infty} e^{i\omega x} u(y) dy, \quad u(y) = \frac{1}{2\pi} \int_{-\infty}^{+\infty} e^{-i\omega x} \hat{u}(\omega) dy, \tag{4}$$

and the Laplace transform pairs

$$\hat{u}(x) = \int_{0}^{\infty} e^{-xt} u(t) dt, \quad u(t) = \int_{c-i\infty}^{c+i\infty} e^{xt} \hat{u}(x) dx. \tag{5}$$

The following are some of the transform results for fractional derivatives:

(i) Atangana-Baleanu derivative in the sense of Caputo

$${}_{a}^{ABC}\mathscr{D}_t^{\alpha} u(t) = \frac{M(\alpha)}{1-\alpha} \int_a^t u'(\xi) E_{\alpha} \left[-\alpha \frac{(t-\xi)^{\alpha}}{1-\alpha} \right] d\xi, \tag{6}$$

with Laplace transform given as

$$\mathscr{L}\left\{ {}_{0}^{ABC}\mathscr{D}_t^{\alpha} u(t) \right\}(s) = \frac{M(\alpha)}{1-\alpha} \frac{s^{\alpha} \mathscr{L}\{u(t)\}(s) - s^{\alpha-1} u(0)}{s^{\alpha} + \alpha(1-\alpha)^{-1}}, \tag{7}$$

where $M(\alpha)$ is the normalized function as defined by Caputo and Fabrizio [10], and E_{α} is the one parameter Mittag-Leffler function defined by the following power series in the entire complex plane

$$E_{\alpha}(z) = \sum_{n=0}^{\infty} \frac{z^n}{\Gamma(\alpha n + 1)}, \quad \alpha > 0, \quad z \in \mathbb{C}, \tag{8}$$

$$\mathscr{L}\left\{ E_{\alpha} \left[-\left(\frac{t}{\xi} \right)^{\alpha} \right] \right\} = \frac{1}{x + \frac{x^{1-\alpha}}{\xi^{\alpha}}} \quad \alpha > 0. \tag{9}$$

Note that

$$E_1(z) = \sum_{n=0}^{\infty} \frac{z^n}{\Gamma(n+1)} = \sum_{n=0}^{\infty} \frac{z^n}{n!} = e^z. \tag{10}$$

(ii) Caputo derivative

$${}_{0}^{C}\mathscr{D}_t^{\alpha} u(t) = \frac{1}{\Gamma(n-\alpha)} \int_0^t \frac{u^{(n)}\xi}{(t-\xi)^{\alpha+1-n}} d\xi, \quad n-1 < \alpha < n, \tag{11}$$

with Laplace transform

$$\mathscr{L}\left\{ {}_{0}^{C}\mathscr{D}_t^{\alpha} u(t) \right\} = x^{\alpha} \hat{u}(x) - [x^{\alpha-1} u(0)]. \tag{12}$$

(iii) Riemann-Liouville derivative

$$\mathscr{D}_t^{\alpha} u(t) = \frac{1}{\Gamma(\alpha)} \frac{d}{dt} \left(\int_0^t \frac{u(\xi)}{(t-\xi)^{1-\alpha} d\xi} \right) \quad 0 < \alpha < 1, \tag{13}$$

with Laplace transform

$$\mathscr{L}\left\{\mathscr{D}_t^{\alpha} u(t)\right\} = x^{\alpha} \hat{u}(x) - \left[\mathscr{D}_t^{\alpha-1} u(t)\right]_{t=0}, \tag{14}$$

and Fourier transform

$$\mathscr{F}\left[\mathscr{D}_y^{\alpha} u(y)\right] = (i\omega)^{\alpha} \hat{u}(\omega). \tag{15}$$

(iv) Atangana-Baleanu fractional derivative in Riemann-Liouville sense

$${}_{a}^{ABR}\mathscr{D}_t^{\alpha}[u(t)] = \frac{M(\alpha)}{1-\alpha} \frac{d}{dt} \int_a^t u(\xi) E_{\alpha}\left[-\alpha \frac{(t-\xi)^{\alpha}}{1-\alpha}\right] d\xi, \tag{16}$$

with Laplace transform given as

$$\mathscr{L}\left\{{}_{0}^{ABR}\mathscr{D}_t^{\alpha} u(t)\right\}(s) = \frac{M(\alpha)}{1-\alpha} \frac{s^{\alpha} \mathscr{L}\{u(t)\}(s)}{s^{\alpha} + \alpha(1-\alpha)^{-1}}. \tag{17}$$

Many other fractional derivatives and their Laplace transforms can be found in classical books [32, 36, 37, 48].

3 Numerical Methods

We start our numerical approximation by first considering a semi-discrete equation formed by a difference method through space. That is, we discretize the spatial derivative by using a finite difference scheme on a uniform mesh on interval $[0, L]$, defined by

$$x_0 = 0 < x_1 < \cdots < x_M, \quad x_m = m\Delta x, \quad \Delta x = L/M.$$

We adopt the second-order central difference method

$$\frac{\partial^2 u(x_i, t)}{\partial x^2} = \frac{u(x_{i-1}, t) - 2u(x_i, t) + u(x_{i+1}, t)}{(\Delta x)^2} + \frac{\partial^4 u(\tau, t)}{\partial x^4} \frac{(\Delta x)^2}{12}, \tag{18}$$

with $\tau \in [x_{i-1}, x_{i+1}]$, provided that function u is sufficiently smooth. Some of the important features to take note about the centered finite difference schemes is that they are symmetric in nature, and possess an even-order of accuracy [38]. The weights of some of the central finite difference methods are well presented in [18].

In addition, we also consider second-order approximations of the Neumann type at the boundary by the centered difference scheme

$$u_x(x,t) = \frac{u(x+\Delta x,t) - u(x-\Delta x,t)}{2\Delta x} + \frac{(\Delta x)^2}{6} u_{xx}(x,t) + \mathscr{O}((\Delta x)^4). \quad (19)$$

By following [9, 31, 45, 46], at the internal mesh points we introduce the discrete operator, say $\mathscr{L}$, such that

$$\mathscr{L}u(x_i,t) = \frac{u(x_{i-1},t) - 2u(x_i,t) + u(x_{i+1},t)}{(\Delta x)^2}, \quad (20)$$

so that (1) becomes

$$\frac{\partial^\alpha u(x_i,t)}{\partial t^\alpha} = \mathscr{L}u(x_i,t) + f(x_i,t) + E(x_i,t), \quad i = 1, 2, \ldots, M-1, \quad (21)$$

where

$$|E(x_i,t)| \le \frac{1}{12} \max_{[0,L]\times[0,T]} \left| \frac{\partial^4 u(x,t)}{\partial x^4} \right| (\Delta x)^2.$$

In compact form, the semi-discretization of equation (1) can be written as

$$\frac{\partial^\alpha u(t)}{\partial t^\alpha} = Lu(t) + f(t), \quad (22)$$

where $u(t) = [u_1(t), u_2(t), \ldots, u_{M-1}(t)]^{\mathrm{T}}$, $u_i(t)$ denotes an approximation of $u_i(x_i, t)$, $f(t) = [f(x_1,t), f(x_2,t), \ldots, f(x_{M-1},t)]^{\mathrm{T}}$, and the tridiagonal matrix L, expressed as

$$\mathbf{L} = \begin{pmatrix} -2 & 1 & & & & \\ 1 & -2 & 1 & & & \\ & & \ddots & \ddots & \ddots & \\ & & & 1 & -2 & 1 \\ & & & & 1 & -2 \end{pmatrix}_{(M-1)\times(M-1)} . \quad (23)$$

If Eq. (19) is used to discretize the boundary conditions, we obtain

$$u_{-1}(t) = u_1(t) \;\text{ and }\; u_{M+1}(t) = u_{M-1}(t). \quad (24)$$

To fully discretize, one only require to approximate the component $u^i(t)$ of (22) by u^i_N. By using the points $t_0 = 0 < t_1 < t_2 < \cdots < t_N = T$, with

$$u^i_N(t) = \sum_k^N \omega_k(t) u^i_N(t_k). \quad (25)$$

Further, we express

$$\omega_k(t) = {}_0\mathscr{D}_t^{\alpha}\omega_k(t),$$

so as to formulate the following approximation

$${}_0\mathscr{D}_t^{\alpha}u^i(t) \implies {}_0\mathscr{D}_t^{\alpha}u_N^i(t) = \sum_k^N \omega_k(t)u_N^i(t_k).$$

Therefore

$${}_0\mathscr{D}_t^{\alpha}u^i(t_j) \implies \sum_{k=0}^N \omega_k(t_j)u_{i,k} = \sum_{k=1}^N \omega_k(t_j)u_{i,k} + \omega_0(t_j)u_0(x_i), \tag{26}$$

for $j = 1, 2, \ldots, N$, $u_{i,k} = u_N^i(t_k) \implies u(x_i, t_k)$ and $u_0(x)$ is the initial function expected to be given.

For the boundary conditions, we employ the above centered difference approximation (19) to obtain (24), and by Eq. (22), we get

$$\begin{aligned}
\sum_{k=0}^N \omega_k(t_j)u_{0,k} &= \frac{2u_{1,j} - 2u_{0,j}}{(\Delta x)^2} + f(x_0, t_j),\\
\sum_{k=0}^N \omega_k(t_j)u_{i,k} &= \frac{u_{i-1,j} - 2u_{i,j} + u_{i+1,j}}{(\Delta x)^2} + f(x_i, t_j),\ i = 1, 2, \ldots, M-1,\\
\sum_{k=0}^N \omega_k(t_j)u_{M,k} &= \frac{2u_{M-1,j} - 2u_{M,j}}{(\Delta x)^2} + f(x_M, t_j),\ j = 1, 2, \ldots, N.
\end{aligned}$$

Hence, the discretized form of problem (1) is represented by the linear system

$$AU + au_0{}^T = \frac{UB}{\Delta x^2} + F, \tag{27}$$

with matrix B defines as

$$\mathbf{B} = \begin{pmatrix} -2 & 1 & & & & & \\ 2 & -2 & 1 & & & & \\ & 1 & -2 & 1 & & & \\ & & \ddots & \ddots & \ddots & & \\ & & & & 1 & -2 & 2 \\ & & & & & 1 & -2 \end{pmatrix}, \tag{28}$$

with

$$(A)_{j,k} = \omega_k(t_j),\ (a)_j = \omega_0(t_j),\ j, k = 1, 2, \ldots, N,$$

and

$$(u_0)_{i+1} = u_0(x_i), (U)_{j,i+1} = u_{i,j}, (F)_{j,i+1} = f(x_i, t_j),\ j = 1, 2, \ldots, N,\ i = 0, 1, \ldots, M.$$

Secondly, we follow [21] and derive a two-step Adams-Bashforth scheme via the Laplace method to approximate (22) in the sense of the Caputo and the Atangana-Baleanu fractional derivatives. This is achieved by taking the Laplace of both sides of (22), to obtain

$$\mathscr{L}\left\{\frac{\partial^\alpha u(x,t)}{\partial t^\alpha}\right\} = \mathscr{L}\left\{\mathbf{L}u(x,t) + \mathbf{F}u(x,t)\right\}. \tag{29}$$

In the sense of Caputo derivative, above expression transforms into

$$\underbrace{\mathscr{L}\left\{\frac{\partial^\alpha u(x,t)}{\partial t^\alpha}\right\}}_{{}_0^C\mathscr{D}_t^\alpha u(t)} = \mathscr{L}\underbrace{\{\mathbf{L}u(x,t) + \mathbf{F}u(x,t)\}}_{\mathscr{G}(u,t)}, \tag{30}$$

where ${}_0^C\mathscr{D}_t^\alpha u(t)$ is the Caputo fractional derivative of order α as given above. Next, the Caputo fractional integral operator is applied on both sides of (30) to get

$$u(t) - u(t_0) = \frac{1}{\Gamma(\alpha)}\int_0^t (t-\xi)^{\alpha-1}\mathscr{G}(u,\xi)d\xi. \tag{31}$$

With $t = t_{n+1}$, one obtains

$$u(t_{n+1}) = u_0 + \frac{1}{\Gamma(\alpha)}\int_0^{t_{n+1}} (t_{n+1}-\xi)^{\alpha-1}\mathscr{G}(u,\xi)d\xi.$$

Likewise, with $t = t_n$ yields

$$u(t_n) = u_0 + \frac{1}{\Gamma(\alpha)}\int_0^{t_n} (t_n-\xi)^{\alpha-1}\mathscr{G}(u,\xi)d\xi,$$

so that

$$u_{n+1} - u_n = \frac{1}{\Gamma(\alpha)}\left\{\int_0^{t_{n+1}} (t_{n+1}-\xi)^{\alpha-1}\mathscr{G}(u,\xi)d\xi - \int_0^{t_n} (t_n-\xi)^{\alpha-1}\mathscr{G}(u,\xi)d\xi\right\}. \tag{32}$$

Bear in mind that with

$$\int_0^{t_{n+1}} (t_{n+1}-\xi)^{\alpha-1}\mathscr{G}(u,\xi)d\xi = \sum_{i=0}^{n}\int_{t_i}^{t_{i+1}} (t_{n+1}-\xi)^{\alpha-1}\mathscr{G}(u,\xi)d\xi,$$

the nonlinear function $\mathscr{G}(u,t)$ in (30) can be approximated using the Lagrange polynomial of the form

$$\mathscr{P}(t) \approx \mathscr{G} = \frac{t-t_{n-1}}{t_n - t_{n-1}}\mathscr{G}_n + \frac{t-t_n}{t_{n-1}-t_n}\mathscr{G}_{n-1}. \tag{33}$$

Conveniently, we can write the fractional integral in (32) as

$$\begin{aligned}
\int_0^{t_{n+1}} (t_{n+1}-\xi)^{\alpha-1}\mathscr{G}(u,\xi)d\xi &= \sum_{i=0}^{n}\int_{t_i}^{t_{i+1}} (t_{n+1}-t)^{\alpha-1}\left(\frac{t-t_{n-1}}{t_n-t_{n-1}}\mathscr{G}_n + \frac{t-t_n}{t_{n-1}-t_n}\mathscr{G}_{n-1}\right)dt,\\
&= \sum_{i=0}^{n}\left\{\frac{\mathscr{G}_n}{t_n-t_{n-1}}\int_{t_i}^{t_{i+1}}(t_{n+1}-t)^{\alpha-1}(t-t_{n-1})dt + \frac{\mathscr{G}_{n-1}}{t_n-t_{n-1}}\int_{t_i}^{t_{i+1}}(t_{n+1}-t)^{\alpha-1}(t-t_n)dt\right\},\\
&= \sum_{i=0}^{n}\left\{\frac{\mathscr{G}_n}{\hbar}\int_{t_i}^{t_{i+1}}(t_{n+1}-t)^{\alpha-1}(t-t_{n-1})dt - \frac{\mathscr{G}_{n-1}}{\hbar}\int_{t_i}^{t_{i+1}}(t_{n+1}-t)^{\alpha-1}(t-t_n)dt\right\}.
\end{aligned} \tag{34}$$

By adopting some change of variables, such as $\tau = t_{n+1}-t,\ t = t_{n+1}-\tau$ and $dt = -d\tau$, we have

$$\begin{aligned}
\int_{t_i}^{t_{i+1}}(t_{n+1}-t)^{\alpha-1}(t-t_{n-1})dt &= \int_{t_{n+1}-t_i}^{t_{n+1}-t_{i+1}} \tau^{\alpha-1}(-\tau+t_{n+1}-t_{n-1})d\tau,\\
&= \frac{1}{\alpha+1}\left\{\tau^{\alpha+1}\right\}_{t_{n+1}-t_i}^{t_{n+1}-t_{i+1}} - \frac{2\hbar}{\alpha}\left\{\tau^{\alpha}\right\}_{t_{n+1}-t_i}^{t_{n+1}-t_{i+1}},\\
&= \frac{1}{\alpha+1}\left[(t_{n+1}-t_{i+1})^{\alpha+1}-(t_{n+1}-t_i)^{\alpha+1}\right] - \frac{2\hbar}{\alpha}\left[(t_{n+1}-t_{i+1})^{\alpha}-(t_{n+1}-t_i)^{\alpha}\right].
\end{aligned} \tag{35}$$

Similarly,

$$\begin{aligned}
\int_{t_i}^{t_{i+1}}(t_{n+1}-t)^{\alpha-1}(t-t_n)dt &= -\int_{t_{n+1}-t_i}^{t_{n+1}-t_{i+1}} \tau^{\alpha-1}(-\tau+t_{n+1}-t_n)d\tau,\\
&= \frac{1}{\alpha+1}\left\{\tau^{\alpha+1}\right\}_{t_{n+1}-t_i}^{t_{n+1}-t_{i+1}} - \frac{\hbar}{\alpha}\left\{\tau^{\alpha}\right\}_{t_{n+1}-t_i}^{t_{n+1}-t_{i+1}},\\
&= \frac{1}{\alpha+1}\left[(t_{n+1}-t_{i+1})^{\alpha+1}-(t_{n+1}-t_i)^{\alpha+1}\right] - \frac{\hbar}{\alpha}\left[(t_{n+1}-t_{i+1})^{\alpha}-(t_{n+1}-t_i)^{\alpha}\right],
\end{aligned} \tag{36}$$

which implies that

$$\begin{aligned}
\int_0^{t_{n+1}}(t_{n+1}-\xi)^{\alpha-1}\mathscr{G}(u,\xi)d\xi = \frac{\mathscr{G}_n}{\hbar}&\left\{\sum_{i=0}^{n}\left[\frac{1}{\alpha+1}\left[(t_{n+1}-t_{i+1})^{\alpha+1}-(t_{n+1}-t_i)^{\alpha+1}\right]\right]\right.\\
&\left.-\frac{2\hbar}{\alpha}\sum_{i=0}^{n}\left(\left[(t_{n+1}-t_{i+1})^{\alpha}-(t_{n+1}-t_i)^{\alpha}\right]\right)\right\}\\
-\frac{\mathscr{G}_{n-1}}{\hbar}&\left\{\sum_{i=0}^{n}\left[\frac{1}{\alpha+1}\left[(t_{n+1}-t_{i+1})^{\alpha+1}-(t_{n+1}-t_i)^{\alpha+1}\right]\right]\right.\\
&\left.-\frac{\hbar}{\alpha}\sum_{i=0}^{n}\left(\left[(t_{n+1}-t_{i+1})^{\alpha}-(t_{n+1}-t_i)^{\alpha}\right]\right)\right\}.
\end{aligned} \tag{37}$$

That is,

$$\int_0^{t_{n+1}} (t_{n+1}-\xi)^{\alpha-1}\mathscr{G}(u,\xi)d\xi = \frac{\mathscr{G}_n}{\hbar}\left[\frac{2\hbar}{\alpha}(t_{n+1}-t_0)^{\alpha} - \frac{1}{\alpha+1}(t_{n+1}-t_0)^{\alpha+1}\right]$$
$$-\frac{\mathscr{G}_{n-1}}{\hbar}\left[\frac{\hbar}{\alpha}(t_{n+1}-t_0)^{\alpha} - \frac{1}{\alpha+1}(t_{n+1}-t_0)^{\alpha+1}\right],$$
$$= \frac{\mathscr{G}_n}{\hbar}\left[\frac{2\hbar(n+1)^{\alpha}\hbar^{\alpha}}{\alpha} - \frac{(n+1)^{\alpha+1}\hbar^{\alpha+1}}{\alpha+1}\right]$$
$$-\frac{\mathscr{G}_{n-1}}{\hbar}\left[\frac{\hbar(n+1)^{\alpha}\hbar^{\alpha}}{\alpha} - \frac{(n+1)^{\alpha+1}\hbar^{\alpha+1}}{\alpha+1}\right]. \quad (38)$$

By simplifying further, we obtain

$$\int_0^{t_{n+1}} (t_{n+1}-\xi)^{\alpha-1}\mathscr{G}(u,\xi)d\xi = \hbar^{\alpha}\left\{\left[\frac{2(n+1)^{\alpha}}{\alpha} - \frac{(n+1)^{\alpha+1}}{\alpha+1}\right]\mathscr{G}_n - \left[\frac{(n+1)^{\alpha}}{\alpha} - \frac{(n+1)^{\alpha+1}}{\alpha+1}\right]\mathscr{G}_{n-1}\right\}. \quad (39)$$

Similarly, we evaluate the second fractional integral in (32) as

$$\int_0^{t_n} (t_n-\xi)^{\alpha-1}\mathscr{G}(u,\xi)d\xi = \frac{\mathscr{G}_n}{\hbar}\sum_{i=0}^{n-1}\int_{t_i}^{t_{i+1}} (t_n-t)^{\alpha-1}(t-t_{n-1})dt$$
$$- \frac{\mathscr{G}_{n-1}}{\hbar}\sum_{i=0}^{n-1}\int_{t_i}^{t_{i+1}} (t_n-t)^{\alpha-1}(t-t_n)dt. \quad (40)$$

Next, we introduce some change of variables parameters as: $\tau = t_n - t,\ t = t_n - \tau,\ dt = -d\tau$, so that

$$\int_0^{t_n} (t_n-\xi)^{\alpha-1}\mathscr{G}(u,\xi)d\xi = \frac{\mathscr{G}_n}{\hbar}\sum_{i=0}^{n-1}\int_{t_n-t_i}^{t_n-t_{i+1}}\left(\tau^{\alpha}-\hbar\tau^{\alpha-1}\right)d\tau - \frac{\mathscr{G}_{n-1}}{\hbar}\sum_{i=0}^{n-1}\int_{t_n-t_i}^{t_n-t_{i+1}}\tau^{\alpha}d\tau,$$
$$= \frac{\mathscr{G}_n}{\hbar}\sum_{i=0}^{n-1}\left[\frac{\tau^{\alpha+1}}{\alpha+1} - \frac{\hbar\tau^{\alpha}}{\alpha}\right]_{t_n-t_i}^{t_n-t_{i+1}} - \frac{\mathscr{G}_{n-1}}{\hbar}\sum_{i=0}^{n-1}\left[\frac{\tau^{\alpha+1}}{\alpha+1}\right]_{t_n-t_i}^{t_n-t_{i+1}},$$
$$= \frac{\mathscr{G}_n}{\hbar}\sum_{i=0}^{n-1}\left[\frac{(t_n-t_{i+1})^{\alpha+1}}{\alpha+1} - \frac{\hbar(t_n-t_{i+1})^{\alpha}}{\alpha} - \frac{(t_n-t_i)^{\alpha+1}}{\alpha+1} - \frac{\hbar(t_n-t_i)^{\alpha}}{\alpha}\right]$$
$$-\frac{\mathscr{G}_{n-1}}{\hbar}\sum_{i=0}^{n-1}\left[\frac{(t_n-t_{i+1})^{\alpha+1}}{\alpha+1} - \frac{(t_n-t_i)^{\alpha+1}}{\alpha+1}\right],$$
$$= \frac{\mathscr{G}_n}{\hbar}\left[\sum_{i=0}^{n-1}\left(\frac{(t_n-t_{i+1})^{\alpha+1}}{\alpha+1} - \frac{(t_n-t_i)^{\alpha+1}}{\alpha+1}\right) - \frac{\hbar}{\alpha}\sum_{i=0}^{n-1}\left((t_n-t_{i+1})^{\alpha} - (t_n-t_i)^{\alpha}\right)\right]$$

$$-\frac{\mathscr{G}_{n-1}}{\hbar}\sum_{i=0}^{n-1}\left(\frac{(t_n-t_{i+1})^{\alpha+1}}{\alpha+1}-\frac{(t_n-t_i)^{\alpha+1}}{\alpha+1}\right),$$

which implied that

$$\int_0^{t_n}(t_n-\xi)^{\alpha-1}\mathscr{G}(u,\xi)d\xi=\frac{\mathscr{G}_n}{\hbar}\left[-\frac{(t_n-t_0)^{\alpha+1}}{\alpha+1}+\frac{\hbar(t_n-t_0)^{\alpha}}{\alpha}\right]+\frac{\mathscr{G}_{n-1}(t_n-t_0)^{\alpha+1}}{\hbar(\alpha+1)}. \tag{41}$$

With $t_n = n\hbar$ and $t_{n+1} = (n+1)\hbar$ yields

$$\int_0^{t_n}(t_n-\xi)^{\alpha-1}\mathscr{G}(u,\xi)d\xi=\frac{\mathscr{G}_n}{\hbar}\left[-\frac{n^{\alpha+1}\hbar^{\alpha+1}}{\alpha+1}+\frac{n^{\alpha}\hbar^{\alpha+1}}{\alpha}\right]+\frac{n^{\alpha+1}\hbar^{\alpha+1}}{\alpha+1}. \tag{42}$$

Collecting the like terms in powers of $\hbar$ gives

$$\int_0^{t_n}(t_n-\xi)^{\alpha-1}\mathscr{G}(u,\xi)d\xi=\hbar^{\alpha}\left[\left(\frac{n^{\alpha}}{\alpha}-\frac{n^{\alpha+1}}{\alpha+1}\right)\mathscr{G}_n+\frac{n^{\alpha+1}}{\alpha+1}\right]. \tag{43}$$

Next, we substitute for both Eqs. (39) and (43) in (32) so that

$$\begin{aligned}u_{n+1}-u_n=\frac{\hbar^{\alpha}}{\Gamma(\alpha)}\Bigg\{&\left[\frac{2(n+1)^{\alpha}}{\alpha}-\frac{(n+1)^{\alpha+1}}{\alpha+1}\right]\mathscr{G}_n\\&-\left[\frac{(n+1)^{\alpha}}{\alpha}-\frac{(n+1)^{\alpha+1}}{\alpha+1}\right]\mathscr{G}_{n-1}-\left(\frac{n^{\alpha}}{\alpha}-\frac{n^{\alpha+1}}{\alpha+1}\right)\mathscr{G}_n-\frac{n^{\alpha+1}}{\alpha+1}\mathscr{G}_{n-1}\Bigg\},\end{aligned}$$

which on simplification results to

$$\begin{aligned}u_{n+1}-u_n=&\frac{\hbar^{\alpha}}{\Gamma(\alpha)}\Bigg\{\left[\frac{2(n+1)^{\alpha}-n^{\alpha}}{\alpha}+\frac{n^{\alpha+1}-(n+1)^{\alpha+1}}{\alpha+1}\right]\mathscr{G}_n\\&-\left[\frac{(n+1)^{\alpha}}{\alpha}+\frac{n^{\alpha+1}-(n+1)^{\alpha+1}}{\alpha+1}\right]\mathscr{G}_{n-1}\Bigg\}.\end{aligned} \tag{44}$$

It should be noted that when the value of α turns to unity in (44), we recover the classical version of the Adams-Bashforth method.

By applying the inverse Laplace to (44), we obtain the following numerical scheme

$$\begin{aligned}u(x,t_{n+1})=&\mathscr{L}^{-1}\Bigg\{u_n+\frac{\hbar^{\alpha}}{\Gamma(\alpha)}\Bigg[\left(\frac{n^{\alpha+1}-(n+1)^{\alpha+1}}{\alpha+1}+\frac{2(n+1)^{\alpha}-n^{\alpha}}{\alpha}\right)\mathscr{G}_n\\&-\left(\frac{n^{\alpha+1}-(n+1)^{\alpha+1}}{\alpha+1}+\frac{(n+1)^{\alpha}}{\alpha}\right)\mathscr{G}_{n-1}\Bigg]\Bigg\}.\end{aligned} \tag{45}$$

Theorem 3.1 *Given the general time-fractional partial differential equation of the form (1). The numerical solution with Laplace Adams-Bashforth scheme is expressed as*

$$u(x,t_{n+1}) = \mathscr{L}^{-1}\left\{u_n + \frac{\hbar^\alpha}{\Gamma(\alpha)}\left[\left(\frac{n^{\alpha+1}-(n+1)^{\alpha+1}}{\alpha+1}+\frac{2(n+1)^\alpha-n^\alpha}{\alpha}\right)\mathscr{G}_n - \left(\frac{n^{\alpha+1}-(n+1)^{\alpha+1}}{\alpha+1}+\frac{(n+1)^\alpha}{\alpha}\right)\mathscr{G}_{n-1}\right]+\mathscr{E}_n^{\alpha\alpha}\right\}, \tag{46}$$

where $\mathscr{E}_n^{\alpha\alpha} < \infty$.

Proof See the work in Refs. [7, 21] for a similar proof.

Also for the Atangana-Baleanu operator when applied to (30), we get

$$u(t) - u(0) = \frac{1-\alpha}{AB(\alpha)}\mathscr{G}(u,t) + \frac{\alpha}{AB(\alpha)\Gamma(\alpha)}\int_0^t \mathscr{G}(u,\xi)(t-\xi)^{\alpha-1}d\xi, \tag{47}$$

with $t = t_{n+1}$, the above expression becomes

$$u(t_{n+1}) = u_0 + \frac{1-\alpha}{AB(\alpha)}\mathscr{G}(t_{n+1}) + \frac{\alpha}{AB(\alpha)\Gamma(\alpha)}\int_0^{t_{n+1}} \mathscr{G}(u,\xi)(t_{n-1}-\xi)^{\alpha-1}d\xi. \tag{48}$$

We let $\hbar = t_{n+1} - t_n,\ n = 0, 1, 2, \ldots$ and $k = x_{j+1} - x_j$, so that if $t = t_n$, we have

$$u_(t_n) = u_0 + \frac{1-\alpha}{AB(\alpha)}\mathscr{G}(t_n) + \frac{\alpha}{AB(\alpha)\Gamma(\alpha)}\int_0^{t_n} \mathscr{G}(u,\xi)(t_n-\xi)^{\alpha-1}d\xi. \tag{49}$$

By subtracting (49) from (48) we obtain

$$u(t_{n+1}) - u(t_n) = \frac{1-\alpha}{AB(\alpha)}\{\mathscr{G}(t_{n+1}) - \mathscr{G}(t_n)\} + \frac{\alpha}{\mathscr{G}(\alpha)}\left\{\int_0^{t_{n+1}} \mathscr{G}(u,\xi)(t_{n+1}-\xi)^{\alpha-1}d\xi - \int_0^{t_n} \mathscr{G}(u,\xi)(t_n-\xi)^{\alpha-1}d\xi\right\}. \tag{50}$$

Next, we apply the Lagrange interpolation on the function $\mathscr{G}(u,\xi)$ to get

$$u(t_{n+1}) - u(t_n) = \frac{\hbar^\alpha}{\Gamma(\alpha)}\left\{\left[\frac{2(n+1)^\alpha - n^\alpha}{\alpha} + \frac{n^{\alpha+1}-(n+1)^{\alpha+1}}{\alpha+1}\right]\mathscr{G}_n - \left[\frac{(n+1)^\alpha}{\alpha} + \frac{n^{\alpha+1}-(n+1)^{\alpha+1}}{\alpha+1}\right]\mathscr{G}_{n-1}\right\} + \frac{1-\alpha}{AB(\alpha)}\left[\mathscr{G}(t_{n+1}) - \mathscr{G}(t_n)\right]. \tag{51}$$

To obtain the numerical solution, we take the inverse Laplace transform of the above equation as

$$
\begin{aligned}
u(t_{n+1}) = \mathscr{L}^{-1}\Bigg\{u(t_n) + \frac{\hbar^{\alpha}}{\Gamma(\alpha)}\Bigg[\Bigg[\frac{2(n+1)^{\alpha}-n^{\alpha}}{\alpha} + \frac{n^{\alpha+1}-(n+1)^{\alpha+1}}{\alpha+1}\Bigg]\mathscr{G}_n \\
-\Bigg[\frac{(n+1)^{\alpha}}{\alpha} + \frac{n^{\alpha+1}-(n+1)^{\alpha+1}}{\alpha+1}\Bigg]\mathscr{G}_{n-1}\Bigg]\Bigg\} \\
+\frac{1-\alpha}{AB(\alpha)}\mathscr{L}^{-1}\left\{\left[\mathscr{G}(t_{n+1})-\mathscr{G}(t_n)\right]\right\}.
\end{aligned}
\tag{52}
$$

In space, Eq. (52) is discretized as

$$
\begin{aligned}
u(x, t_{n+1}) = \Bigg\{u(x_j, t_n) + \frac{\hbar^{\alpha}}{\Gamma(\alpha)}\Bigg[\Bigg[\frac{2(n+1)^{\alpha}-n^{\alpha}}{\alpha} + \frac{n^{\alpha+1}-(n+1)^{\alpha+1}}{\alpha+1}\Bigg]\mathscr{G}_j^n \\
-\Bigg[\frac{(n+1)^{\alpha}}{\alpha} + \frac{n^{\alpha+1}-(n+1)^{\alpha+1}}{\alpha+1}\Bigg]\mathscr{G}_j^{n-1}\Bigg]\Bigg\} \\
+\frac{1-\alpha}{AB(\alpha)}\left[\mathscr{G}_j^{n+1}-\mathscr{G}_j^{n}\right],
\end{aligned}
\tag{53}
$$

which is the two-step Laplace Adams-Bashforth scheme.

4 Main Equations and Numerical Experiments

Here, we consider some illustrative examples of time-fractional reaction-diffusion-like problems involving the numerical methods discussed in above section. For the numerical computations, we apply the fractional two-steps Adams-Bashforth solver in the commercial software package (MATLAB) to solve the time-fractional reaction-diffusion equations in one and two dimensions at some instances of fractional index denoted by α.

4.1 Diffusion-Like Equation with Atangana-Baleanu Derivative

In this section, we first consider the classical diffusion-like equation in a modified one dimensional form

$$
{}_{0}^{ABC}\mathscr{D}_t^{\alpha} u(x,t) = dLu(x,t) + F(u(x,t)), \quad -L \le x \le L, \ t > 0, \ \alpha \in (0,1),
\tag{54}
$$

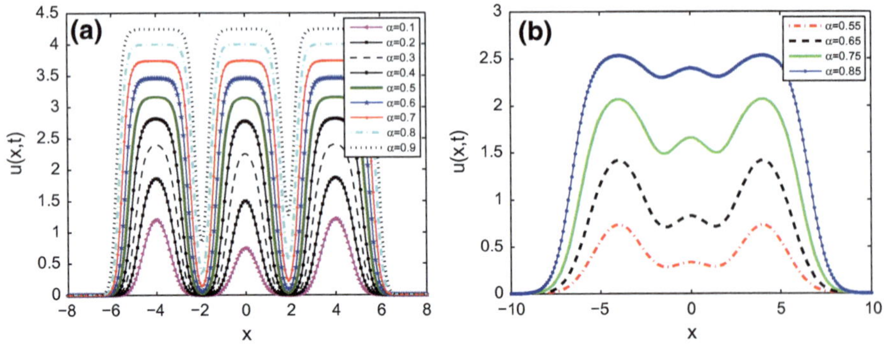

Fig. 1 One dimensional solution of fractional diffusion-like equation (54) for different values of α, with $\mu = 0.8$, $\kappa = 1.0$ and $d = 0.1$. Simulation runs for $t = 1$

where $d > 0$ is a diffusion constant, $Lu(x, t) = \frac{\partial^2 u(x,t)}{\partial x^2}$ and $F(u(x, t))$ represents a nonlinear function given here as $\mu u(x, t)\left(1 - \frac{u(x,t)}{\kappa}\right)$. Parameters μ and κ are the respective growth rate and carrying capacity of the species. This type of equation has been used in the study of nuclear reactor and flame propagation [17, 35, 47].

The above equation is numerically solved using the fractional Adams-Bashforth scheme as presented in Sect. 3. The right-hand side containing the second-order approximation is given as follows:

$$\begin{aligned}\frac{\partial^2 u(x,t)}{\partial x^2} &= \frac{1}{2}\left(\frac{u_{i+1}^{j+1} - 2u_i^{j+1} + u_{i-1}^{j+1}}{2(\Delta x)^2} + \frac{u_{i+1}^{j} - 2u_i^{j} + u_{i-1}^{j}}{2(\Delta x)^2}\right)\\ u(x,t) &= \frac{1}{2}\left(u_i^{j+1} - u_i^j\right).\end{aligned} \tag{55}$$

Equation (54) is solved using the periodic boundary condition clamped at both ends of the domain interval $-\infty \le L \le \infty$, subject to initial condition

$$u_0(x) = 3\exp(-20(x+4)^2) + 2.05\exp(-10(x-4)^2) + \exp(-20(x)^2). \tag{56}$$

In Fig. 1, we simulate with $\mu = 0.8$, $\kappa = 1.0$ and $d = 0.1$ for different values of $\alpha \in (0, 1)$ on different domain of lengths $L = 8$ and $L = 10$ for plots (a) and (b) respectively. It should be mentioned that in attempt to provide enough room for the waves patterns to propagate, the infinite domain is truncated at $L > 0$.

In two dimensions, we let $Lu(\cdot .t) = \left(\frac{\partial^2 u}{\partial x^2} + \frac{\partial^2 u}{\partial y^2}\right)$, $(x, y) \in \Omega = (a \le x, y \le b)$ for $\Omega \subset \mathbf{R}^{n\times n}$, $u(x, y, 0) = u_0(x, y)$ and $\frac{\partial u}{\partial x}(a, x, t) = \frac{\partial u}{\partial x}(b, x, t) = 0$ with similar expression for y-direction. The diffusion coefficients $d = diag(d_1, d_2, \ldots, d_n)$. We discretize in space with mesh $(x_i, y_j)(a + i \times \hbar_x, a + j \times \hbar_y)$, where $\hbar_x = (b - a)/(N_x + 1)$, $\hbar_y = (b - a)/(N_y + 1)$, for $0 \le i \le N_x + 1$ and $0 \le j \le N_y + 1$. We

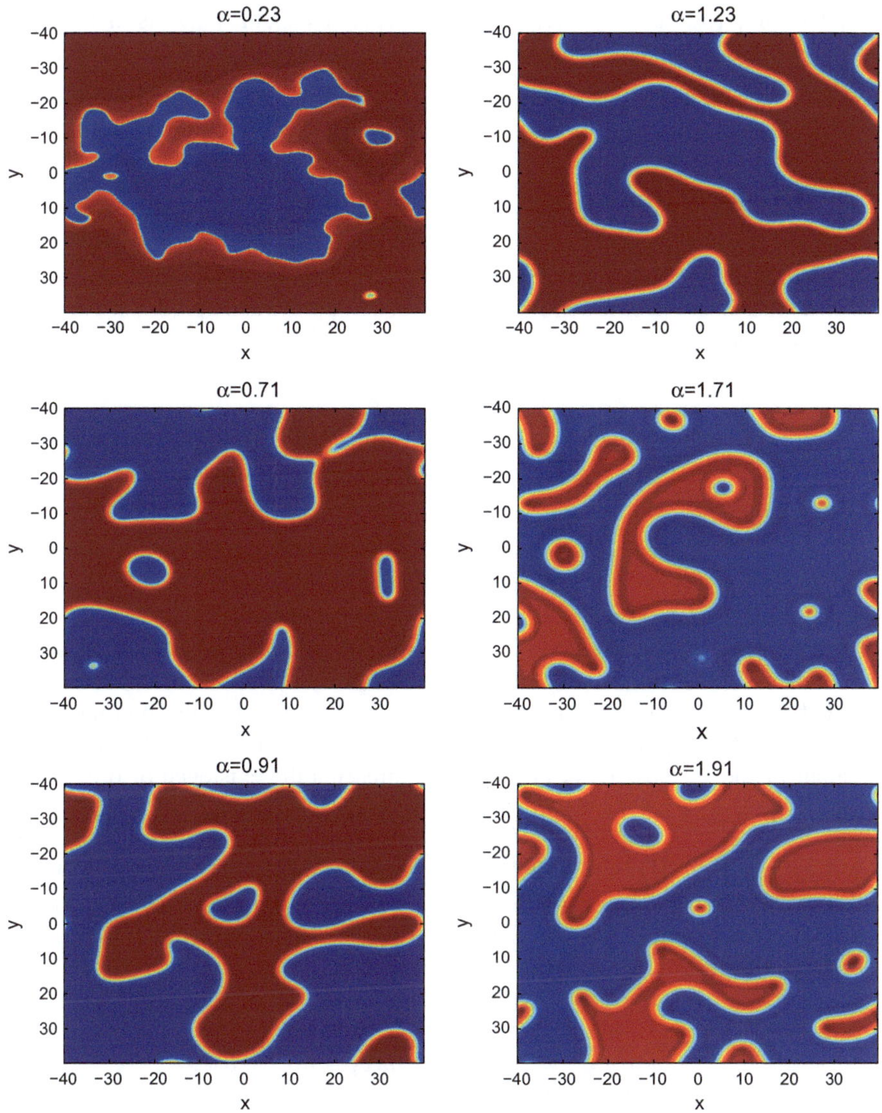

Fig. 2 Two dimensional results for fractional diffusion-like equation (54) for subdiffusive (column-1 $0 < \alpha < 1$) and superdiffusive (column-2 $1 < \alpha < 2$) scenarios. Parameters are: $\mu = 0.5$, $\kappa = 2.0$, $d = 0.1$. Simulation runs for $t = 150$

follow the finite difference technique discussed in [39] to approximate the second-order partial derivatives in two dimensions.

In 2D we obtained the results in Fig. 2 with random initial condition and parameters $d = 0.1$, $\kappa = 2$, $\mu = 0.5$, $\Delta t = 0.25$, $\hbar = 1/4$. Columns-1 and -2 correspond to subdiffusive ($\alpha \in (0, 1)$) and superdiffusive ($\alpha \in (1, 2)$) cases respectively. Apart from the spiral (oscillatory) waves obtained here, other spatial patterns such as spots and stripes are possible depending on the choice of the initial conditions and other parameter values.

4.2 Biological System

The biological system considered in this work models the three species food web which defines inter-specific competition between two preys and a predator. The scaled model is of the form

$$\begin{aligned}\frac{du}{dt} &= f_u(u,v,w) = u - u^2 - uv - uw,\\ \frac{dv}{dt} &= f_v(u,v,w) = -a_1 v + b_1 uv - b_2 v^2 - b_3 vw,\\ \frac{dw}{dt} &= f_w(u,v,w) - a_2 w + c_1 uw + c_2 uv - c_3 w^2,\end{aligned} \tag{57}$$

where a_1, a_2, b_i and c_i for $i = 1, 2, 3$ are positive parameters.

The essence of linear stability is to correctly guide in the choice of parameters when simulating the full reaction-diffusion system. Linear stability analysis at $f_u(u,v,w) = f_v(u,v,w) = f_w(u,v,w) = 0$ shows that the biological system (57) has five equilibrium points $E_0 = (0,0,0)$ which corresponds to the total washout of the species, $E_2 = (1,00)$ that represents the existence of prey u only with $v = w = 0$, $E_2 = \left(\frac{c_3+a_2}{c_1+c_3}, 0, \frac{c_3-a_2}{c_1+c_3}\right)$ which shows no existence of prey v and $E_3 = \left(\frac{b_2+a_1}{b_1+b_2}, \frac{b_1-a_1}{b_1+b_2}, 0\right)$ which corresponds to the predator w extinction. We are not interested in the above four equilibrium points. Hence, we consider the last point that represents the existence of the three species. The coexistence point that is biologically meaningful is denoted by $E^* = (u^*, v^*, w^*)$ where

$$\begin{aligned}u^* &= \frac{b_2c_3 + b_3c_2 + (c_2+c_3)a_1 + (b_2-b_3)a_2}{b_2c_3 + b_3c_2 + b_1c_3 - b_3c_1 + b_1c_2 + b_2c_1},\\ v^* &= \frac{b_1c_3 + b_3c_1 - (b_1+c_3)a_1 + (b_1+b_3)a_2}{b_2c_3 + b_3c_2 + b_1c_3 - b_3c_1 + b_1c_2 + b_2c_1},\\ w^* &= \frac{b_1c_2 + b_2c_1 + (c_1-c_2)a_1 - (b_1+b_2)a_2}{b_2c_3 + b_3c_2 + b_1c_3 - b_3c_1 + b_1c_2 + b_2c_1}.\end{aligned} \tag{58}$$

The community matrix of (57) at point E^* is given by the Jacobian

$$\mathscr{J}(E^*) = \begin{pmatrix} -u^* & -u^* & -u^* \\ b_1v^* & -b_2v^* & -b_3v^* \\ c_1w^* & c_2w^* & -c_3w^* \end{pmatrix}_{(u^*,v^*,w^*)}, \tag{59}$$

with characteristic equation

$$\lambda^3 + \mathscr{Q}_1\lambda^2 + \mathscr{Q}_2\lambda + \mathscr{Q}_3 = 0, \tag{60}$$

where $\mathscr{Q}_1 = u^* + b_2v^* + c_3w^*$, $\mathscr{Q}_2 = (b_1 + b_2)u^*v^* + (b_2c_3 + b_3c_2)v^*w^* + (c_1 + c_3)u^*w^*$ and $\mathscr{Q}_3 = (b_2c_3 + b_3c_2 + b_1c_3 - b_3c_1 + b_1c_2 + b_2c_1)u^*v^*w^*$. We can see that $\mathscr{Q}_1$, $\mathscr{Q}_1\mathscr{Q}_2 - \mathscr{Q}_3$ and $\mathscr{Q}_3$ are positive. Hence, by adopting the Routh's stability criterion, obviously, the coexistence state is unconditionally stable.

In on attempt to examine the effect of non-local and nonsingular kernel into the definition of fractional derivative, we first model (57) in the form of (1) and replace the classical time derivative in Eq. (57) with the Atangana-Baleanu fractional derivative in Caputo sense to obtain

$$\begin{aligned}
{}_0^{ABC}\mathscr{D}_t^{\alpha}u(t) &= d_1\frac{\partial^2 u(x,t)}{\partial x^2} + u - u^2 - uv - uw,\\
{}_0^{ABC}\mathscr{D}_t^{\alpha}v(t) &= d_2\frac{\partial^2 v(x,t)}{\partial x^2} - a_1v + b_1uv - b_2v^2 - b_3vw,\\
{}_0^{ABC}\mathscr{D}_t^{\alpha}w(t) &= d_3\frac{\partial^2 w(x,t)}{\partial x^2} - a_2w + c_1uw + c_2uv - c_3w^2.
\end{aligned} \tag{61}$$

In the experiment, we first observed the behaviour of the species in the absence of diffusion, that is, when $d_i = 0,\ i = 1, 2, 3$. We observed in Fig. 3 that the three species will coexist for period $t > 0 \in T$ as displayed in the second panel. Emergence of the attractor further shows the coexistence of the species. Parameter values used are given in the figure caption.

One dimensional results for system (61) in the presence of diffusion at two instances of fractional power α are displayed in Fig. 4. Plots (a-d) and (e-h) corresponds to $\alpha = 0.91$ and $\alpha = 0.97$ respectively.

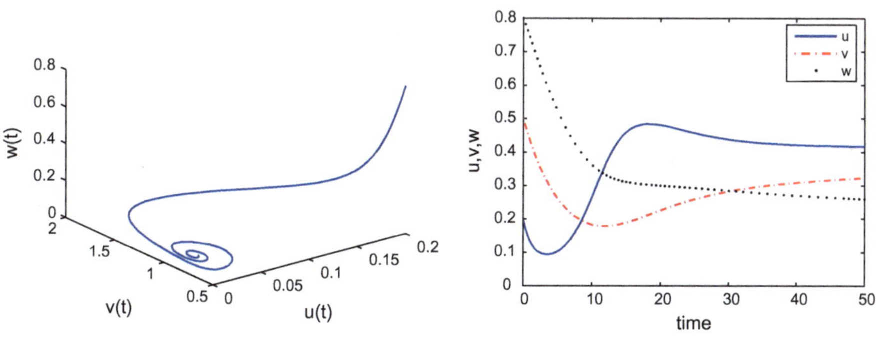

Fig. 3 Solution of the biological system (61) for $\alpha = 0.89$ at $d_i = 0, i = 1, 2, 3$. Other parameters are: $(u_0, v_0, w_0) = (0.2, 0.5, 0.8), a_1 = 0.05, a_2 = 0.09, b_1 = 0.3, b_2 = 0.15, b_3 = 0.09, c_1 = 0.18, c_2 = 0.1$ and simulation time $t = 50$

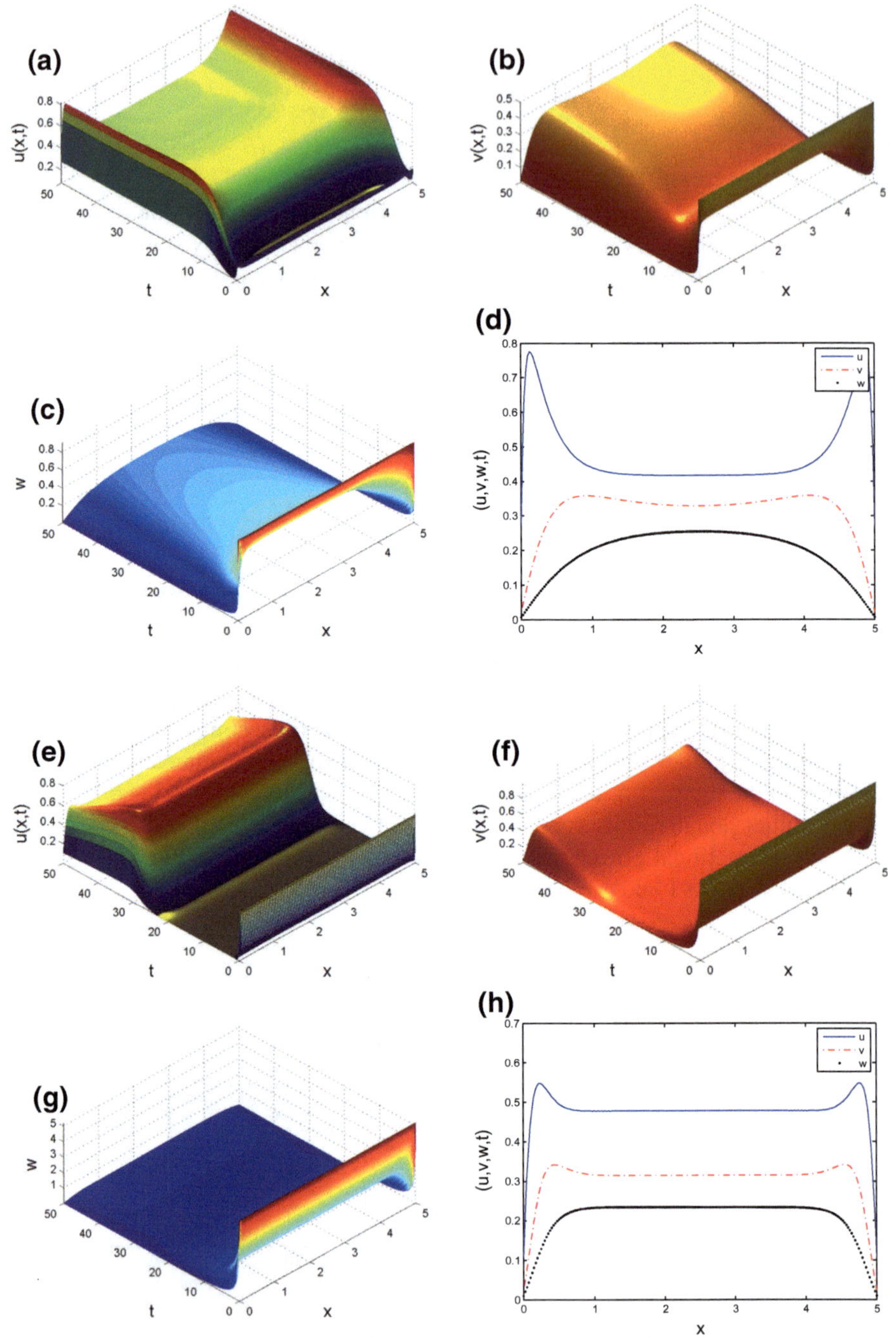

Fig. 4 One dimensional experiment for system (61) with Atangana-Baleanu derivative at $d_1 = 0.007$, $d_2 = 0.001$, $d_3 = 0.002$, $\alpha = 0.91$ for plots (**a**–**d**) and $\alpha = 0.97$ for plots (**e**–**h**). Other parameters are given in Fig. 3

4.3 The Volta's System

The Volta's system displays classic chaotic behavior. The three parameter model consider here is given as

$$\begin{aligned}\frac{du}{dt} &= -u(t) - \phi v(t) - v(t)w(t),\\ \frac{dv}{dt} &= -v(t) - \psi u(t) - u(t)w(t),\\ \frac{dw}{dt} &= \varphi w(t) + u(t) + v(t) + 1,\end{aligned} \tag{62}$$

where ϕ, ψ and φ are the dimensionless positive parameters.

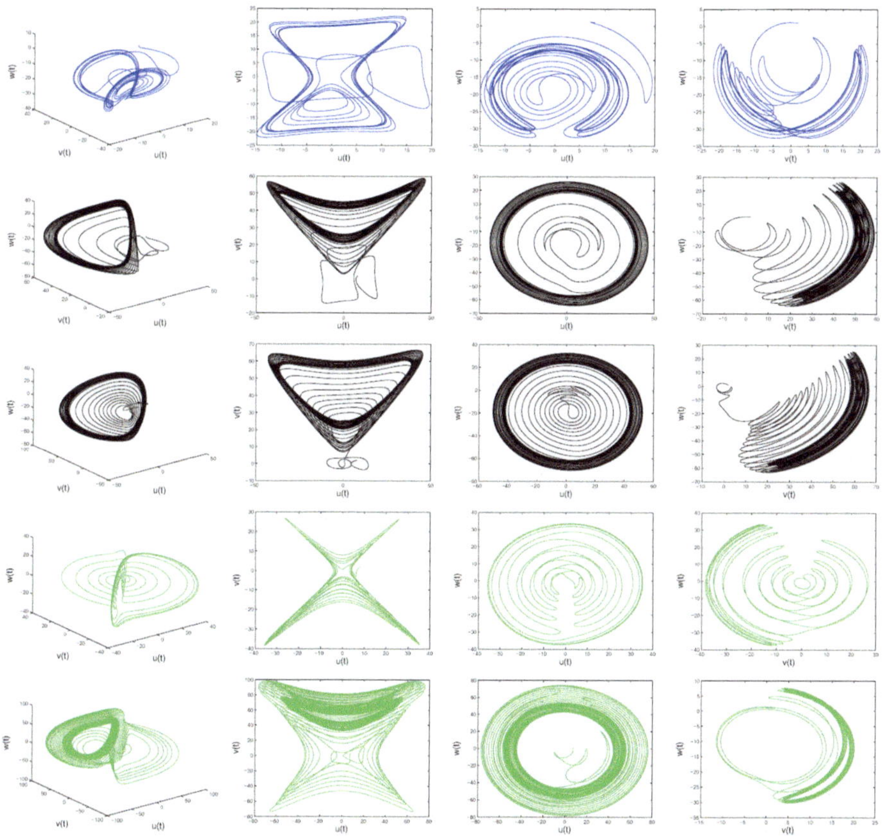

Fig. 5 Numerical simulation of system (63) for different values of α, using the Atangana-Baleanu fractional derivative (rows 1–3) and the Caputo operator (rows 4 and 5). Rows 1–4 correspond to $\alpha = 0.55$, $\alpha = 0.90$, $\alpha = 0.95$, $\alpha = 0.90$ and $\alpha = 0.95$ respectively

We examine the effect of combining power law with the Mittag-Leffler kernel by replacing the classical time derivative in (62) with both the Caputo and Atangana-Baleanu fractional derivatives of order $\alpha \in [0, 1]$. For instance, By using the Atangana-Baleanu operator in the sense of Caputo, the above classical Volta's system becomes

$$\begin{aligned} {}_{0}^{ABC}\mathscr{D}_{t}^{\alpha}u(t) &= -u(t) - \phi v(t) - v(t)w(t), \\ {}_{0}^{ABC}\mathscr{D}_{t}^{\alpha}v(t) &= -v(t) - \psi u(t) - u(t)w(t), \\ {}_{0}^{ABC}\mathscr{D}_{t}^{\alpha}w(t) &= \varphi w(t) + u(t) + v(t) + 1. \end{aligned} \tag{63}$$

Figure 5 shows some numerical simulations obtained at different instances of $\alpha \in [0, 1]$. Rows 1–3 are obtained using the Atangana-Baleanu derivative in Caputo sense for $\alpha = 0.55, \alpha = 0.90$ and $\alpha = 0.95$ respectively. Similarly, we apply the Caputo operator to get the results displayed in Rows 4 and 5 of Fig. 5 for the values $\alpha = 0.90$ and $\alpha = 0.95$. It should be mentioned that the fractional derivative in this system exhibits some chaotic scenarios that are completely missing in the classical model.

5 Conclusion

We introduced a two-step Adams-Bashforth method to solve a range of time-fractional reaction-diffusion problems. The classical derivatives in the problem are replaced with either the Caputo derivative which follows the power law or the Atangana-Baleanu operator which obeys the Mittag-Leffler law to formulate a fractional derivative with no-singular and non-local kernel. The proposed method is flexible and easy to use. Some illustrative examples of time-fractional reaction-diffusion problems are given to justify the robustness of the derivative. The results obtained in one and two dimensions for different values of α in subdiffusion and superdiffusion scenarios are in good agreement with the existing cases. Application of the Atangana-Baleanu derivative to model space-fractional reaction-diffusion equation is left for future research.

References

1. Atangana, A.: Derivative with a New Parameter: Theory, Methods and Applications. Academic Press, New York (2016)
2. Atangana, A.: Fractional Operators with Constant and Variable Order with Application to Geo-Hydrology. Academic Press, New York (2017)
3. Atangana, A., Baleanu, D.: New fractional derivatives with nonlocal and non-singular kernel: theory and application to heat transfer model. Therm. Sci. **20**, 763–769 (2016)

4. Atangana, A., Gómez-Aguilar, J.F.: Hyperchaotic behaviour obtained via a nonlocal operator with exponential decay and Mittag-Leffler laws. Chaos Solitons Fractals **102**, 285–294 (2017)
5. Atangana, A., Gómez-Aguilar, J.F.: Fractional derivatives with no-index law property: application to chaos and statistics. Chaos Solitons Fractals **114**, 516–535 (2018)
6. Atangana, A., Koca, I.: Chaos in a simple nonlinear system with Atangana-Baleanu derivatives with fractional order. Chaos Solitons Fractals **89**, 447–454 (2016)
7. Atangana, A., Owolabi, K.M.: New numerical approach for fractional differential equations. Math. Model. Nat. Phenom. **13**(1), 1–19 (2018)
8. Bueno-Orovio, A., Kay, D., Burrage, K.: Fourier spectral methods for fractional-in-space reaction-diffusion equations. BIT Numer. Math. **54**, 937–954 (2014)
9. Burrage, K., Cardone, A., D'Ambrosio, R., Paternoster, B.: Numerical solution of time fractional diffusion systems. Appl. Numer. Math. **116**, 82–94 (2017)
10. Caputo, M., Fabrizio, M.: A new definition of fractional derivative without singular kernel. Prog. Fract. Differ. Appl. **1**, 73–85 (2015)
11. Caputo, M., Fabrizio, M.: Applications of new time and spatial fractional derivatives with exponential kernels. Prog. Fract. Differ. Appl. **2**, 1–11 (2016)
12. Cetinkaya, A., Klymaz, O.: The solution of the time-fractional diffusion equation by the generalized differential transform method. Math. Comput. Model. **57**, 2349–2354 (2013)
13. Coronel-Escamilla, A., Gómez-Aguilar, J.F., Baleanu, D., Córdova-Fraga, T., Escobar-Jiménez, R.F., Olivares-Peregrino, V.H., Qurashi, M.M.A.l.: Bateman-Feshbach tikochinsky and Caldirola–Kanai oscillators with new fractional differentiation. Entropy **19**(2), 1–21 (2017)
14. Coronel-Escamilla, A., Gómez-Aguilar, J.F., Torres, L., Escobar-Jiménez, R.F., Valtierra-Rodríguez, M.: Synchronization of chaotic systems involving fractional operators of Liouville-Caputo type with variable-order. Phys. A: Stat. Mech. Appl. **487**, 1–21 (2017)
15. Coronel-Escamilla, A., Gómez-Aguilar, J.F., Torres, L., Escobar-Jiménez, R.F.: A numerical solution for a variable-order reaction-diffusion model by using fractional derivatives with non-local and non-singular kernel. Phys. A: Stat. Mech. Appl. **491**, 406–424 (2018)
16. Cuahutenango-Barro, B., Taneco-Hernández, M.A., Gómez-Aguilar, J.F.: On the solutions of fractional-time wave equation with memory effect involving operators with regular kernel. Chaos Solitons Fractals **115**, 283–299 (2018)
17. Fisher, R.A.: The wave of advance of advantageous genes. Ann. Eugen. **7**, 353–369 (1937)
18. Fornberg, B.: Calculation of weights in finite difference formulas. SIAM Rev. **40**, 685–691 (1998)
19. Gafiychuk, V.V., Datsko, B.Y.: Pattern formation in a fractional reaction diffusion system. Phys. A **365**, 300–306 (2006)
20. Ghanbari, B., Gómez-Aguilar, J.F.: Modeling the dynamics of nutrient-phytoplankton-zooplankton system with variable-order fractional derivatives. Chaos Solitons Fractals **116**, 114–120 (2018)
21. Gnitchogna, R., Atangana, A.: New two step Laplace Adam-Bashforth method for integer a noninteger order partial differential equations. Numer. Methods Partial Differ. Equs. **1**, 1–19 (2017)
22. Gómez-Aguilar, J.F.: Novel analytical solutions of the fractional Drude model. Optik **168**, 728–740 (2018)
23. Gómez-Aguilar, J.F., Atangana, A.: New insight in fractional differentiation: power, exponential decay and Mittag-Leffler laws and applications. Eur. Phys. J. Plus **132**, 1–13 (2017)
24. Gómez-Aguilar, J.F., Dumitru, B.: Fractional transmission line with losses. Zeitschrift für Naturforschung A **69**(10–11), 539–546 (2014)
25. Gómez-Aguilar, J.F., Torres, L., Yépez-Martínez, H., Baleanu, D., Reyes, J.M., Sosa, I.O.: Fractional Liénard type model of a pipeline within the fractional derivative without singular kernel. Adv. Differ. Equs. **2016**(1), 1–17 (2016)
26. Gómez-Aguilar, J.F., Escobar-Jiménez, R.F., López-López, M.G., Alvarado-Martínez, V.M.: Atangana-Baleanu fractional derivative applied to electromagnetic waves in dielectric media. J. Electromagn. Waves Appl. **30**(15), 1937–1952 (2016)

27. Gómez-Aguilar, J.F., Atangana, A., Morales-Delgado, J.F.: Electrical circuits RC, LC, and RL described by Atangana-Baleanu fractional derivatives. Int. J. Circ. Theor. Appl. **1**, 1–22 (2017)
28. Gómez-Aguilar, J.F., López-López, M.G., Alvarado-Martínez, V.M., Baleanu, D., Khan, H.: Chaos in a cancer model via fractional derivatives with exponential decay and Mittag-Leffler law. Entropy **19**(12), 1–16 (2017)
29. Hilfer, R.: Applications of Fractional Calculus in Physics. World Scientific, Singapore (2000)
30. Jiang, H., Liu, F., Turner, I., Burrage, K.: Analytical solutions for the multi-term time-fractional diffusion-wave/diffusion equations in a finite domain. Comput. Math. Appl. **64**, 3377–3388 (2012)
31. Kassam, A.K., Trefethen, L.N.: Fourth-order time-stepping for stiff PDEs. SIAM J. Sci. Comput. **26**, 1214–1233 (2005)
32. Kilbas, A.A., Srivastava, H.M., Trujillo, J.J.: Theory and Applications of Fractional Differential Equations. Elsevier, Netherlands (2006)
33. Morales-Delgado, V.F., Taneco-Hernández, M.A., Gómez-Aguilar, J.F.: On the solutions of fractional order of evolution equations. Eur. Phys. J. Plus **132**(1), 1–17 (2017)
34. Morales-Delgado, V.F., Gómez-Aguilar, J.F., Kumar, S., Taneco-Hernández, M.A.: Analytical solutions of the Keller-Segel chemotaxis model involving fractional operators without singular kernel. Eur. Phys. J. Plus **133**(5), 1–20 (2018)
35. Murray, J.D.: Mathematical Biology I: Spatial Models and Biomedical Applications. Springer, Berlin (2003)
36. Oldham, K.B., Spanier, J.: The Fractional Calculus: Theory and Applications of Differentiation and Integration to Arbitrary Order. Dover Publication, New York (2006)
37. Ortigueira, M.D.: Fractional Calculus for Scientists and Engineers. Springer, New York (2011)
38. Owolabi, K.M.: Second or fourth-order finite difference operators, which one is most effective? Int. J. Stat. Math. **1**, 44–54 (2014)
39. Owolabi, K.M.: Robust IMEX schemes for solving two-dimensional reaction-diffusion models. Int. J. Nonlinear Sci. Numer. Simul. **16**, 271–284 (2015)
40. Owolabi, K.M.: Mathematical analysis and numerical simulation of patterns in fractional and classical reaction-diffusion systems. Chaos Solitons Fractals **93**, 89–98 (2016)
41. Owolabi, K.M.: Numerical solution of diffusive HBV model in a fractional medium. Springer Plus **2016**, 1–19 (2016)
42. Owolabi, K.M.: Robust and adaptive techniques for numerical simulation of nonlinear partial differential equations of fractional order. Commun. Nonlinear Sci. Numer. Simuls. **44**, 304–317 (2017)
43. Owolabi, K.M.: Mathematical modelling and analysis of two-component system with Caputo fractional derivative order. Chaos Solitons Fractals **103**, 544–554 (2017)
44. Owolabi, K.M., Atangana, A.: Analysis and application of new fractional Adams-Bashforth scheme with Caputo-Fabrizio derivative. Chaos Solitons Fractals **105**, 111–119 (2017)
45. Owolabi, K.M., Patidar, K.C.: Higher-order time-stepping methods for time-dependent reaction-diffusion equations arising in biology. Appl. Math. Comput. **240**, 30–50 (2014)
46. Owolabi, K.M., Patidar, K.C.: Numerical solution of singular patterns in one-dimensional Gray-Scott-like models. Int. J. Nonlinear Sci. Numer. Simul. **15**, 437–462 (2014)
47. Owolabi, K.M., Patidar, K.C.: Solution of pattern waves for diffusive Fisher-like non-linear equations with adaptive methods. Int. J. Nonlinear Sci. Numer. Simul. **17**, 291–304 (2016)
48. Podlubny, I.: Fractional Differential Equations: An Introduction to Fractional Derivatives, Fractional Differential Equations, to Methods of Their Solution and some of Their Applications. Academic Press, San Diego, Calif, USA (1999)
49. Tateishi, A.A., Ribeiro, H.V., Lenzi, E.K.: The role of fractional time-derivative operators on anomalous diffusion. Front. Phys. **5**, 1–9 (2017)

Heat Transfer Analysis in Ethylene Glycol Based Molybdenum Disulfide Generalized Nanofluid via Atangana–Baleanu Fractional Derivative Approach

Farhad Ali, Muhammad Saqib, Ilyas Khan and Nadeem Ahmad Sheikh

Abstract At the end of 2016, Atangana and Baleanu introduced a new definition for fractional derivatives, namely Atangana–Baleanu fractional derivatives with the non-singular and non-local kernel. The idea of Atangana–Baleanu was used by several authors for various types of fractional problems. However, for heat transfer problem, this idea is rarely used in particular when nanofluid is considered. Based on this motivation, this chapter aims to study the flow of ethylene glycol based Molybdenum disulfide generalized nanofluid (EGMDGN) over an isothermal vertical plate. A fractional model with non-singular and non-local kernel, Atangana–Baleanu fractional derivatives is developed in the form of partial differential equations along with appropriate initial and boundary conditions. Molybdenum disulfide nanoparticles of spherical shape are suspended in Ethylene Glycol (EG) taken as conventional base fluid. The exact solutions are developed for velocity and temperature profiles via the Laplace transform technique. In a limiting sense, the obtained solutions are reduced to fractional Newtonian ($\beta \to \infty$), classical Casson fluid ($\alpha \to 1$) and classical Newtonian nanofluids. The influence of various pertinent parameters is analyzed in various plots and discussed physically.

Keywords Fractional calculus · Atangana–Baleanu fractional derivative · Heat transfer model

F. Ali · M. Saqib · N. Ahmad Sheikh
Computational Analysis Research Group, Ton Duc Thang University, Ho Chi Minh City, Vietnam

F. Ali · M. Saqib · I. Khan (✉) · N. Ahmad Sheikh
Faculty of Mathematics and Statistics, Ton Duc Thang University, Ho Chi Minh City, Vietnam
e-mail: ilyaskhan@tdt.edu.vn

F. Ali · M. Saqib · N. Ahmad Sheikh
Department of Mathematics, City University of Science and Information Technology, Peshawar 25000, Pakistan

J. F. Gómez et al. (eds.), *Fractional Derivatives with Mittag-Leffler Kernel*, Studies in Systems, Decision and Control 194,
https://doi.org/10.1007/978-3-030-11662-0_13

Nomenculture

Py	- The yield stress of the non-Newtonian fluid.
π	- The product of the component of deformation rate itself.
π	- The critical value of the product.
μ_r	- Plastic dynamic viscosity.
u	- Velocity of the fluid.
T	- Temperature of the fluid.
g	- Acceleration due to gravity.
C_p	- Specific heat at a constant pressure.
k_f	- Thermal conductivity of the base fluid.
T_∞	- Fluid temperature far away from the plate.
q	- Laplace transforms parameter.
v_f	- Kinematic viscosity of the base fluid.
μ_f	- Dynamic viscosity.
ρ_f	- Density of the base fluid.
ρ_s	- The density of the solid.
U	- The amplitude of the velocity.
β_T	- The volumetric coefficient of thermal expansion.
B_0	- External Magnetic Field.
ρ_{nf}	- Nanofluids density.
μ_{nf}	- Dynamic viscosity of nanofluid.
σ_{nf}	- The electrical conductivity of nanofluid.
β	- The material parameter of Casson fluid.
$(\beta_T)_{nf}$	- Thermal expansion coefficient of nanofluid.
$(\rho c_p)_{nf}$	- Specific heat capacity of nanofluid.
k_{nf}	- The thermal conductivity of nanofluid.
M	- Magnetic parameter.
G_r	- Thermal Grasshof number.
P_r	- Prandtl number.
Nu_x	- Nusselt number.
ϕ	- Nanoparticles volume fraction.
α	- Fractional order/fractional parameter.

1 Introduction

With the evolution of thermal science and engineering, it is of great concern to promote devices for the flow of microscale liquid that have a high surface to volume ratio and compactness as compared to the conventional flow system. These thermal devices are the requirement to enhance heat transfer in various diverse industries including microelectronics, chemical engineering, aerospace, manufacturing and transportation [1]. But the energy enhancement using such devices leads to an unacceptable increase in the heat transport systems. It is strongly recognized by many investigators that the key factor that affects the efficiency of heat transport is the thermal conductivity of the conventional fluids such as water, alcohol, ethylene glycol and kerosene-oil etc., [2–5]. Maxwell [6] initiated an efficient method to enhance the thermal conductivity of the conventional fluids by suspending micro-sized or milli-sized solid particles into them. However, he grasped that the large milli-sized or micro-sized solid particles originate some scientific problems, i.e. (i) clogging micro-channel of devices, (ii) sedimentation of large particles, (iii) abrasion of surfaces, (iv) erosion of pipelines and (v) increasing drop in pressure [7]. Based on these problems, in 1995, Choi [8] conferred the idea of enhancing thermal conductivity by adding nanometer-sized nanoparticles (metal oxides, metals, polymers, carbon nanotubes or silica) in conventional base fluids (water, alcohol, ethylene glycol and oil etc.) referred to nanofluids. The studies of Das et al. [9], Wang and Mujumdar [10–12] and Buongiorno [13] explained the mechanism that results the enhancement of the heat transport in nanofluids. Later on, Ulhaq et al. [14], Shahzad et al. [15], Khan et al. [16], Wakif et al. [17], Sheikholeslami [18, 19], Aman et al. [20] and Ali et al. [21] enriched the literature of nanofluids.

However, there are still many questions that need to be answered. For example which type of nanoparticles should be suspended in which type of base fluid to get the maximum heat transfer rate. What should be the size, shape, structure etc. of each nanoparticles? Therefore, several types of nanoparticles have been used by the researchers in their experimental or theoretical studies. Among the different kinds of nanoparticles, one of them is called Molybdenum disulfide nanoparticles. These nanoparticles are rarely used in the literature particularly in theoretical work. MoS_2 has a large band gaps structure which is closed to the structure of graphene, due to this reason MoS_2 has been applied at large scale in logic circuits and amplifier devices [22–25]. Moreover, the thermo-physical properties of Molybdenum disulfide such as capabilities of lubrication, thermal conductivity, and heat capacity can be used in mechanical applications [26]. The lubricant graphite and Molybdenum disulfide powders were coated with copper to reinforce their bonding to the copper particles in the composites during sintering [27]. Mao et al. [28] investigated the tribological properties of nanofluids used in minimum quantity lubrication (MQL) grading.

Recently, the first theoretical study on Molybdenum disulfide nanofluid was carried out by Sharidan et al. [29] in a conference paper. Khan et al. [30] studied the effect of magnetic field on Molybdenum disulfide nanofluids in presence of thermal radiation. Khan [31] investigated the effects of different shapes MoS_2 nanoparticles in water based nanofluid. Nonetheless, in the discussed literature, mostly viscous nanofluids model with the ordinary derivatives is considered whereas the present study focused on non-Newtonian nanofluid model namely, the Casson nanofluid model with the fractional derivative approach. It is well accepted that fractional derivatives are more suitable than ordinary derivatives to describe a physical phenomenon [32–48, 57]. But there are some discrepancies in the application of Riemann–Liouville, Caputo and Caputo–Fabrizio fractional derivatives which are recently addressed by Atangana et al. [49–54]. At the end of 2016, Atangana and Baleanu introduced a new definition [53] namely, Atangana–Baleanu fractional derivatives with the non-singular and non-local kernel. Inspired from this new approach of fractional derivatives, in the present study, we adopted the Atangana–Baleanu fractional derivatives approach for the generalization of Casson nanofluid for the first time in the literature. In this study, Ethylene glycol (EG) is chosen as a base fluid and Molybdenum disulfide nanoparticles are dispersed into them in order to enhance the heat transfer in fluid flow. The governing equations of the model are solved via the Laplace transform technique to obtain the exact solutions. The exact solutions obtained here are of great importance because these solutions can be used as a benchmark by experimentalists and numerical solvers to check the accuracy of their results.

2 Mathematical Formulation

Let us consider the unsteady flow of MDGCNF over a flat plate with constant wall temperature. MoS_2 nanoparticles are suspended in EG taken as conventional base fluid. The flow is assumed to be along the x-direction. Initially, both the fluid and plate are at rest with uniform temperature T_∞. At $t = 0^+$, the plate starts motion with constant velocity as shown in Fig. 1.

For incompressible Casson fluid the rheological equation is given as the following [55]:

$$\tau_{ij} = \begin{cases} 2\left(\mu_\gamma + \frac{p_y}{\sqrt{2\pi}}\right)e_{ij}, & \pi > \pi_c \\ 2\left(\mu_\gamma + \frac{p_y}{\sqrt{2\pi_c}}\right)e_{ij}, & \pi_c < \pi. \end{cases} \tag{1}$$

By using the Boussinesq's approximation the free convective flow of Casson nanofluid is governed by [21]:

Fig. 1 Geometry of the flow

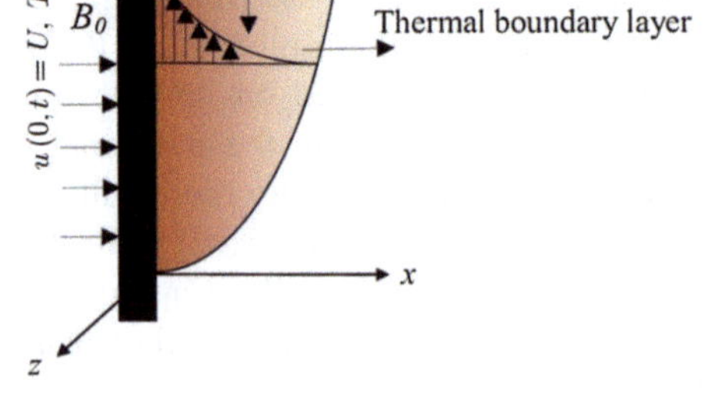

Table 1 Thermo-physical properties of nanofluids

Materials	ρ (kgm^{-3})	c_p (kg^{-1}K^{-1})	k (Wm^{-1}K^{-1})	$\beta \times 10^{-5}$ (K^{-1})	σ (s/m)
E.G	1115	2386	0.2599	3.41×10^{-8}	1.07×10^{-8}
Mos_2	5.06×10^3	397.21	904.4	2.8424	2.09×10^{-4}

$$\rho_{nf}\frac{\partial u}{\partial t} = \mu_{nf}\left(1+\frac{1}{\beta}\right)\frac{\partial^2 u}{\partial t^2} - \sigma_{nf} B_0^2 u + (\rho\beta_T)_{nf} g(T - T_\infty), \tag{2}$$

$$(\rho c_p)_{nf}\frac{\partial T}{\partial t} = k_{nf}\frac{\partial^2 u}{\partial t^2}, \tag{3}$$

subject to the following conditions:

$$\begin{aligned} u(y,0) &= 0, \qquad T(y,0) = T_\infty, \\ u(0,t) &= U, \qquad T(0,t) = T_w, \\ u(\infty,t) &= 0, \qquad T(\infty,t) = T_\infty. \end{aligned} \tag{4}$$

The nanoparticles expressions restricted to spherical shape [21] are used and the thermo-physical properties of the materials are given in Table 1:

Introducing the following dimensionless variables:

$$v = \frac{u}{U}, \qquad \xi = \frac{U}{\nu_f}, \qquad \tau = \frac{U^2}{\nu_f}, \qquad \theta = \frac{T - T_\infty}{T_w - T_\infty},$$

into Eqs. (2)–(4), we get

$$\frac{\partial v}{\partial \tau} = \frac{1}{\gamma_0}\frac{\partial^2 v}{\partial \xi^2} - M_0 v + Gr_0\theta, \tag{5}$$

$$\delta_0\frac{\partial \theta}{\partial \tau} = \frac{\partial^2 \theta}{\partial \xi^2}, \tag{6}$$

$$\begin{aligned} v(\xi,0) &= 0, \qquad \theta(\xi,0) = 0, \\ v(0,\tau) &= 1, \qquad \theta(0,\tau) = 1, \\ v(\infty,\tau) &= 0, \qquad \theta(\infty,\tau) = 0, \end{aligned} \tag{7}$$

where

$$\gamma_0 = b\gamma, \quad \frac{1}{\gamma} = 1 + \frac{1}{\beta}, \quad M_0 = \frac{1}{Re}(1-\phi)^{2.5}M, \quad Gr_0 = \phi_2 G_r, \quad \delta_0 = \frac{Pr\phi_3}{\lambda_{nf}},$$

$$b = (1-\phi)^{2.5}\left[(1-\phi) + \phi\left(\frac{\rho_s}{\rho_f}\right)\right], \quad \phi_1 = 1 + \frac{3(\sigma-1)\phi}{(\sigma+2)-(\sigma-1)\phi},$$

$$\phi_2 = \frac{(1-\phi)\rho_f - \phi\rho_s\left(\frac{\beta_{T_f}}{\beta_{T_s}}\right)}{\rho_{nf}}, \quad \phi_3 = (1-\phi) + \phi\left(\frac{(\rho c_p)_s}{(\rho c_p)_f}\right), \quad \lambda_{nf} = \frac{k_{nf}}{k_f},$$

$$\frac{k_{nf}}{k_f} = \frac{(k_s + 2k_f) - 2\varphi(k_f - k_s)}{(k_s + 2k_f) + \varphi(k_f - k_s)},$$

$$M = \frac{\nu\sigma_f B_0^2}{\rho_f U^2}, \quad Gr = \frac{\nu g\beta_{Tf}}{U^3}(T_w - T_\infty), \quad Pr = \frac{(\mu c_p)_f}{k_f}.$$

In order to develop a generalized Casson nanofluid model, we have replaced partial derivatives with respect to τ by Atangana–Baleanu fractional operator of order α, which leads Eqs. (5) and (6) to the following form:

$$D_\tau^\alpha v(\xi,\tau) = \frac{1}{\gamma_0}\frac{\partial^2 v}{\partial \xi^2} - M_0 v + Gr_0\theta, \tag{8}$$

$$\delta_0 D_\tau^\alpha \theta(\xi,\tau) = \frac{\partial^2\theta(\xi,\tau)}{\partial\xi^2}, \tag{9}$$

where $D_\tau^\alpha(\cdot)$ is the Atangana–Baleanu time fractional operator defined by [53]:

$${}^{ABC}D_\tau^\alpha(f(\tau)) = \frac{M(\alpha)}{1-\alpha}\int_0^\tau E_\alpha\left[-\alpha\frac{(t-\tau)^\alpha}{1-\alpha}\right]f'(\tau)d\tau; \qquad 0<\alpha<1. \tag{10}$$

where $M(\alpha)$ is a normalization function such that $M(0) = M(1) = 1$.

3 Solution of the Problem

In the virtue of Eq. (7), the Laplace transformation of Eqs. (8) and (9) is given as the following:

$$\frac{d^2\bar{\theta}(\xi,q)}{d\xi^2} - \frac{\delta_1 q^\alpha}{q^\alpha + a_1}\bar{\theta}(\xi,q) = 0, \tag{11}$$

$$\frac{d^2\bar{v}(\xi,q)}{d\xi^2}-\left[\frac{a_2q^\alpha+a_3}{q^\alpha+a_1}\right]\bar{v}(\xi,q)=-Gr_1\bar{\theta}(\xi,q), \tag{12}$$

where

$$\delta_1=\delta_0a_0,\qquad a_1=\alpha a_0,\qquad a_0=\frac{1}{1-\alpha},\qquad a_2=\gamma_0(a_0+M_0),$$

$$a_3=M_0\gamma_0a_1,\qquad Gr_1=Gr_0\gamma 0,$$

along with

$$\bar{\theta}(0,q)=\frac{1}{q},\qquad \bar{\theta}(\infty,q)=0, \tag{13}$$

$$\bar{v}(0,q)=\frac{1}{q},\qquad \bar{v}(\infty,q)=0, \tag{14}$$

The solution of Eqs. (11)–(14) is evaluated as

$$\bar{\theta}(\xi,q)=\frac{1}{q^{1-\alpha}}\bar{\Phi}(\xi,\sqrt{\delta_1},q,0,0,a_1), \tag{15}$$

$$\bar{v}(\xi,q)=\frac{1}{q^{1-\alpha}}\Big\{a_6\bar{\Phi}(\xi\sqrt{a_2},q,a_8,a_5,a_1)-a_9\bar{\Phi}(\xi\sqrt{a_2},q,0,a_5,a_1)-$$

$$-a_6\bar{\Phi}(\xi\sqrt{\delta_1},q,a_8,0,a_1)+a_7\bar{\Phi}(\xi\sqrt{\delta_1},q,0,0,a_1)\Big\}, \tag{16}$$

where

$$a_5=\frac{a_3}{a_2},\qquad a_6=\frac{Gr_1(a_3+a_1a_4)}{a_3a_4},\qquad a_7=\frac{Gr_1a_1}{a_3},\qquad a_8=\frac{a_3}{a_4},\qquad a_9=a_7-1.$$

Inverting the Laplace transforms of Eqs. (15) and (16), we get:

$$\theta(\xi,\tau)=h(\tau,\alpha)\Phi(\xi\sqrt{\delta_1},\tau,0,0,a_1), \tag{17}$$

$$v(\xi,\tau)=h(\tau,\alpha)*\Big\{a_6\Phi(\xi\sqrt{a_2},\tau,a_8,a_5,a_1)-a_9\Phi(\xi\sqrt{a_2},\tau,0,a_5,a_1)-$$

$$-a_6\Phi(\xi\sqrt{\delta_1},\tau,a_8,0,a_1)+a_7\Phi(\xi\sqrt{\delta_1},\tau,0,0,a_1)\Big\}, \tag{18}$$

where

$$\bar{\Phi}(y,q,a,b,c)=\frac{1}{q^{\alpha}+a}\exp\left[-y\sqrt{\frac{q^{\alpha}+b}{q^{\alpha}+c}}\right],$$

and the inverse Laplace transform of $\bar{\theta}(y,q,a,b,c)$ is given by [56]:

$$\Phi(y,\tau,a,b,c)=\frac{1}{\pi}\int_0^{\infty}\int_0^{\infty}\Phi_1(y,u,a,b,c)\exp(-\tau r-ur^{\alpha}\cos(\alpha\pi))\sin(ur^{\alpha}\sin(\alpha\pi))drdu,$$

$$h(\tau,\alpha)=\frac{1}{t^{\alpha}\Gamma(1-\alpha)},$$

$$\bar{\Phi}_1(y,q,a,b,c)=\frac{1}{q+a}\exp\left[-y\sqrt{\frac{q+b}{q+c}}\right],$$

$$\Phi_1(y,\tau,a,b,c)=\exp(-a\tau-y)-$$

$$-\frac{y\sqrt{b-c}}{2\sqrt{\pi}}\int_0^{\infty}\int_0^{\tau}\frac{\exp(-at)}{\sqrt{\tau}}\exp\left[at-c\tau-\frac{y^2}{4u}-u\right]I_1(2\sqrt{(b-c)u\tau})d\tau du,$$

where $I_1(\cdot)$ is the Bessel function of the first kind.

3.1 *Fractional Newtonian Fluid with Nanoparticles*

By making $\beta\to\infty\Rightarrow\gamma_0\to1$, Eq. (18) reduces to the following expression for the velocity of fractional Newtonian nanofluid

$$v(\xi,\tau)=h(\tau,\alpha)\Big\{a_6\Phi(\xi\sqrt{a_2},\tau,a_8,a_5,a_1)-a_9\Phi(\xi\sqrt{a_2},\tau,0,a_5,a_1)-$$

$$-a_6\Phi(\xi\sqrt{\delta_1},\tau,a_8,0,a_1)+a_7\Phi(\xi\sqrt{\delta_1},\tau,0,0,a_1)\Big\}, \tag{19}$$

where

$$a_2=a_0+M_0,\qquad a_3=M_0a_1,\qquad a_6=\frac{Gr_0(a_3+a_1a_4)}{a_3a_4},\qquad a_7=\frac{Gr_0a_1}{a_3}.$$

3.2 Classical Casson Nanofluid

For $\alpha = 1$:

$$\begin{aligned}
\lim_{\alpha\to 1} D_\tau^\alpha(\xi,\tau) &= \lim_{\alpha\to 1} L^{-1}\Big[L\Big\{D_\tau^\alpha v(\xi,\tau)\Big\}\Big],\\
&= L^{-1}\left\{\lim_{\alpha\to 1}\frac{q^\alpha \bar{v}(\xi,q)-v(\xi,0)}{(1-\alpha)q^\alpha+\alpha}\right\} = L^{-1}\Big\{q\bar{v}(\xi,q)-v(\xi,0)\Big\},\\
&= L^{-1}\Big[L\Big\{v'(\xi,\tau)\Big\}\Big] = v'(\xi,\tau).
\end{aligned}$$

$$\begin{aligned}
v(\xi,\tau) &= \tfrac{1}{2}\Big[\exp(-\xi\sqrt{M_0\gamma_0})\mathrm{erfc}\Big(\tfrac{\xi}{2\sqrt{\tau}}-\sqrt{M_0\gamma_0\tau}\Big)+\exp(\xi\sqrt{M_0\gamma_0})\mathrm{erfc}\Big(\tfrac{\xi}{2\sqrt{\tau}}+\sqrt{M_0\gamma_0\tau}\Big)\Big]+\\
&+\tfrac{Gr_0}{2M_0\gamma_0}\Big[\exp(-\xi\sqrt{M_0\gamma_0})\mathrm{erfc}\Big(\tfrac{\xi}{2\sqrt{\tau}}-\sqrt{M_0\gamma_0\tau}\Big)+\exp(\xi\sqrt{M_0\gamma_0})\mathrm{erfc}\Big(\tfrac{\xi}{2\sqrt{\tau}}+\sqrt{M_0\gamma_0\tau}\Big)\Big]-\\
&-\frac{Gr_0\exp\Big(\frac{M_0\gamma_0}{\delta_1}\tau\Big)}{2M_1}\Big[\exp\Big(-\xi\sqrt{M_0\gamma_0\Big(1-\tfrac{1}{\delta_1}\Big)}\Big)\mathrm{erfc}\Big(\tfrac{\xi}{2\sqrt{\tau}}-\sqrt{M_0\gamma_0\Big(1-\tfrac{1}{\delta_1}\Big)\tau}\Big)+\\
&+\exp\Big(\xi\sqrt{M_0\gamma_0\Big(1-\tfrac{1}{\delta_1}\Big)}\Big)\mathrm{erfc}\Big(\tfrac{\xi}{2\sqrt{\tau}}+\sqrt{M_0\gamma_0\Big(1-\tfrac{1}{\delta_1}\Big)\tau}\Big)\Big]-\tfrac{Gr_0}{M_0\gamma_0}\mathrm{erfc}\Big(\tfrac{\xi\sqrt{\delta_0}}{2\sqrt{\tau}}\Big)+\\
&+\frac{Gr_0\exp\Big(\frac{M_0\gamma_0}{\delta_1}\tau\Big)}{2M_0}\Big[\exp\Big(-a\sqrt{\delta_0\tfrac{M_0\gamma_0}{\delta_1}}\Big)\mathrm{erfc}\Big(\tfrac{\xi\sqrt{\delta_0}}{2\sqrt{\tau}}-\sqrt{\tfrac{M_0\gamma_0}{\delta_1}\tau}\Big)+\\
&+\exp\Big(a\sqrt{\delta_0\tfrac{M_0\gamma_0}{\delta_1}}\Big)\mathrm{erfc}\Big(\tfrac{\xi\sqrt{\delta_0}}{2\sqrt{\tau}}+\sqrt{\tfrac{M_0\gamma_0}{\delta_1}\tau}\Big)\Big].
\end{aligned} \tag{20}$$

which are the corresponding classical solutions of Eq. (18).

3.3 Classical Newtonian Fluid with Nanoparticles

By making $\beta \to \infty \Rightarrow \gamma_0 \to 1$ and $\alpha = 1$, Eq. (20) reduces to the following expression for the velocity of classical Newtonian nanofluid

$$\begin{aligned}
v(\xi,\tau) &= \frac{1}{2}\Big[\exp(-\xi\sqrt{M_0})\mathrm{erfc}\Big(\frac{\xi}{2\sqrt{\tau}}-\sqrt{M_0\tau}\Big)+\exp(\xi\sqrt{M_0})\mathrm{erfc}\Big(\frac{\xi}{2\sqrt{\tau}}+\sqrt{M_0\tau}\Big)\Big]+\\
&+\frac{Gr_0}{2M_0}\Big[\exp(-\xi\sqrt{M_0})\mathrm{erfc}\Big(\frac{\xi}{2\sqrt{\tau}}-\sqrt{M_0\tau}\Big)+\exp(\xi\sqrt{M_0})\mathrm{erfc}\Big(\frac{\xi}{2\sqrt{\tau}}+\sqrt{M_0\tau}\Big)\Big]-\\
&-\frac{Gr_0\exp\Big(\frac{M_0}{\delta_1}\tau\Big)}{2M_1}\Big[\exp\Big(-\xi\sqrt{M_0\Big(1-\frac{1}{\delta_1}\Big)}\Big)\mathrm{erfc}\Big(\frac{\xi}{2\sqrt{\tau}}-\sqrt{M_0\Big(1-\frac{1}{\delta_1}\Big)\tau}\Big)+\\
&+\exp\Big(\xi\sqrt{M_0\Big(1-\frac{1}{\delta_1}\Big)}\Big)\mathrm{erfc}\Big(\frac{\xi}{2\sqrt{\tau}}+\sqrt{M_0\Big(1-\frac{1}{\delta_1}\Big)\tau}\Big)\Big]-\frac{Gr_0}{M_0}\mathrm{erfc}\Big(\frac{\xi\sqrt{\delta_0}}{2\sqrt{\tau}}\Big)+
\end{aligned}$$

$$+\frac{Gr_0 \exp\left(\frac{M_0}{\delta_1}\tau\right)}{2M_0}\left[\exp\left(-a\sqrt{\delta_0\frac{M_0}{\delta_1}}\right)\operatorname{erfc}\left(\frac{\xi\sqrt{\delta_0}}{2\sqrt{\tau}}-\sqrt{\frac{M_0}{\delta_1}\tau}\right)\right. \tag{21}$$
$$\left.+\exp\left(a\sqrt{\delta_0\frac{M_0}{\delta_1}}\right)\operatorname{erfc}\left(\frac{\xi\sqrt{\delta_0}}{2\sqrt{\tau}}+\sqrt{\frac{M_0}{\delta_1}\tau}\right)\right].$$

4 Nusselt Number

The dimensionless expression for the Nusselt number is given by:

$$Nu_x = -\frac{(k_s + 2k_f) - 2\varphi(k_f - k_s)}{(k_s + 2k_f) + \varphi(k_f - k_s)} \frac{\partial\theta(\xi,\tau)}{\partial\xi}\bigg|_{\xi=0}. \tag{22}$$

5 Parametric Study

A fractional model for the flow of MDGCNF over an isothermal vertical plate is studied. The coupled partial differential equations with Atangana–Baleanu time-fractional derivatives are solved analytically by the Laplace transform method. The solutions for fractional Newtonian nanofluid, classical Casson nanofluid and classical Newtonian nanofluid are presented as special cases. The influence of different embedded parameters, such as α, ϕ, β, M, t and Gr are shown graphically.

Figures 2, 3, 4, 5 and 6 depict the effect of fractional parameter α on $v(\xi,\tau)$. Clearly, increasing values of α decreases $v(\xi,\tau)$. Moreover, classical velocity ($\alpha = 1$) is less than fractional velocity. In classical case, ($\alpha = 1$) the boundary layer is stronger and thicker which reduces the fluids velocity.

Figure 2 is plotted in order to compare Casson nanofluid ($\phi = 0.04$) with customary Casson fluid ($\phi = 0.00$). It is noticed that the velocity of customary Casson fluid is higher than the velocity of Casson nanofluid. Physically, it is due to the higher viscosity of Casson nanofluid as compared to customary Casson fluid.

The effect of ϕ of Molybdenum disulfide nanoparticles on the nanofluid velocity is presented in Fig. 3. It is cleared from this figure that $v(\xi,\tau)$ is a decreasing function of ϕ. This is true because when ϕ increases, the fluid becomes more viscous which leads to decrease the velocity of nanofluid. ϕ is also responsible for increase in the thermal conductivity of the nanofluid.

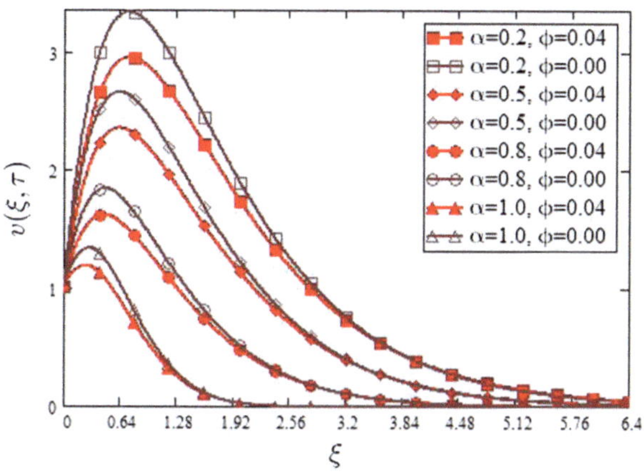

Fig. 2 Comparison of Casson nanofluid with customary Casson fluid when $\beta = 2.5$, $Re = 1$, $\phi = 0.02$, $M = 2$, $Gr = 30$ and $t = 0.10$

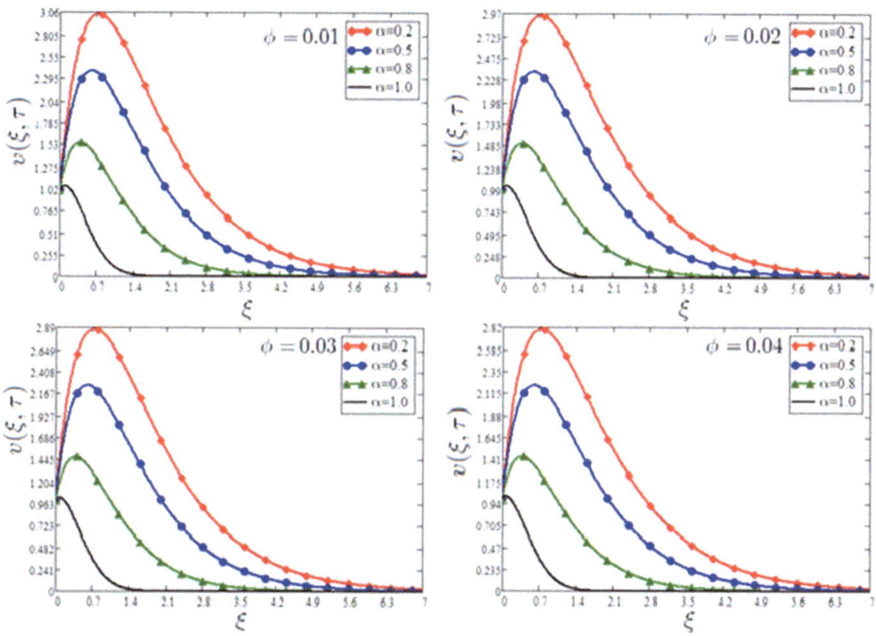

Fig. 3 Variation in the velocity profile for different values of ϕ when $\beta = 2.5$, $Re = 1$, $\phi = 0.02$, $M = 2$, $Gr = 30$ and $t = 0.10$

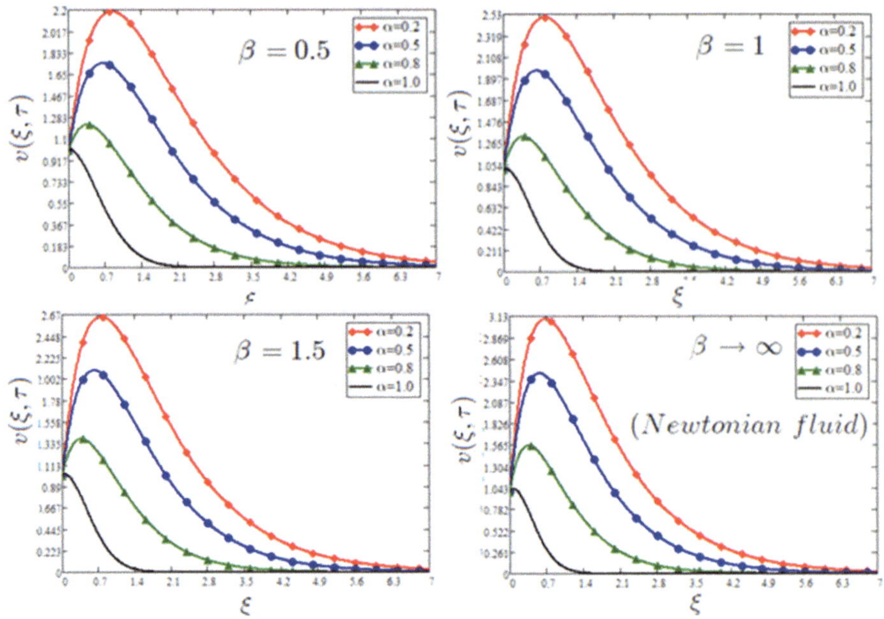

Fig. 4 Variation in the velocity profile for different values of β when $Re = 1$, $\phi = 0.02$, $M = 2$, $Gr = 30$ and $t = 0.10$

Table 2 Variation in Nusselt number for various parameters

α	σ	t	Nu_x
0.2	0.02	0.2	0.707
0.5	0.02	0.2	0.831
0.2	0.04	0.2	0.724
0.2	0.02	1	0.642

Figure 4 depicts the variation in velocity distribution for different values of β. For large values of β, the fluid velocity increases. This is physically possible, when β increases, the momentum boundary layer significantly decreases. As a result, the nanofluid velocity increases. Furthermore, for $\beta \to \infty \Rightarrow \gamma_0 \to 1$, fluid behaves as Newtonian nanofluid. The influence of M is presented in Fig. 5. From this figure it is noticed that the nanofluid velocity decreases when M is increased. M is a dimensionless number whose increasing values enhances the Lorentz force which retards the nanofluid velocity. The velocity profile for different values of Gr is plotted

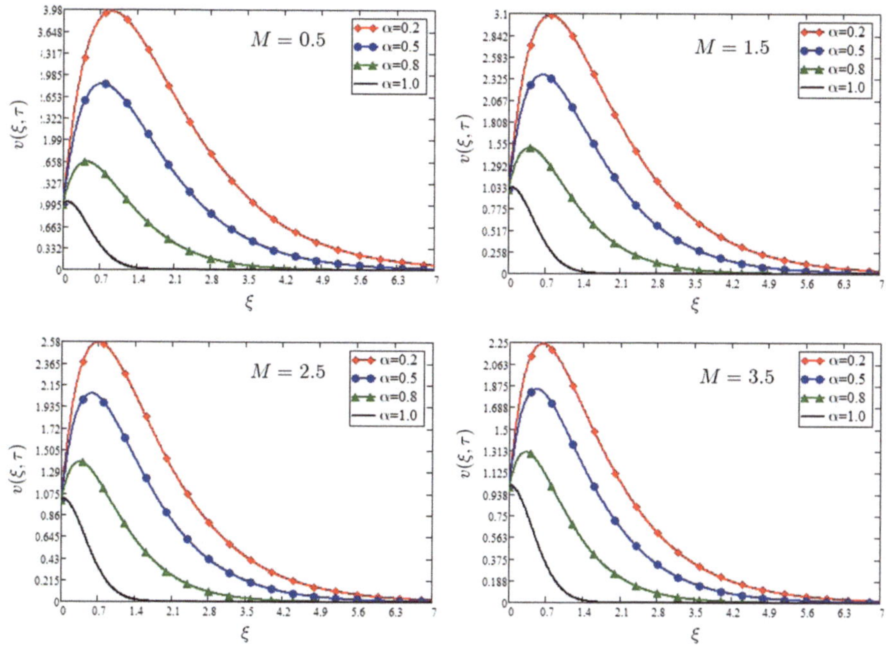

Fig. 5 Variation in the velocity profile for different values of M when $\beta = 2.5$, $Re = 1$, $\phi = 0.02$, $Gr = 30$ and $t = 0.10$

in Fig. 6. It is observed that the velocity of nanofluid increases with increasing values of Gr. Physically, Gr increases the gradient of temperature as a result the buoyancy force enhances and the velocity of nanofluid increases. Table 2 exhibits variation in Nu_x. For large values of α and ϕ, the rate of heat transfer increases whereas it decreases with increase in τ. Large values of ϕ tend to more nanoparticle per unit volume which enhance the thermal conductivity of the fluid.

6 Concluding Remarks

Using the Atangana–Baleanu time-fractional derivatives, a model has been developed for the free convection flow of generalized nanofluid with MoS_2 nanoparticles. The exact solutions are obtained via the Laplace transform method. The key findings of the present study have been summarized as follows:

- Increasing values of ϕ,M decreases the nanofluid velocity.
- The nanofluid velocity enhances with enhancement in β and Gr.
- For large values of α,ϕ the rate of heat transfer increases.
- The rate of heat transfer decreases with increase in τ.

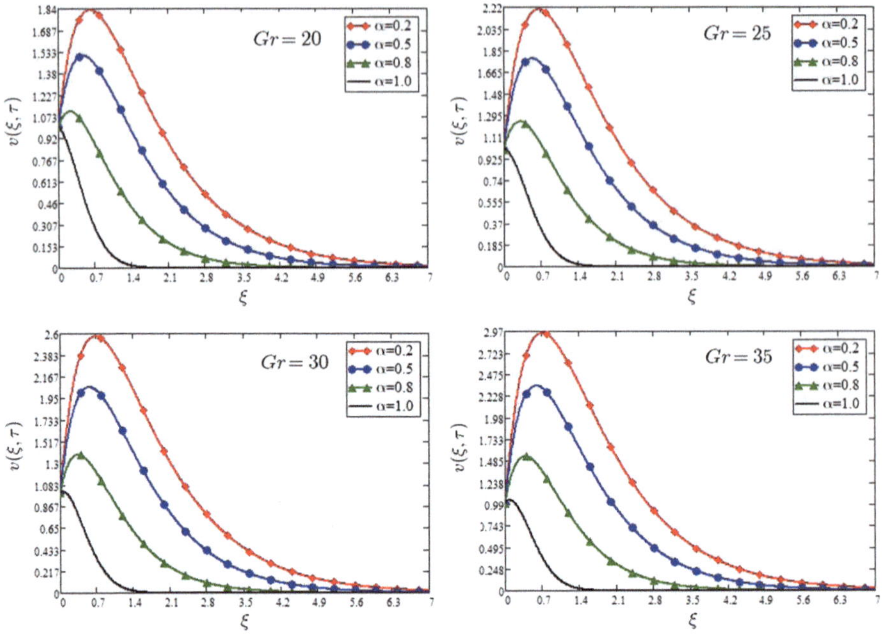

Fig. 6 Variation in the velocity profile for different values of Gr when $\beta = 2.5$, $Re = 1$, $\phi = 0.02$, $M = 2$ and $t = 0.10$

References

1. Li, Y., Tung, S., Schneider, E., Xi, S.: A review on development of nanofluid preparation and characterization. Powder Technol. **196**(2), 89–101 (2009)
2. Bashirnezhad, K., Bazri, S., Safaei, M.R., Goodarzi, M., Dahari, M., Mahian, O., Wongwises, S.: Viscosity of nanofluids: a review of recent experimental studies. Int. Commun. Heat Mass Transf. **73**, 114–123 (2016)
3. Mohamoud, M.J., Singh, T., Mahmoud, S.E., Koc, M., Samara, A., Isaifan, R.J., Atieh, M.A.: Critical review on nanofluids: preparation, characterization and applications. J. Nanomater. **1**, 1–22 (2016). (Hindawy)
4. Öztop, H.F., Estellé, P., Yan, W.M., Al-Salem, K., Orfi, J., Mahian, O.: A brief review of natural convection in enclosures under localized heating with and without nanofluids. Int. Commun. Heat Mass Transf. **60**, 37–44 (2015)
5. Kasaeian, A., Azarian, R.D., Mahian, O., Kolsi, L., Chamkha, A.J., Wongwises, S., Pop, I.: Nanofluid flow and heat transfer in porous media: a review of the latest developments. Int. J. Heat Mass Transf. **107**, 778–791 (2017)
6. Maxwell, J.C., Garnett, W., Pesic, P.: An Elementary Treatise on Electricity. Courier Corporation (2005)
7. Gul, A., Khan, I., Shafie, S., Khalid, A., Khan, A.: Heat transfer in MHD mixed convection flow of a ferrofluid along a vertical channel. PloS One **10**(11), 1–14 (2015)
8. Choi, S.U.: Enhancing thermal conductivity of fluids with nanoparticles. ASME-Publications-Fed **231**, 99–106 (1995)
9. Das, S.K., Choi, S.U., Yu, W., Pradeep, T.: Nanofluids: Science and Technology. Wiley, New York (2007)

10. Wang, X.Q., Mujumdar, A.S.: Heat transfer characteristics of nanofluids: a review. Int. J. Therm. Sci. **46**(1), 1–19 (2007)
11. Ding, Y., Chen, H., Wang, L., Yang, C.Y., He, Y., Yang, W., Huo, R.: Heat transfer intensification using nanofluids. KONA Powder Part. J. **25**, 23–38 (2007)
12. Wang, X.Q., Mujumdar, A.S.: A review on nanofluids-part II: experiments and applications. Braz. J. Chem. Eng. **25**(4), 631–648 (2008)
13. Buongiorno, J.: Convective transport in nanofluids. J. Heat Transf. **128**(3), 240–250 (2006)
14. Haq, R.U., Shahzad, F., Al-Mdallal, Q.M.: MHD pulsatile flow of engine oil based carbon nanotubes between two concentric cylinders. Results Phys. **7**, 57–68 (2017)
15. Shahzad, F., Haq, R.U., Al-Mdallal, Q.M.: Water driven Cu nanoparticles between two concentric ducts with oscillatory pressure gradient. J. Mol. Liq. **224**, 322–332 (2016)
16. Khan, U., Ahmed, N., Mohyud-Din, S.T.: Heat transfer effects on carbon nanotubes suspended nanofluid flow in a channel with non-parallel walls under the effect of velocity slip boundary condition: a numerical study. Neural Comput. Appl. **28**(1), 37–46 (2017)
17. Wakif, A., Boulahia, Z., Sehaqui, R.: Numerical analysis of the onset of longitudinal convective rolls in a porous medium saturated by an electrically conducting nanofluid in the presence of an external magnetic field. Results Phys. **1**, 1–21 (2017)
18. Sheikholeslami, M., Vajravelu, K.: Forced convection heat transfer in Fe_3O_4-ethylene glycol nanofluid under the influence of Coulomb force. J. Mol. Liq. **233**, 203–210 (2017)
19. Sheikholeslami, M., Hayat, T., Alsaedi, A.: Numerical simulation of nanofluid forced convection heat transfer improvement in existence of magnetic field using lattice Boltzmann method. Int. J. Heat Mass Transf. **108**, 1870–1883 (2017)
20. Aman, S., Khan, I., Ismail, Z., Al-Mdallal, Q.M.: Heat transfer enhancement in free convection fow of CNTs Maxwell nanofluids with four different types of molecular liquids. Sci. Rep. **7**(2445), 1–13 (2017)
21. Ali, F., Gohar, M., Khan, I.: MHD flow of water-based Brinkman type nanofluid over a vertical plate embedded in a porous medium with variable surface velocity, temperature and concentration. J. Mol. Liq. **223**, 412–419 (2016)
22. Wang, H., Yu, L., Lee, Y.H., Shi, Y., Hsu, A., Chin, M.L., Palacios, T.: Integrated circuits based on bilayer MoS2 transistors. Nano Lett. **12**(9), 4674–4680 (2012)
23. Das, S., Chen, H.Y., Penumatcha, A.V., Appenzeller, J.: High performance multilayer MoS_2 transistors with scandium contacts. Nano Lett. **13**(1), 100–105 (2012)
24. Radisavljevic, B., Radenovic, A., Brivio, J., Giacometti, I.V., Kis, A.: Single-layer MoS_2 transistors. Nat. Nanotechnol. **6**(3), 147–150 (2011)
25. Castellanos-Gómez, A., Poot, M., Steele, G.A., Van der Zant, H.S., Agraït, N., Rubio-Bollinger, G.: Mechanical properties of freely suspended semiconducting graphene-like layers based on MoS_2. Nanoscale Res. Lett. **7**(1), 1–8 (2012)
26. Winer, W.O.: Molybdenum disulfide as a lubricant: a review of the fundamental knowledge. Wear **10**(6), 422–452 (1967)
27. Kato, H., Takama, M., Iwai, Y., Washida, K., Sasaki, Y.: Wear and mechanical properties of sintered copper-tin composites containing graphite or molybdenum disulfide. Wear **255**(1), 573–578 (2003)
28. Mao, C., Huang, Y., Zhou, X., Gan, H., Zhang, J., Zhou, Z.: The tribological properties of nanofluid used in minimum quantity lubrication grinding. Int. J. Adv. Manuf. Technol. **71**(5–8), 1221–1228 (2014)
29. Shafie, S., Gul, A., Khan, I.: Molybdenum disulfide nanoparticles suspended in water-based nanofluids with mixed convection and flow inside a channel filled with saturated porous medium. In: Proceedings of the 2nd International Conference on Mathematics, Engineering and Industrial Applications (icomeia2016), vol. 1775, no. 1, pp. 1–6. AIP Publishing (2016)
30. Khan, I., Gul, A., Shafie, S.: Effects of magnetic field on molybdenum disulfide nanofluids in mixed convection flow inside a channel filled with a saturated porous medium. J. Porous Media **20**(5), 1–14 (2017)
31. Khan, I.: Shape effects of MoS_2 nanoparticles on MHD slip flow of molybdenum disulphide nanofluid in a porous medium. J. Mol. Liq. **233**, 442–451 (2017)

32. Saqib, M., Ali, F., Khan, I., Sheikh, N.A., Jan, S.A.A.: Exact solutions for free convection flow of generalized Jeffrey fluid: a Caputo-Fabrizio fractional model. Alex. Eng. J. **1**, 1–14 (2017)
33. Hristov, J.: Derivatives with non-singular kernels from the Caputo-Fabrizio definition and beyond: appraising analysis with emphasis on diffusion models. Front. Fract. Calc. **1**, 235–295 (2017). (Sharjah: Bentham Science Publishers)
34. Cuahutenango-Barro, B., Taneco-Hernández, M.A., Gómez-Aguilar, J.F.: On the solutions of fractional-time wave equation with memory effect involving operators with regular kernel. Chaos, Solitons & Fractals **115**, 283–299 (2018)
35. Morales-Delgado, V.F., Taneco-Hernández, M.A., Gómez-Aguilar, J.F.: On the solutions of fractional order of evolution equations. Eur. Phys. J. Plus **132**(1), 1–17 (2017)
36. Saad, K.M., Gómez-Aguilar, J.F.: Coupled reaction-diffusion waves in a chemical system via fractional derivatives in Liouville-Caputo sense. Rev. Mex. Fís **64**(5), 539–547 (2018)
37. Ali, F., Saqib, M., Khan, I., Sheikh, N.A.: Application of Caputo-Fabrizio derivatives to MHD free convection flow of generalized Walters'-B fluid model. Eur. Phys. J. Plus **131**(10), 1–13 (2016)
38. Sheikh, N.A., Ali, F., Saqib, M., Khan, I., Jan, S.A.A.: A comparative study of Atangana-Baleanu and Caputo-Fabrizio fractional derivatives to the convective flow of a generalized Casson fluid. Eur. Phys. J. Plus **132**(1), 1–15 (2017)
39. Hristov, J.: Derivation of the fractional Dodson equation and beyond: transient diffusion with a non-singular memory and exponentially fading-out diffusivity. Prop. Fract. Differ. Appl. **3**(4), 255–270 (2017)
40. Coronel-Escamilla, A., Gómez-Aguilar, J.F., Baleanu, D., Córdova-Fraga, T., Escobar-Jiménez, R.F., Olivares-Peregrino, V.H., Qurashi, M.M.A.: Bateman-Feshbach tikochinsky and Caldirola–Kanai oscillators with new fractional differentiation. Entropy **19**(2), 1–21 (2017)
41. Hristov, J.: Steady-state heat conduction in a medium with spatial non-singular fading memory: derivation of Caputo-Fabrizio space-fractional derivative with Jeffrey's kernel and analytical solutions. Therm. Sci. **1**, 115–125 (2016)
42. Hristov, J.: Transient heat diffusion with a non-singular fading memory: from the Cattaneo constitutive equation with Jeffrey's kernel to the Caputo-Fabrizio time-fractional derivative. Therm. Sci. **20**(2), 757–762 (2016)
43. Saad, K.M., Gómez-Aguilar, J.F.: Analysis of reaction diffusion system via a new fractional derivative with non-singular kernel. Phys. A: Stat. Mech. Appl. **509**, 703–716 (2018)
44. Gómez-Aguilar, J.F., López-López, M.G., Alvarado-Martínez, V.M., Reyes-Reyes, J., Adam-Medina, M.: Modeling diffusive transport with a fractional derivative without singular kernel. Phys. A: Stat. Mech. Appl. **447**, 467–481 (2016)
45. Gómez-Aguilar, J.F., Miranda-Hernández, M., López-López, M.G., Alvarado-Martínez, V.M., Baleanu, D.: Modeling and simulation of the fractional space-time diffusion equation. Commun. Nonlinear Sci. Numer. Simul. **30**(1–3), 115–127 (2016)
46. Alegría-Zamudio, M., Escobar-Jiménez, R.F., Gómez-Aguilar, J.F.: Fault tolerant system based on non-integers order observers: application in a heat exchanger. ISA Trans. **80**, 286–296 (2018)
47. Gómez-Aguilar, J.F.: Space-time fractional diffusion equation using a derivative with nonsingular and regular kernel. Phys. A: Stat. Mech. Appl. **465**, 562–572 (2017)
48. Ali, F., Jan, S.A.A., Khan, I., Gohar, M., Sheikh, N.A.: Solutions with special functions for time fractional free convection flow of Brinkman-type fluid. Eur. Phys. J. Plus **131**(9), 1–13 (2016)
49. Atangana, A., Baleanu, D.: New fractional derivatives with nonlocal and non-singular kernel: theory and application to heat transfer model. Therm. Sci. **20**(2), 763–769 (2016)
50. Atangana, A., Gómez-Aguilar, J.F.: Decolonisation of fractional calculus rules: breaking commutativity and associativity to capture more natural phenomena. Eur. Phys. J. Plus **133**, 1–22 (2018)
51. Gómez-Aguilar, J.F., Atangana, A.: New insight in fractional differentiation: power, exponential decay and Mittag-Leffler laws and applications. Eur. Phys. J. Plus **132**(1), 1–13 (2017)
52. Atangana, A., Gómez-Aguilar, J.F.: Fractional derivatives with no-index law property: application to chaos and statistics. Chaos, Solitons & Fractals **114**, 516–535 (2018)

53. Gómez-Aguilar, J.F., Atangana, A.: Fractional derivatives with the power-law and the Mittag-Leffler kernel applied to the nonlinear Baggs-Freedman model. Fractal Fract. **2**(1), 1–10 (2018)
54. Atangana, A., Koca, I.: Chaos in a simple nonlinear system with Atangana-Baleanu derivatives with fractional order. Chaos, Solitons & Fractals **1**, 1–8 (2016)
55. Casson, N.: A Flow Equation for Pigment-Oil Suspensions of the Printing Ink Type. Pergamon Press, London (1959)
56. Aghili, A.: Solution to time fractional Couette flow. **3**, 1–9 (2017). (in other words)
57. Gómez-Aguilar, J.F., Escobar-Jiménez, R.F., Olivares-Peregrino, V.H., Benavides-Cruz, M., Calderón-Ramon, C.: Nonlocal electrical diffusion equation. Int. J. Mod. Phys. C **27**(01), 1–16 (2016)

Atangana–Baleanu Derivative with Fractional Order Applied to the Gas Dynamics Equations

Sunil Kumar, Amit Kumar, J. J. Nieto and B. Sharma

Abstract We apply the new Atangana–Baleanu derivative in Caputo sense to study gas dynamics equations of arbitrary order using modified homotopy analysis transform method (MHATM). Atangana and Baleanu suggested an interesting fractional operator in 2016 which is based on the exponential kernel. An alternative framework of MHATM with Atangana–Baleanu derivative is presented and the modified Gas dynamics equations are solved numerically and analytically using aforesaid the method. Illustrative examples are included to demonstrate the validity and applicability of the presented technique with new Atangana–Baleanu derivative.

Keywords Fractional calculus · Atangana–Baleanu fractional derivative · Gas dynamics equations

1 Introduction

In the past few decades, the remarkable achievement of fractional calculus in diverse fields of engineering has been gradually realized. The fractional differential equations have becomes a very important topic of many scholars and scientists. Anomalous diffusion behaviors have been observed in several systems such as in polymers, biopolymers, organisms, liquid crystals, fractals and percolation clusters, proteins and ecosystems. Some of the relevant applications are given in the books [1–6]. The

S. Kumar (✉) · B. Sharma
Department of Mathematics, National Institute of Technology,
Jamshedpur 831014, Jharkhand, India
e-mail: skumar.math@nitjsr.ac.in

A. Kumar
Department of Mathematics, Balarampur College Purulia,
Balarampur, Purulia 723143, West Bengal, India

J. J. Nieto
Department of Statistics, Mathematical Analysis and Optimization,
University of Santiago de Compostela, 15782 Santiago de Compostela, Spain

J. F. Gómez et al. (eds.), *Fractional Derivatives with Mittag-Leffler Kernel*,
Studies in Systems, Decision and Control 194,
https://doi.org/10.1007/978-3-030-11662-0_14

famous Caputo fractional derivative first computes an ordinary derivative followed by a fractional integral to achieve the desired order of fractional derivative, and the Riemann–Liouville fractional derivative is computed in the reverse order and both derivatives are based on singular kernel. In 2016, an interesting and new derivatives without singular kernel were introduced by Atangana and Baleanu [7]. This paper explores the relevancy of new derivatives applied in the field of applied modeling. Throughout this paper the time-fractional derivative is considered in Atangana–Baleanu sense [8] .

It is well known that several physical phenomena are described by non-linear differential equations (both ODEs and PDEs) and the solutions of that type of differential equations play an important role to study the behavior of models. Therefore, the study of the many analytical and numerical methods used for solving the non-linear differential equations is a very important topic for the analysis of engineering practical problems [9–20]. The gas dynamics equation is the mathematical expressions of conservation laws which exist in engineering practices such as conservation of mass, conservation of momentum, conservation of energy etc. The nonlinear equations of ideal gas dynamics are applicable for three types of nonlinear waves like shock fronts, rarefactions and contact discontinuities. The different types of gas dynamics equations in physics have been solved by several authors [21–31]. Recently, Kumar and Rashidi [32] have obtained the solutions for the homogenous and non-homogenous time fractional gas dynamics equation using homotopy analysis transform method with Caputo derivative.

The homotopy analysis method (HAM) was first proposed and applied by Liao [33–37] based on homotopy, a fundamental concept in topology and differential geometry. The HAM is based on the construction of a homotopy which continuously deforms an initial guess approximation to the exact solution of the given problem. An auxiliary linear operator is chosen to construct the homotopy and an auxiliary parameter is used to control the region of convergence of the solution series, which is not possible in other methods like Adomian's decomposition method [38, 39], homotopy asymptotic method [40–43] and harmonic balance method [44, 45]. The homotopy analysis method provides greater flexibility in choosing initial approximations and auxiliary linear operators with Caputo derivative [46–48]. The main goal of the present study is to find the semi analytic solution of the fractional Gas dynamics equations using MHATM and the time fractional derivative is considered in Atangana–Baleanu sense. To the best knowledge of the authors, the modified form of the homotopy analysis transform method is not working with Atangana–Baleanu sense.

The rest of the paper is systematized as follows: Basic definitions of new Atangana–Baleanu derivative is presented in Sect. 2. In Sect. 3, Basic idea of MHATM with new derivative is presented. Convergence analysis of the proposed technique is presented in Sect. 4. Next, application of MHATM with new derivative through few examples is presented in Sect. 5. Finally, Sect. 6 concludes the output of the whole paper.

2 History of Fractional Order Derivatives Without Singular Kernel

The old versions of Riemann–Liouville and Caputo derivatives of fractional order are respectively defined as

$$D_x^{\alpha}u(x,t) = \frac{1}{\Gamma(1-\alpha)}\frac{d}{dx}\int_0^x (x-t)^{-\alpha}u(x,t)dt \qquad 0 < \alpha \leq 1, \tag{1}$$

and

$${}^{c}D_x^{\alpha}u(x,t) = \frac{1}{\Gamma(1-\alpha)}\int_0^x (x-t)^{-\alpha}\frac{d}{dt}u(x,t)dt, \tag{2}$$

Definition 1 (*Caputo–Fabrizio derivative with fractinal order(CFFD)*): Let $u(x,t)$ be a function in $H^1(a;b)$; $b > a$; $\alpha \in (0,1)$ then the CFFD is defined as [2]:

$${}^{CF}D_t^{\alpha}u(x,t) = \frac{M(\alpha)}{(1-\alpha)}\int_a^t u'(x,\tau)e^{-\frac{\alpha(1-\tau)}{1-\alpha}}d\tau. \tag{3}$$

Without loss of generality we can put $a = 0$ to have

$${}^{CF}D_t^{\alpha}u(x,t) = \frac{M(\alpha)}{(1-\alpha)}\int_0^t u'(x,\tau)e^{-\frac{\alpha(1-\tau)}{1-\alpha}}d\tau, \tag{4}$$

where $M(\alpha)$ is a normalization function such that $M(0) = M(1) = 1$. But for the function that does not belong to $H^1(a;b)$, we defined its Caputo–Fabrizio fractional as

$${}^{CF}D_t^{\alpha}u(x,t) = \frac{\alpha M(\alpha)}{(1-\alpha)}\int_a^t (u(x,t)-u(x,\tau))e^{-\frac{\alpha(1-\tau)}{1-\alpha}}d\tau. \tag{5}$$

Definition 2 (*Atangana–Baleanu derivative in Caputo sense*): Let $u(x,t)$ be a function in $H^1(a;b)$; $b > a$; $\alpha \in (0,1)$ then the Atangana–Baleanu derivative in Caputo sense is defined as [7]:

$${}^{ABC}D_t^{\alpha}u(x,t) = \frac{M(\alpha)}{(1-\alpha)}\int_a^t u'(x,\tau)E_{\alpha}\left[-\alpha\frac{(t-\tau)^{\alpha}}{1-\alpha}\right]d\tau. \tag{6}$$

Where M has the same properties as in Caputo–Fabrizio case. The above definition will be helpful to real world problem and also will have great advantage when using Laplace transform to solve some physical problems.

Definition 3 The Laplace transform of the Atangana–Baleanu derivative in Caputo sense is given by [7]

$$L[{}^{ABC}D_t^{\alpha}u(x,t)](s) = \frac{M(\alpha)}{(1-\alpha)}\frac{F(s)s^{\alpha}-s^{\alpha-1}u(x,0)}{s^{\alpha}+\frac{\alpha}{1-\alpha}}.$$

3 Basic Idea of the Modified Homotopy Analysis Transform Method with New Atangana–Baleanu Derivative in Caputo Sense

The basic concept of MHATM using new fractional derivative is discussed through the following general form of the fractional differential equation:

$$ {}^{ABC}D_t^{(\alpha+n)}u(x,t) + R[x]u(x,t) + M[x]u(x,t) = g(x,t),\ t > 0,\ x \in \mathbb{R},\ 0 < \alpha \leq 1, \tag{7}$$

where $R[x]$ is a general linear operator in x, $N[x]$ is a general non-linear operator in x, $g(x,t)$ is a continuous function. For simplicity we ignore all boundary or initial conditions, which can be treated in the similar way.

According to the methodology, applying Laplace transform to both sides of the Eq. (7), we get

$$L[{}^{ABC}D_t^{(\alpha+n)}u(x,t) + R[x]u(x,t) + M[x]u(x,t)] = L[g(x,t)]. \tag{8}$$

Next, by using the property of Laplace transform we get

$$L[u(x,t)] - \left(\frac{1}{s^{n+1}}\right)\sum_{k=0}^{n} s^{n-k}u^k(x,0) + \left(\frac{s+\alpha-\alpha s}{s^{n+2}}\right)L\big[R[x] + M[x] - g(x,t)\big] = 0. \tag{9}$$

Now let us define a non-linear operator as

$$N[\phi(x,t;q) = L[\phi(x,t;q)] - \frac{1}{s^{n+1}}\sum_{k=0}^{n} s^{n-k}u^k + \left(\frac{s+\alpha-\alpha s}{s}\right)L\big[R[x] + M[x] - g(x,t)\big] = 0, \tag{10}$$

where $q \in [0,1]$, be an embedding parameter and $\phi(x,t;q)$ is the real function of x, t and q. By means of generalizing the zeroth order deformation equation is given by

$$(1-q)L[\phi(x,t;q) - u_0(x,t)] = \hbar q H(x,t) M[\phi(x,t;q)], \tag{11}$$

where $\hbar$ is a nonzero auxiliary parameter which helps us to increase the convergence results, $H(x,t)$ an auxiliary function, $u_0(x,t)$ is an initial guess of $u(x,t)$ and $\phi(x,t;q)$ is an unknown function. It is important that one has great freedom to choose auxiliary parameters in MHATM. This freedom plays an important role in establishing the keystone of validity and flexibility of MHATM. Obviously, when $q = 0$ and $q = 1$, it holds

$$\phi(x,t;0) = u_0(x,t) \quad \text{and} \quad \phi(x,t;1) = u(x,t)$$

respectively. Thus, as q increases from 0 and 1, the solution varies from the initial guess $u_0(x,t)$ to the solution $u(x,t)$. Expanding $\phi(x,t;q)$ in Taylor's series with respect to q, we have

$$\phi(x,t;q) = u_0(x,t) + \sum_{m=1}^{\infty} q^m u_m(x,t), \tag{12}$$

where $u_m(x,t) = \frac{1}{m!}\frac{\partial^m \phi(x,t;q)}{\partial q^m}\Big|_{q=0}$. Let us define a vector as

$$\mathbf{u}_n = \{u_0(x,t), u_1(x,t), \ldots, u_n(x,t)\}.$$

Differentiating equation (11) m times with respect to q and the setting $q = 0$ and finally dividing by $m!$ we obtain mth order deformation equation

$$L[u_{m-1}(x,t) - \chi_m u_{m-1}(x,t)] = \hbar q H(x,t) R_m(\mathbf{u}_{m-1}, x, t), \tag{13}$$

where

$$R_m(\mathbf{u}_{m-1}, x, t) = \frac{1}{(m-1)!}\frac{\partial^{m-1}}{\partial q^{m-1}} M(\phi(x,t;q))\Big|_{q=0}, \tag{14}$$

and $\chi_m = \begin{cases} 0, & m \le 1, \\ 1 & m > 1 \end{cases}.$

Operating inverse Laplace transform both sides on Eq. (13), we get

$$u_{m-1}(x,t) = \chi_m u_{m-1}(x,t) + \hbar q L^{-1}[H(x,t) R_m(\mathbf{u}_{m-1}, x, t)]. \tag{15}$$

For convenience point of view, the present nonlinear operator form has been modified in the homotopy analysis transforms method i.e., the nonlinear term $M[x,t]u(x,t)$ is expanded in terms of homotopy polynomials [49] as

$$M[u(x,t)] = M\left(\sum_{k=0}^{m-1} u_m(x,t)\right) = \sum_{m=0}^{\infty} P_m u^m. \tag{16}$$

From Eq. (15), we calculate the terms $u_m(x,t)$, for $m \ge 1$. Therefore, the series solution of the (7) can be obtained as,

$$u_m(x,t) = \sum_{m=0}^{\infty} u_m(x,t)\ . \tag{17}$$

Often, one will be interested in a truncated series of the form $\sum_{m=0}^{k} u_m(x,t)$ as this serves as an approximation to the exact solution.

4 Convergence Analysis of MHATM with New Atangana–Baleanu Derivative in Caputo Sense

The convergence analysis of MHATM with new Atangana–Baleanu derivative in Caputo sense is given through the following Theorem.

Theorem 1 *The series solution* $\sum_{m=0}^{k} u_m(x,t)$ *defined above converges if for a pre-assigned positive* ε *there exist a natural number* k *such that* $\left|u_{m+p}\right| < \varepsilon, \forall m \geq k$ *and* $p = 1, 2, 3 \ldots$

Proof Define the sequence of function $\{g_m\}_{m=0}^{\infty}$ as follows:

$$\begin{aligned} g_0 &= u_0 \\ g_1 &= u_0 + u_1 \\ g_2 &= u_0 + u_1 + u_2 \\ &\cdots \\ g_m &= u_0 + u_1 + u_2 + \cdots + u_m. \end{aligned}$$

We have to show that $\{g_m\}_{m=0}^{\infty}$ converges uniformly on $\Re$. From the assumption of Theorem, we have

$$\left|g_{m+p} - g_m\right| = \left|u_{m+p}\right| < \varepsilon \forall\ m \geq k \text{ and } p = 1, 2, 3, \ldots \tag{18}$$

Therefore, by the well-known Cauchy criterion, we have the sequence $\{g_m\}_{m=0}^{\infty}$ converges uniformly on $\Re$, and hence, series solution $\sum_{m=0}^{k} u_m(x,t)$ converges.

5 Application of HATM with New Atangana–Baleanu Derivative in Caputo Sense to Time Fractional Gas Dynamics Equations

Example 1 In this example, we consider the following homogeneous time fractional gas dynamics equation [31, 32]:

$$\frac{\partial^\alpha u}{\partial t^\alpha} + u\frac{\partial u}{\partial x} - u(1-u) = 0, \quad t > 0, \quad x \in \mathbb{R}, \quad 0 < \alpha \leq 1, \tag{19}$$

with initial condition $u(x,0) = e^{-x}$, and the solution $u(x,t) = e^{-x+t}$ is an exact solution of standard gas dynamics equation [32], that is for $\alpha = 1$. Applying Laplace transform to both sides of Eq. (19) and after using the differential property of Laplace transform , we get

$$\left(\frac{s}{s+\alpha(1-s)}\right)[L\{u(x,t)\} - u(x,0)] + L[uu_x - u + u^2] = 0. \tag{20}$$

On simplifying and operating inverse Laplace transform on both sides, we get

$$u(x,t) - e^{-x} + L^{-1}\left[\left(\frac{s+\alpha(1-s)}{s}\right) L(uu_x - u + u^2)\right] = 0. \tag{21}$$

We consider the linear operator as

$$T[\phi(x,t;q)] = L[\phi(x,t;q)], \tag{22}$$

with property $T[c] = 0$, where c is constant. Here non-linear operator defined as

$$N[\phi(x,t;q)] = L[\phi(x,t;q)] - s^{-1}e^{-x} + \left[\frac{s+\alpha(1-s)}{s}\right] L[\phi\phi_x - \phi + \phi^2]. \tag{23}$$

Using the above definition, with assumption $H(x,t) = 1$, we construct the zeroth-order deformation equation:

$$(1-q)T[\phi(x,t;q) - u_0(x,t)] = q\hbar N[\phi(x,t;q)]. \tag{24}$$

Obviously, when $q = 0$ and $q = 1$, $\phi(x,t;0) = u_0(x,t)$, and $\phi(x,t;1) = u(x,t)$ respectively. Thus, we obtain the mth-order deformation equation:

$$T[u_m(x,t) - \chi_m u_{m-1}(x,t)] = \hbar R_m(\mathbf{u}_{m-1}, x, t). \tag{25}$$

Operating inverse Laplace transform we get

$$u_m(x,t) = \chi_m u_{m-1}(x,t) + L^{-1}[\hbar q R_m(\mathbf{u}_{m-1}, x, t)], \tag{26}$$

where

$$R_m(\mathbf{u}_{m-1}, x, t) = L[u_{m-1}] - (1-\chi_m)e^{-x} + \left[\frac{s+\alpha(1-s)}{s}\right] L\left[P_m^1 - u_m + P_m^2\right]. \tag{27}$$

Here P_m^1 and P_m^2 are homotopy polynomials given by

$$\begin{aligned} P_m^1 &= \frac{1}{\Gamma(m+1)}\left[\frac{\partial^m}{\partial q^m} N\Big[(q\phi(x,t;q))(q\Phi(x,t;q))_x\Big]\right]_{q=0}, \\ P_m^2 &= \frac{1}{\Gamma(m+1)}\left[\frac{\partial^m}{\partial q^m} N\Big[\Big((q\phi(x,t;q))(q\Phi(x,t;q))\Big)\Big]\right]_{q=0}. \end{aligned} \tag{28}$$

Now the solution of the mth order Eq. (26) is given as

$$u_{(x,t)} = (\chi_m + \hbar)u_{m-1} - \hbar(1-\chi_m)e^{-x} + \hbar L^{-1}\left[\left(\frac{s+\alpha(1-s)}{s}\right) L\left[P_m^1 - u_m + P_m^2\right]\right]. \tag{29}$$

Using the initial value $u_0(x,t) = u(x,0) = e^{-x}$ and the iterative scheme (29), we get the following values:

$$u_1(x,t) = -e^{-x}\hbar(1-\alpha+\alpha t),$$
$$u_2(x,t) = -e^{-x}\hbar(1+\hbar)(1-\alpha+\alpha t) + e^{-x}\hbar^2\left[(1-\alpha+\alpha t)^2 - \frac{t^2\alpha^2}{2}\right],$$
$$u_3(x,t) = -e^{-x}\hbar(1+\hbar)(1-\alpha+\alpha t) + e^{-x}\hbar^2\left[(1-\alpha+\alpha t)^2 - \frac{t^2\alpha^2}{2}\right] - e^{-x}\hbar^2 t$$
$$(t\alpha-\alpha^2-\alpha)(1-2\hbar+3\hbar\alpha) + e^{-x}\hbar^2(\alpha-1)^2(1+\alpha\hbar) - \frac{1}{6}e^{-x}\hbar^3t^3\alpha^3.$$

In a similar manner, we can find the rest of the terms for $m \geq 4$, and the final solution can be obtained as:

$$u(x,t) = u_0(x,t) + \sum_{m=0}^{\infty} u_m(x,t)$$
$$= e^{-x} - e^{-x}\hbar(1-\alpha+\alpha t) - e^{-x}\hbar(1+\hbar)(1-\alpha+\alpha t) + e^{-x}\hbar^2\left[(1-\alpha+\alpha t)^2 - \frac{t^2\alpha^2}{2}\right]$$
$$- e^{-x}\hbar(1+\hbar)(1-\alpha+\alpha t) + e^{-x}\hbar^2\left[(1-\alpha+\alpha t)^2 - \frac{t^2\alpha^2}{2}\right] - e^{-x}\hbar^2 t$$
$$(t\alpha-\alpha^2-\alpha)(1-2\hbar+3\hbar\alpha) + e^{-x}\hbar^2(\alpha-1)^2(1+\alpha\hbar) - \frac{1}{6}e^{-x}\hbar^3t^3\alpha^3 + \cdots$$

In particular, if we take $\alpha = 1, \hbar = -1$ the above solution reduced to the exact solution e^{-x+t} [32].

Next the obtained results are verified through the different graphical representation. Figure 1 represents comparisons of the 4th order approximate solution of the function $u(x,t)$ for different values of the fractional derivative α with the known exact solution. It can be noted that there exist a very good agreement between them.

To validate the efficiency and accuracy of the analytical scheme, we give the absolute error curve $E_4 = |u(x,t) - u_4(x,t)|$. Figure 2 shows that our solution obtained by the proposed methods converges very rapidly to the exact solutions in only 4th order approximations.

The behavior of the approximate analytical solution for different fractional Brownian motions $\alpha = 0.7, \alpha = 0.8, \alpha = 0.9$ and standard motions, i.e., $\alpha = 1$ is shown in Fig. 3. It is seen in Fig. 3 that the solution obtained by MHATM with new derivative increase very rapidly with the increases in t at the value of $x = 1$.

Figure 4 shows the $\hbar-$ curves obtained from the 4th order approximation solution. In our theory, it is obvious from Fig. 4 that the acceptable range of auxiliary parameter $\hbar$ is $-2.00 \leq \hbar < 0$ and the valid region of convergence corresponds to the line segment nearly parallel to the horizontal axis.

The comparison of the results between the proposed method with the exact solution, consequently the absolute error is presented in Table 1. Tabulated data shows that our approximate solution is also very near to the exact solution.

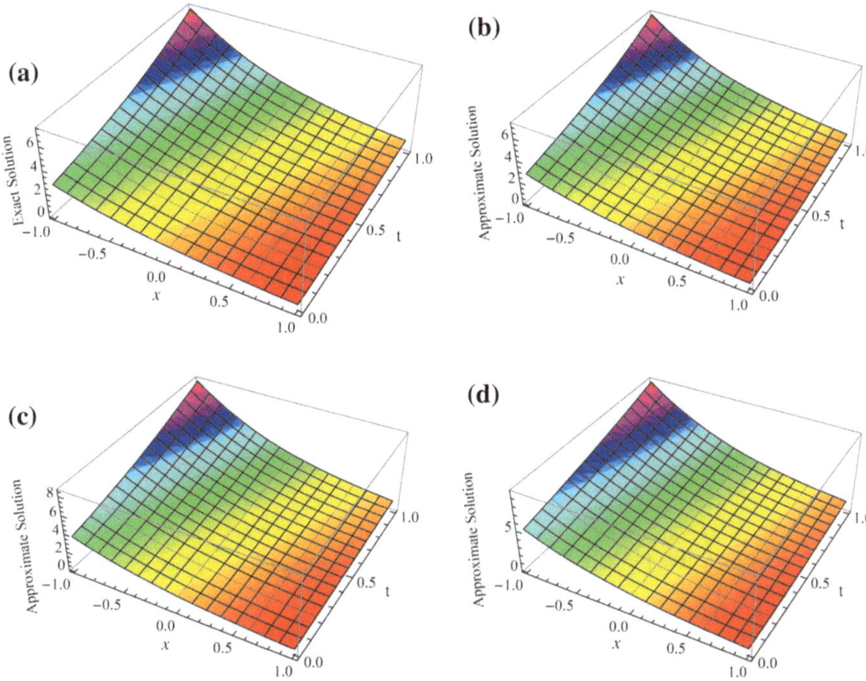

Fig. 1 The surface graph of the exact solution $u(x,t)$ and the 4th order approximate solution $u_4(x,t)$: (a) $u(x,t)$ when $\alpha = 1$, (b) $u_4(x,t)$ when $\alpha = 1$, (c) $u_4(x,t)$ when $\alpha = 0.75$, (d) $u_4(x,t)$ when $\alpha = 0.5$

Fig. 2 Plot of absolute error $E_4(u) = |u(x,t) - u_4(x,t)|$ when $\alpha = 1$

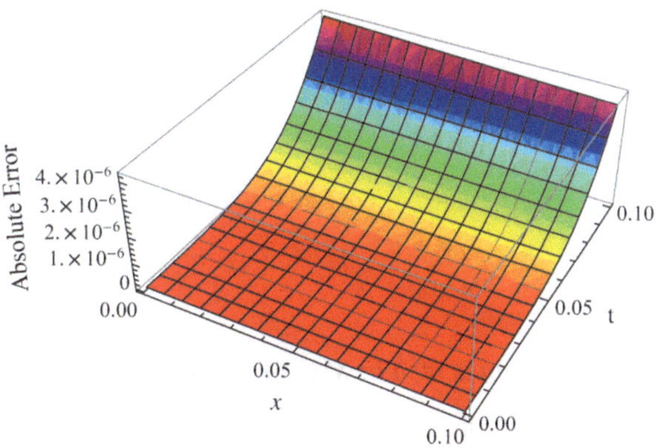

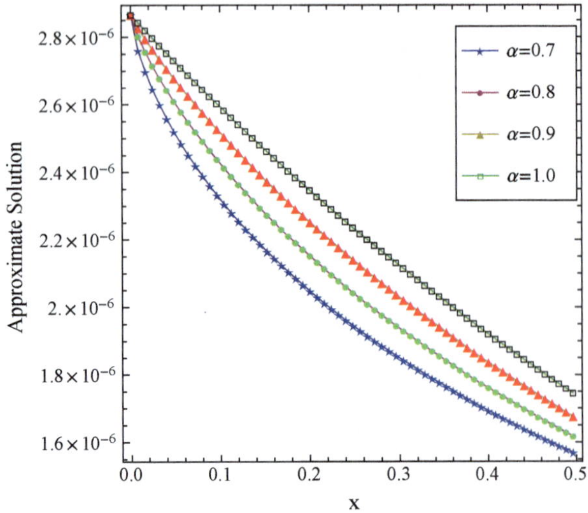

Fig. 3 Plot of $u(x,t)$ verses x time for different values of α at $t=1$ and $\hbar=-1$

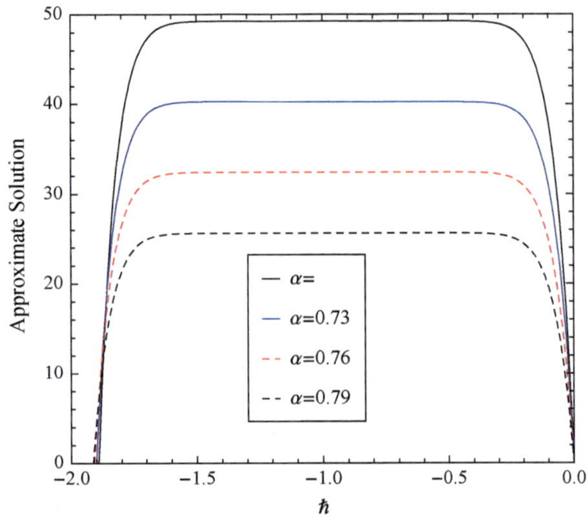

Fig. 4 Plot of $\hbar-$ curve for different values of α at $x=0.5$ and $t=0.01$

Optimal Values of $\hbar$ in MHATM with New Derivative

At the mth- order of approximation, one can define the exact square residual error as

$$\Delta_m^u = \int_0^1 \int_0^1 \left(N\left[\sum_{i=0}^{m} u_i(x,t) \right] \right)^2 dx\, dt, \tag{30}$$

where $N[u(x,t)] = \frac{\partial^\beta u}{\partial t^\alpha} + au\frac{\partial u}{\partial x} + b\frac{\partial^3 u}{\partial x^3}$. However, the exact square residual error Δ_m defined by (30) needs too much CPU time to calculate, even if the order of

Table 1 The absolute error in the solution of fractional gas dynamic equation using MHATM with new derivative at different points of x and t when $\alpha = 1$

(x,t)	Exact solution	Approximation solution	$\lvert u_{exact} - u_{MHATM} \rvert$
(0.1,0.1)	0.00498777	0.00498777	2.03442×10^{-16}
(0.1,0.2)	0.00498801	0.00498801	3.26508×10^{-15}
(0.1,0.3)	0.00498826	0.00498826	1.6523×10^{-14}
(0.2,0.1)	0.00495082	0.00495082	1.89671×10^{-16}
(0.2,0.2)	0.00495131	0.00495131	3.05813×10^{-15}
(0.2,0.3)	0.0049518	0.0049518	1.54813×10^{-14}
(0.3,0.1)	0.00488989	0.00488989	1.54813×10^{-16}
(0.3,0.2)	0.00489062	0.00489062	2.72966×10^{-15}
(0.3,0.3)	0.00489134	0.00489134	1.38293×10^{-14}

Table 2 L_2 and L_∞ error norm for fractional gas dynamic equation using MHATM with new derivative at various points x for $\alpha = 1$

x	L_2 error norm	L_∞ error norm
0.1	1.43210×10^{-15}	2.03442×10^{-16}
0.2	1.65243×10^{-15}	1.89671×10^{-16}
0.3	1.98623×10^{-15}	1.54813×10^{-16}

Table 3 Optimal value of $\hbar$

Order of approx.	Optimal value of $\hbar$ for $\alpha = 1$	Optimal value of $\hbar$ for $\alpha = 0.9$	Value of E_m for $\alpha = 1$	Value of E_m for $\alpha = 0.9$
1	−0.93090	−0.921927	3.45691×10^{-5}	2.56713×10^{-5}
2	−0.92271	−0.95123	3.36789×10^{-6}	3.23478×10^{-6}
3	−0.96235	−0965467	1.56732×10^{-7}	4.87612×10^{-7}

approximation is not very high. Thus, to overcome this disadvantage i.e., to decrease the CPU time, we introduced here the so-called averaged residual error defined by

$$E_m^u = \frac{1}{k_1^2} \sum_{j=1}^{k_1} \sum_{l=1}^{k_1} \left(N \left[\sum_{i=0}^{m} u_i(j\Delta x, l\Delta t) \right] \right)^2, \tag{31}$$

where $\Delta x = \frac{1}{40k_1}$, $\Delta t = \frac{1}{40k_2}$, $k_1 = k_2 = 5$ for gas dynamic equation. The optimal value of $\hbar$ can be obtained by means of minimizing the so called averaged residual error E_m defined by (31), corresponding to the nonlinear algebraic equations $\frac{\partial E_m^u}{\partial \hbar} = 0$ (Tables 2 and 3).

Above the table shows the selection of the values of $\hbar$ as well as the averaged residual error E_m for the different order of approximations. Here we see that there is a great freedom to choose the auxiliary parameters $\hbar$.

Example 2 In this example, we consider the following homogeneous nonlinear time-fractional gas dynamics equation [31, 32]:

$$\frac{\partial^\alpha u}{\partial t^\alpha} + u\frac{\partial u}{\partial x} - u(1-u)\log a = 0, \tag{32}$$

with the initial condition $u(x, 0) = a^{-x}$, and the solution $u(x, t) = a^{t-x}$ is the exact solution for $\alpha = 1$ [32].

Now, applying the aforesaid technique as Example 1, we define a nonlinear operator as

$$N[\phi(x,t;q)] = L[\phi(x,t;q)] - \frac{1}{s}a^{-x} + \frac{s+\alpha(1-s)}{s}L\Big[\phi(x,t;q)\phi_x(x,t;q) - \phi(x,t;q)\log a + \phi^2(x,t;q)\log a\Big]. \tag{33}$$

Therefore the mth order deformation equation is

$$T[u_m(x,t) - \chi_m u_{m-1}(x,t)] = \hbar R_m(\mathbf{u}_{m-1}, x, t). \tag{34}$$

Operating inverse Laplace transform we get

$$u_m(x,t) = \chi_m u_{m-1}(x,t) + L^{-1}[\hbar q R_m(\mathbf{u}_{m-1}, x, t)], \tag{35}$$

where

$$R_m(\mathbf{u}_{m-1}, x, t) = L[u_{m-1}] - (1-\chi_m)a^{-x} + \hbar L^{-1}\left[\frac{s+\alpha(1-s)}{s}L[P_m^1 - u_m\log a + \log a P_m^2]\right]. \tag{36}$$

Now the solution of the mth order of Eq. (35) is given as

$$u_m(x,t) = (\chi_m + \hbar)u_{m-1} - \hbar(1-\chi_m)a^{-x} + \hbar L^{-1}\left[\frac{s+\alpha(1-s)}{s}L[P_m^1 - u_m\log a + \log a P_m^2]\right]. \tag{37}$$

Using the initial approx $u_0(x,t) = u(x,0) = a^{-x}$ and the iterative scheme (37), we get

$$
\begin{aligned}
u_1(x,t) &= -a^{-x}\hbar(1-\alpha+\alpha t)\log a,\\
u_2(x,t) &= -a^{-x}\hbar(1+\hbar)(1-\alpha+\alpha t)\log a + a^{-x}\hbar^2\left[(1-\alpha+\alpha t)^2-\frac{t^2\alpha^2}{2}\right](\log a)^2,\\
u_3(x,t) &= -a^{-x}\hbar(1+\hbar)^2(1-\alpha+\alpha t)\log a + a^{-x}\hbar^2\left[(1-\alpha+\alpha t)^2-\frac{t^2\alpha^2}{2}\right](\log a)^2 - a^{-x}\hbar^2 t\\
&\quad (t\alpha-\alpha^2-\alpha)(1-2h+3\hbar\alpha)(\log a) + a^{-x}\hbar^2(\alpha-1)^2(1+\alpha\hbar) - \frac{1}{6}a^{-x}\hbar^3 t^3\alpha^3(\log a)^3.
\end{aligned}
$$

In a similar manner, we can find the rest of the terms for $m \geq 4$, and the final solution can be obtained as:

$$
u(x,t) = u_0(x,t) + \sum_{m=0}^{\infty} u_m(x,t).
$$

In particular, if we take $\alpha = 1$, $\hbar = -1$ the above obtained solution is reduced to the exact solution a^{t-x} [32].

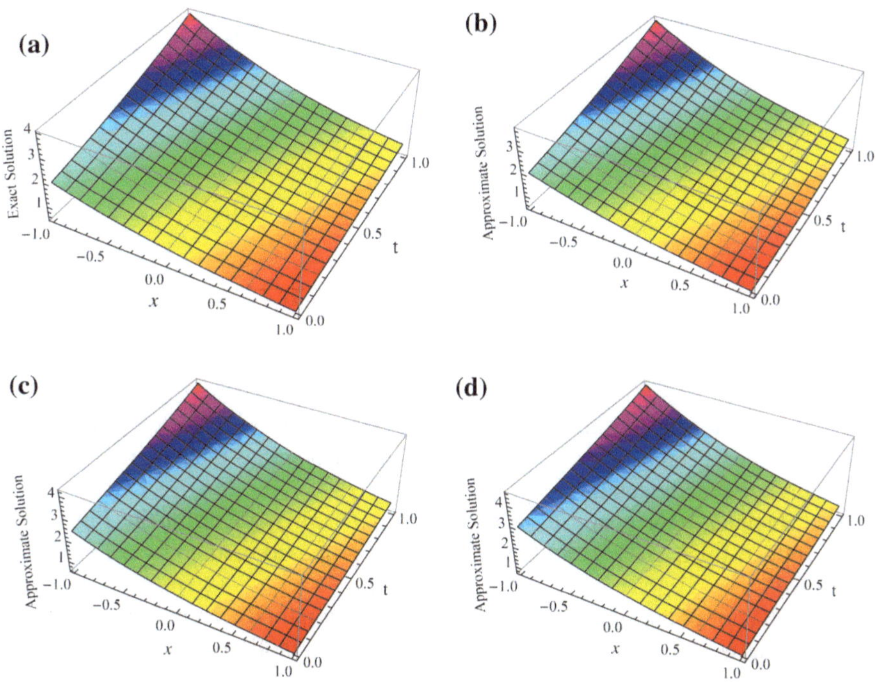

Fig. 5 The surface graph of the exact solution $u(x,t)$ and the 4th order approximate solution $u_4(x,t)$: (a) $u(x,t)$ when $\alpha = 1$, (b) $u_4(x,t)$ when $\alpha = 1$, (c) $u_4(x,t)$ when $\alpha = 0.75$, (d) $u_4(x,t)$ when $\alpha = 0.5$

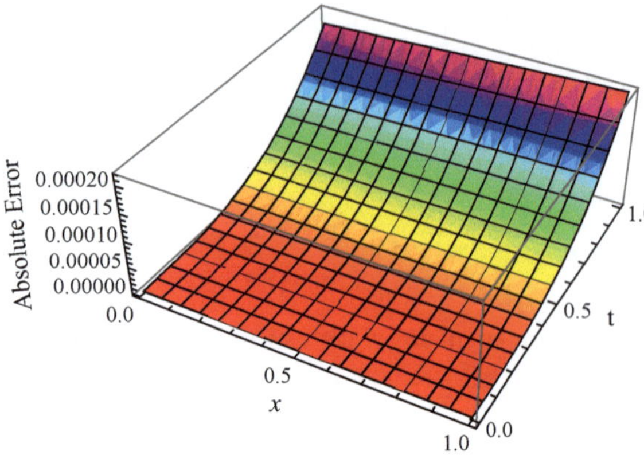

Fig. 6 Plot of absolute error $E_4(u) = |u(x,t) - u_4(x,t)|$ when $\alpha = 1$

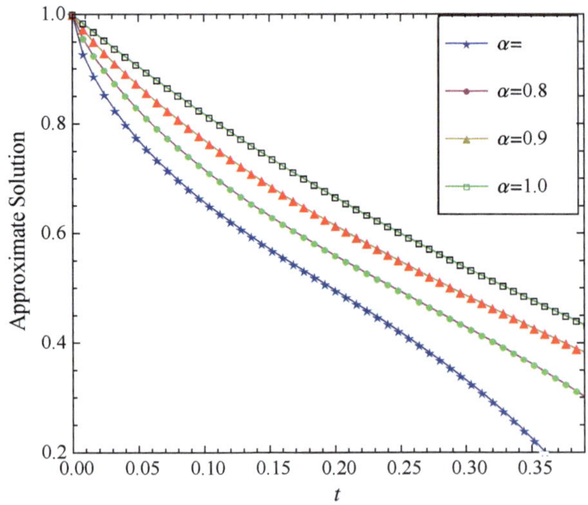

Fig. 7 Plot of $u(x,t)$ verses x time for different values of α at $t = 1$ and $\hbar = -1$

Figure 5 represents comparisons of the 4th order approximate solution with the exact solution. It can be noted that despite of different values of α, there exist a very good agreement between them.

The absolute error curve $E_4 = |u(x,t) - u_4(x,t)|$ is given in the Fig. 6. This figure reflects the accuracy and flexibility of the proposed method by using the new derivative.

The behavior of the approximate analytical solution for different fractional Brownian motions $\alpha = 0.7$, $\alpha = 0.8$, $\alpha = 0.9$ and standard motions, i.e., $\alpha = 1$ is shown in Fig. 7.

Figure 8 shows the $\hbar-$ curves obtained from the 4th order approximation solution. In our theory, it is obvious from Fig. 8 that the acceptable range of auxiliary parameter $\hbar$ is $-2.00 \leq \hbar < 0$ and the valid region of convergence corresponds to the line segment nearly parallel to the horizontal axis.

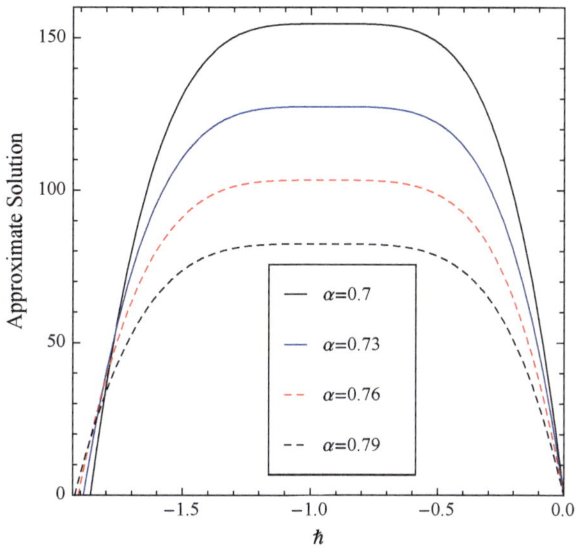

Fig. 8 Plot of $\hbar-$ curve for different values of α at $x = 0.5$ and $t = 0.01$

6 Conclusion

In this manuscript, a new derivative with non-singular kernel was discussed for fractional Gas dynamics equations. An efficient and accurate method based on the MHATM method is proposed for solving the nonlinear Gas dynamics equations with new Caputo–Fabrizio derivative. Few numerical examples were given to demonstrate the validity and applicability of the proposed method. The results show that the MHATM method with new fractional derivative is simple and accurate. In fact by selecting few terms, excellent numerical results are obtained.

Acknowledgements The first author Sunil Kumar would like to acknowledge the financial support received from the National Board for Higher Mathematics. Department of Atomic Energy, Government of India (Approval No. 2/48(20)/2016/NBHM(R.P.)/R and D II/1014).

References

1. Kilbas, A.A., Srivastava, H.M., Trujillo, J.J.: Theory and Applications of Fractional Differential Equations. Elsevier (North-Holland) Science Publishers, Amsterdam (2006)
2. Podlubny, I.: Fractional Differential Equations: An Introduction to Fractional Derivatives, Fractional Differential Equations, to Methods of their Solution and Some of their Applications. Academic Press, San Diego, Calif, USA (1999)
3. Sabatier, J., Agrawal, O.P., Tenreiro Machado, J.A.: Advances in Fractional Calculus: Theoretical Developments and Applications in Physics and Engineering. Springer, Dordrecht (2007)
4. Miller, K.S., Ross, B.: An Introduction to the Fractional Calculus and Fractional Differential Equations. Wiley, New York (1993)

5. Yang, X.J., Baleanu, D., Srivastava, H.M.: Local Fractional Integral Transform and their Applications. Academic Press, New York (2015)
6. Kilbas, A.A., Srivastava, H.M., Trujillo, J.J.: Theory and Applications of Fractional Differential Equations. Elsevier, Amsterdam (2006)
7. Atangana, A., Baleanu, D.: New fractional derivatives with nonlocal and non-singular kernal: theory and applications to heat transfer model. Therm. Sci. (2016)
8. Atangana, A., Koca, I.: Chaos in a simple nonlinear system with Atangana-Baleanu derivatives with fractional order. Chaos, Solitons & Fractals **89**, 447–454 (2016)
9. Atangana, A., Gómez-Aguilar, J.F.: A new derivative with normal distribution kernel: theory, methods and applications. Phys. A: Stat. Mech. Appl. **476**, 1–14 (2017)
10. Gómez-Aguilar, J.F., Atangana, A.: New insight in fractional differentiation: power, exponential decay and Mittag-Leffler laws and applications. Eur. Phys. J. Plus **132**(1), 1–13 (2017)
11. Atangana, A., Gómez-Aguilar, J.F.: Decolonisation of fractional calculus rules: breaking commutativity and associativity to capture more natural phenomena. Eur. Phys. J. Plus **133**, 1–22 (2018)
12. Gómez-Aguilar, J.F.: Irving-Mullineux oscillator via fractional derivatives with Mittag-Leffler kernel. Chaos, Solitons & Fractals **95**(35), 1–7 (2017)
13. Cuahutenango-Barro, B., Taneco-Hernández, M.A., Gómez-Aguilar, J.F.: On the solutions of fractional-time wave equation with memory effect involving operators with regular kernel. Chaos, Solitons & Fractals **115**, 283–299 (2018)
14. Saad, K.M., Gómez-Aguilar, J.F.: Coupled reaction-diffusion waves in a chemical system via fractional derivatives in Liouville-Caputo sense. Rev. Mex. Fís **64**(5), 539–547 (2018)
15. Yépez-Martínez, H., Gómez-Aguilar, J.F.: Numerical and analytical solutions of nonlinear differential equations involving fractional operators with power and Mittag-Leffler kernel. Math. Model. Nat. Phenom. **13**(1), 1–13 (2018)
16. Coronel-Escamilla, A., Gómez-Aguilar, J.F., Torres, L., Escobar-Jiménez, R.F., Valtierra-Rodríguez, M.: Synchronization of chaotic systems involving fractional operators of Liouville-Caputo type with variable-order. Phys. A: Stat. Mech. Appl. **487**, 1–21 (2017)
17. Coronel-Escamilla, A., Gómez-Aguilar, J.F., Torres, L., Escobar-Jiménez, R.F.: A numerical solution for a variable-order reaction-diffusion model by using fractional derivatives with nonlocal and non-singular kernel. Phys. A: Stat. Mech. Appl. **491**, 406–424 (2018)
18. Gómez-Aguilar, J.F., Escobar-Jiménez, R.F., López-López, M.G., Alvarado-Martínez, V.M.: Atangana-Baleanu fractional derivative applied to electromagnetic waves in dielectric media. J. Electromagn. Waves Appl. **30**(15), 1937–1952 (2016)
19. Saad, K.M., Gómez-Aguilar, J.F.: Analysis of reaction diffusion system via a new fractional derivative with non-singular kernel. Phys. A: Stat. Mech. Appl. **509**, 703–716 (2018)
20. Coronel-Escamilla, A., Gómez-Aguilar, J.F., Baleanu, D., Córdova-Fraga, T., Escobar-Jiménez, R.F., Olivares-Peregrino, V.H., Qurashi, M.M.A.: Bateman-Feshbach tikochinsky and Caldirola-Kanai oscillators with new fractional differentiation. Entropy **19**(2), 1–21 (2017)
21. Jafari, H., Chun, C., Seifi, S., Saeidy, M.: Analytical solution for nonlinear gas dynamics equation by homotopy analysis method. Appl. Appl. Math. **4**(1), 149–154 (2009)
22. Mohamad-Jawad, A.J., Petkovic, M.D., Biswas, A.: Applications of He's principles to partial differential equations. Appl. Math. Comput. **217**, 7039–7047 (2011)
23. Bogomolov, S.V.: Quasi gas dynamics equations. Mat. Modelirovanie **21**(12), 145–151 (2009)
24. Evans, D.J., Bulut, H.: A new approach to the gas dynamics equation: an application of the decomposition method. Int. J. Comput. Math. **79**(7), 817–822 (2002)
25. Steger, J.L., Warming, R.F.: Flux vector splitting of the inviscid gas dynamic equations with application to finite-difference methods. J. Comput. Phys. **40**(2), 263–293 (1981)
26. Aziz, A., Anderson, D.: The use of pocket computer in gas dynamics. Comput. Educ. **9**(1), 41–56 (1985)
27. Rasulov, M., Karaguler, T.: Finite difference scheme for solving system equation of gas dynamics in a class of discontinuous function. Appl. Math. Comput. **143**(1), 145–164 (2003)
28. Liu, T.P.: Nonlinear Waves in Mechanics and Gas Dynamics. Defense Technical Information Center, Maryland University, College Park Department of Mathematics (1990)

29. Biazar, J., Eslami, M.: Differential transform method for nonlinear fractional gas dynamics equation. Int. J. Phys. Sci. **6**(5), 1–12 (2011)
30. Das, S., Kumar, R.: Approximate analytical solutions of fractional gas dynamics. Appl. Math. Comput. **217**(24), 9905–9915 (2011)
31. Kumar, S., Kocak, H., Yildirim, A.: A fractional model of gas dynamics equations and its analytical approximate solution using Laplace transform. Z. Nat. **67a**, 388–396 (2012)
32. Kumar, S., Rashidi, M.M.: New method for fractional gas dynamics in shock fronts. Comput. Phys. Commun. **185**, 1947–1954 (2014)
33. Liao, S,J.: The proposed homotopy analysis technique for the solution of nonlinear problems, Ph.D. thesis, Shanghai Jiao Tong University (1992)
34. Liao, S.J.: Beyond Perturbation: Introduction to the Homotopy Analysis Method. CRC Press, Chapman and Hall, Boca Raton (2003)
35. Liao, S.J.: On the homotopy analysis method for nonlinear problems. Appl. Math. Comput. **147**, 499–513 (2004)
36. Liao, S.J.: Notes on the homotopy analysis method: some definition and theorems. Commun. Nonlinear Sci. Numer. Simul. **14**, 983–997 (2009)
37. Liao, S.J.: Homotopy analysis method: a new analytical technique for nonlinear problem. Commun. Nonlinear Sci. Numer. Simul. **2**, 95–100 (1997)
38. Wu, G.-C.: Adomian decomposition method for non-smooth initial value problems. Math. Comput. Model. **54**, 2104–2108 (2011)
39. Wazwaz, A.M.: The combined Laplace transform-Adomian decomposition method for handling nonlinear Volterra integro-differential equations. Appl. Math. Comput. **216**(4), 1304–1309 (2010)
40. Pandey, R.K., Singh, O.P., Baranwal, V.K., Tripathi, M.P.: An analytic solution for the space-time fractional advection-dispersion equation using the optimal homotopy asymptotic method. Comput. Phys. Commun. **183**, 2098–2106 (2012)
41. Golbabai, A., Fardi, M., Sayevand, K.: Application of the optimal homotopy asymptotic method for solving a strongly nonlinear oscillatory system. Math. Comput. Model. **58**(11–12), 1837–1843 (2013)
42. Gómez-Aguilar, J.F., Yépez-Martínez, H., Torres-Jiménez, J., Córdova-Fraga, T., Escobar-Jiménez, R.F., Olivares-Peregrino, V.H.: Homotopy perturbation transform method for nonlinear differential equations involving to fractional operator with exponential kernel. Adv. Differ. Equ. **2017**(1), 1–18 (2017)
43. Morales-Delgado, V.F., Gómez-Aguilar, J.F., Yépez-Martínez, H., Baleanu, D., Escobar-Jiménez, R.F., Olivares-Peregrino, V.H.: Laplace homotopy analysis method for solving linear partial differential equations using a fractional derivative with and without kernel singular. Adv. Differ. Equ. **2016**(1), 1–16 (2016)
44. Chen, Y.M., Liu, J.K., Meng, G.: Relationship between the homotopy analysis method and harmonic balance method. Commun. Nonlinear Sci. Numer. Simul. **15**, 2017–2025 (2010)
45. Liao, H.: Piece wise constrained optimization harmonic balance method for predicting the limit cycle oscillations of an airfoil with various nonlinear structures. J. Fluid Struct. **55**, 324–346 (2015)
46. Kumar, S., Kumar, A., Odibat, Z.: A nonlinear fractional model to describe the population dynamics of two interacting species. Math. Methods Appl. Sci. **40**(11), 4134–4148 (2017)
47. Kumar, S., Kumar, A., Argyros, I.K.: A new analysis for the Keller Segel model of fractional order. Numer. Algorithms **75**(1), 213–228 (2017)
48. Li, C., Kumar, A., Kumar, S., Yang, X.J.: On the approximate solution of nonlinear time-fractional KdV equation via modified homotopy analysis Laplace transform method. J. Nonlinear Sci. Appl. **9**, 5463–5470 (2016)
49. Odibat, Z., Bataineh, A.S.: An adaptation of homotopy analysis method for reliable treatment of strongly nonlinear problems: construction of homotopy polynomials. Math. Methods Appl. Sci. **38**, 991–1000 (2015)

New Direction of Atangana–Baleanu Fractional Derivative with Mittag-Leffler Kernel for Non-Newtonian Channel Flow

Muhammad Saqib, Ilyas Khan and Sharidan Shafie

Abstract This book chapter higlights a new direction of Atangana–Baleanu fractional derivative to channel flow of non-Newtonian fluids. Because the idea to apply fractional derivatives with Mittag-Leffler kernel is a quite new direction for non-Newtonian fluids when flow is in a parallel plate channel. This new and inreresting fractional derivative launched by Atangana and Baleanu with a new fractional operator namely, Atangana–Baleanu fractional operator with Mittag-Leffler function as the kernel of integration has attracted the interest of the researchers. Because this new operator is an efficient tool to model complex and real-world problems. Therefore, this chapter deals with modeling and solution of generalized magnetohydrodynamic (MHD) flow of Casson fluid in a microchannel. The microchannel is taken of infinite length in the vertical direction and of finite width in the horizontal direction. The flow is modeled in terms of a set of partial differential equations involving Atangana–Baleanu time fractional operator with physical initial and boundary conditions. The partial differential equations are transformed to ordinary differential equations via fractional Laplace transformation and solved for exact solutions. To explore the physical significance of various pertinent parameters, the solutions are numerically computed and plotted in different graphs with a physical explanation. The results obtained here may have useful industrial and engineering applications.

Keywords Fractional calculus · Atangana–Baleanu fractional derivative · Non-Newtonian fluid

---

M. Saqib · S. Shafie
Department of Mathematical Sciences, Faculty of Science,
Universiti Teknologi Malaysia, 81310 Skudai, Johor Bahru, Malaysia

I. Khan (✉)
Faculty of Mathematics and Statistics, Ton Duc Thang University,
Ho Chi Minh City, Vietnam
e-mail: ilyaskhan@tdt.edu.vn

J. F. Gómez et al. (eds.), *Fractional Derivatives with Mittag-Leffler Kernel*,
Studies in Systems, Decision and Control 194,
https://doi.org/10.1007/978-3-030-11662-0_15

Nomenclature

u - Velocity of the fluid.
t - Time.
γ - Casson fluid parameter.
T - Temperature of the fluid.
U_0 - Amplitude of the velocity.
g - Acceleration due to gravity.
C_p - Specific heat at a constant pressure.
k_1 - Thermal conductivity of the fluid.
ν - Kinematic viscosity of the fluid.
ν - Dimensionless velocity of the fluid.
B_0 - External magnetic field.
γ_0 - Dimensionless Casson fluid parameter.
ω - Frequency of oscillation.
p_y - Yield stress of the non-Newtonian fluid.
ϕ - Product of the component of deformation rate itself.
ϕ_c - Critical value of product.
β_T - Volumetric coefficient of thermal expansion.
k_ϵ - Mean absorption coefficient.
μ - Dynamic viscosity.
ρ - Fluid density.
σ_1 - Stephen-Boltzmann constant.
q - Laplace transforms parameter.
q_r - Thermal radiation.
G_r - Thermal Grashof number.
α - Fractional order/Fractional parameter.
$N(\alpha)$ - Normalization function.
k - Permeability of porous medium.
ϕ - Porosity of porous medium.
Pr - Prandtl number.
Pr_{eff} - Effective Prandtl number.
K - Permeability of porous medium.
K_{eff} - Effective permeability.
τ - Shear stress.
e_{ij} - $(i, j)^{th}$ component of deformation rate.

1 Introduction

Recently, it is recognized by many researchers that fractional calculus is an efficient tool to describe complex and real-world problems including viscoelastic materials, rheology, fluid flow, diffusive transport, probability, electrical networks and electromagnetic theory [1–14]. It is worth mentioning here that the real world problems obey three leading mathematical laws based on, exponential decay function, power function and the generalized Mittag-Leffler function [15–20]. Now, there is a discussion to distinguish that which fractional operator is suitable to describe the complex and real-world problems. According to some applied mathematicians, the Caputo fractional operator is an efficient tool to describe such problems because it allows the usual initial conditions in the application of integral transform [21]. This argument is since the researchers want to see the fractional operator like a classical one. On the other hand, some researchers have agreed that the Riemann–Liouville operator is more suitable because while using the Laplace transform, one gets the initial conditions with a fractional exponent which is considered more realistic [22, 23]. However, it is pointed out that both the Riemann–Liouville and Caputo operators have some discrepancies. The kernel of Caputo operator is a singular for $t = \tau$; in the Riemann–Liouville case, the derivative of a constant is not zero [24, 25]. To overcome this shortfall, Caputo and Fabrizio developed a fractional operator using exponential decay law with non-singular but local kernel [26]. This new fractional operator is considered as a filter in the literature and criticized by many researchers [9, 27, 28]. Recently, Atangana and Baleanu introduced new fractional operators namely, Atanaga–Baleanu operators in Riemann–Liouville and Caputo sense [29]. The kernel of these operators is based on the generalized Mittag-Leffler function which is non-singular and non-local. These operators are more suitable to describe the complex and real-world problems because the non-locality of the kernel gives a better description of the memory within the structure with different scale [30]. Furthermore, the Atangana–Baleanu operators satisfied all the mathematical principle under the scope of fractional calculus [31, 32]. The idea of Atangana–Baleanu fractional derivatives was applied by Sheikh et al. [33, 34] for the first time in the literature to study the convective fluid problems. As discussed above, many known mathematicians, physicists and engineers contributed to the theory of fractional derivatives, yet the idea of fractional derivatives particularly of Atangana–Baleanu fractional derivatives, needs to be further explored.

The complex and real-world problems, particularly the flow problems in industry and engineering are non-Newtonian in nature such as chocolates, polymers, paints, varnishes, toothpaste, jelly, coffey and honey etc. Such types of fluids exhibit viscoelastic behavior and possess memory. The governing equations of these fluids are challengeable to handle even with classical derivatives. Under some assumptions, the numerical and approximate solutions are available in the literature, but exact analytical solutions are rare. However, classical derivatives are not suitable to interpret the memory effect of these fluids [35, 36]. Among different types of non-Newtonian fluids, the flow of Casson fluid is considered in a vertical microchannel embedded porous medium due to industrial and engineering applications which include space systems, glass blowing, polymer suspension, high-power-density chips, supercomputers, electronics, micro electro mechanical systems (MEMS), micro opto electro mechanical systems (MOEMS), material processing operations, paper production extraction of plastic sheets and many other [37–39]. Whereas, for better interpretation of viscoelastic behavior and memory effect the fractional operator of Atangana and Baleanu in Caputo sense is considered. The fractional Laplace transformation is utilized to obtain exact analytical solutions. The obtained solutions are computed numerically and plotted in various graphs with physical interpretation.

2 Mathematical Formulation

Let consider the MHD flow of a generalized Casson fluid in a vertical microchannel. The channel consists two parallel plates at a distance d. The fluid flows in x-directions whereas, the y-axis is chosen perpendicular to the flow direction. At the time, $t = 0$, the fluid is at rest and the temperature of the plate at $y = 0$ is T_0. After $t = 0^+$, the temperature of the plate at $y = d$ is raised from T_0 to T_w which enhances the buoyancy force. The physical configuration is shown in Fig. 1 and the constitutive equation of Casson fluid is given in [40]

$$\tau_{ij} = \begin{cases} 2\Big(\mu_\gamma + \frac{p_y}{\sqrt{2\pi}}\Big)e_{ij}, & \pi > \pi_c \\ 2\Big(\mu_\gamma + \frac{p_y}{\sqrt{2\pi_c}}\Big)e_{ij}, & \pi_c < \pi, \end{cases} \tag{1}$$

where

$$\gamma = \frac{\mu_B\sqrt{2\pi_c}}{P_y}. \tag{2}$$

The flow of incompressible Casson fluid, under the influence uniform magnetic field and porous medium along with heat transfer is governed by [40]

$$\frac{\partial u(y,t)}{\partial t} = \nu\left(1+\frac{1}{\gamma}\right)\frac{\partial^2 u(y,t)}{\partial y^2} - \left(\frac{\sigma B_0^2}{\rho} + \left(1+\frac{1}{\gamma}\right)\frac{\nu\phi}{k}\right)u(y,t) + gB_T(T(y,t)-T_0), \tag{3}$$

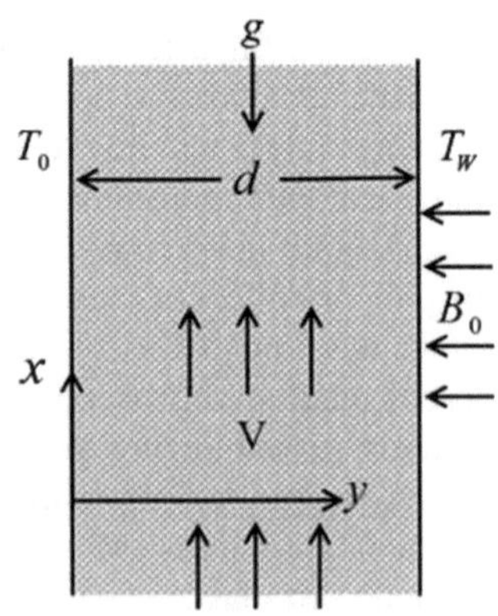

Fig. 1 Schematic diagram for vertical microchannel

$$\rho c_p \frac{\partial T(y,t)}{\partial t} = k_1 \frac{\partial^2 T(y,t)}{\partial y^2} - \frac{\partial q_r}{\partial y}, \tag{4}$$

associated with the following physical initial and boundary conditions:

$$u(y,0) = 0, \qquad T(y,0) = T_0, \forall y \geq 0, \tag{5}$$

$$u(0,t) = 0, \qquad T(0,t) = T_0,$$

$$u(d,t) = 0, \qquad T(d,t) = T_w, \qquad t > 0. \tag{6}$$

The radiative flux is given as under [40]

$$q_r = -\frac{4\sigma_1}{3k_\epsilon}\frac{\partial T^4}{\partial y}, \tag{7}$$

T^4 is expended about T_0 using Taylor series as

$$T^4 \cong 4TT_0^3 - 3T_0^4. \tag{8}$$

Incorporating Eqs. (7) and (8) in Eq. (4), the energy equation becomes:

$$\rho c_p \frac{\partial T(y,t)}{\partial t} = k_1 \left(1 + \frac{16\sigma T_0}{k_1 k_\epsilon}\right) \frac{\partial^2 T(y,t)}{\partial y^2}. \tag{9}$$

Introducing the following dimensionless variables

$$v = \frac{d}{v_f}u, \qquad \xi = \frac{y}{d}, \qquad \tau = \frac{v_f}{d^2}t, \qquad \theta = \frac{T - T_0}{T_w - T_0},$$

into Eqs. (3), (5), (6) and Eq. (9), we get:

$$\frac{\partial v(\xi,\tau)}{\partial \tau} = \frac{1}{\gamma_0}\frac{\partial^2 v(\xi,\tau)}{\partial y^2} - k_{eff} v(\xi,\tau) + G_r \theta(\xi,\tau), \tag{10}$$

$$Pr_{eff} \frac{\partial \theta(\xi,\tau)}{\partial \tau} = \frac{\partial^2 \theta(\xi,\tau)}{\partial \xi^2}, \tag{11}$$

$$v(\xi,0) = 0, \qquad \theta(\xi,0) = T_0, \forall \xi \geq 0, \tag{12}$$

$$v(0,\tau) = 0, \qquad \theta(0,t) = 0,$$

$$u(1,t) = 0, \qquad \theta(d,t) = 1, \qquad t > 0, \tag{13}$$

with

$$\gamma_0 = \frac{\gamma}{\gamma + 1}, \qquad M = \frac{\sigma B_0^2 d^2}{\mu}, \qquad \frac{1}{K} = \frac{\phi d^2}{k}, K_{eff} = M + \frac{1}{\gamma_0 K},$$

$$G_r = \frac{d^3 g \beta_T (T_w - T_0)}{\nu}, \qquad Pr = \frac{\mu C_p}{k}, \qquad R = \frac{16\sigma T_0^3}{3kk_1}, \qquad Pr_{eff} = \frac{Pr}{R+1}.$$

The classical derivatives $\frac{\partial f(.,.)}{\partial \tau}$ are replaced by $D_\tau^\alpha f(.,.)$ via Atangana–Baleanu fractional operator into Eqs. (10) and (12) to generate the generalized model as:

$$D_\tau^\alpha v(\xi, \tau) = \frac{1}{\gamma_0} \frac{\partial^2 v(\xi, \tau)}{\partial y^2} - k_{eff} v(\xi, \tau) + G_r \theta(\xi, \tau), \tag{14}$$

$$Pr_{eff} D_\tau^\alpha \theta(\xi, \tau) = \frac{\partial^2 \theta(\xi, \tau)}{\partial \xi^2}. \tag{15}$$

The Atangana–Baleanu fractional operator with non-Singular and non-local kernel based on Mittag-Leffler function is defined by [29]:

$$D_\tau^\alpha f(\xi, \tau) = \frac{N(\alpha)}{1-\alpha} \int_0^\tau E_\alpha \left(-\alpha \frac{(\tau - t)^\alpha}{1-\alpha} \right) f'(\xi, \tau) dt; \qquad \text{for} \qquad 0 < \alpha < 1, \tag{16}$$

here

$$E_\alpha(-t^\alpha) = \sum_{k=0}^{\infty} \frac{(-t)^{\alpha k}}{\Gamma(\alpha k + 1)},$$

is the Mittag-Leffler function.

3 Exact Analytical Solutions

Here the fractionalized Laplace transformation is applied to obtain solutions for velocity and temperature distributions.

3.1 Solution of the Energy Equation

Applying fractional Laplace transformation on Eq. (15) using Eq. (16) and boundary conditions from (13) yield to the following:

$$\frac{d^2\bar{\theta}(\xi,q)}{d\xi^2} - Pr_{eff}a_0\frac{q^\alpha}{q^\alpha + a_1}\bar{\theta}(\xi,\tau) = 0, \tag{17}$$

$$\bar{\theta}(0,q) = 0, \qquad \bar{\theta}(1,q) = \frac{1}{q}, \tag{18}$$

where $a_0 = \frac{1}{1-\alpha}$ and $a_1 = a_0\alpha$.

The solution of Eq. (17) using the transform boundary conditions from Eq. (18) is given by:

$$\bar{\theta}(\xi,q) = \frac{1}{q}\frac{\sinh \xi\sqrt{\frac{Pr_{eff}a_0q^\alpha}{q^\alpha+a_1}}}{\sinh\sqrt{\frac{Pr_{eff}a_0q^\alpha}{q^\alpha+a_1}}}. \tag{19}$$

To obtain the inverse Laplace transform Eq. (19) can be written in following form:

$$\begin{aligned}\bar{\theta}(\xi,q) =& \frac{1}{q^{1-\alpha}}\left(\sum_{n=0}^{\infty}\frac{1}{q^\alpha}\exp\left(-(1+2n-\xi)\right)\sqrt{\frac{Pr_{eff}a_0q^\alpha}{q^\alpha+b_1}} - \right.\\ &\left.\sum_{n=0}^{\infty}\frac{1}{q^\alpha}\exp\left(-(1+2n+\xi)\right)\sqrt{\frac{Pr_{eff}a_0q^\alpha}{q^\alpha+b_1}}\right).\end{aligned} \tag{20}$$

Upon inverting the Laplace transform, Eq. (20) Yield to the following form:

$$\theta(\xi,\tau) = h(\tau) * \left(\sum_{n=0}^{\infty}\Phi((1+2n-\xi),\tau;Pr_{eff}a_0,b_1) - \sum_{n=0}^{\infty}\Phi((1+2n+\xi),\tau;Pr_{eff}a_0,b_1)\right), \tag{21}$$

here $*$ represents convolution product and the functions $h(\tau)$ and $\Phi(\eta,t,a,b)$ are presented in Appendix (A1)–(A3).

3.2 Solution of Momentum Equation

In the virtue of Eq. (16) the fractional Laplace transformation of Eq. (14) and of the corresponding boundary conditions (13) is calculated as follow:

$$\frac{d^2\bar{v}(\xi,q)}{d\xi^2} - \left(\frac{\gamma_0a_0q^\alpha}{q^\alpha+a_1} + \gamma_0K_{eff}\right)\bar{v}(\xi,\tau) = -\frac{\gamma_0G_r}{q}\frac{\sinh\xi\sqrt{\frac{Pr_{eff}a_0q^\alpha}{q^\alpha+a_1}}}{\sinh\sqrt{\frac{Pr_{eff}a_0q^\alpha}{q^\alpha+a_1}}}, \tag{22}$$

for

$$\bar{v}(0,q) = 0, \qquad \bar{v}(1,q) = 0. \tag{23}$$

The analytical solution of Eq. (22) using (23) is given by:

$$\bar{v}(\xi,q)=\frac{\gamma_0 G_r(q^\alpha+a_1)}{a_4q^\alpha+a_3}\frac{1}{q^{1-\alpha}}\frac{\sinh\xi\sqrt{\frac{a_2q^\alpha+a_3}{q^\alpha+a_1}}}{q^\alpha\sinh\sqrt{\frac{a_2q^\alpha+a_3}{q^\alpha+a_1}}}$$

$$-\frac{\gamma_0 G_r(q^\alpha+a_1)}{a_4q^\alpha+a_3}\frac{1}{q^{1-\alpha}}\frac{\sinh\xi\sqrt{\frac{Pr_{eff}a_0q^\alpha}{q^\alpha+a_1}}}{q^\alpha\sinh\sqrt{\frac{Pr_{eff}a_0q^\alpha}{q^\alpha+a_1}}}, \tag{24}$$

where $a_2=\gamma_0(a_0+K_{eff})$, $a_3=\gamma_0a_1K_{eff}$, $a_4=a_0Pr_{eff}-a_2$.

Equation (24) is further simplified as

$$\bar{v}(\xi,q)=a_6\left(\frac{q^\alpha}{q^\alpha+a_5}+\frac{a_1}{q^\alpha+a_5}\right)\frac{1}{q^{1-\alpha}}\times\left[\sum_{n=0}^{\infty}\frac{1}{q^\alpha}\exp\left(-(1+2n-\xi)\frac{a_2q^\alpha+a_3}{q^\alpha+a_1}\right)\right.$$

$$\left.-\sum_{n=0}^{\infty}\frac{1}{q^\alpha}\exp\left(-(1+2n+\xi)\frac{a_2q^\alpha+a_3}{q^\alpha+a_1}\right)\right]-a_6\left(\frac{q^\alpha}{q^\alpha+a_5}+\frac{a_1}{q^\alpha+a_5}\right)\frac{1}{q^{1-\alpha}}$$

$$\times\left[\sum_{n=0}^{\infty}\frac{1}{q^\alpha}\exp\left(-(1+2n-\xi)\frac{a_2q^\alpha+a_3}{q^\alpha+a_1}\right)-\sum_{n=0}^{\infty}\frac{1}{q^\alpha}\exp\left(-(1+2n+\xi)\frac{a_2q^\alpha+a_3}{q^\alpha+a_1}\right)\right], \tag{25}$$

where $a_5=\frac{a_3}{a_4}$, $a_6=\frac{\gamma_0G_r}{a_4}$.

Taking the inverse Laplace Transform of Eq. (25) yield to the following form:

$$\bar{v}(\xi,\tau)=h(\tau)*[a_6(R_{\alpha,\alpha}(-a_5,\tau)+a_1F_\alpha(a_5,\tau))]*\left(\sum_{n=0}^{\infty}\Psi((1+2n-\xi),\tau;a_2,a_2,a_1)\right.$$

$$\left.-\sum_{n=0}^{\infty}\Psi((1+2n+\xi),\tau;a_2,a_2,a_1)\right)-h(\tau)*[a_6(R_{\alpha,\alpha}(-a_5,\tau)+a_1F_\alpha(a_5,\tau))]$$

$$*\left(\sum_{n=0}^{\infty}\Phi((1+2n-\xi),\tau;Pr_{eff},a_0,b_1)-\sum_{n=0}^{\infty}\Phi((1+2n+\xi),\tau;\tau;Pr_{eff},a_0,b_1)\right). \tag{26}$$

The functions $h(\tau)$ and $\Phi(\eta,t,a,b)$ are previously defined whereas, the new function $\Psi(\eta,t;a,b,c)$ is presented in Appendix (A4) and (A5). Here $R_{\alpha,v}(.,.)$ and $F_\alpha(.,.)$ are the Lorenzo and Hartleys' and Robotnov and Hartleys' functions, respectively. These special functions can be defined as [8]:

$$R_{\alpha,v}(-m,t)=\left(\frac{q^v}{q^\alpha+m}\right)=\sum_{n=0}^{\infty}\frac{(-m)^nt^{(n+1)\alpha-1-v}}{\Gamma[(n+1)\alpha-v]}, \tag{27}$$

$$F_\alpha(-m,t)=\left(\frac{1}{q^\alpha+m}\right)=\sum_{n=0}^{\infty}\frac{(-m)^n t^{(n+1)\alpha-1}}{\Gamma[(n+1)\alpha]}. \tag{28}$$

It is worth mentioning here that Eqs. (21) and (26) clearly satisfies all the imposed initial and boundary conditions which validate our results.

4 Parametric Studies Through Graphs

This section deals with the physical influences of the various pertinent parameter on velocity and temperature distributions. The results obtained for velocity distribution are presented in figures. The influence of α, γ_0, M, K, G_r and of P_r on the velocity profile is discussed physically. Moreover, the results for velocity and temperature are compared in tabular form.

Figures 2a, b are plotted to study the impact of α on the fluid velocity. These figures clearly show that the physical significance of α on the flow is very strong. The velocity increases when α is increased. The classical velocity ($\alpha = 1$) is maximum in case of large time i.e. $t = 1$. This effect is because of the fact that the thickness momentum boundary layer is minimum as compared to the thermal boundary layer in this case. However, Fig. 2b justifies that this effect reverses for small time i.e. $t = 0.1$.

Figure 3 depicts variation in velocity profile due to variation in γ_0 and the rest of parameters are kept constant. The increase in γ_0 enhances viscous forces as compared to thermal forces that tend to decrease fluid velocity. Figure 4 presents the effect of M on velocity profile. The velocity decrease with increase in M.

Large value of M physically, this is true due to the fact M strengthen the Lorentz forces which tends to retard the velocity.

The influence of K is shown in Fig. 5. It is explored that velocity increases for larger values of K. This effect can be physically justified as, when K is increased, the drag forces became weaker consequently the velocity is increased. The influence of G_r on the velocity distribution is displayed in Fig. 6. The increase in G_r causes an increase in the buoyancy forces due to which the fluid velocity increases. The effect of P_r on velocity profile is illustrated in Fig. 7. The P_r is of great physical significance because it shows the ratio of vicious and thermal forces. Greater the values of P_r, stronger will be the viscous forces compare to the thermal forces and thereby the fluid velocity decreases.

Tables 1 and 2 show the variation in velocity and temperature distributions due to α. Clearly, both velocity and temperature distributions increase with an increase in α. Moreover, these tables show that, our solutions satisfy the imposed initial and boundary conditions

Fig. 2 Influence of α on $v(\xi, \tau)$ when $\gamma_0 = 1.5$, $M = 0.5$, $K = 1.5$, $G_r = 10$, $Pr = 10$, $R = 1.5$ and $\omega t = \frac{\pi}{2}$

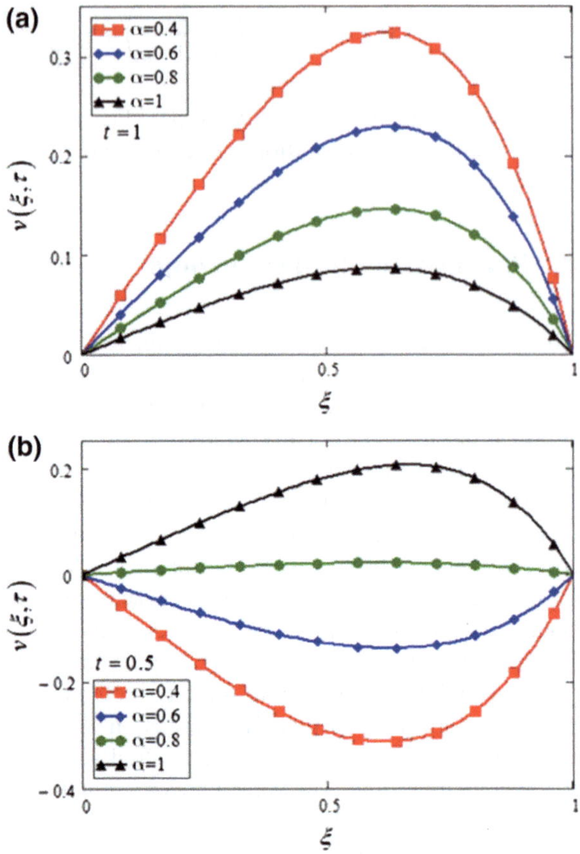

Fig. 3 Influence of γ_0 on $v(\xi, \tau)$ when $\alpha = 0.5$, $M = 0.5$, $K = 1.5$, $G_r = 10$, $Pr = 10$, $R = 1.5$, $t = 1$ and $\omega t = \frac{\pi}{2}$

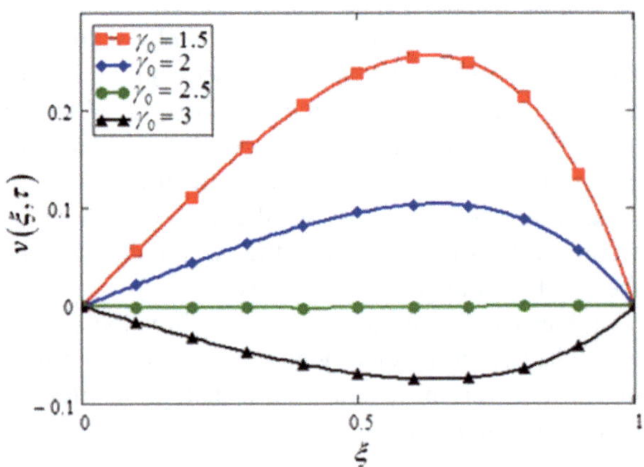

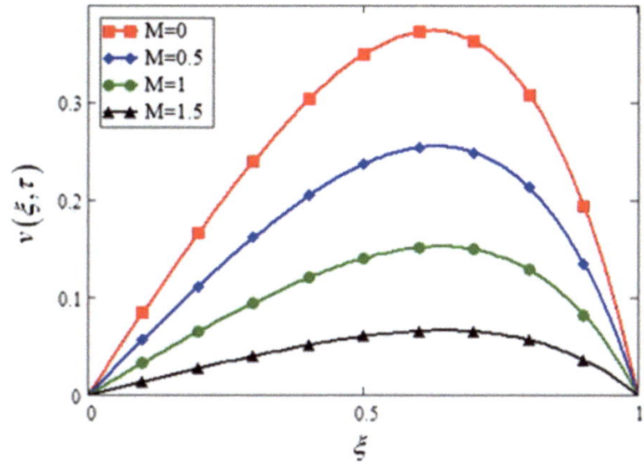

Fig. 4 Influence of M on $v(\xi, \tau)$ when $\alpha = 0.5$, $\gamma_0 = 1.5$, $K = 1.5$, $G_r = 10$, $Pr = 10$, $R = 1.5$, $t = 1$ and $\omega t = \frac{\pi}{2}$

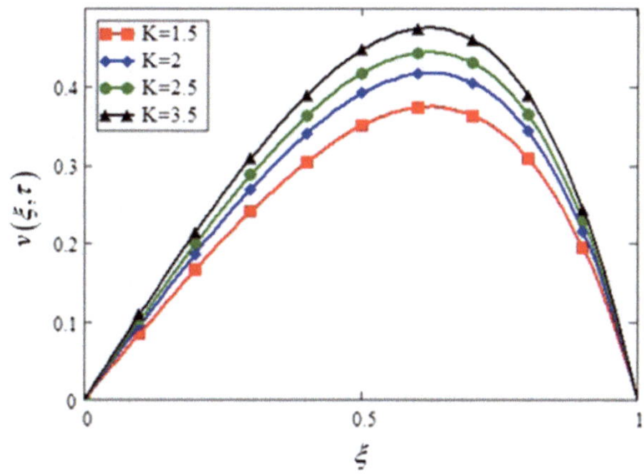

Fig. 5 Influence of K on $v(\xi, \tau)$ when $\alpha = 0.5$, $\gamma_0 = 1.5$, $M = 0.5$, $G_r = 10$, $Pr = 10$, $R = 1.5$, $t = 1$ and $\omega t = \frac{\pi}{2}$

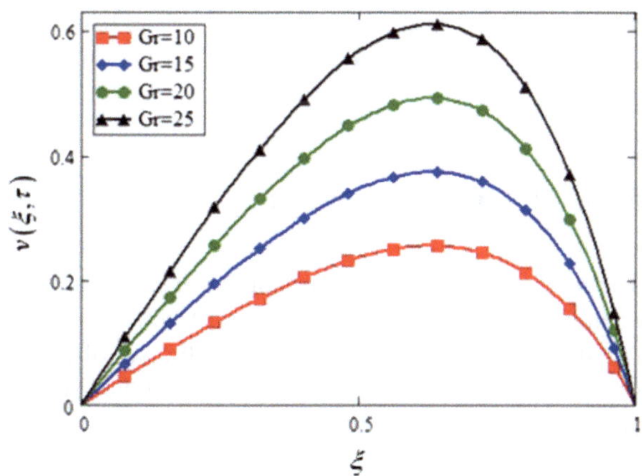

Fig. 6 Influence of G_r on $v(\xi, \tau)$ when $\alpha = 0.5$, $M = 0.5$, $K = 1.5$, $\gamma_0 = 1.5$, $Pr = 10$, $R = 1.5$, $t = 1$ and $\omega t = \frac{\pi}{2}$

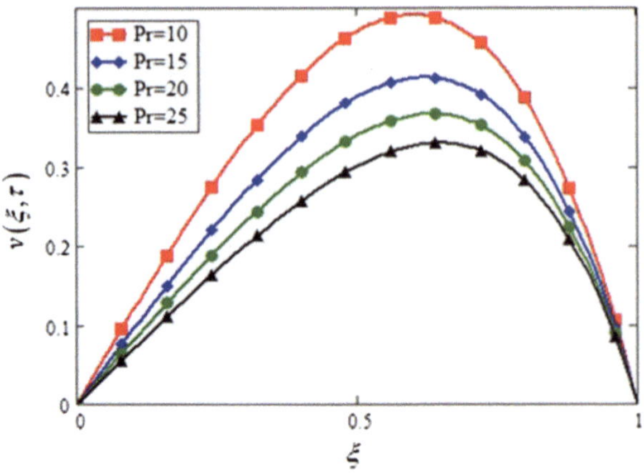

Fig. 7 Influence of Pr on $v(\xi, \tau)$ when $\alpha = 0.5$, $\gamma_0 = 1.5$, $M = 0.5$, $G_r = 10$, $K = 1.5$, $R = 1.5$, $t = 1$ and $\omega t = \frac{\pi}{2}$

Table 1 Variation in velocity distribution due to α

	$\alpha = 0.2$	$\alpha = 0.4$	$\alpha = 0.6$	$\alpha = 0.8$	$\alpha = 1$
ξ	Velocity distribution	Velocity distribution	Velocity distribution	Velocity distribution	Velocity distribution
0	0	0	0	0	0
0.1	0.1190	0.0740	0.0460	0.0420	0.0720
0.2	0.2320	0.1440	0.0890	0.0830	0.1400
0.3	0.3330	0.2090	0.1300	0.1190	0.1910
0.4	0.4160	0.2640	0.1640	0.1480	0.2330
0.5	0.4710	0.3040	0.1900	0.1680	0.2640
0.6	0.4920	0.3230	0.2040	0.1750	0.2650
0.7	0.4670	0.3410	0.1990	0.1650	0.2390
0.8	0.3870	0.2660	0.1710	0.1360	0.1880
0.9	0.2360	0.1670	0.1080	0.0820	0.1140
1.0	−0.00000008	−0.00000008	−0.00000008	−0.00000008	−0.00000008

5 Concluding Remarks

The time fractional derivative approach with a non-singular and non-local kernel of Atangana and Baleanu is utilized to fractionalize the flow. The fractional Laplace transform technique is used to solve the set of partial differential equations. Parametric studies are carried out through graphs. From this study, the major key points are summarized as follow:

- The fractional parameter α is of physical significance affecting viscous, as well as, buoyancy forces.
- The velocity increases with increase in α for long time whereas, this effect reverses for a shorter time.

Table 2 Variation in temperature distribution due to α

	$\alpha = 0.2$	$\alpha = 0.4$	$\alpha = 0.6$	$\alpha = 0.8$	$\alpha = 1$
ξ	Temperature distribution	Temperature distribution	Temperature distribution	Temperature distribution	Temperature distribution
0	0	0	0	0	0
0.1	0.0580	0.0670	0.0800	0.0930	0.1000
0.2	0.1190	0.1360	0.1620	0.1870	0.2000
0.3	0.1840	0.2090	0.2460	0.1810	0.3000
0.4	0.2550	0.2870	0.3340	0.3770	0.3990
0.5	0.3360	0.3730	0.4260	0.4750	0.4890
0.6	0.4290	0.4680	0.5240	0.5750	0.5970
0.7	0.5380	0.5760	0.6290	0.6770	0.6980
0.8	0.6660	0.6980	0.7430	0.7820	0.8030
0.9	0.8180	0.8380	0.8660	0.8890	0.9100
1.0	0.99999994	0.99999994	0.99999994	0.99999994	0.99999994

- The velocity enhances with the enhancement of G_r and K.
- A decreasing behavior of velocity is observed while increasing γ_0, M and Pr.
- The temperature profile is increased when α is increased.

Acknowledgements The authors would like to acknowledge Ministry of Higher Education (MOHE) and Research Management Centre-UTM, Universiti Teknologi Malaysia UTM for the financial support through vote numbers 15H80 and 13H74 for this research.

Appendix-A

$$\Phi_1(\eta, t; a, b) = \mathscr{L}\left[\frac{1}{q}\exp\left(-\eta\left(\sqrt{\frac{aq}{q+b}}\right)\right)\right] = 1 - \frac{2a}{\pi}\int_0^\infty \frac{\sin(\eta s)}{s(a+s^2)}\exp\left(-\frac{bts^2}{a+s^2}\right)ds, \quad \text{(A1)}$$

$$\Phi(\eta, t; a, b) = \mathscr{L}\left[\frac{1}{q}\exp\left(-\eta\left(\sqrt{\frac{aq^\alpha}{q^\alpha+b}}\right)\right)\right]$$

$$= \frac{1}{\pi}\int_0^\infty\int_0^\infty \Phi_1(\eta, u; a, b)\exp(-\tau - r - ur^\alpha\cos\alpha\pi)(ur^\alpha\sin\alpha\pi)drdu, \quad \text{(A2)}$$

$$h(\tau) = \mathscr{L}\left(\frac{1}{q^{1-\alpha}}\right) = \frac{1}{\tau^\alpha\Gamma(1-\alpha)}, \quad \text{(A3)}$$

$$\Psi_1(\eta, t; a, b, c) = \mathscr{L}\left[\frac{1}{q}\exp\left(-\eta\sqrt{\frac{aq+b}{q+c}}\right)\right.$$

$$= \exp(-\eta\sqrt{a}) - \frac{\sqrt{b-ac}}{2\sqrt{\pi}}\int_0^\infty\int_0^\tau \frac{1}{\sqrt{t}}\exp\left(-ct - \frac{\eta}{4u} - au\right) I_1(2\sqrt{(b-ac)ut})dtdu, \quad \text{(A4)}$$

$$\Psi(\eta, t; a, b, c) = \mathscr{L}\left[\frac{1}{q}\exp\left(-\eta\sqrt{\frac{aq^\alpha+b}{q^\alpha+c}}\right)\right.$$

$$= \frac{1}{\pi}\int_0^\infty\int_0^\infty \Psi_1(\eta, u; a, b, c)\exp(-\tau - r - ur^\alpha\cos\alpha\pi)(ur^\alpha\sin\alpha\pi)drdu. \quad \text{(A5)}$$

References

1. Ali, F., Saqib, M., Khan, I., Sheikh, N.A.: Application of Caputo-Fabrizio derivatives to MHD free convection flow of generalized Walters'-B fluid model. Eur. Phys. J. Plus **131**(10), 1–10 (2016)
2. Ali, F., Sheikh, N.A., Khan, I., Saqib, M.: Solutions with Wright function for time fractional free convection flow of Casson fluid. Arab. J. Sci. Eng. **42**(6), 2565–2572 (2017)
3. Gómez-Aguilar, J.F., Torres, L., Yépez-Martínez, H., Baleanu, D., Reyes, J.M., Sosa, I.O.: Fractional Liénard type model of a pipeline within the fractional derivative without singular kernel. Adv. Differ. Equ. **2016**(1), 1–17 (2016)
4. Gómez-Aguilar, J.F., Dumitru, B.: Fractional transmission line with losses. Z. für Naturforschung A **69**(10–11), 539–546 (2014)
5. Gómez-Aguilar, J.F., Yépez-Martínez, H., Escobar-Jiménez, R.F., Astorga-Zaragoza, C.M., Reyes-Reyes, J.: Analytical and numerical solutions of electrical circuits described by fractional derivatives. Appl. Math. Model. **40**(21–22), 9079–9094 (2016)
6. Morales-Delgado, V.F., Taneco-Hernández, M.A., Gómez-Aguilar, J.F.: On the solutions of fractional order of evolution equations. Eur. Phys. J. Plus **132**(1), 1–17 (2017)
7. Alegría-Zamudio, M., Escobar-Jiménez, R.F., Gómez-Aguilar, J.F.: Fault tolerant system based on non-integers order observers: application in a heat exchanger. ISA Trans. **80**, 286–296 (2018)
8. Ali, F., Sheikh, N.A., Khan, I., Saqib, M.: Magnetic field effect on blood flow of Casson fluid in axisymmetric cylindrical tube: a fractional model. J. Magn. Magn. Mater. **423**, 327–336 (2017)
9. Cuahutenango-Barro, B., Taneco-Hernández, M.A., Gómez-Aguilar, J.F.: On the solutions of fractional-time wave equation with memory effect involving operators with regular kernel. Chaos Solitons Fractals **115**, 283–299 (2018)
10. Coronel-Escamilla, A., Gómez-Aguilar, J.F., Baleanu, D., Córdova-Fraga, T., Escobar-Jiménez, R.F., Olivares-Peregrino, V.H., Qurashi, M.M.A.: Bateman-Feshbach tikochinsky and Caldirola-Kanai oscillators with new fractional differentiation. Entropy **19**(2), 1–21 (2017)
11. Dalir, M., Bashour, M.: Applications of fractional calculus. Appl. Math. Sci. **4**(21), 1021–1032 (2010)
12. Coronel-Escamilla, A., Gómez-Aguilar, J.F., Baleanu, D., Escobar-Jiménez, R.F., Olivares-Peregrino, V.H., Abundez-Pliego, A.: Formulation of Euler-Lagrange and Hamilton equations involving fractional operators with regular kernel. Adv. Differ. Equ. **2016**(1), 1–21 (2016)

13. Gómez-Aguilar, J.F., Yépez-Martínez, H., Escobar-Jiménez, R.F., Astorga-Zaragoza, C.M., Morales-Mendoza, L.J., González-Lee, M.: Universal character of the fractional space-time electromagnetic waves in dielectric media. J. Electromagn. Waves Appl. **29**(6), 727–740 (2015)
14. Singh, J., Kumar, D., Hammouch, Z., Atangana, A.: A fractional epidemiological model for computer viruses pertaining to a new fractional derivative. Appl. Math. Comput. **316**, 504–515 (2018)
15. Atangana, A., Owolabi, K.M.: New numerical approach for fractional differential equations. Math. Model. Nat. Phenom. **13**(1), 1–13 (2018)
16. Gómez-Aguilar, J.F., Atangana, A.: New insight in fractional differentiation: power, exponential decay and Mittag-Leffler laws and applications. Eur. Phys. J. Plus **132**(1), 1–19 (2017)
17. Atangana, A., Gómez-Aguilar, J.F.: Fractional derivatives with no-index law property: application to chaos and statistics. Chaos Solitons Fractals **114**, 516–535 (2018)
18. Coronel-Escamilla, A., Gómez-Aguilar, J.F., López-López, M.G., Alvarado-Martínez, V.M., Guerrero-Ramírez, G.V.: Triple pendulum model involving fractional derivatives with different kernels. Chaos Solitons Fractals **91**, 248–261 (2016)
19. Caputo, M.: Linear models of dissipation whose Q is almost frequency independent. Ann. Geophys. **19**(4), 383–393 (1966)
20. Caputo, M.: Linear models of dissipation whose Q is almost frequency independent–II. Geophys. J. Int. **13**(5), 529–539 (1967)
21. Atangana, A., Gómez-Aguilar, J.F.: Numerical approximation of Riemann-Liouville definition of fractional derivative: from Riemann-Liouville to Atangana-Baleanu. Numer. Methods Part. Differ. Equ. **34**(5), 1502–1523 (2018)
22. Bakkyaraj, T., Sahadevan, R.: Invariant analysis of nonlinear fractional ordinary differential equations with Riemann-Liouville fractional derivative. Nonlinear Dyn. **80**(1–2), 447–455 (2015)
23. Sousa, E., Li, C.: A weighted finite difference method for the fractional diffusion equation based on the Riemann-Liouville derivative. Appl. Numer. Math. **90**, 22–37 (2015)
24. Saqib, M., Ali, F., Khan, I., Sheikh, N.A., Jan, S.A.A.: Exact solutions for free convection flow of generalized Jeffrey fluid: a Caputo-Fabrizio fractional model. Alex. Eng. J. **1**, 1–10 (2017)
25. Sheikh, N.A., Ali, F., Khan, I., Saqib, M.: A modern approach of Caputo-Fabrizio time-fractional derivative to MHD free convection flow of generalized second-grade fluid in a porous medium. Neural Comput. Appl. **30**(6), 1865–1875 (2018)
26. Caputo, M., Fabrizio, M.: A new definition of fractional derivative without singular kernel. Progr. Fract. Differ. Appl **1**(2), 1–13 (2015)
27. Atangana, A., Gómez-Aguilar, J.F.: Hyperchaotic behaviour obtained via a nonlocal operator with exponential decay and Mittag-Leffler laws. Chaos Solitons Fractals **102**, 285–294 (2017)
28. Atangana, A., Gómez-Aguilar, J.F.: A new derivative with normal distribution kernel: theory, methods and applications. Phys. A Stat. Mech. Appl. **476**, 1–14 (2017)
29. Atangana, A., Baleanu, D.: New fractional derivatives with nonlocal and non-singular kernel: theory and application to heat transfer model. Therm. Sci. **20**(2), 763–769 (2016)
30. Al-Salti, FAMN, Karimov E., Initial and Boundary Value Problems for Fractional differential equations involving Atangana-Baleanu Derivative. arXiv:1706.00740, 2017
31. Atangana, A., Alqahtani, R.T.: Modelling the spread of river blindness disease via the caputo fractional derivative and the beta-derivative. Entropy **18**(2), 1–18 (2016)
32. Atangana, A., Baleanu, D.: Caputo-Fabrizio derivative applied to groundwater flow within confined aquifer. J. Eng. Mech. **143**(5), 1–18 (2017)
33. Sheikh, N.A., Ali, F., Saqib, M., Khan, I., Jan, S.A.A.: A comparative study of Atangana-Baleanu and Caputo-Fabrizio fractional derivatives to the convective flow of a generalized Casson fluid. Eur. Phys. J. Plus **132**(1), 1–15 (2017)
34. Sheikh, N.A., Ali, F., Saqib, M., Khan, I., Jan, S.A.A., Alshomrani, A.S., Alghamdi, M.S.: Comparison and analysis of the Atangana-Baleanu and Caputo-Fabrizio fractional derivatives for generalized Casson fluid model with heat generation and chemical reaction. Results Phys. **7**, 789–800 (2017)

35. Jan, S.A.A., Ali, F., Sheikh, N.A., Khan, I., Saqib, M., Gohar, M.: Engine oil based generalized brinkman-type nano-liquid with molybdenum disulphide nanoparticles of spherical shape: Atangana-Baleanu fractional model. Numer. Methods Part. Differ. Equ. **34**(5), 1472–1488 (2018)
36. Saqib, M., Khan, I., Shafie, S.: Application of Atangana-Baleanu fractional derivative to MHD channel flow of CMC-based-CNT's nanofluid through a porous medium. Chaos Solitons Fractals **116**, 79–85 (2018)
37. Narahari, M., Pendyala, R.: Exact solution of the unsteady natural convective radiating gas flow in a vertical channel. AIP Conf. Proc. **1557**(1), 121–124 (2013)
38. Seth, G.S., Sharma, R., Kumbhakar, B.: Effects of Hall current on unsteady MHD convective Couette flow of heat absorbing fluid due to accelerated movement of one of the plates of the channel in a porous medium. J. Porous Media **19**(1), 13–30 (2016)
39. Singh, A.K., Gholami, H.R., Soundalgekar, V.M.: Transient free convection flow between two vertical parallel plates. Heat Mass Transf. **31**(5), 329–331 (1996)
40. Saqib, M., Ali, F., Khan, I., Sheikh, N.A.: Heat and mass transfer phenomena in the flow of Casson fluid over an infinite oscillating plate in the presence of first-order chemical reaction and slip effect. Neural Comput. Appl. **30**(7), 2159–2172 (2018)

Exact Solutions for the Liénard Type Model via Fractional Homotopy Methods

V. F. Morales-Delgado, J. F. Gómez-Aguilar, L. Torres, R. F. Escobar-Jiménez and M. A. Taneco-Hernandez

Abstract In this chapter, we present the solution for a Liénard type model of a pipeline expressed by Liouville–Caputo and Atangana-Baleanu-Caputo fractional order derivatives. For this model, new approximated analytical solutions are derived by using the Laplace homotopy perturbation method and the modified homotopy analysis transform method. Both the efficiency and the accuracy of the method are verified by comparing the obtained solutions versus the exact analytical solution.

Keywords Fractional calculus · Atangana–Baleanu fractional derivative · Liénard type model

1 Introduction

Many dynamical phenomena can be represented by the Liénard equation, such as biological, mechanical, electronic and electrical systems [1–10]. Concerning to the topic of the present paper, in [11–13], the authors proposed a Liénard type model for representing the fluid dynamics of a pipeline; the motivation to formulate the model was its application on parameter identification. However, a solution for this model was not presented, which is a gap that should be fulfilled. The use of derivatives

V. F. Morales-Delgado · M. A. Taneco-Hernandez
Facultad de Matemáticas, Universidad Autónoma de Guerrero,
Av. Lázaro Cárdenas S/N, Chilpancingo, Guerrero, Mexico

J. F. Gómez-Aguilar (✉)
CONACYT-Tecnológico Nacional de México, Centro Nacional de Investigación y Desarrollo Tecnológico, Interna del Internado, Palmira, 62490 Cuernavaca, Morelos, México
e-mail: jgomez@cenidet.edu.mx

L. Torres
CONACYT-Instituto de Ingeniería, Universidad Nacional Autónoma de México, México City, México

R. F. Escobar-Jiménez
Tecnológico Nacional de México, Centro Nacional de Investigación y Desarrollo Tecnológico, Interna del Internado, Palmira, 62490 Cuernavaca, Morelos, México

J. F. Gómez et al. (eds.), *Fractional Derivatives with Mittag-Leffler Kernel*, Studies in Systems, Decision and Control 194,
https://doi.org/10.1007/978-3-030-11662-0_16

with fractional order sometimes is more convenient since models with this kind of derivatives are more general in comparison with classical order models. In the recent decades several physical problems have been represented mathematically by fractional derivatives. These representations have offered excellent results for modelling real world problems [14–21]. A physical interpretation of equations with fractional derivatives with respect to time is connected with the memory effects. The fractional derivatives include an integral operator of which kernel function (power, exponential o Mittag-leffler type) is a memory function that involves non-local interaction [22–26].

It is very difficult to solve highly nonlinear differential equations of fractional order. In this context, many powerful methods for finding exact solutions have been developed to study the solutions of nonlinear FPDE's, for instance, Hermite colocation method [27], invariant subspace method [28], optimal homotopy asymptotic method [29], Adomian decomposition methods [30], homotopy analysis Sumudu transform method [31], homotopy decomposition method [32], the homotopy perturbation transform method [33–35], homotopy–Padé technique [36], the fractional homotopy analysis transform method [37], and Chebyshev operational collocation method [38].

The homotopy analysis method transforms a problem into an infinite number of linear problems without using the perturbation techniques; this method employs the concept of the homotopy from topology to generate a convergent series solution [39, 40]. The Laplace homotopy perturbation method (LHPM) is a combination of the homotopy analysis method proposed by Liao [41] and the Laplace transform [42–45]. The modified homotopy analysis transform method (MHATM) was proposed in [46]; this method is an analytical technique based on the combination of the homotopy analysis method and Laplace transform with homotopy polynomial. In [46] considering the Liouville–Caputo fractional derivative, the authors developed the MHATM method with homotopy polynomial for solving the time-fractional K-S equation. A convergence analysis of MHATM was obtained by the proposed method and verified through different graphical representations.

In this chapter, we examine the fractional Lienard's type model of a pipeline with the aid of the LHPM and the MHATM using the fractional operators of Liouville–Caputo and Atangana–Baleanu in Liouville–Caputo sense. The model considered is expressed in terms of the flow rate or in terms of the pressure head as required. The organization of this article is as follows: Sect. 2 presents the literature related to Lienard's type model of a pipeline. In Sects. 3 and 4, LHPM and MHATM are applied to Lienard's model via Liouville–Caputo and Atangana–Baleanu–Caputo fractional order derivatives. Finally in Sect. 5, conclusions are presented.

2 Preliminaries

Chaudhry in [47], described the momentum and continuity equations that govern the dynamics of a fluid in a horizontal pipeline, which are given by

$$\frac{\partial Q(z,t)}{\partial t} + gA_r \frac{\partial H(z,t)}{\partial z} + \frac{f}{2\phi A_r} Q(z,t)|Q(z,t)| = 0, \tag{1}$$

$$\frac{\partial H(z,t)}{\partial t} + \frac{b^2}{gA_r} \frac{\partial Q(z,t)}{\partial z} = 0, \tag{2}$$

where $(t, z) \in (0, L) \times (0, \infty)$ are the spatial coordinates measured in meters and the temporal coordinates measured in seconds coordinates respectively; L is the length of the pipe $[m]$, $H(z, t)$ is the pressure head $[m]$, $Q(z, t)$ is the flow rate $[m^3/s]$, b is the wave speed in the fluid $[m/s]$, g is the gravitational acceleration $[m/s^2]$, A_r is the cross-sectional area of the pipe $[m^2]$, ϕ is the inside diameter of the pipe $[m]$, and f is the Darcy–Weisbach friction factor [48]. The above equations consider that convective changes in the velocity are negligible, as well as, both the liquid density and the cross-sectional area are constant.

The authors in [47, 50], presented the linearized version of Eqs. (1) and (2). The linear version is given by

$$\frac{\partial h(z,t)}{\partial z} + \frac{1}{A_r} \frac{\partial q(z,t)}{\partial t} + \frac{f q_0}{2g\phi A_r^2} q(z,t) = 0, \tag{3}$$

$$\frac{\partial q(z,t)}{\partial z} + \frac{gA_r}{b^2} \frac{\partial h(z,t)}{\partial t} = 0, \tag{4}$$

where q_0 and h_0 are the flow and pressure in equilibrium, $q(z, t)$ and $h(z, t)$ are the flow rate and pressure head around the equilibrium (q_0, h_0), respectively. The physical parameters of the pipeline can be redefined in terms of electrical parameters as follows

$$L = \frac{1}{A_r}, \qquad C = \frac{gA_r}{b^2}, \qquad R = \frac{f q_0}{2g\phi A_r^2}, \tag{5}$$

such that Eqs. (3) and (4) can be rewritten as

$$\frac{\partial h(z,t)}{\partial z} + L \frac{\partial q(z,t)}{\partial t} + Rq(z,t) = 0, \tag{6}$$

$$\frac{\partial q(z,t)}{\partial z} + C \frac{\partial h(z,t)}{\partial t} = 0. \tag{7}$$

The above mentioned equations are the Telegrapher's equations without the admittance term G [51]. In an electrical transmission line, if $G \to \infty$ means that there is a short circuit in the line. In a pipeline, the meaning of this admittance term is a leak, hence the following equations represent a pipeline with a leak at the point z_L

$$\frac{\partial h(z,t)}{\partial z} + L \frac{\partial q(z,t)}{\partial t} + Rq(z,t) = 0, \tag{8}$$

$$\frac{\partial q(z,t)}{\partial z} + C \frac{\partial h(z,t)}{\partial t} + G\delta(z - z_L)u(t - t_L)h(z_L, t) = 0, \tag{9}$$

where G is given by

$$G = \alpha \lambda h_{L0}^{\chi-1},$$

with h_{L0} as the equilibrium pressure at the leak position and χ is a constant that depends on the geometry of the leak. For round holes, $\chi = 1$.

By differentiating Eqs. (6) and (7) and applying some algebraic manipulation, we obtain the following equations

$$\frac{\partial^2 h(z,t)}{\partial z^2} = LC\frac{\partial^2 h(z,t)}{\partial t^2} + (RC + GL)\frac{\partial h(z,t)}{\partial t} + GRh(z,t), \tag{10}$$

$$\frac{\partial^2 q(z,t)}{\partial z^2} = LC\frac{\partial^2 q(z,t)}{\partial t^2} + (RC + GL)\frac{\partial q(z,t)}{\partial t} + GRq(z,t). \tag{11}$$

Using the Liénard transform [52] in terms of the flow we get

$$\Phi : \left(x_1(t)\ x_2(t),\right) \to \left(q(z,t)\ \dot{q}(z,t) + F(q(z,t))\right), \tag{12}$$

or in terms of the pressure head

$$\Phi : \left(x_1(t)\ x_2(t),\right) \to \left(h(z,t)\ \dot{h}(z,t) + F(h(z,t))\right), \tag{13}$$

the couple of Eqs. (6) and (7) becomes

$$\begin{aligned} \frac{\partial x_1(z,t)}{\partial t} - x_2(z,t) + \left(\frac{RC+GL}{LC}\right) x_1(z,t) &= 0, \\ \frac{\partial x_2(z,t)}{\partial t} + \left(\frac{GR}{LC}\right) x_1(z,t) - \left(\frac{1}{LC}\right)\frac{\partial^2 x_1(z,t)}{\partial z^2} &= 0. \end{aligned} \tag{14}$$

The above equations correspond to the Liénard Model of a fluid transmission line. In the next section, we apply the Laplace homotopy perturbation method and the modified homotopy analysis transform method for obtaining the new approximated analytical solutions of the system (14).

3 Implementation of the Laplace Homotopy Perturbation Method via Liouville–Caputo and Atangana–Baleanu–Caputo Fractional Order Derivatives

Case 1. Liénard model of a fluid transmission line via Liouville–Caputo fractional order derivative.

Usually, several authors replace the time integer derivative operator with fractional ones on a purely mathematical basis omitting the dimensionality of the fractional equation. The dimension mismatch of the fractional systems can be mathematically corrected introducing an auxiliary parameter Ξ in the following way [53]

$$\frac{d}{dt} \to \frac{1}{\Xi^{1-\alpha}} \cdot {}_0\mathscr{D}_t^{\alpha}, \qquad m-1<\alpha\leq m, \qquad m\in M=1,2,3,\ldots \tag{15}$$

where α represents the order of the fractional temporal operator and Ξ has the dimension of seconds. The parameter Ξ characterizes the existence of fractional structures which emerge from the non-local behaviour of the system in time (these components change the time constant of the system) [53]. The authors of [54] used the Planck time, $t_p = 5.39106 \times 10^{-44}$ s, for preserving the dimensional compatibility. Considering [54] the Ξ parameter corresponds to the t_p for developing our calculations. Following this idea, we consider the following fractional equations in Liouville–Caputo sense

$$ {}_0^C\mathscr{D}_t^{\alpha} x_1(z,t) - t_p^{1-\alpha} x_2(z,t) + t_p^{1-\alpha}\left(\frac{RC+GL}{LC}\right)x_1(z,t)=0, \tag{16}$$

$$ {}_0^C\mathscr{D}_t^{\alpha} x_2(z,t) + t_p^{1-\alpha}\left(\frac{RG}{LC}\right)x_1(z,t) - t_p^{1-\alpha}\left(\frac{1}{LC}\right)\frac{\partial^2 x_1(z,t)}{\partial z^2}=0, \tag{17}$$

$$ \frac{\partial^k x_i(z,0)}{\partial z^k} = x_{i,k}(z,0), \;\; k=0,1,\ldots,n-1, \tag{18}$$

$$ x_i(0,t) = x_{i,0}(t). \tag{19}$$

where ${}_0^C\mathscr{D}_t^{\alpha}$ is the Liouville–Caputo operator (C) defined as follows [23]

$$ {}_0^C\mathscr{D}_t^{\alpha} u(x,t) = \frac{1}{\Gamma(n-\alpha)}\int_0^t (t-\theta)^{n-\alpha-1}u^n(x,\theta)d\theta, \qquad n-1<\alpha<n, \tag{20}$$

where u^n is the derivative of integer nth order of $u(x,t)$, $n=1,2,\ldots\in N$ and $n-1<\alpha\leq n$.

If $0<\alpha\leq 1$, then we define the Laplace transform for the Liouville–Caputo fractional derivative as follows

$$\mathscr{L}\left[{}_0^C\mathscr{D}_t^{\alpha} u(x,t)\right](s) = s^{\alpha}\mathscr{L}[u(x,t)](s) - s^{\alpha-1}[u(x,0)]. \tag{21}$$

Solution. Applying Laplace transform (21) to Eqs. (16) and (17) we have

$$\begin{aligned} X_1(z,s) &= \frac{x_1(z,0)}{s} + t_p^{1-\alpha}\frac{1}{s^{\alpha}}\left[X_2(z,s) - \left(\frac{RC+GL}{LC}\right)X_1(z,s)\right], \\ X_2(z,s) &= \frac{x_2(z,0)}{s} + t_p^{1-\alpha}\frac{1}{s^{\alpha}}\left[-\left(\frac{RG}{LC}\right) + \left(\frac{1}{LC}\right)\frac{\partial^2}{\partial z^2}\right]X_1(z,s). \end{aligned} \tag{22}$$

Applying the LHPM to above equation then

$$\sum_{i=0}^{\infty} p^i\ X_{1,i}(z,s) = \frac{x_1(z,0)}{s} + p\, t_p^{1-\alpha}\, \frac{1}{s^{\alpha}} \Big[\sum_{i=0}^{\infty} p^i X_{2,i}(z,s) - \Big(\frac{RC+GL}{LC}\Big) \sum_{i=0}^{\infty} p^i X_{1,i}(z,s)\Big],$$
$$\sum_{i=0}^{\infty} p^i X_{2,i}(z,s) = \frac{x_2(z,0)}{s} + p\, t_p^{1-\alpha}\, \frac{1}{s^{\alpha}} \Big[-\Big(\frac{RG}{LC}\Big) + \Big(\frac{1}{LC}\Big)\frac{\partial^2}{\partial z^2}\cdot\Big] \sum_{i=0}^{\infty} p^i X_{1,i}(z,s). \quad (23)$$

Comparing terms we obtain

$$\begin{aligned}
p^0 : X_{1,0}(z,s) &= \frac{x_1(z,0)}{s},\\
X_{2,0}(z,s) &= \frac{x_2(z,0)}{s},\\
p^1 : X_{1,1}(z,s) &= t_p^{1-\alpha}\, \frac{1}{s^{\alpha}} \Big[X_{2,0}(z,s) - \Big(\frac{RC+GL}{LC}\Big) X_{1,0}(z,s)\Big],\\
X_{2,1}(z,s) &= t_p^{1-\alpha}\, \frac{1}{s^{\alpha}} \Big[-\Big(\frac{RG}{LC}\Big) + \Big(\frac{1}{LC}\Big)\frac{\partial^2}{\partial z^2}\Big] X_{1,0}(z,s),\\
&\ \vdots\\
p^{n+1} : X_{1,n+1}(z,s) &= t_p^{1-\alpha}\, \frac{1}{s^{\alpha}} \Big[X_{2,n}(z,s) - \Big(\frac{RC+GL}{LC}\Big) X_{1,n}(z,s)\Big],\\
X_{2,n+1}(z,s) &= t_p^{1-\alpha}\, \frac{1}{s^{\alpha}} \Big[-\Big(\frac{RG}{LC}\Big) + \Big(\frac{1}{LC}\Big)\frac{\partial^2}{\partial z^2}\Big] X_{1,n}(z,s),
\end{aligned} \quad (24)$$

when $p \to 1$, the approximated solutions of Eqs. (16) and (17) are given by

$$\begin{aligned}
H_{1,n}(z,s) &= \sum_{i=0}^{n} X_{1,i}(z,s),\\
H_{2,n}(z,s) &= \sum_{i=0}^{n} X_{2,i}(z,s).
\end{aligned} \quad (25)$$

Finally, applying inverse Laplace transform we get

$$\begin{aligned}
x_1(z,t) &= \mathscr{L}^{-1}\big[H_{1,n}(z,s)\big],\\
x_2(z,t) &= \mathscr{L}^{-1}\big[H_{2,n}(z,s)\big].
\end{aligned} \quad (26)$$

Considering the fractional equations (16) and (17) with initial conditions

$$\begin{aligned}
x_1(z,0) &= \Big(\frac{2\sqrt{LC}(CR+GL)}{(CR-GL)}\Big)\ \sin\Big(\frac{(CR-GL)}{2\sqrt{LC}} z\Big),\\
x_2(z,0) &= \Big(\frac{(CR+GL)}{\sqrt{LC}(CR-GL)}\Big)\ \sin\Big(\frac{(CR-GL)}{2\sqrt{LC}} z\Big).
\end{aligned} \quad (27)$$

we get

$$
\begin{aligned}
p^0 : x_{1,0}(z,t) &= x_1(z,0),\\
x_{2,0}(z,t) &= x_2(z,0),\\
p^1 : x_{1,1}(z,s) &= \mathscr{L}^{-1}\left\{t_p^{1-\alpha}\,\frac{1}{s^\alpha}\left[X_{2,0}(z,s)-\left(\frac{RC+GL}{LC}\right)X_{1,0}(z,s)\right]\right\},\\
&= \frac{t_p^{1-\alpha}\big(CR+GL\big)\,t^\alpha}{\Gamma(\alpha+1)\,\sqrt{LC}\big(GL-CR\big)}\,\sin\left(\frac{(CR-GL)}{2\sqrt{LC}}z\right),\\
x_{2,1}(z,t) &= \mathscr{L}^{-1}\left\{t_p^{1-\alpha}\,\frac{1}{s^\alpha}\left[-\left(\frac{RG}{LC}\right)+\left(\frac{1}{LC}\right)\frac{\partial^2}{\partial z^2}\right]X_{1,0}(z,s)\right\},\\
&= \frac{t_p^{1-\alpha}\big(CR+GL\big)^2\,t^\alpha}{2\,\Gamma(\alpha+1)\,\sqrt{(LC)^3}\big(GL-CR\big)}\,\sin\left(\frac{(CR-GL)}{2\sqrt{LC}}z\right),\\
p^1 : x_{1,2}(z,t) &= \mathscr{L}^{-1}\left\{t_p^{1-\alpha}\,\frac{1}{s^\alpha}\left[X_{2,1}(z,s)-\left(\frac{RC+GL}{LC}\right)X_{1,1}(z,s)\right]\right\},\\
&= -\frac{t_p^{2(1-\alpha)}\big(CR+GL\big)^2\,t^{2\alpha}}{2\,\Gamma(2\alpha+1)\,\sqrt{(LC)^3}\big(GL-CR\big)}\,\sin\left(\frac{(CR-GL)}{2\sqrt{LC}}z\right),\\
x_{2,2}(z,t) &= \mathscr{L}^{-1}\left\{t_p^{1-\alpha}\,\frac{1}{s^\alpha}\left[-\left(\frac{RG}{LC}\right)+\left(\frac{1}{LC}\right)\frac{\partial^2}{\partial z^2}\right]X_{1,1}(z,s)\right\},\\
&= -\frac{t_p^{2(1-\alpha)}\big(CR+GL\big)^3\,t^{2\alpha}}{4\,\Gamma(2\alpha+1)\,\sqrt{(LC)^5}\big(GL-CR\big)}\,\sin\left(\frac{(CR-GL)}{2\sqrt{LC}}z\right),\\
&\;\;\vdots
\end{aligned}
\tag{28}
$$

The approximated solution is given by

$$
\begin{aligned}
x_{1,n}(z,t) &= \sum_{i=0}^{n} x_{1,i}(z,t),\\
&= \left[-\left(\frac{2\sqrt{LC}(CR+GL)}{(CR-GL)}\right)+\frac{t_p^{1-\alpha}\big(CR+GL\big)\,t^\alpha}{\Gamma(\alpha+1)\,\sqrt{LC}\big(GL-CR\big)}\right.\\
&\quad\left.-\frac{t_p^{2(1-\alpha)}\big(CR+GL\big)^2\,t^{2\alpha}}{2\,\Gamma(2\alpha+1)\,\sqrt{(LC)^3}\big(GL-CR\big)}+\cdots\right]\sin\left(\frac{(CR-GL)}{2\sqrt{LC}}z\right).
\end{aligned}
$$

$$x_{2,n}(z,t) = \sum_{i=0}^{n} x_{2,i}(z,t),$$
$$= \left[-\left(\frac{(CR+GL)}{\sqrt{LC}(CR-GL)} \right) + \frac{t_p^{1-\alpha}(CR+GL)^2\, t^{\alpha}}{2\,\Gamma(\alpha+1)\,\sqrt{(LC)^3}(GL-CR)} \right.$$
$$\left. - \frac{t_p^{2(1-\alpha)}(CR+GL)^3\, t^{2\alpha}}{4\,\Gamma(2\alpha+1)\,\sqrt{(LC)^5}(GL-CR)} + \cdots \right] \sin\left(\frac{(CR-GL)}{2\sqrt{LC}} z \right). \tag{29}$$

If $n \to \infty$, we get

$$x_{1,n}(z,t) = \left[\frac{2\,\Gamma(2\alpha+1)\,\sqrt{(LC)^3} t_p^{\alpha} \exp\left(\frac{(CR+GL)\, t^{\alpha} t_p}{-\Gamma(2\alpha+1) LC\, t_p^{\alpha}} \right)}{\Gamma(2\alpha+1) LC t_p^{\alpha}(GL-CR)} \right] \sin\left(\frac{(CR-GL)}{2\sqrt{LC}} z \right). \tag{30}$$

$$x_{2,n}(z,t) = \left[\frac{2\,\Gamma(\alpha+1)\,\sqrt{LC} t_p^{\alpha} \exp\left(\frac{(CR+GL)\, t^{\alpha} t_p}{-2\,\Gamma(\alpha+1) LC\, t_p^{\alpha}} \right)}{2\,\Gamma(\alpha+1)\, LC t_p^{\alpha}(GL-CR)} \right] \sin\left(\frac{(CR-GL)}{2\sqrt{LC}} z \right). \tag{31}$$

Example

Consider the following values $C = 4.8240$, $L = 12.3453$ and $R = 6.414$ arbitrarily chosen. Figure 1a–d show the numerical simulations of the Eqs. (30) and (31) for $\alpha = 1$ and $\alpha = 0.9$.

Case 2. Liénard model of a fluid transmission line via Atangana–Baleanu–Caputo fractional order derivative.

In this case, we consider the following fractional equations in ABC sense

$${}_0^{ABC}\mathscr{D}_t^{\alpha} x_1(z,t) - t_p^{1-\alpha} x_2(z,t) + t_p^{1-\alpha} \left(\frac{RC+GL}{LC} \right) x_1(z,t) = 0, \tag{32}$$

$${}_0^{ABC}\mathscr{D}_t^{\alpha} x_2(z,t) + t_p^{1-\alpha} \left(\frac{RG}{LC} \right) x_1(z,t) - t_p^{1-\alpha} \left(\frac{1}{LC} \right) \frac{\partial^2 x_1(z,t)}{\partial z^2} = 0, \tag{33}$$

$$\frac{\partial^k x_i(z,0)}{\partial z^k} = x_{i,k}(z,0), \quad k = 0, 1, \ldots, n-1. \tag{34}$$

$$x_i(0,t) = x_{i,0}(t). \tag{35}$$

where t_p corresponds to the Planck time and ${}_0^{ABC}\mathscr{D}_t^{\alpha}$ is the fractional derivative with generalized Mittag-Leffler law in Liouville–Caputo sense (ABC) defined as follows [26]

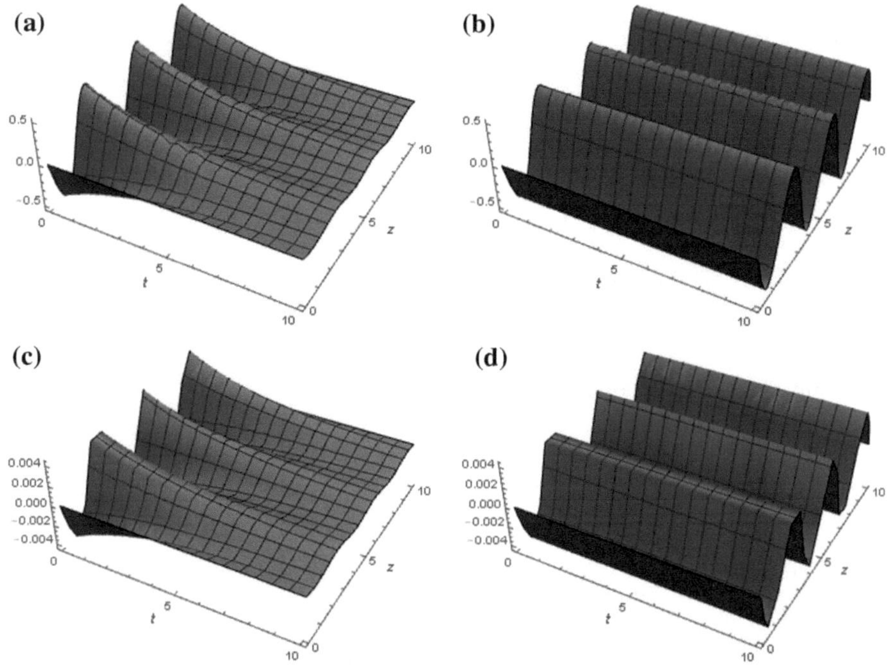

Fig. 1 Numerical simulation for the approximate solutions given by Eqs. (30) and (31). In **a** and **b** Eq. (30) for $\alpha = 1$ and $\alpha = 0.9$, respectively. In **c** and **d** Eq. (31) for $\alpha = 1$ and $\alpha = 0.9$, respectively

$$ {}_{0}^{ABC}\mathscr{D}_t^{\alpha} u(x,t) = \frac{B(\alpha)}{1-\alpha} \int_0^t E_{\alpha}\left[-\alpha \frac{(t-\theta)^{\alpha}}{n-\alpha}\right] u^n(x,\theta) d\theta, \tag{36} $$

where $B(\alpha)$ is a normalization function, $B(0) = B(1) = 1$. This fractional operator uses the Mittag-Leffler law as nonsingular and nonlocal kernel.

If $0 < \alpha \leq 1$, then we define the Laplace transform for the Atangana–Baleanu fractional derivative as follows

$$ \mathscr{L}\left[{}_{0}^{ABC}D_t^{\alpha} u(x,t)\right](s) = \left(\frac{s^{\alpha}\mathscr{L}\left[u(x,t)\right](s) - s^{\alpha-1}\left[u(x,0)\right]}{s^{\alpha}(1-\alpha)+\alpha}\right). \tag{37} $$

Solution. Applying Laplace transform (37) to Eqs. (32) and (33) we have

$$ \begin{aligned} X_1(z,s) &= \frac{x_1(z,0)}{s} + t_p^{1-\alpha}\,\frac{(1-\alpha)s^{\alpha}+\alpha}{B(\alpha)s^{\alpha}}\left[X_2(z,s) - \left(\frac{RC+GL}{LC}\right)X_1(z,s)\right], \\ X_2(z,s) &= \frac{x_2(z,0)}{s} + t_p^{1-\alpha}\,\frac{(1-\alpha)s^{\alpha}+\alpha}{B(\alpha)s^{\alpha}}\left[-\left(\frac{RG}{LC}\right) + \left(\frac{1}{LC}\right)\frac{\partial^2}{\partial z^2}\right]X_1(z,s). \end{aligned} \tag{38} $$

Applying the LHPM to above mentioned equation then

$$\sum_{i=0}^{\infty} p^i\, X_{1,i}(z,s) = \frac{x_1(z,0)}{s} + p\, t_p^{1-\alpha}\, \frac{(1-\alpha)s^\alpha + \alpha}{B(\alpha)s^\alpha}$$
$$\times \left[\sum_{i=0}^{\infty} p^i X_{2,i}(z,s) - \left(\frac{RC+GL}{LC}\right) \sum_{i=0}^{\infty} p^i X_{1,i}(z,s) \right],$$
$$\sum_{i=0}^{\infty} p^i X_{2,i}(z,s) = \frac{x_2(z,0)}{s} + p\, t_p^{1-\alpha}\, \frac{(1-\alpha)s^\alpha + \alpha}{B(\alpha)s^\alpha}$$
$$\times \left[-\left(\frac{RG}{LC}\right) + \left(\frac{1}{LC}\right)\frac{\partial^2}{\partial z^2} \right] \sum_{i=0}^{\infty} p^i X_{1,i}(z,s). \tag{39}$$

Comparing terms we obtain

$$p^0 : X_{1,0}(z,s) = \frac{x_1(z,0)}{s},$$
$$X_{2,0}(z,s) = \frac{x_2(z,0)}{s},$$
$$p^1 : X_{1,1}(z,s) = t_p^{1-\alpha}\, \frac{(1-\alpha)s^\alpha + \alpha}{B(\alpha)s^\alpha} \left[X_{2,0}(z,s) - \left(\frac{RC+GL}{LC}\right) X_{1,0}(z,s) \right],$$
$$X_{2,1}(z,s) = t_p^{1-\alpha}\, \frac{(1-\alpha)s^\alpha + \alpha}{B(\alpha)s^\alpha} \left[-\left(\frac{RG}{LC}\right) + \left(\frac{1}{LC}\right)\frac{\partial^2}{\partial z^2} \right] X_{1,0}(z,s),$$
$$\vdots$$
$$p^{n+1} : X_{1,n+1}(z,s) = t_p^{1-\alpha}\, \frac{(1-\alpha)s^\alpha + \alpha}{B(\alpha)s^\alpha} \left[X_{2,n}(z,s) - \left(\frac{RC+GL}{LC}\right) X_{1,n}(z,s) \right],$$
$$X_{2,n+1}(z,s) = t_p^{1-\alpha}\, \frac{(1-\alpha)s^\alpha + \alpha}{B(\alpha)s^\alpha} \left[-\left(\frac{RG}{LC}\right) + \left(\frac{1}{LC}\right)\frac{\partial^2}{\partial z^2} \right] X_{1,n}(z,s), \tag{40}$$

when $p \to 1$, the approximated solutions of Eqs. (32) and (33) are given by

$$H_{1,n}(z,s) = \sum_{i=0}^{n} X_{1,i}(z,s),$$
$$H_{2,n}(z,s) = \sum_{i=0}^{n} X_{2,i}(z,s). \tag{41}$$

Finally, applying inverse Laplace transform we get

$$x_1(z,t) = \mathscr{L}^{-1}\big[H_{1,n}(z,s)\big],$$
$$x_2(z,t) = \mathscr{L}^{-1}\big[H_{2,n}(z,s)\big]. \tag{42}$$

Considering the fractional equations (32) and (33) with initial conditions

$$x_1(z,0)=\left(\frac{2\sqrt{LC}(CR+GL)}{(CR-GL)}\right)\sin\left(\frac{(CR-GL)}{2\sqrt{LC}}z\right),$$
$$x_2(z,0)=\left(\frac{(CR+GL)}{\sqrt{LC}(CR-GL)}\right)\sin\left(\frac{(CR-GL)}{2\sqrt{LC}}z\right). \tag{43}$$

we get

$$\begin{aligned}
p^0 &: x_{1,0}(z,t)=x_1(z,0),\\
&\quad x_{2,0}(z,t)=x_2(z,0),\\
p^1 &: x_{1,1}(z,s)=\mathscr{L}^{-1}\left\{t_p^{1-\alpha}\,\frac{(1-\alpha)s^\alpha+\alpha}{B(\alpha)s^\alpha}\left[X_{2,0}(z,s)-\left(\frac{RC+GL}{LC}\right)X_{1,0}(z,s)\right]\right\},\\
&=\frac{t_p^{1-\alpha}(CR+GL)\,(\alpha\,t^\alpha+(1-\alpha)\Gamma(\alpha+1))}{B(\alpha)\,\Gamma(\alpha+1)\,\sqrt{LC}(GL-CR)}\sin\left(\frac{(CR-GL)}{2\sqrt{LC}}z\right),\\
x_{2,1}(z,t)&=\mathscr{L}^{-1}\left\{t_p^{1-\alpha}\,\frac{(1-\alpha)s^\alpha+\alpha}{B(\alpha)s^\alpha}\left[-\left(\frac{RG}{LC}\right)+\left(\frac{1}{LC}\right)\frac{\partial^2}{\partial z^2}\right]X_{1,0}(z,s)\right\},\\
&=\frac{t_p^{1-\alpha}(CR+GL)^2\,(\alpha\,t^\alpha+(1-\alpha)\Gamma(\alpha+1))}{2\,B(\alpha)\,\Gamma(\alpha+1)\,\sqrt{(LC)^3}(GL-CR)}\sin\left(\frac{(CR-GL)}{2\sqrt{LC}}z\right),\\
p^1 &: x_{1,2}(z,t)=\mathscr{L}^{-1}\left\{t_p^{1-\alpha}\,\frac{(1-\alpha)s^\alpha+\alpha}{B(\alpha)s^\alpha}\left[X_{2,1}(z,s)-\left(\frac{RC+GL}{LC}\right)X_{1,1}(z,s)\right]\right\},\\
&=-\frac{t_p^{2(1-\alpha)}(CR+GL)^2\,((1-\alpha)^2\Gamma(\alpha+1)\Gamma(2\alpha+1)+2\alpha(1-\alpha)\Gamma(2\alpha+1)\,t^\alpha+\alpha^2\Gamma(\alpha+1)t^{2\alpha})}{2\,B(\alpha)^2\,\Gamma(\alpha+1)\,\Gamma(2\alpha+1)\,\sqrt{(LC)^3}(GL-CR)}\\
&\times\sin\left(\frac{(CR-GL)}{2\sqrt{LC}}z\right),
\end{aligned}$$

$$\begin{aligned}
x_{2,2}(z,t)&=\mathscr{L}^{-1}\left\{t_p^{1-\alpha}\,\frac{(1-\alpha)s^\alpha+\alpha}{B(\alpha)s^\alpha}\left[-\left(\frac{RG}{LC}\right)+\left(\frac{1}{LC}\right)\frac{\partial^2}{\partial z^2}\right]X_{1,1}(z,s)\right\},\\
&=-\frac{t_p^{2(1-\alpha)}(CR+GL)^3\,((1-\alpha)^2\Gamma(\alpha+1)\Gamma(2\alpha+1)+2\alpha(1-\alpha)\Gamma(2\alpha+1)\,t^\alpha+\alpha^2\Gamma(\alpha+1)t^{2\alpha})}{4\,B(\alpha)^2\,\Gamma(\alpha+1)\,\Gamma(2\alpha+1)\,\sqrt{(LC)^5}(GL-CR)}\\
&\times\sin\left(\frac{(CR-GL)}{2\sqrt{LC}}z\right),\\
&\vdots
\end{aligned} \tag{44}$$

If $n\to\infty$, we get

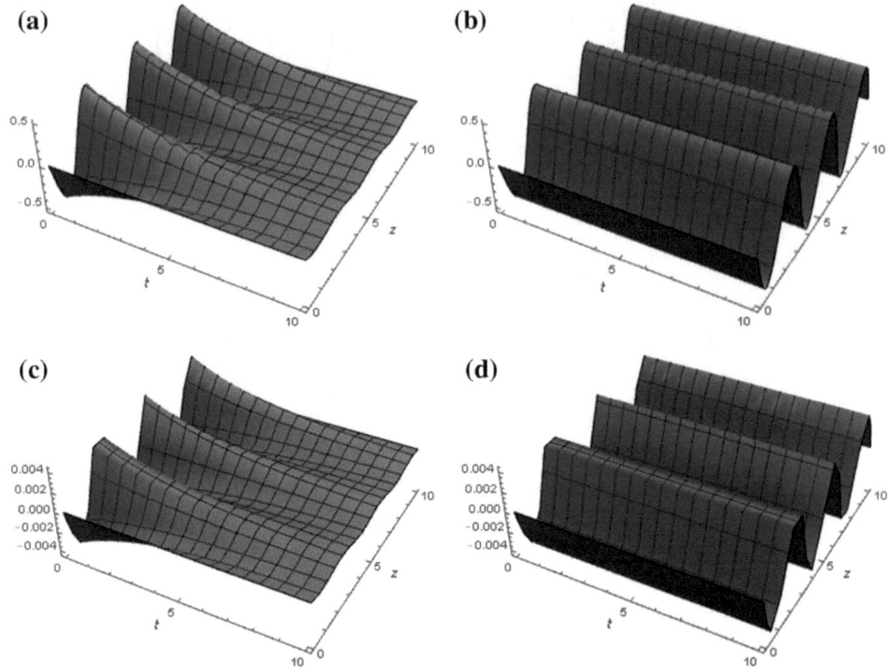

Fig. 2 Numerical simulation for the approximate solutions given by Eqs. (45) and (46). In **a** and **b** Eq. (45) for $\alpha = 1$ and $\alpha = 0.9$, respectively. In **c** and **d** Eq. (46) for $\alpha = 1$ and $\alpha = 0.9$, respectively

$$x_1(z,t) = \left[\frac{2\,\Gamma(\alpha+1)\,\Gamma(2\alpha+1)\,\sqrt{(LC)^3}t_p^{\alpha}\exp\left(\frac{(CR+GL)\,\alpha\,t^{\alpha}t_p}{\left(-\Gamma(\alpha+1)\,\Gamma(2\alpha+1)LC\,t_p^{\alpha}-(1-\alpha)\Gamma(\alpha+1)\,t_p\,(GL+CR)\right)}\right)}{\left(\Gamma(\alpha+1)\,\Gamma(2\alpha+1)LCt_p^{\alpha}+(1-\alpha)\Gamma(\alpha+1)t_p\right)(GL-CR)} \right] \times$$
$$\sin\left(\frac{(CR-GL)}{2\sqrt{LC}}z\right). \tag{45}$$

$$x_2(z,t) = \left[\frac{2\,\Gamma(\alpha+1)\,\sqrt{LC}t_p^{\alpha}\exp\left(\frac{(CR+GL)\,\alpha\,t^{\alpha}t_p}{\left(-2\,\Gamma(\alpha+1)LC\,t_p^{\alpha}-(1-\alpha)\Gamma(\alpha+1)t_p\,(GL+CR)\right)}\right)}{\left(2\,\Gamma(\alpha+1)\,LCt_p^{\alpha}+(1-\alpha)\Gamma(\alpha+1)(GL+CR)t_p\right)(GL-CR)} \right]$$
$$\times\sin\left(\frac{(CR-GL)}{2\sqrt{LC}}z\right). \tag{46}$$

Example

Consider the following values $C = 4.8240$, $L = 12.3453$ and $R = 6.414$ arbitrarily chosen. Figure 2a–d show the numerical simulations of the Eqs. (45) and (46) for $\alpha = 1$ and $\alpha = 0.9$.

4 Implementation of the Modified Homotopy Analysis Transform Method via Liouville–Caputo and Atangana–Baleanu–Caputo Fractional Order Derivatives

Case 1. Liénard model of a fluid transmission line via Liouville–Caputo fractional order derivative.

In this section, we consider the MHATM [46] for solving the Liénard type model of a fluid transmission line considering Liouville–Caputo and Atangana–Baleanu–Caputo fractional order derivatives.

In this case, we consider the following fractional equations in Liouville–Caputo sense

$$ {}_{0}^{C}\mathscr{D}_t^{\alpha} x_1(z,t) - t_p^{1-\alpha} x_2(z,t) + t_p^{1-\alpha}\left(\frac{RC+GL}{LC}\right) x_1(z,t) = 0, \tag{47} $$

$$ {}_{0}^{C}\mathscr{D}_t^{\alpha} x_2(z,t) + t_p^{1-\alpha}\left(\frac{RG}{LC}\right) x_1(z,t) - t_p^{1-\alpha}\left(\frac{1}{LC}\right)\frac{\partial^2 x_1(z,t)}{\partial z^2} = 0, \tag{48} $$

$$ x_1(z,0) = \left(\frac{2\sqrt{LC}(CR+GL)}{(CR-GL)}\right)\sin\left(\frac{(CR-GL)}{2\sqrt{LC}}z\right), \tag{49} $$

$$ x_2(z,0) = \left(\frac{(CR+GL)}{\sqrt{LC}(CR-GL)}\right)\sin\left(\frac{(CR-GL)}{2\sqrt{LC}}z\right), \tag{50} $$

where t_p corresponds to the Planck time and ${}_{0}^{C}\mathscr{D}_t^{\alpha}$ is the fractional derivative of Liouville–Caputo type [23].

Solution. Applying Laplace transform (21) to Eqs. (47) and (48), and taking initial conditions (49) and (50), we get

$$ \begin{aligned} X_1(z,s) &= \frac{x_1(z,0)}{s} + t_p^{1-\alpha}\frac{1}{s^{\alpha}}\left[X_2(z,s) - \left(\frac{RC+GL}{LC}\right)X_1(z,s)\right], \\ X_2(z,s) &= \frac{x_2(z,0)}{s} + t_p^{1-\alpha}\frac{1}{s^{\alpha}}\left[-\left(\frac{RG}{LC}\right) + \left(\frac{1}{LC}\right)\frac{\partial^2}{\partial z^2}\right]X_1(z,s). \end{aligned} \tag{51} $$

applying the inverse Laplace transform to Eq. (51), we obtain

$$ \begin{aligned} x_1(z,t) &= x_1(z,0) + \mathscr{L}^{-1}\left\{t_p^{1-\alpha}\frac{1}{s^{\alpha}}\mathscr{L}\left[x_2(z,t) - \left(\frac{RC+GL}{LC}\right)x_1(z,t)\right]\right\}, \\ x_2(z,t) &= x_2(z,0) + \mathscr{L}^{-1}\left\{t_p^{1-\alpha}\frac{1}{s^{\alpha}}\mathscr{L}\left[-\left(\frac{RG}{LC}\right) + \left(\frac{1}{LC}\right)\frac{\partial^2}{\partial z^2}\right]x_1(z,t)\right\}. \end{aligned} \tag{52} $$

We choose the linear operator as

$$\mathfrak{F}\big[\phi_j(z,t;q)\big] = \mathscr{L}\big[\phi_j(z,t;q)\big],\ j=1,2.$$

With property $\mathfrak{F}(c) = 0$, where c is a constant. Next, defining the system of non-linear operator as

$$\begin{aligned} N\big[\phi(z,t;q)\big] &= \mathscr{L}\big[\phi(z,t;q)\big] - \left(\frac{2\sqrt{LC}(CR+GL)}{(CR-GL)}\right) \sin\left(\frac{(CR-GL)}{2\sqrt{LC}}z\right) \\ &\quad - t_p^{1-\alpha}\,\frac{1}{s^{\alpha}}\,\mathscr{L}\left[\Phi - \left(\frac{RC+GL}{LC}\right)\phi\right], \end{aligned} \tag{53}$$

$$\begin{aligned} N\big[\Phi(z,t;q)\big] &= \mathscr{L}\big[\Phi(z,t;q)\big] - \left(\frac{(CR+GL)}{\sqrt{LC}(CR-GL)}\right) \sin\left(\frac{(CR-GL)}{2\sqrt{LC}}z\right) \\ &\quad - t_p^{1-\alpha}\,\frac{1}{s^{\alpha}}\,\mathscr{L}\left[-\left(\frac{RG}{LC}\right) + \left(\frac{1}{LC}\right)\frac{\partial^2}{\partial z^2}\right]\phi. \end{aligned} \tag{54}$$

Now, we construct the so-called zeroth-order deformation equation in the following manner

$$(1-q)\,\mathfrak{F}[\phi_j(z,t;q) - u_0(z,t)] = q\,\hbar\,N[\phi_j(z,t;q)],\quad j=1,2.$$

when $q=0$ and $q=1$, we have

$$\phi_j(z,t;0) = u_0(z,t),\quad \phi_j(z,t;1) = u(z,t),\quad j=1,2. \tag{55}$$

Thus, we obtain the mth-order deformation equations as

$$\begin{aligned} \mathscr{L}[x_{1_m}(z,t) - \chi_m x_{1_{m-1}}(z,t)] &= \hbar\,R_m\big(\vec{x}_{1_{m-1}}, z, t\big), \\ \mathscr{L}[x_{2_m}(z,t) - \chi_m x_{2_{m-1}}(z,t)] &= \hbar\,R_m\big(\vec{x}_{2_{m-1}}, z, t\big), \end{aligned} \tag{56}$$

applying the inverse Laplace transform to Eq. (56) we get

$$x_{1_m}(z,t) = \chi_m x_{1_{m-1}}(z,t) + \hbar\,R_m\big(\vec{x}_{1_{m-1}}, z, t\big), \tag{57}$$

$$x_{2_m}(z,t) = \chi_m x_{2_{m-1}}(z,t) + \hbar\,R_m\big(\vec{x}_{2_{m-1}}, z, t\big), \tag{58}$$

where

$$\begin{aligned} R_m\big(\vec{x}_{1_{m-1}}, z, t\big) &= \mathscr{L}[x_{1_{m-1}}(z,t)] - (1-\chi_m)\left(\frac{2\sqrt{LC}(CR+GL)}{(CR-GL)}\right) \sin\left(\frac{(CR-GL)}{2\sqrt{LC}}z\right) \\ &\quad - t_p^{1-\alpha}\,\frac{1}{s^{\alpha}}\,\mathscr{L}\left[x_{2_{m-1}} - \left(\frac{RC+GL}{LC}\right)x_{1_{m-1}}\right], \\ R_m\big(\vec{x}_{2_{m-1}}, z, t\big) &= \mathscr{L}[x_{2_{m-1}}(z,t)] - (1-\chi_m)\left(\frac{(CR+GL)}{\sqrt{LC}(CR-GL)}\right) \sin\left(\frac{(CR-GL)}{2\sqrt{LC}}z\right) \end{aligned}$$

$$- t_p^{1-\alpha} \frac{1}{s^\alpha} \mathscr{L}\left[-\left(\frac{RG}{LC}\right)+\left(\frac{1}{LC}\right)\frac{\partial^2}{\partial z^2}\right] x_{1_{m-1}}. \tag{59}$$

Now, the solution of mth-order deformation equations (56) are given as

$$\begin{aligned}
x_{1_m}(z,t) &= (\chi_m+\hbar)x_{1_{m-1}} - \hbar(1-\chi_m)\left(\frac{2\sqrt{LC}(CR+GL)}{(CR-GL)}\right)\sin\left(\frac{(CR-GL)}{2\sqrt{LC}}z\right)\\
&\quad - \hbar\,\mathscr{L}^{-1}\left\{t_p^{1-\alpha}\frac{1}{s^\alpha}\mathscr{L}\left[x_{2_{m-1}} - \left(\frac{RC+GL}{LC}\right)x_{1_{m-1}}\right]\right\},\\
x_{2_m}(z,t) &= (\chi_m+\hbar)x_{2_{m-1}} - \hbar(1-\chi_m)\left(\frac{(CR+GL)}{\sqrt{LC}(CR-GL)}\right)\sin\left(\frac{(CR-GL)}{2\sqrt{LC}}z\right)\\
&\quad - \hbar\,\mathscr{L}^{-1}\left\{t_p^{1-\alpha}\frac{1}{s^\alpha}\mathscr{L}\left[-\left(\frac{RG}{LC}\right)+\left(\frac{1}{LC}\right)\frac{\partial^2}{\partial z^2}\right]x_{1_{m-1}}\right\}.
\end{aligned} \tag{60}$$

Taking initial conditions and the iterative scheme (60), we obtain the following iterations

$$\begin{aligned}
x_{1,0}(z,t) &= x_1(z,0),\\
x_{2,0}(z,t) &= x_2(z,0),\\
x_{1,1}(z,t) &= -\frac{t_p^{1-\alpha}(CR+GL)\,t^\alpha\,\hbar}{\Gamma(\alpha+1)\sqrt{LC}(GL-CR)}\sin\left(\frac{(CR-GL)}{2\sqrt{LC}}z\right),\\
x_{2,1}(z,t) &= -\frac{t_p^{1-\alpha}(CR+GL)^2\,t^\alpha\,\hbar}{2\,\Gamma(\alpha+1)\sqrt{(LC)^3}(GL-CR)}\sin\left(\frac{(CR-GL)}{2\sqrt{LC}}z\right),\\
x_{1,2}(z,t) &= -\frac{t_p^{1-\alpha}(CR+GL)\,t^\alpha\,\hbar(1+\hbar)}{\Gamma(\alpha+1)\sqrt{LC}(GL-CR)}\sin\left(\frac{(CR-GL)}{2\sqrt{LC}}z\right),\\
&\quad - \frac{t_p^{2(1-\alpha)}(CR+GL)^2\,t^{2\alpha}\,\hbar^2}{2\,\Gamma(2\alpha+1)\sqrt{(LC)^3}(GL-CR)}\sin\left(\frac{(CR-GL)}{2\sqrt{LC}}z\right),\\
x_{2,2}(z,t) &= -\frac{t_p^{1-\alpha}(CR+GL)^2\,t^\alpha\,\hbar(1+\hbar)}{2\,\Gamma(\alpha+1)\sqrt{(LC)^3}(GL-CR)}\sin\left(\frac{(CR-GL)}{2\sqrt{LC}}z\right),\\
&\quad - \frac{t_p^{2(1-\alpha)}(CR+GL)^3\,t^{2\alpha}\,\hbar^2}{4\,\Gamma(2\alpha+1)\sqrt{(LC)^5}(GL-CR)}\sin\left(\frac{(CR-GL)}{2\sqrt{LC}}z\right),\\
&\ \ \vdots
\end{aligned} \tag{61}$$

Finally the solutions of Eqs. (47) and (48) are given by

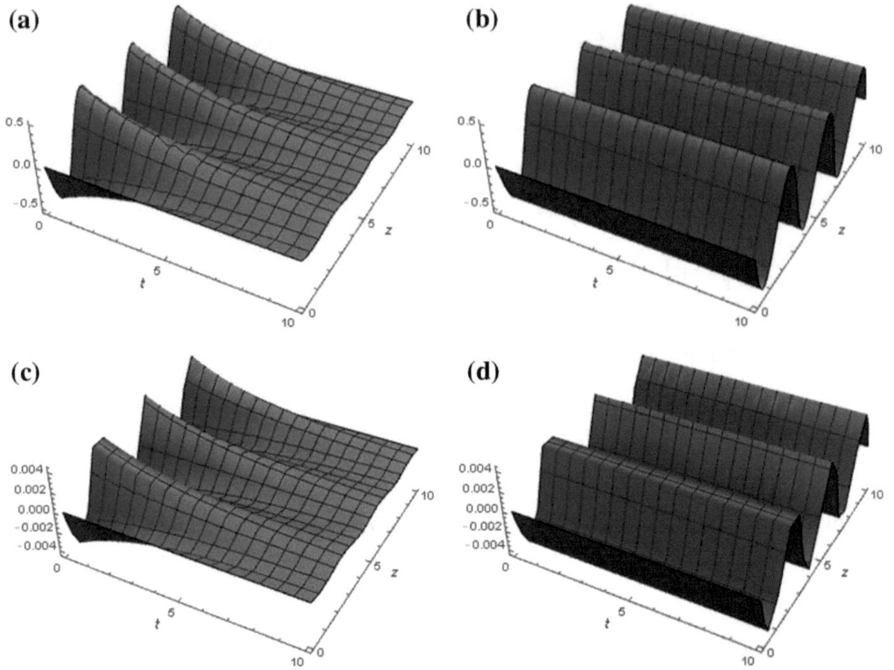

Fig. 3 Numerical simulation for the approximate solutions given by Eqs. (62) and (63). In **a** and **b** Eq. (62) for $\alpha = 1$ and $\alpha = 0.9$, respectively. In **c** and **d** Eq. (63) for $\alpha = 1$ and $\alpha = 0.9$, respectively

$$x_1(z,t) = \sum_{m=0}^{\infty} x_{1,m}(z,t), \tag{62}$$

$$x_2(z,t) = \sum_{m=0}^{\infty} x_{2,m}(z,t). \tag{63}$$

Example

Consider the following values $C = 4.8240$, $L = 12.3453$ and $R = 6.414$ arbitrarily chosen. Figure 3a–d show the numerical simulations of the Eqs. (62) and (63) for $\alpha = 1$ and $\alpha = 0.9$.

Case 2. Liénard model of a fluid transmission line via Atangana–Baleanu–Caputo fractional order derivative.

In this case, we consider the following fractional equations in ABC sense

$${}_{0}^{ABC}\mathscr{D}_t^{\alpha} x_1(z,t) - t_p^{1-\alpha} x_2(z,t) + t_p^{1-\alpha}\left(\frac{RC+GL}{LC}\right) x_1(z,t) = 0, \tag{64}$$

$${}_{0}^{ABC}\mathscr{D}_t^{\alpha} x_2(z,t) + t_p^{1-\alpha}\left(\frac{RG}{LC}\right) x_1(z,t) - t_p^{1-\alpha}\left(\frac{1}{LC}\right)\frac{\partial^2 x_1(z,t)}{\partial z^2} = 0, \tag{65}$$

$$x_1(z,0) = \left(\frac{2\sqrt{LC}(CR+GL)}{(CR-GL)}\right) \sin\left(\frac{(CR-GL)}{2\sqrt{LC}} z\right), \tag{66}$$

$$x_2(z,0) = \left(\frac{(CR+GL)}{\sqrt{LC}(CR-GL)}\right) \sin\left(\frac{(CR-GL)}{2\sqrt{LC}} z\right), \tag{67}$$

where t_p corresponds to the Planck time and ${}_0^{ABC}\mathscr{D}_t^\alpha$ is the fractional derivative with generalized Mittag-Leffler law in Liouville–Caputo sense [26].
Solution. Applying the Laplace transform (37) to the Eqs. (64) and (65), we obtain

$$\begin{aligned} X_1(z,s) &= \frac{x_1(z,0)}{s} + t_p^{1-\alpha} \frac{(1-\alpha)s^\alpha + \alpha}{B(\alpha)s^\alpha} \left[X_2(z,s) - \left(\frac{RC+GL}{LC}\right) X_1(z,s)\right], \\ X_2(z,s) &= \frac{x_2(z,0)}{s} + t_p^{1-\alpha} \frac{(1-\alpha)s^\alpha + \alpha}{B(\alpha)s^\alpha} \left[-\left(\frac{RG}{LC}\right) + \left(\frac{1}{LC}\right)\frac{\partial^2}{\partial z^2}\right] X_1(z,s). \end{aligned} \tag{68}$$

applying the inverse Laplace transform to Eq. (68), we obtain

$$\begin{aligned} x_1(z,t) &= x_1(z,0) + \mathscr{L}^{-1}\left\{t_p^{1-\alpha} \frac{(1-\alpha)s^\alpha + \alpha}{B(\alpha)s^\alpha} \mathscr{L}\left[x_2(z,t) - \left(\frac{RC+GL}{LC}\right) x_1(z,t)\right]\right\}, \\ x_2(z,t) &= x_2(z,0) + \mathscr{L}^{-1}\left\{t_p^{1-\alpha} \frac{(1-\alpha)s^\alpha + \alpha}{B(\alpha)s^\alpha} \mathscr{L}\left[-\left(\frac{RG}{LC}\right) + \left(\frac{1}{LC}\right)\frac{\partial^2}{\partial z^2}\right] x_1(z,t)\right\}. \end{aligned} \tag{69}$$

We choose the linear operator as

$$\mathfrak{F}[\phi_j(z,t;q)] = \mathscr{L}[\phi_j(z,t;q)], \ j = 1, 2.$$

With property $\mathfrak{F}(c) = 0$, where c is constant. Next, defining the system of nonlinear operator as

$$\begin{aligned} N[\phi(z,t;q)] = \mathscr{L}[\phi(z,t;q)] &- \left(\frac{2\sqrt{LC}(CR+GL)}{(CR-GL)}\right) \sin\left(\frac{(CR-GL)}{2\sqrt{LC}} z\right) \\ &- t_p^{1-\alpha} \frac{(1-\alpha)s^\alpha + \alpha}{B(\alpha)s^\alpha} \mathscr{L}\left[\Phi - \left(\frac{RC+GL}{LC}\right) \phi\right], \end{aligned} \tag{70}$$

$$\begin{aligned} N[\Phi(z,t;q)] = \mathscr{L}[\Phi(z,t;q)] &- \left(\frac{(CR+GL)}{\sqrt{LC}(CR-GL)}\right) \sin\left(\frac{(CR-GL)}{2\sqrt{LC}} z\right) \\ &- t_p^{1-\alpha} \frac{(1-\alpha)s^\alpha + \alpha}{B(\alpha)s^\alpha} \mathscr{L}\left[-\left(\frac{RG}{LC}\right) + \left(\frac{1}{LC}\right)\frac{\partial^2}{\partial z^2}\right] \phi. \end{aligned} \tag{71}$$

Now, we construct the so-called zeroth-order deformation equation in the following manner

$$(1-q)\,\mathfrak{F}[\phi_j(z,t;q)-u_0(z,t)] = q\,\hbar\,N[\phi_j(z,t;q)],\quad j=1,2. \tag{72}$$

when $q=0$ and $q=1$, we have

$$\phi_j(z,t;0)=u_0(z,t),\quad \phi_j(z,t;1)=u(z,t),\quad j=1,2.$$

Thus, we obtain the mth-order deformation equations as

$$\begin{aligned}
\mathscr{L}[x_{1_m}(z,t)-\chi_m x_{1_{m-1}}(z,t)] &= \hbar\,R_m\big(\vec{x_{1_{m-1}}},z,t\big),\\
\mathscr{L}[x_{2_m}(z,t)-\chi_m x_{2_{m-1}}(z,t)] &= \hbar\,R_m\big(\vec{x_{2_{m-1}}},z,t\big),
\end{aligned} \tag{73}$$

applying the inverse Laplace transform to the Eq. (73) we get

$$x_{1_m}(z,t)=\chi_m x_{1_{m-1}}(z,t)+\hbar\,R_m\big(\vec{x_{1_{m-1}}},z,t\big), \tag{74}$$

$$x_{2_m}(z,t)=\chi_m x_{2_{m-1}}(z,t)+\hbar\,R_m\big(\vec{x_{2_{m-1}}},z,t\big), \tag{75}$$

where

$$\begin{aligned}
R_m\big(\vec{x_{1_{m-1}}},z,t\big) &= \mathscr{L}[x_{1_{m-1}}(z,t)]-(1-\chi_m)\left(\frac{2\sqrt{LC}(CR+GL)}{(CR-GL)}\right)\sin\left(\frac{(CR-GL)}{2\sqrt{LC}}z\right)\\
&\quad - t_p^{1-\alpha}\,\frac{(1-\alpha)s^\alpha+\alpha}{B(\alpha)s^\alpha}\,\mathscr{L}\left[x_{2_{m-1}}-\left(\frac{RC+GL}{LC}\right)x_{1_{m-1}}\right],\\
R_m\big(\vec{x_{2_{m-1}}},z,t\big) &= \mathscr{L}[x_{2_{m-1}}(z,t)]-(1-\chi_m)\left(\frac{(CR+GL)}{\sqrt{LC}(CR-GL)}\right)\sin\left(\frac{(CR-GL)}{2\sqrt{LC}}z\right)\\
&\quad - t_p^{1-\alpha}\,\frac{(1-\alpha)s^\alpha+\alpha}{B(\alpha)s^\alpha}\,\mathscr{L}\left[-\left(\frac{RG}{LC}\right)+\left(\frac{1}{LC}\right)\frac{\partial^2}{\partial z^2}\right]x_{1_{m-1}}.
\end{aligned} \tag{76}$$

Now, the solution of mth-order deformation equations (73) are given as

$$\begin{aligned}
x_{1_m}(z,t) &= (\chi_m+\hbar)x_{1_{m-1}}-\hbar(1-\chi_m)\left(\frac{2\sqrt{LC}(CR+GL)}{(CR-GL)}\right)\sin\left(\frac{(CR-GL)}{2\sqrt{LC}}z\right)\\
&\quad -\hbar\,\mathscr{L}^{-1}\left\{t_p^{1-\alpha}\,\frac{(1-\alpha)s^\alpha+\alpha}{B(\alpha)s^\alpha}\,\mathscr{L}\left[x_{2_{m-1}}-\left(\frac{RC+GL}{LC}\right)x_{1_{m-1}}\right]\right\},\\
x_{2_m}(z,t) &= (\chi_m+\hbar)x_{2_{m-1}}-\hbar(1-\chi_m)\left(\frac{(CR+GL)}{\sqrt{LC}(CR-GL)}\right)\sin\left(\frac{(CR-GL)}{2\sqrt{LC}}z\right)\\
&\quad -\hbar\,\mathscr{L}^{-1}\left\{t_p^{1-\alpha}\,\frac{(1-\alpha)s^\alpha+\alpha}{B(\alpha)s^\alpha}\,\mathscr{L}\left[-\left(\frac{RG}{LC}\right)+\left(\frac{1}{LC}\right)\frac{\partial^2}{\partial z^2}\right]x_{1_{m-1}}\right\}.
\end{aligned} \tag{77}$$

Taking initial conditions and the iterative scheme (77), we obtain the following iterations

$$p^0 : x_{1,0}(z,t) = x_1(z,0),$$
$$x_{2,0}(z,t) = x_2(z,0),$$
$$p^1 : x_{1,1}(z,s) = -\frac{t_p^{1-\alpha}(CR+GL)\left(\alpha\, t^{\alpha} + (1-\alpha)\Gamma(\alpha+1)\right)\hbar}{B(\alpha)\,\Gamma(\alpha+1)\,\sqrt{LC}(GL-CR)} \sin\left(\frac{(CR-GL)}{2\sqrt{LC}}z\right),$$
$$x_{2,1}(z,t) = -\frac{t_p^{1-\alpha}(CR+GL)^2\left(\alpha\, t^{\alpha} + (1-\alpha)\Gamma(\alpha+1)\right)\hbar}{2\,B(\alpha)\,\Gamma(\alpha+1)\,\sqrt{(LC)^3}(GL-CR)} \sin\left(\frac{(CR-GL)}{2\sqrt{LC}}z\right),$$
$$p^1 : x_{1,2}(z,t) = -\frac{t_p^{1-\alpha}(CR+GL)\left(\alpha\, t^{\alpha} + (1-\alpha)\Gamma(\alpha+1)\right)\hbar(1+\hbar)}{B(\alpha)\,\Gamma(\alpha+1)\,\sqrt{LC}(GL-CR)} \sin\left(\frac{(CR-GL)}{2\sqrt{LC}}z\right),$$

$$-\frac{t_p^{2(1-\alpha)}(CR+GL)^2\left((1-\alpha)^2\Gamma(\alpha+1)\Gamma(2\alpha+1) + 2\alpha(1-\alpha)\Gamma(2\alpha+1)\,t^{\alpha} + \alpha^2\Gamma(\alpha+1)t^{2\alpha}\right)\hbar^2}{2\,B(\alpha)^2\,\Gamma(\alpha+1)\,\Gamma(2\alpha+1)\,\sqrt{(LC)^3}(GL-CR)}$$
$$\times \sin\left(\frac{(CR-GL)}{2\sqrt{LC}}z\right),$$
$$x_{2,2}(z,t) = -\frac{t_p^{1-\alpha}(CR+GL)^2\left(\alpha\, t^{\alpha} + (1-\alpha)\Gamma(\alpha+1)\right)\hbar(1+\hbar)}{2\,B(\alpha)\,\Gamma(\alpha+1)\,\sqrt{(LC)^3}(GL-CR)} \sin\left(\frac{(CR-GL)}{2\sqrt{LC}}z\right),$$

$$-\frac{t_p^{2(1-\alpha)}(CR+GL)^3\left((1-\alpha)^2\Gamma(\alpha+1)\Gamma(2\alpha+1) + 2\alpha(1-\alpha)\Gamma(2\alpha+1)\,t^{\alpha} + \alpha^2\Gamma(\alpha+1)t^{2\alpha}\right)\hbar^2}{4\,B(\alpha)^2\,\Gamma(\alpha+1)\,\Gamma(2\alpha+1)\,\sqrt{(LC)^5}(GL-CR)}$$

$$\times \sin\left(\frac{(CR-GL)}{2\sqrt{LC}}z\right),$$
$$\vdots \tag{78}$$

Finally, the solutions of the Eqs. (64) and (65) are given by

$$x_1(z,t) = \sum_{m=0}^{\infty} x_{1,m}(z,t), \tag{79}$$

$$x_2(z,t) = \sum_{m=0}^{\infty} x_{2,m}(z,t). \tag{80}$$

Example

Consider the following values $C = 4.8240$, $L = 12.3453$ and $R = 6.414$ arbitrarily chosen. Figure 4a–d show the numerical simulations of the Eqs. (79) and (80) for $\alpha = 1$ and $\alpha = 0.9$.

5 Conclusions

In this chapter, the LHPM and MHATM have been applied to solve the fractional Liénard type model of a pipeline via Liouville–Caputo and Atangana–Baleanu–Caputo fractional order derivatives. The main aim of this chapter was to provide the series solution of the time-fractional Liénard model by using these fractional methods. Approximate analytical solutions and numerical simulations were obtained for some parameters. We found that there is a very good approximation between our solutions and the exact solution in the limit when $\alpha \to 1$. From the above discussion we concluded that the present methods are reliable.

Both methods presented are an important mathematical tool for developing scientific work in various areas of the natural sciences. The two definitions of fractional operators considered in this work could be applied conveniently depending on the nature and the phenomenological behavior of the system under consideration. The illustrative examples showed that both methods are easy to use and represent an effective tool to numerically solve fractional partial differential equations.

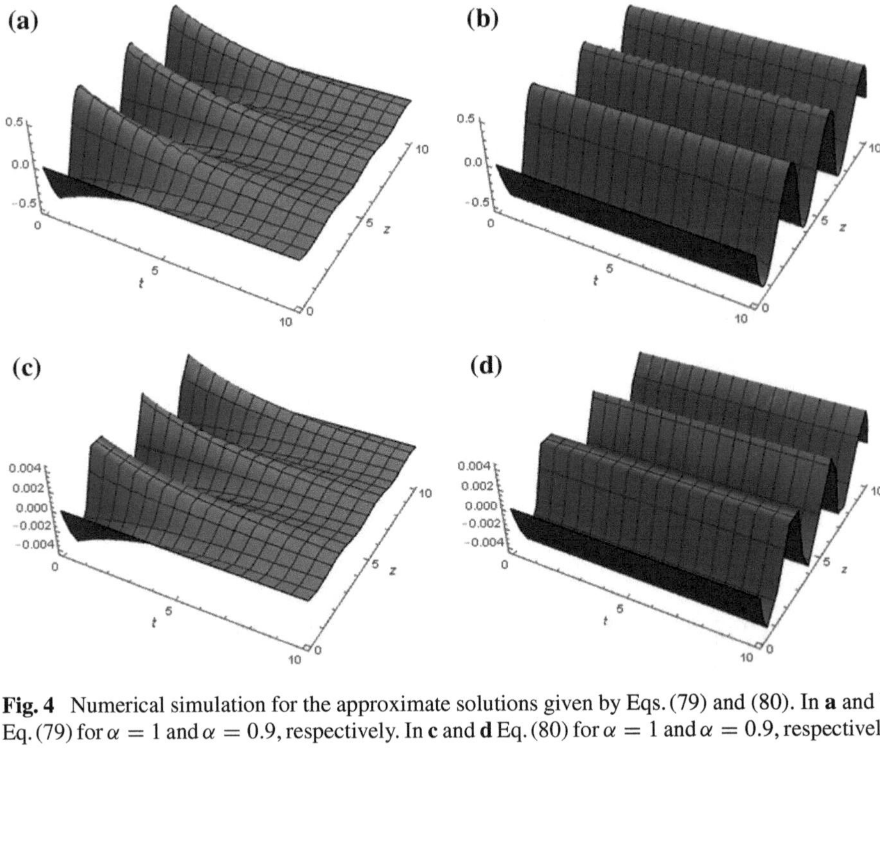

Fig. 4 Numerical simulation for the approximate solutions given by Eqs. (79) and (80). In **a** and **b** Eq. (79) for $\alpha = 1$ and $\alpha = 0.9$, respectively. In **c** and **d** Eq. (80) for $\alpha = 1$ and $\alpha = 0.9$, respectively

Acknowledgements José Francisco Gómez Aguilar and Lizeth Torres acknowledges the support provided by CONACyT: Cátedras CONACyT para jóvenes investigadores 2014. José Francisco Gómez Aguilar, Lizeth Torres, Ricardo Fabricio Escobar Jiménez and Marco Antonio Taneco Hernández acknowledges the support provided by SNI-CONACyT.

References

1. Kong, D.: Explicit exact solutions for the Liénard equation and its applications. Phys. Lett. A **196**, 301–306 (1995)
2. Kudryashov, N.A., Sinelshchikov, D.I.: On the connection of the quadratic Liénard equation with an equation for the elliptic functions. Regul. Chaotic Dyn. **20**(4), 486–496 (2015)
3. Nowak, W., Geiyer, D., Das, T.: Absolute Stability analysis using the Liénard equation: a study derived from control of fuel cell ultracapacitor hybrids. J. Dyn. Syst. Meas. Control. **138**(3), 1–22 (2016)
4. Sinha, M., Dörfler, F., Johnson, B.B., Dhople, S.V.: Synchronization of Liénard-type oscillators in uniform electrical networks. In: American Control Conference, vol. 1. IEEE, pp. 4311–4316 (2016)
5. Martins, R.M., Mereu, A.C.: Limit cycles in discontinuous classical Liénard equations. Nonlinear Anal. R. World Appl. **20**, 67–73 (2014)
6. Harko, T., Liang, S.D.: Exact solutions of the Liénard-and generalized Liénard-type ordinary nonlinear differential equations obtained by deforming the phase space coordinates of the linear harmonic oscillator. J. Eng. Math. **98**(1), 93–111 (2016)
7. Kudryashov, N.A., Sinelshchikov, D.I.: On the integrability conditions for a family of Liénard-type equations. Regul. Chaotic Dyn. **21**(5), 548–555 (2016)
8. Feng, Z.: On explicit exact solutions for the Liénard equation and its applications. Phys. Lett. A **239**, 50–56 (2002)
9. Matinfar, M., Hosseinzadeh, H., Ghanbari, M.: A numerical implementation of the variational iteration method for the Liénard equation. World J. Model. Simul. **4**, 205–210 (2008)
10. Matinfar, M., Mahdavi, M., Raeisy, Z.: Exact and numerical solution of Liénard's equation by the variational homotopy perturbation method. J. Inf. Comput. Sci. **6**(1), 73–80 (2011)
11. Torres, L., Besancon, G., Verde, C.: Liénard type model of fluid flow in pipelines: application to estimation. In: 2015 12th International Conference on Electrical Engineering, Computing Science and Automatic Control, vol. 1. IEEE, pp. 1–6 (2015)
12. Torres, L., Aguiñaga, J.A.D., Besancon, G., Verde, C., Begovich, O.: Equivalent Liénard-type models for a fluid transmission line. Comptes Rendus Mécanique **344**(8), 582–595 (2016)
13. Jiménez, J., Torres, L., Rubio, I., Sanjuan, M.: Auxiliary signal design and Liénard-type models for identifying pipeline parameters. Modeling and Monitoring of Pipelines and Networks, vol. 1, pp. 99–124. Springer International Publishing, Berlin (2017)
14. Singh, J., Kumar, D., Qurashi, M.A., Baleanu, D.: Analysis of a new fractional model for damped Bergers' equation. Open Phys. **15**(1), 35–41 (2017)
15. Hristov, J.: Space-fractional diffusion with a potential power-law coefficient: transient approximate solution. Progr. Fract. Differ. Appl. **3**(1), 19–39 (2017)
16. Owolabi, K.M., Atangana, A.: Numerical simulations of chaotic and complex spatiotemporal patterns in fractional reaction-diffusion systems. Comput. Appl. Math. **1**, 1–24 (2017)
17. Owolabi, K.M., Atangana, A.: Mathematical analysis and numerical simulation of two-component system with non-integer-order derivative in high dimensions. Adv. Differ. Equ. **1**, 1–24 (2017)
18. Kumar, S., Kumar, A., Odibat, Z.M.: A nonlinear fractional model to describe the population dynamics of two interacting species. Math. Methods Appl. Sci. **1**, 1–15 (2017)
19. Ali, F., Sheikh, N.A., Khan, I., Saqib, M.: Solutions with Wright function for time fractional free convection flow of Casson fluid. Arab. J. Sci. Eng. **42**(6), 2565–2572 (2017)

20. Atangana, A., Baleanu, D., Alsaedi, A.: Analysis of time-fractional Hunter-Saxton equation: a model of neumatic liquid crystal. Open Phys. **14**(1), 145–149 (2016)
21. Atangana, A.: A novel model for the lassa hemorrhagic fever: deathly disease for pregnant women. Neural Comput. Appl. **26**(8), 1895–1903 (2015)
22. Baleanu, D., Diethelm, K., Scalas, E., Trujillo, J.J.: Fractional Calculus Models and Numerical Methods. Series on Complexity, Nonlinearity and Chaos. World Scientific, Singapore (2012)
23. Podlubny, I.: An introduction to fractional derivatives, fractional differential equations, to methods of their solution and some of their applications. Fractional Differential Equations. Academic, San Diego (1999)
24. Atangana, A., Secer, A.: A note on fractional order derivatives and table of fractional derivatives of some special functions. Abstr. Appl. Anal. **1**, 1–21 (2013)
25. Caputo, M., Fabrizio, M.: A new definition of fractional derivative without singular Kernel. Progr. Fract. Differ. Appl. **1**(2), 73–85 (2015)
26. Atangana, A., Baleanu, D.: New fractional derivatives with nonlocal and non-singular Kernel: theory and application to heat transfer model. Therm. Sci. **20**(2), 763–769 (2016)
27. Pirim, N.A., Ayaz, F.: A new technique for solving fractional order systems: Hermite collocation method. Appl. Math. **7**(18), 1–12 (2016)
28. Choudhary, S., Daftardar-Gejji, V.: Invariant subspace method: a tool for solving fractional partial differential equations (2016). arXiv:1609.04209
29. Hamarsheh, M., Ismail, A.I., Odibat, Z.: An analytic solution for fractional order Riccati equations by using optimal homotopy asymptotic method. Appl. Math. Sci. **10**(23), 1131–1150 (2016)
30. Ghorbani, A.: Beyond Adomain's polynomials: He's polynomials. Chaos Solitons Fractals **39**, 1486–1492 (2009)
31. Rathore, S., Kumar, D., Singh, J., Gupta, S.: Homotopy analysis Sumudu transform method for nonlinear equations. Int. J. Ind. Math. **4**(4), 301–314 (2012)
32. Atangana, A., Alabaraoye, E.: Solving a system of fractional partial differential equations arising in the model of HIV infection of CD4+ cells and attractor one-dimensional Keller-Segel equations. Adv. Differ. Equ. **94**, 1–14 (2013)
33. Vahidi, J.: The combined Laplace-homotopy analysis method for partial differential equations. J. Math. Comput. Sci.-JMCS **16**(1), 88–102 (2016)
34. Atangana, A.: Extension of the Sumudu homotopy perturbation method to an attractor for onedimensional Keller-Segel equations. Appl. Math. Model. **39**, 2909–2916 (2015)
35. Pandey, R.K., Mishra, H.K.: Homotopy analysis Sumudu transform method for time-fractional third order dispersive partial differential equation. Adv. Comput. Math. **1**, 1–19 (2016)
36. Abbasbandy, S., Shivanian, E., Vajravelu, K., Kumar, S.: A new approximate analytical technique for dual solutions of nonlinear differential equations arising in mixed convection heat transfer in a porous medium. Int. J. Numer. Methods Heat Fluid Flow **27**(2), 486–503 (2017)
37. Kumar, D., Agarwal, R.P., Singh, J.: A modified numerical scheme and convergence analysis for fractional model of Liénard's equation. J. Comput. Appl. Math. **1**, 1–14 (2017)
38. Singh, H.: Solution of fractional Liénard equation using Chebyshev operational matrix method. Nonlinear Sci. Lett. A **8**(4), 397–404 (2017)
39. Jafari, H., Seifi, S.: Homotopy analysis method for solving linear and nonlinear fractional diffusion-wave equation. Commun. Nonlinear Sci. Numer. Simul. **14**(5), 2006–2012 (2009)
40. Jafari, H., Seifi, S.: Solving a system of nonlinear fractional partial differential equations using homotopy analysis method. Commun. Nonlinear Sci. Numer. Simul. **14**(5), 1962–1969 (2009)
41. Liao, S.J.: A kind of approximate solution technique which does not depend upon small parameters (II): an application in fluid mechanics. Int. J. Nonlinear Mech. **32**(5), 815–822 (1997)
42. Kumar, S., Kumar, A., Odibat, Z.M.: A nonlinear fractional model to describe the population dynamics of two interacting species. Math. Methods Appl. Sci. **40**(11), 4134–4148 (2017)
43. Morales-Delgado, V.F., Gómez-Aguilar, J.F., Yépez-Martínez, H., Baleanu, D., Escobar-Jiménez, R.F., Olivares-Peregrino, V.H.: Laplace homotopy analysis method for solving linear partial differential equations using a fractional derivative with and without kernel singular. Adv. Differ. Equ. **1**, 1–17 (2016)

44. Gómez-Aguilar, J.F., Torres, L., Yépez-Martínez, H., Baleanu, D., Reyes, J.M., Sosa, I.O.: Fractional Liénard type model of a pipeline within the fractional derivative without singular kernel. Adv. Differ. Equ. **1**, 1–13 (2016)
45. Li, C., Kumar, A., Kumar, S., Yang, X.J.: On the approximate solution of nonlinear time-fractional KdV equation via modified homotopy analysis Laplace transform method. J. Nonlinear Sci. Appl. **9**, 5463–5470 (2016)
46. Kumar, S., Kumar, A., Argyros, I.K.: A new analysis for the Keller-Segel model of fractional order. Numer. Algorithms **75**(1), 213–228 (2017)
47. Chaudhry, M.H.: Applied Hydraulic Transients, pp. 426–431. Van Nostrand Reinhold, New York (1979)
48. Brown, G.O.: The history of the Darcy-Weisbach equation for pipe flow resistance. Environ. Water Resour. Hist. **1**, 34–43 (2003)
49. Ferrante, M., Brunone, B., Meniconi, S.: Leak detection in branched pipe systems coupling wavelet analysis and a lagrangian model. J. Water Supply Res. Technol.-AQUA **58**(2), 95–106 (2009)
50. Wylie, E.B., Streeter, V.L., Suo, L.: Fluid Transients in Systems. Prentice Hall, Englewood Cliffs (1993)
51. Gómez-Aguilar, J.F., Baleanu, D.: Solutions of the telegraph equations using a fractional calculus approach. Proc. Rom. Acad. A **15**, 27–34 (2014)
52. Liénard, A.: Etude des oscillations entretenues. Rev. Gén. Électr. **23**, 901–954 (1928)
53. Gómez-Aguilar, J.F., Rosales-García, J.J., Bernal-Alvarado, J.J., Córdova-Fraga, T., Guzmán-Cabrera, R.: Fractional mechanical oscillators. Rev. Mex. Fis. **58**, 524–537 (2012)
54. Calik, A.E.: Investigation of electrical RC circuit within the framework of fractional calculus. Rev. Mex. Fis. **61**, 58–63 (2015)

Model of Coupled System of Fractional Reaction-Diffusion Within a New Fractional Derivative Without Singular Kernel

K. M. Saad, J. F. Gómez-Aguilar, A. Atangana and R. F. Escobar-Jiménez

Abstract In this chapter, the approximate analytical solutions of a new reaction-diffusion fractional time model are studied. For this analysis is used the p-homotopy transform method based on different kernels (power, exponential and Mittag-Leffler). The system nonlinearities are addressed by the Adomian polynomials. The system convergence is studied by determining the interval of the convergence by $\hbar$-curves, as well as, searching for the optimal value of $\hbar$ which minimize the residual error. Therefore, the optimal $\hbar$ value is calculated to estimate the order β error. At the end of the chapter, we explained the obtained behavior by plotting the solutions in 3D. The results are accurate.

Keywords Fractional calculus · Atangana–Baleanu fractional derivative · Cubic isothermal auto-catalytic chemical system

K. M. Saad
Department of Mathematical Sciences, University of South Africa,
Florida 0003, South Africa

J. F. Gómez-Aguilar (✉)
CONACYT-Tecnológico Nacional de México, Centro Nacional de Investigación y Desarrollo Tecnológico, Interna del Internado, Palmira, 62490 Cuernavaca, Morelos, México
e-mail: jgomez@cenidet.edu.mx

A. Atangana
Faculty of Natural and Agricultural Sciences, Institute of Groundwater Studies,
University of Free State, Bloemfontein 9301, South Africa

R. F. Escobar-Jiménez
Tecnológico Nacional de México, Centro Nacional de Investigación y Desarrollo Tecnológico,
Interna del Internado, Palmira, 62490 Cuernavaca, Morelos, México

J. F. Gómez et al. (eds.), *Fractional Derivatives with Mittag-Leffler Kernel*,
Studies in Systems, Decision and Control 194,
https://doi.org/10.1007/978-3-030-11662-0_17

1 Introduction

In this chapter, we study the reaction-diffusion traveling waves governed by cubic auto-catalysis which can be initiated in a coupled isothermal chemical system. We hypothesized that reactions occur along semipermeable membrane interfaces commonly occurs on one interface (region I) being cubic auto-catalysis (1)

$$U + V \rightarrow 2V(rate\ r_1 uv^2), \tag{1}$$

with the step of the linear termination (2)

$$V \rightarrow W(rate\ r_2 v), \tag{2}$$

where u and v correspond to the concentrations of the reactant U and auto-catalyst V, respectively, the $r_i (i = 1, 2)$ represents the rate constants and W is some inert product of reaction. On the other interface (region II), the reaction is assumed to be just the autocatalytic step (1) with the same rate r_1. Both regions are considered to be coupled by a linear diffusive interchange of the auto-catalytic V. This yields to the following system

$$\frac{\partial u_1}{\partial \eta} = \frac{\partial^2 u_1}{\partial \xi^2} - u_1 v_1^2 + \nu(u_2 - u_1), \tag{3}$$

$$\frac{\partial v_1}{\partial \eta} = \frac{\partial^2 v_1}{\partial \xi^2} + u_1 v_1^2 - \kappa v_1, \tag{4}$$

$$\frac{\partial u_2}{\partial \eta} = \frac{\partial^2 u_2}{\partial \xi^2} - u_2 v_2^2 + \nu(u_1 - u_2), \tag{5}$$

$$\frac{\partial v_2}{\partial \eta} = \frac{\partial^2 v_2}{\partial \xi^2} + u_2 v_2^2, \tag{6}$$

with the boundary conditions

$$u_i(0, \eta) = u_i(\ell, \eta) = 1, \quad v_i(0, \eta) = v_i(\ell, \eta) = 0, \tag{7}$$

where ν represents the coupling between (I) and (II) and κ represents the strength of the auto-catalyst decay. For more details see [1].

In the recent decades, fractional calculus has been used to model real-world problems in many fields, such as: engineering, technology and science. Many definitions of fractional order derivatives have been introduced, including: Riemann–Liouville (RL), Grünwald–Letnikov (GL) and the Liouville–Caputo (LC) [2–5]. The RL derivative involve the convolution of a power-law kernel and a given function [3]. It was constructed using the fundamental relations of the RL fractional integral [2].

The LC fractional derivative involve the convolution of power law function and the local derivative of a given function [3]. More recently, Caputo and Fabrizio (CF) in [4] proposed a new fractional derivative based on the exponential decay law which is a generalized power law function [5–11]. Atangana and Baleanu (AB) proposed a fractional derivative with non-local kernel based on the Mittag-Leffler function. With this fractional derivative, we can describe complex physical problems which follows at the same time exponential decay law and power law [12–18].

In this chapter, our aim is to use p-homotopy analysis transform method (p-HATM) based on LC, CF and AB operators to evaluate the approximate solutions of the fractional time reaction-diffusion system (TFRDS). This chapter is divided as follows: In second section we apply the p-HATM on the TFRDS with LC, CF and AB. In section three, we evaluate the approximate solutions and discuss the numerical simulation. Finally, in section four, we present the conclusion.

2 New p-HATM Solutions

In this section, we use the p-HATM on TFRDS applying LC, CF and AB operators.

2.1 *Application of Liouville–Caputo Fractional-Order Derivative on TFRDS*

In this section, the p-HATM is used applying the LC operator. Now, we develop a new TFCAR by replacing D_η by the ${}_0^{LC}D_\eta^\beta$. The set of the Eqs. (3)–(6) become

$${}_0^{LC}D_\eta^\beta u_1 = u_{1,\xi\xi} - u_1 v_1^2 + \nu(u_2 - u_1), \tag{8}$$

$${}_0^{LC}D_\eta^\beta v_1 = v_{1,\xi\xi} + u_1 v_1^2 - \kappa v_1, \tag{9}$$

$${}_0^{LC}D_\eta^\beta u_2 = u_{2,\xi\xi} - u_2 v_2^2 + \nu(u_1 - u_2), \tag{10}$$

$${}_0^{LC}D_\eta^\beta v_2 = v_{2,\xi\xi} + u_2 v_2^2. \tag{11}$$

The ${}_0^{LC}D_\eta^\beta$ is given by

$${}_0^{C}D_\eta^\beta(f(\eta)) = J^{m-\beta}D^m f(\eta) = \frac{1}{\Gamma(m-\beta)}\int_0^\eta (\eta - t)^{m-\beta-1} f^{(m)}(t)\,dt,$$

for $m - 1 < \beta \leq m$, $m \in \mathbb{N}$, $t > 0$, $f \in C_\mu^m$, $\mu \geq -1$. We set the initial conditions as

$$u_i(\xi,0)=1-\sum_{n=1}^{\infty}a_i\cos(0.5(\ell-2\xi)\lambda)\sin(\lambda\ell/2), \tag{12}$$

$$v_i(\xi,0)=\sum_{n=1}^{\infty}b_i\cos(0.5(\ell-2\xi)\lambda)\sin(\lambda\ell/2), \tag{13}$$

where $\lambda=\frac{n\pi}{\ell}$, $i=1,2$.

We have from [19–27] that p-HAM is based on continuous mapping

$$u_i(\xi,\eta)\to\phi_i(\xi,\eta;p),\quad v_i(\xi,\eta)\to\psi_i(\xi,\eta;p), \tag{14}$$

such that, as the embedding parameter p increases from 0 to $1/n$; $\phi_i(\xi,\eta;p)$, $\psi_i(\xi,\eta;p)$ varies from the initial iteration to the exact solution.

Now, we define the nonlinear operators as

$$\begin{aligned}\mathscr{N}_1(\phi_1(\xi,\eta;p))&=\mathscr{L}(\phi_1(\xi,\eta;p))-\frac{1}{s}u_1(\xi,0)-\Omega^{(\cdot)}(s,\beta)\mathscr{L}\Big(\phi_{1,\xi\xi}(\xi,\eta;p)\\&\quad-\phi_1(\xi,\eta;p)\psi_1^2(\xi,\eta;p)-\nu(\phi_2(\xi,\eta;p)-\phi_1(\xi,\eta;p)\Big),\\ \mathscr{M}_1(\psi_1(\xi,\eta;p))&=\mathscr{L}(\psi_1(\xi,\eta;p))-\frac{1}{s}v_1(\xi,0)-\Omega^{(\cdot)}(s,\beta)\mathscr{L}\Big(\psi_{1,\xi\xi}(\xi,\eta;p)\\&\quad-\kappa\psi_1(\xi,\eta;p)+\phi_1(\xi,\eta;p)\psi_1^2(\xi,\eta;p)\Big),\\ \mathscr{N}_2(\phi_2(\xi,\eta;p))&=\mathscr{L}(\phi_2(\xi,\eta;p))-\frac{1}{s}u_2(\xi,0)-\Omega^{(\cdot)}(s,\beta)\mathscr{L}\Big(\phi_{2,\xi\xi}(\xi,\eta;p)\\&\quad-\phi_2(\xi,\eta;p)\psi_2^2(\xi,\eta;p)-\nu(\phi_1(\xi,\eta;p)-\phi_2(\xi,\eta;p)\Big),\\ \mathscr{M}_2(\psi_2(\xi,\eta;p))&=\mathscr{L}(\psi_2(\xi,\eta;p))-\frac{1}{s}v_2(\xi,0)-\Omega^{(\cdot)}(s,\beta)\mathscr{L}\Big(\psi_{2,\xi\xi}(\xi,\eta;p)\\&\quad+\phi_2(\xi,\eta;p)\psi_2^2(\xi,\eta;p)\Big),\end{aligned} \tag{15}$$

where $\Omega^{LC}(s,\beta)=s^{-\beta}$.

By using p, we can develop the following equations

$$\begin{aligned}(1-np)\mathscr{L}(\phi_i(\xi,\eta;p)-u_{i,0}(\xi,\eta))&=p\hbar_iH(\xi,\eta)\mathscr{N}_i(\phi_i(\xi,\eta;p)),\\(1-np)\mathscr{L}(\psi_i(\xi,\eta;p)-v_{i,0}(\xi,\eta))&=p\hbar_iH(\xi,\eta)\mathscr{M}_i(\psi_i(\xi,\eta;p)),\end{aligned} \tag{16}$$

with initial conditions

$$\phi_i(\xi,0;p)=u_{i,0}(\xi,0),\quad \psi_i(\xi,0;p)=v_{i,0}(\xi,0),$$

where $\hbar_i \neq 0$ is the auxiliary parameter and $H(\xi, \eta) \neq 0$ is the auxiliary function. We expand $\phi_i(\xi, \eta; p)$ and $\psi_i(\xi, \eta; p)$ in series form by utilizing the Taylor theorem w. r. t. p, and obtain the following equation

$$\begin{aligned}
\phi_i(\xi, \eta; p) &= u_{i,0}(\xi, \eta) + \sum_{m=1}^{\infty} u_{i,m}(\xi, \eta) p^m, \\
\psi_i(\xi, \eta; p) &= v_{i,0}(\xi, \eta) + \sum_{m=1}^{\infty} v_{i,m}(\xi, \eta) p^m,
\end{aligned} \tag{17}$$

where

$$\begin{aligned}
u_{i,m}(\xi, \eta) &= \frac{1}{m!} \frac{\partial^m \phi_i(\xi, \eta; p)}{\partial p^m}|_{p=0}, \\
v_{i,m}(\xi, \eta) &= \frac{1}{m!} \frac{\partial^m \psi_i(\xi, \eta; p)}{\partial p^m}|_{p=0}.
\end{aligned} \tag{18}$$

The series (17) become after let $p = \frac{1}{n}$ as

$$\begin{aligned}
u_i(\xi, \eta) &= u_{i,0}(\xi, \eta) + \sum_{m=1}^{\infty} u_{i,m}(\xi, \eta) \left(\frac{1}{n}\right)^m, \\
v_i(\xi, \eta) &= v_{i,0}(\xi, \eta) + \sum_{m=1}^{\infty} v_{i,m}(\xi, \eta) \left(\frac{1}{n}\right)^m.
\end{aligned} \tag{19}$$

From [28, 29], we can build the mth-order deformation equation as follows

$$\begin{aligned}
&\mathscr{L}(u_{i,m}(\xi, \eta) - \mathscr{X}_m u_{i,(m-1)}(\xi, \eta)) = \hbar_1 H(\xi, \eta) RU_i, \\
&\mathscr{L}(v_{i,m}(\xi, \eta) - \mathscr{X}_m v_{i,(m-1)}(\xi, \eta)) = \hbar_i H(\xi, \eta) RV_i,
\end{aligned} \tag{20}$$

with initial conditions $u_m(\xi, 0) = 0$ and $v_m(\xi, 0) = 0$, for $m > 1$

$$\begin{aligned}
R_{u_1}^{(\cdot)} &= \mathscr{L}\left(u_{1,(m-1)}(\xi, \eta)\right) - \frac{1}{s} u_1(\xi, 0) \left(1 - \frac{\mathscr{X}_m}{n}\right) \\
&- \Omega^{(\cdot)}(s, \beta) \mathscr{L}\Big(u_{1,(m-1),\xi\xi}(\xi, t) - N_1(u_{1,(m-1)}(\xi, \eta), v_{1,(m-1)}(\xi, \eta)) \\
&+ \nu(u_{2,(m-1)}(\xi, \eta) - u_{1,(m-1)}(\xi, \eta))\Big)
\end{aligned} \tag{21}$$

$$\begin{aligned}
R_{v_1}^{(\cdot)} &= \mathscr{L}_{(m-1)}\left(v_{1,(m-1)}(\xi, \eta)\right) - \frac{1}{s} v_1(\xi, 0) \left(1 - \frac{\mathscr{X}_m}{n}\right) \\
&- \Omega^{(\cdot)}(s, \beta) \mathscr{L}\Big(v_{1,(m-1),\xi\xi}(\xi, \eta) + N_1(u_{1,(m-1)}(\xi, \eta), v_{1,(m-1)}(\xi, \eta)) \\
&- \kappa v_{1,(m-1)}(\xi, \eta)\Big).
\end{aligned} \tag{22}$$

$$R_{u_2}^{(\cdot)} = \mathscr{L}\left(u_{2,(m-1)}(\xi,\eta)\right) - \frac{1}{s}u_2(\xi,0)\left(1-\frac{\mathscr{X}_m}{n}\right)$$
$$- \Omega^{(\cdot)}(s,\beta)\mathscr{L}\Big(u_{2,(m-1),\xi\xi}(\xi,t) - N_2(u_{1,(m-1)}(\xi,\eta), v_{1,(m-1)}(\xi,\eta))$$
$$+ \nu(u_{1,(m-1)}(\xi,\eta) - u_{2,(m-1)}(\xi,\eta))\Big) \tag{23}$$

$$R_{v_2}^{(\cdot)} = \mathscr{L}_{(m-1)}\left(v_{2,(m-1)}(\xi,\eta)\right) - \frac{1}{s}v_2(\xi,0)\left(1-\frac{\mathscr{X}_m}{n}\right)$$
$$- \Omega^{(\cdot)}(s,\beta)\mathscr{L}\Big(v_{2,(m-1),\xi\xi}(\xi,\eta) + N_2(u_{1,(m-1)}(\xi,\eta), v_{1,(m-1)}(\xi,\eta))\Big). \tag{24}$$

$$N_1(u_{1,(m-1)}(\xi,\eta), v_{1,(m-1)}(\xi,\eta)) = u_{1,(m-1)}(\xi,\eta)v_{1,(m-1)}^2(\xi,\eta) \tag{25}$$
$$= \sum_{k=0}^{m-1} A_k \tag{26}$$

$$N_2(u_{2,(m-1)}(\xi,\eta), v_{2,(m-1)}(\xi,\eta)) = u_{2,(m-1)}(\xi,\eta)v_{2,(m-1)}^2(\xi,\eta) \tag{27}$$
$$= \sum_{k=0}^{m-1} B_k \tag{28}$$

where

$$A_m = \frac{1}{m!}\left[\frac{d^m}{d\lambda^m}N_1(u_{1,(m-1)}(\xi,\eta), v_{1,(m-1)}(\xi,\eta))\right]_{\lambda=0}, \tag{29}$$

$$B_m = \frac{1}{m!}\left[\frac{d^m}{d\lambda^m}N_2(u_{2,(m-1)}(\xi,\eta), v_{2,(m-1)}(\xi,\eta))\right]_{\lambda=0}, \tag{30}$$

where A_m and B_m are called the Adomian polynomials [30–32].

If we put $\mathscr{L}$ = Laplace transform and then inverse Laplace transform= $\mathscr{L}^{-1}$, the Eqs. (20)–(20) become

$$u_{i,m} = \mathscr{X}_m u_{i,(m-1)} + \hbar_i \mathscr{L}^{-1} R_{u_i}^{(\cdot)},$$

$$v_{i,m} = \mathscr{X}_m v_{i,(m-1)} + \hbar_i \mathscr{L}^{-1} R_{v_i}^{(\cdot)}. \tag{31}$$

2.2 *Application of Caputo–Fabrizio Fractional-Order Derivative on TFRDS*

By replacing ${}_0^C D_\eta^\beta(\cdot)$ into (8)–(11) by ${}_0^{CFC} D_\eta^\beta(\cdot)$, We obtain the TFRDS with Caputo–Fabrizio operator in Liouville–Caputo sense [4]

$$ {}_0^{CF}D_\eta^\beta u_1(\xi,\eta) - u_{1,\xi\xi}(\xi,\eta) + u_1(\xi,\eta)v_1^2(\xi,\eta) - \nu(u_2(\xi,\eta) - u_1(\xi,\eta)) = 0, \tag{32} $$

$$ {}_0^{CF}D_\eta^\beta v_1(\xi,\eta) - v_{1,\xi\xi}(\xi,\eta) - u_1(\xi,\eta)v_1^2(\xi,\eta) + \kappa v_1(\xi,\eta) = 0, \tag{33} $$

$$ {}_0^{CF}D_\eta^\beta u_2(\xi,\eta) - u_{2,\xi\xi}(\xi,\eta) + u_2(\xi,\eta)v_2^2(\xi,\eta) - \nu(u_1(\xi,\eta) - u_2(\xi,\eta)) = 0, \tag{34} $$

$$ {}_0^{CF}D_\eta^\beta v_2(\xi,\eta) - v_{2,\xi\xi}(\xi,\eta) - u_2(\xi,\eta)v_2^2(\xi,\eta) = 0, \tag{35} $$

where ${}_0^{CF}D_\eta^\beta(\cdot)$ is given by

$$ {}_0^{CF}D_\eta^\beta(\cdot) = \frac{M(\beta)}{1-\beta}\int_0^\eta \exp\left(\frac{-\beta(\eta-t)}{1-\beta}\right) D(\cdot)dt, $$

where $M(\beta)$ is a normalization function such that $M(0) = M(1) = 1$.

The definition of the Laplace transform for the CF fractional derivative is given by

$$ \mathscr{L}\left[{}_0^{CF}\mathscr{D}_t^\beta(\cdot)\right](s) = M(\beta)\left(\frac{s\mathscr{L}[u(\xi,\eta)](s) - u(\xi,0)}{s+\beta(1-s)}\right).,\ 0 < \beta \leq 1. \tag{36} $$

The Eqs. (32)–(35) become after applying the Laplace transform (36) on both ends as follows

$$ \frac{M(\beta)(s\mathscr{L}(u_1(\xi,\eta)) - u_1(\xi,0))}{(s+\beta(1-s))} = \mathscr{L}\Big(u_{1,\xi\xi}(\xi,\eta) - u_1(\xi,\eta)v_1^2(\xi,\eta) + \nu(u_2(\xi,\eta) - u_1(\xi,\eta))\Big), \tag{37} $$

$$ \frac{M(\beta)(s\mathscr{L}(v_1(\xi,\eta)) - v_1(\xi,0))}{(s+\beta(1-s))} = \mathscr{L}\left(v_1\xi\xi(\xi,\eta) + u_1\xi,\eta)v_1^2(\xi,\eta) - \kappa v_1(\xi,\eta)\right), \tag{38} $$

$$ \frac{M(\beta)(s\mathscr{L}(u_2(\xi,\eta)) - u_2(\xi,0))}{(s+\beta(1-s))} = \mathscr{L}\Big(u_{2,\xi\xi}(\xi,\eta) - u_2(\xi,\eta)v_2^2(\xi,\eta) + \nu(u_1(\xi,\eta) - u_2(\xi,\eta))\Big), \tag{39} $$

$$ \frac{M(\beta)(s\mathscr{L}(v_2(\xi,\eta)) - v_2(\xi,0))}{(s+\beta(1-s))} = \mathscr{L}\left(v_{2,\xi\xi}(\xi,\eta) + u_2\xi,\eta)v_2^2(\xi,\eta)\right). \tag{40} $$

As in Sect. 2.1, we follow the same procedure and we obtain

$$ u_{i,m} = \mathscr{X}_m u_{1,(m-1)} + \hbar\mathscr{L}^{-1}R_{u_i}^{CFC}, \tag{41} $$

$$v_{i,m} = \mathscr{X}_m v_{1,(m-1)} + \hbar \mathscr{L}^{-1} R_{v_i}^{CFC}, \tag{42}$$

where $R_{u_i}^{CF}$ and $R_{v_i}^{CF}$ are given by (21)–(24) with $\Omega^{CF}(s,\beta) = -\frac{\beta(1-s)+s}{sM(\beta)}$.

2.3 Application of Atangana–Baleanu Fractional-Order Derivative on TFRDS

By replacing ${}_0^C D_\eta^\beta(\cdot)$ into (8)–(11) by ${}_0^{AB} D_\eta^\beta(\cdot)$, we obtain the TFRDS with AB operator in LC sense [5]

$${}_0^{AB} D_\eta^\beta u_1(\xi,\eta) - u_{1,\xi\xi}(\xi,\eta) + u_1(\xi,\eta)v_1^2(\xi,\eta) - \nu(u_2(\xi,\eta) - u_1(\xi,\eta)) = 0, \tag{43}$$

$${}_0^{AB} D_\eta^\beta v_1(\xi,\eta) - v_{1,\xi\xi}(\xi,\eta) - u_1(\xi,\eta)v_1^2(\xi,\eta) + \kappa v_1(\xi,\eta) = 0, \tag{44}$$

$${}_0^{AB} D_\eta^\beta u_2(\xi,\eta) - u_{2,\xi\xi}(\xi,\eta) + u_2(\xi,\eta)v_2^2(\xi,\eta) - \nu(u_1(\xi,\eta) - u_2(\xi,\eta)) = 0, \tag{45}$$

$${}_0^{AB} D_\eta^\beta v_2(\xi,\eta) - v_{2,\xi\xi}(\xi,\eta) - u_2(\xi,\eta)v_2^2(\xi,\eta) = 0, \tag{46}$$

where ${}_0^{AB} D_\eta^\beta(.)$ is given by

$${}_0^{AB} D_\eta^\beta(\cdot) = \frac{M(\beta)}{1-\beta}\int_0^\eta E_\beta\left(\frac{-\beta(\eta-t)}{1-\beta}\right) D(\cdot)dt,$$

where $E_\beta(z) = \sum_{k=0}^{\infty} \frac{z^k}{\Gamma(\beta k+1)}$ is Mittag-Leffler function and $M(\beta)$ is defined above.

The definition of the Laplace transform for the AB fractional derivative is given by

$$\mathscr{L}\left[{}_0^{AB} D_t^\beta(\cdot)\right](s) = M(\beta)\left(\frac{s^\beta \mathscr{L}[u(\xi,\eta)](s) - s^{\beta-1}[u(\xi,0)]}{s^\beta(1-\beta)+\beta}\right), 0 < \beta \le 1. \tag{47}$$

The Eqs. (43)–(46) become after applying the Laplace transform (47) on both ends as follows

$$\frac{M(\beta)\left(\mathscr{L}(u_1(\xi,\eta))s^\beta - u_1(\xi,0)s^{\beta-1}\right)}{(1-\beta)\left(\frac{\beta}{1-\beta}+s^\beta\right)} = \mathscr{L}\Big(u_{1,\xi\xi}(\xi,\eta) - u_1(\xi,\eta)v_1^2(\xi,\eta) + \nu(u_2(\xi,\eta) - u_1(\xi,\eta))\Big), \tag{48}$$

$$\frac{M(\beta)\left(\mathscr{L}(v_1(\xi,\eta))s^{\beta}-v_1(\xi,0)s^{\beta-1}\right)}{(1-\beta)\left(\frac{\beta}{1-\beta}+s^{\beta}\right)}=\mathscr{L}\left(v_1\xi\xi(\xi,\eta)+u_1\xi,\eta)v_1^2(\xi,\eta)-\kappa v_1(\xi,\eta)\right), \tag{49}$$

$$\frac{M(\beta)\left(\mathscr{L}(u_2(\xi,\eta))s^{\beta}-u_2(\xi,0)s^{\beta-1}\right)}{(1-\beta)\left(\frac{\beta}{1-\beta}+s^{\beta}\right)}=\mathscr{L}\Big(u_{2,\xi\xi}(\xi,\eta)-u_2(\xi,\eta)v_2^2(\xi,\eta) + \nu(u_1(\xi,\eta)-u_2(\xi,\eta))\Big), \tag{50}$$

$$\frac{M(\beta)\left(\mathscr{L}(v_2(\xi,\eta))s^{\beta}-v_2(\xi,0)s^{\beta-1}\right)}{(1-\beta)\left(\frac{\beta}{1-\beta}+s^{\beta}\right)}=\mathscr{L}\left(v_{2,\xi\xi}(\xi,\eta)+u_2\xi,\eta)v_2^2(\xi,\eta)\right). \tag{51}$$

As in Sect. 2.1, we follow the same procedure and we obtain

$$u_{i,m}=\mathscr{X}_m u_{1,(m-1)}+\hbar\mathscr{L}^{-1}R_{u_i}^{CFC}, \tag{52}$$

$$v_{i,m}=\mathscr{X}_m v_{1,(m-1)}+\hbar\mathscr{L}^{-1}R_{v_i}^{CFC}, \tag{53}$$

where $R_{u_i}^{AB}$ and $R_{v_i}^{AB}$ are given by (21)–(24) with $\Omega^{AB}(s,\beta)=\frac{s^{-\beta-1}\left(\beta s^{\beta+1}-s^{\beta+1}-\beta s\right)}{M}$.

3 Numerical Results

In this section, we compute the first approximation for the three operators listed above. Also, we evaluate the interval of convergence by plotting the $\hbar$-curves and the optimal values of $\hbar$ with ARE. Furthermore, we evaluate the residual function for computing the accurate and efficient for LC, CF and AB operators, respectively. We set the initial approximation as

$$u_{i,0}(\xi,\eta)=u_{i,0}(\xi,0),\quad v_{i,0}(\xi,\eta)=v_{i,0}(\xi,0). \tag{54}$$

For $m=1$, we obtain the first approximation as follows

$$\begin{aligned}u_{1,1}(\xi,\eta)=\hbar_{u_1}\mathscr{L}^{-1}\Bigg(&\mathscr{L}\left(u_{1,0}(\xi,\eta)\right)-\frac{1}{s}u_1(\xi,0)\left(1-\frac{\mathscr{X}_1}{n}\right)\\&+\Omega^{(\cdot)}(s,\beta)\mathscr{L}\left(u_{1,0,\xi\xi}(\xi,\eta)-A_0\right)\Big)\\&+\nu(u_{2,0}(\xi,\eta)-u_{1,0}(\xi,\eta)),\end{aligned} \tag{55}$$

$$v_{1,1}(\xi,\eta)=\hbar_{v_1}\mathscr{L}^{-1}\Bigg(\mathscr{L}\left(v_{1,0}(\xi,\eta)\right)-\frac{1}{s}v(\xi,0)\left(1-\frac{\mathscr{X}_m}{n}\right)$$
$$+\,\Omega^{(\cdot)}(s,\beta)\mathscr{L}\left(v_{1,0,\xi\xi}(\xi,\eta)-\kappa v_{1,0}(\xi,\eta)+A_0\right)\Bigg), \tag{56}$$

$$u_{2,1}(\xi,\eta)=\hbar_{u_2}\mathscr{L}^{-1}\left(\mathscr{L}\left(u_{2,0}(\xi,\eta)\right)-\frac{1}{s}u_2(\xi,0)\left(1-\frac{\mathscr{X}_1}{n}\right)\right.$$
$$+\,\Omega^{(\cdot)}(s,\beta)\mathscr{L}\left(u_{2,0,\xi\xi}(\xi,\eta)-B_0\right))$$
$$+\,\nu(u_{1,0}(\xi,\eta)-u_{2,0}(\xi,\eta)), \tag{57}$$

$$v_{2,1}(\xi,\eta)=\hbar_{v_2}\mathscr{L}^{-1}\Bigg(\mathscr{L}\left(v_{2,0}(\xi,\eta)\right)-\frac{1}{s}v(\xi,0)\left(1-\frac{\mathscr{X}_m}{n}\right)$$
$$+\,\Omega^{(\cdot)}(s,\beta)\mathscr{L}\left(v_{2,0,\xi\xi}(\xi,\eta)+B_0\right)\Bigg). \tag{58}$$

Therefore, we obtain the first approximation for LC, CF and AB by taking $\Omega^{(\cdot)}(s,\beta) = \Omega^{LC}(s,\beta)$, $\Omega^{CF}(s,\beta)$ and $\Omega^{ABC}(s,\beta)$, respectively.

We can evaluate the rest of the approximations by the similar manner. We therefore have p-HATM solutions via three operators LC, CF and AB as

$$u_i(\xi,\eta)=u_{i,0}(\xi,\eta)+\sum_{j=1}^{m}\frac{u_{i,j}(\xi,\eta)}{n^j}, \tag{59}$$

$$v_i(\xi,\eta)=v_{i,0}(\xi,\eta)+\sum_{j=1}^{m}\frac{v_{i,j}(\xi,\eta)}{n^j}. \tag{60}$$

Figure 1 shows the numerical solutions for $u_{i,\eta}(\xi,0)$, $v_{i,\eta}(\xi,0)$ against $\hbar$, taking $n=5$, $\kappa=0.1$, $\nu=0.2$, $\ell=100$, $\xi=4$, $\eta=0$, $a_1=0.001$, $b_1=0.001$, $a_2=0.002$ and $b_2=0.002$. We plot the $\hbar$-curves of 4-terms of p-HATM solutions for (8)–(11), (32)–(35) and (43)–(46) with the aim to observe the intervals of convergence. It can be observed from this figure that the straight line that parallels to the $\hbar$-axis is in fact the interval of the convergence to be determined [23]. However, it does not give us the optimal value of $\hbar$ to make the convergence faster and more accurate, so estimate the residual error, then we minimize this error. The following steps illustrate this [33–42].

$$E_{u_i}(\hbar)=\frac{1}{(S+1)(J+1)}\sum_{s=0}^{S}\sum_{j=0}^{J}\left[\mathscr{N}_i\left(\sum_{k=0}^{m}u_{i,k}\left(\frac{10s}{S},\frac{10j}{J}\right)\right)\right]^2, \tag{61}$$

$$E_{v_i}(\hbar) = \frac{1}{(S+1)(J+1)} \sum_{s=0}^{S}\sum_{j=0}^{J}\left[\mathscr{M}_i\left(\sum_{k=0}^{m} v_{i,k}\left(\frac{10s}{S}, \frac{10j}{J}\right)\right)\right]^2, \tag{62}$$

corresponding to a nonlinear algebraic equations

$$\frac{dE_{u_i}(\hbar)}{d\hbar} = 0, \tag{63}$$

$$\frac{dE_{v_i}(\hbar)}{d\hbar} = 0. \tag{64}$$

The Eqs. (61)–(62) are called the average residual error (ARE). Figure 2 shows the ARE for the LC, CF and AB operators. This Figure shows the $E_{u_i}(\hbar)$ and $E_{v_i}(\hbar)$ for 4-terms obtained with p-HATM. We set into (61)–(62) $S = 10$ and $J = 10$ with $\kappa = 0.1$, $\nu = 0.2$, $\ell = 10$, $a_1 = 0.001$, $b_1 = 0.001$, $a_2 = 0.002$ and $b_2 = 0.002$. To get the optimal values of $\hbar$ we use the command "Minimize" of Mathematica for plotting the AVE against $\hbar$. From Fig. 2, we observe the ARE of order 10^{-6}–10^{-9}. This observation assures that the p-HATM solutions of LC, CF, and AB converge quickly. While an exact solution of the fractional time derivative system exists for $\beta = 1$, there is no such exact solution for $0 < \beta < 1$. So, we define the residual error

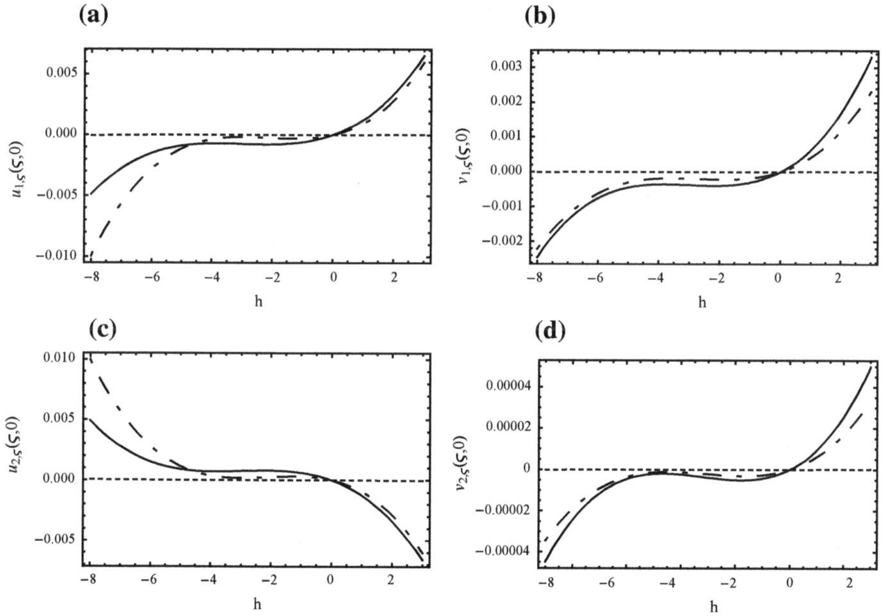

Fig. 1 $\hbar$-curves via LC (solid line), CF (dashed line) and AB (dashed-doted-dashed line) operators with $\beta = 0.5, n = 5, \xi = 4, \eta = 0, \kappa = 0.1, \nu = 0.2, \ell = 10, a_1 = 0.001, b_1 = 0.001, a_2 = 0.002, b_2 = 0.002$

function of thep-HATM solution as

$$REF_0^{(\cdot)} = D_\eta^\beta u_i - u_{i,\xi\xi} + u_i v_i^2 + (-1)^i(u_1 - u_2), \tag{65}$$

$$REF_0^{(\cdot)} = D_\eta^\beta v_i - v_{i,\xi\xi} - u_i v_i^2 + \kappa(2 - i)\kappa v_1, \tag{66}$$

where u_i and v_i are the p-HATM solution (59)–(60). Now, we substitute u_i and v_i from (59)–(60) into (65)–(66) and then we can plot the REF, as in Fig. 2. Figure 3 shows the residual error function for the p-HATM solution (59)–(60) for $\eta = 20, n = 5, \hbar = -3, \kappa = 0.1, \nu = 0.2, \ell = 100, a_1 = 0.001, b_1 = 0.001, a_2 = 0.002, b_2 = 0.002$. From this figure, it can be seen that the p-HATM gives accurate solutions of the fractional time derivatives LC, CF and AB.

Finally, we plot the p-HATM solutions for LC, CF and ABC for different values of β. Figures 4, 5, 6 and 7 show the behavior of the TFRDS with three operators for $\beta = 1, 0.6$, and 3. These figures show that the approximate solutions symmetry about $\xi = \ell/2$ and go down when β increases for u_1, u_2 and up for v_1, and v_2.

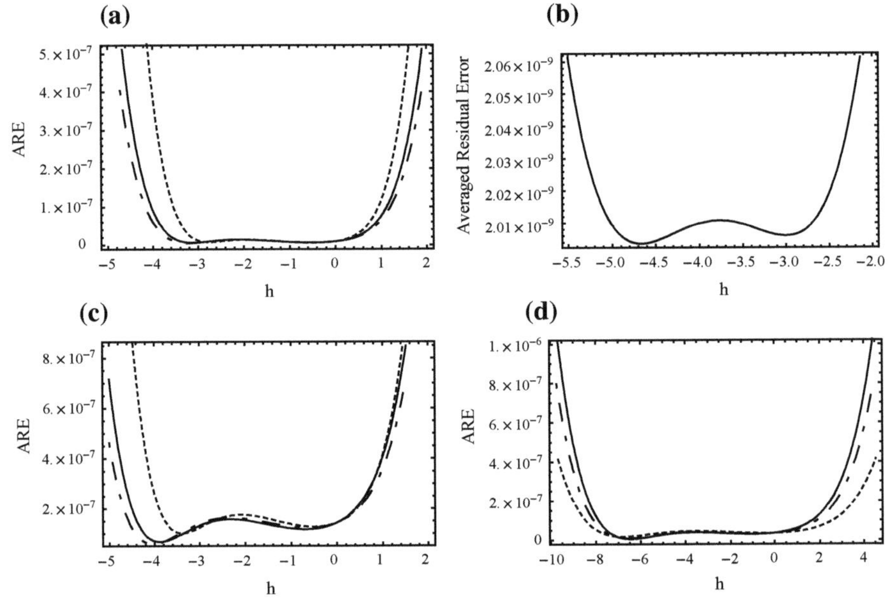

Fig. 2 Plotting the ARE for 4-terms of p-HATM solutions for LC (solid line), CF (dashed line) and AB (dashed-dot-dashed line) with $\beta = 0.6, n = 5, 0 \leq \eta \leq 10, 0 \leq x \leq 10, k = 0.1, \nu = 0.2, \ell = 10, a_1 = 0.001, b_1 = 0.001, a_2 = 0.002, b_2 = 0.002$

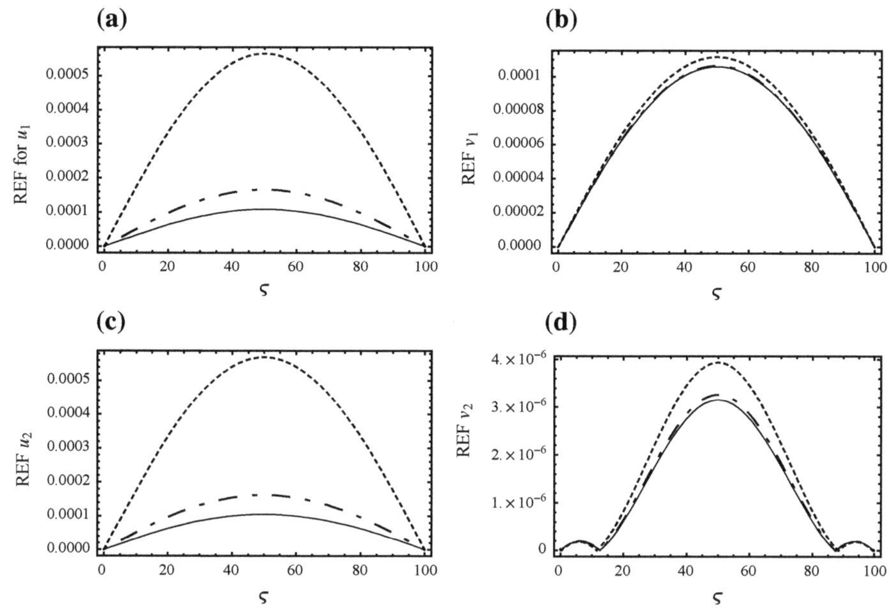

Fig. 3 The plotting of the REF (65)–(66) for LC (solid line), CF (dashed line) and AB (dashed-dot-dashed line) for $\eta = 20, n = 5, \hbar = -3, k = 0.1, \nu = 0.2, \ell = 100, a_1 = 0.001, b_1 = 0.001, a_2 = 0.002, b_2 = 0.002$

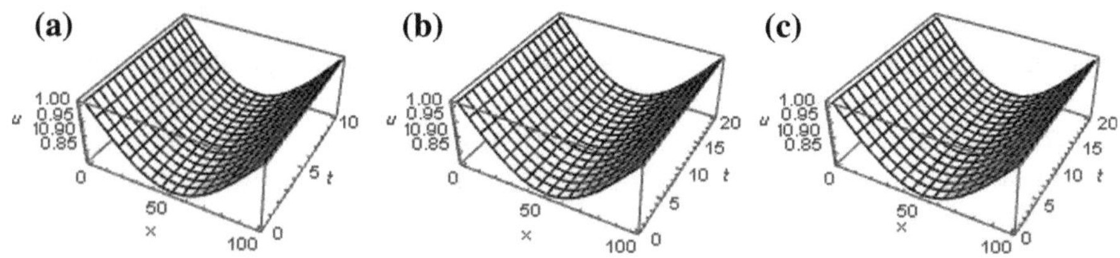

Fig. 4 The plot of 4-terms of p-HATM solutions for $u_1(\xi, \eta)$ with $\beta = 0.7, n = 5, k = 0.001, \nu = 0.2, \ell = 100, \hbar = -0.1, \ell = 100, a_1 = 0.1, b_1 = 0.09, a_2 = 0.2, b_2 = 0.1$. **a** LC, **b** CF and **c** AB

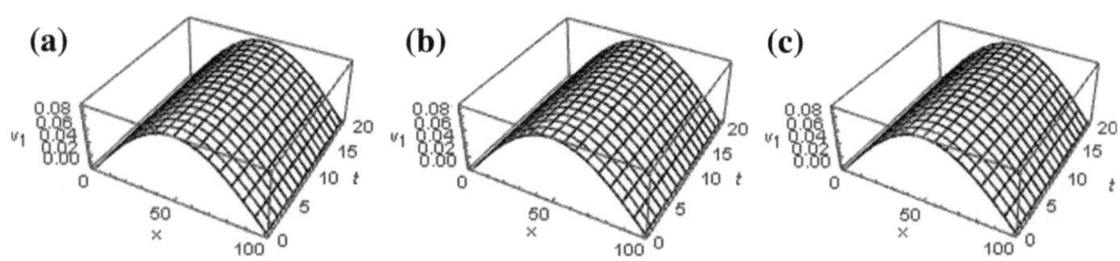

Fig. 5 The plot of 4-terms of p-HATM solutions for $v_1(\xi, \eta)$ with $\beta = 0.7, n = 5, k = 0.001, \nu = 0.2, \ell = 100, \hbar = -0.1, \ell = 100, a_1 = 0.1, b_1 = 0.09, a_2 = 0.2, b_2 = 0.1$. **a** LC, **b** CF and **c** AB

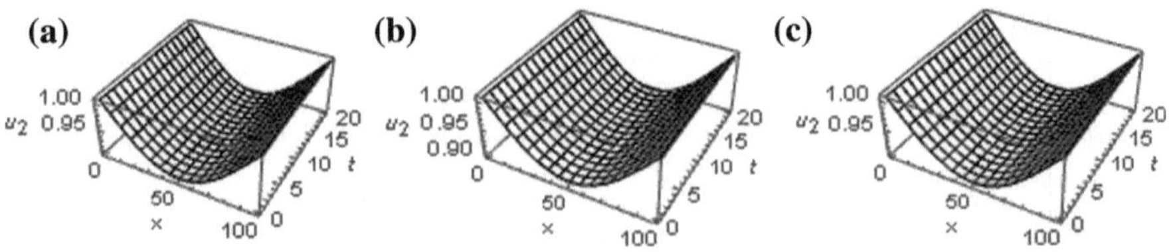

Fig. 6 The plot of 4-terms of p-HATM solutions for $u_2(\xi, \eta)$ with $\beta = 0.7, n = 5, k = 0.001, \nu = 0.2, \ell = 100, \hbar = -0.1, \ell = 100, a_1 = 0.1, b_1 = 0.09, a_2 = 0.2, b_2 = 0.1$. **a** LC, **b** CF and **c** AB

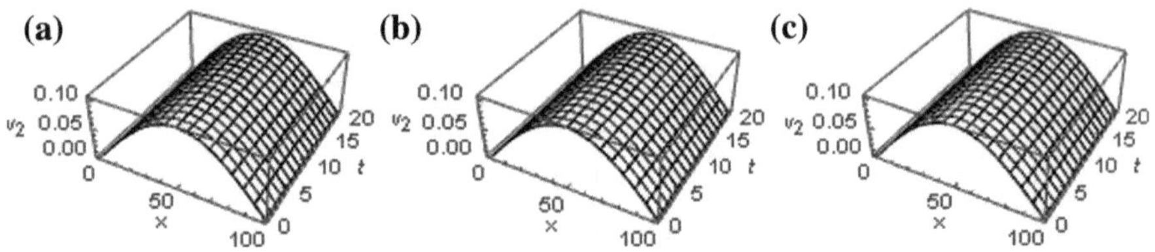

Fig. 7 The plot of 4-terms of p-HATM solutions for $v_2(\xi, \eta)$ with $\beta = 0.7, n = 5, k = 0.001, \nu = 0.2, \ell = 100, \hbar = -0.1, \ell = 100, a_1 = 0.1, b_1 = 0.09, a_2 = 0.2, b_2 = 0.1$. **a** LC, **b** CF and **c** AB

4 Conclusion

In this chapter, the p-HATM was used to compute the approximate solutions of TFCIACS using the LC, CF and AB operators. LC, CF and AB fractional-order derivatives present alternative solutions of the TFCIACS. We expanded the nonlinear terms using Adomian polynomial. We evaluated the order of the error using the ARE and REF and we found the small order. Also, the interval of the convergence of p-HATM and optimal value of $\hbar$ were compute. The numerical simulations showed that the proposed methodology is accurate.

Acknowledgements José Francisco Gómez Aguilar acknowledges the support provided by CONACyT: Cátedras CONACyT para jóvenes investigadores 2014.

References

1. Merkin, J.H., Needham, D.J., Scott, S.K.: Coupled reaction-diffusion waves in an isothermal autocatalytic chemical system. IMA J. Appl. Math. **50**, 43–76 (1993)
2. Podlubny, I.: An introduction to fractional derivatives, fractional differential equations, to methods of their solution and some of their applications. Fractional Differential Equations. Academic Press, San Diego (1999)
3. Caputo, M., Mainardi, F.: A new dissipation model based on memory mechanism. Pure Appl. Geophys. **91**, 134–147 (1971)
4. Caputo, M., Fabrizio, M.: A new definition of fractional derivative without singular Kernel. Progr. Fract. Differ. Appl. **1**(2), 73–85 (2015)

5. Atangana, A., Baleanu, D.: New fractional derivatives with nonlocal and non-singular Kernel. Theory and application to heat transfer model. Therm. Sci. **20**(2), 763–769 (2016)
6. Alsaedi, A., Baleanu, D., Etemad, S., Rezapour, S.: On coupled systems of time-fractional differential problems by using a new fractional derivative. J. Funct. Spaces **1**, 1–16 (2016)
7. Atangana, A., Nieto, J.J.: Numerical solution for the model of RLC circuit via the fractional derivative without singular kernel. Adv. Mech. Eng. **7**, 1–7 (2015)
8. Atangana, A.: On the new fractional derivative and application to nonlinear fishers reaction diffusion equation. App. Math. Comput. **273**, 948–956 (2016)
9. Atangana, A., Badr, S.T.A.: Analysis of the Keller-Segel model with a fractional derivative without singular Kernel. Entropy **17**, 4439–4453 (2015)
10. Atangana, A., Badr, S.T.A.: Extension of the RLC electrical circuit to fractional derivative without singular Kernel. Adv. Mech. Eng. **7**(6), 1–6 (2015)
11. Atangana, A., Badr, S.T.A.: New model of groundwater flowing within a confine aquifer: application of Caputo-Fabrizio derivative. Arab. J. Geosci. **9**(1), 1–8 (2016)
12. Djida, J.D., Atangana, A., Area, I.: Numerical computation of a fractional derivative with non-local and non-singular Kernel. Math. Model. Nat. Phenom. **12**(3), 4–13 (2017)
13. Koca, I.: Analysis of Rubella disease model with non-local and non-singular fractional derivatives. Int. J. Optim. Control. Theor. Appl. (IJOCTA) **8**(1), 17–25 (2017)
14. Abro, K.A., Hussain, M., Baig, M.M.: An analytic study of molybdenum disulfide nanofluids using the modern approach of Atangana-Baleanu fractional derivatives. Eur. Phys. J. Plus **132**(10), 1–14 (2017)
15. Khan, A., Abro, K.A., Tassaddiq, A., Khan, I.: Atangana-Baleanu and Caputo-Fabrizio analysis of fractional derivatives for heat and mass transfer of second grade fluids over a vertical plate: a comparative study. Entropy **19**(8), 1–12 (2017)
16. Alkahtani, B.S.T., Atangana, A., Koca, I.: Novel analysis of the fractional Zika model using the Adams type predictor-corrector rule for non-singular and non-local fractional operators. J. Nonlinear Sci. Appl. **10**(6), 3191–3200 (2017)
17. Singh, J., Kumar, D., Baleanu, D.: On the analysis of chemical kinetics system pertaining to a fractional derivative with Mittag-Leffler type Kernel. Chaos Interdiscip. J. Nonlinear Sci. **27**(10), 1–13 (2017)
18. Kumar, D., Singh, J., Baleanu, D.: Analysis of regularized long-wave equation associated with a new fractional operator with Mittag-Leffler type Kernel. Phys. A Stat. Mech. Appl. **492**, 155–167 (2018)
19. El-Tawil, M.A., Huseen, S.N.: The q-homotopy analysis method (q-ham). Int. J. Appl. Math. Mech. **8**(15), 51–75 (2012)
20. Huseen, S.N., Grace, S.R., El-Tawil, M.A.: The optimal q-homotopy analysis method (Oq-HAM). Int. J. Comput. Technol. **11**(8), 2859–2866 (2013)
21. Iyiola, O.S.: q-Homotopy analysis method and application to fingero-imbibition phenomena in double phase flow through porous media. Asian J. Curr. Eng. Math. **2**(4), 283–286 (2013)
22. Liao, S.-J.: The proposed homotopy analysis technique for the solution of nonlinear problems. Ph.D. thesis, Shanghai Jiao Tong University (1992)
23. Liao, S.-J.: Beyond Perturbation: Introduction to the Homotopy Analysis Method. Chapman and Hall/CRC Press, Boca Raton (2003)
24. Liao, S.-J.: On the homotopy analysis method for nonlinear problems. Appl. Math. Comput. **147**, 499–513 (2004)
25. Liao, S.-J.: Comparison between the homotopy analysis method and homotopy perturbation method. Appl. Math. Comput. **169**, 1186–1194 (2005)
26. Mohamed, M.S., Hamed, Y.S.: Solving the convection-diffusion equation by means of the optimal q-homotopy analysis method (Oq-HAM). Results Phys. **6**, 20–25 (2016)
27. Saad, K.M., AL-Shomrani, A.A.: An application of homotopy analysis transform method for Riccati differential equation of fractional order. J. Fract. Calc. Appl. **7**, 61–72 (2016)
28. Kumar, D., Singh, J., Baleanu, D.: A new analysis for fractional model of regularized long-wave equation arising in ion acoustic plasma waves. Math. Methods Appl. Sci. **40**(15), 5642–5653 (2017)

29. Saad, K.M., AL-Shareef, E.H., Mohamed, M.S., Yang, X.J.: Optimal q-homotopy analysis method for time-space fractional gas dynamics equation. Eur. Phys. J. Plus **132**(1), 1–11 (2017)
30. Khan, Y., Austin, F.: Application of the Laplace decomposition method to nonlinear homogeneous and non-homogenous advection equations. Z. Nat. Sect. A **65**, 849–853 (2010)
31. Elsaid, A.: Fractional differential transform method combined with the Adomian polynomials. Appl. Math. Comput. **218**, 6899–6911 (2012)
32. Yin, X.B., Kumar, S., Kumar, D.: A modified homotopy analysis method for solution of fractional wave equations. Adv. Fract. Dyn. Mech. Eng. **7**(12), 1–8 (2015)
33. Abbasbandy, S., Jalili, M.: Determination of optimal convergence-control parameter value in homotopy analysis method. Numer. Algorithms **64**(4), 593–605 (2013)
34. Abbasbandy, S., Shivanian, E.: Predictor homotopy analysis method and its application to some nonlinear problems. Commun. Nonlinear Sci. Numer. Simulat. **16**, 2456–2468 (2011)
35. Abo-Dahab, S.M., Mohamed, M.S., Nofal, T.A.: A one step optimal homotopy analysis method for propagation of harmonic waves in nonlinear generalized magnetothermoelasticity with two relaxation times under influence of rotation. Abstract and Applied Analysis, vol. 1, pp. 1–15. Hindawi Publishing Corporation, Cairo (2013)
36. Ghanbari, M., Abbasbandy, S., Allahviranloo, T.: A new approach to determine the convergence-control parameter in the application of the homotopy analysis method to systems of linear equations. Appl. Comput. Math. **12**(3), 355–364 (2013)
37. Gondal, M.A., Arife, A.S., Khan, M., Hussain, I.: An efficient numerical method for solving linear and nonlinear partial differential equations by combining homotopy analysis and transform method. World Appl. Sci. J. **14**(12), 1786–1791 (2011)
38. Liao, S.-J.: An optimal homotopy-analysis approach for strongly nonlinear differential equations. Commun. Nonlinear Sci. Numer. Simulat. **15**(8), 2003–2016 (2010)
39. Singh, J., Kumar, D., Swroop, R.: Numerical solution of time- and space-fractional coupled burgers equations via homotopy algorithm. Alex. Eng. J. **55**(2), 1753–1763 (2016)
40. Singh, J., Kumar, D., Swroop, R., Kumar, S.: An efficient computational approach for time-fractional Rosenau-Hyman equation. Neural Comput. Appl. **45**, 192–204 (2017)
41. Srivastava, H.M., Kumar, D., Singh, J.: An efficient analytical technique for fractional model of vibration equation. Appl. Math. Model. **45**, 192–204 (2017)
42. Yamashita, M., Yabushita, K., Tsuboi, K.: An analytic solution of projectile motion with the quadratic resistance law using the homotopy analysis method. J. Phys. A Math. Gen. **40**, 8403–8416 (2007)

Upwind-Based Numerical Approximation of a Space-Time Fractional Advection-Dispersion Equation for Groundwater Transport Within Fractured Systems

A. Allwright and A. Atangana

Abstract Modelling groundwater transport in fractured aquifer systems is complex due to the uncertainty associated with delineating the specific fractures along which water and potential contaminants could be transported. The resulting uncertainty in modelled contaminant movement has implications for the protection of the environment, where inadequate mitigation or remediation measures could be employed. To improve the governing equation for groundwater transport modelling, the Atangana–Baleanu in Caputo sense (ABC) fractional derivative is applied to the advection-dispersion equation with a focus on the advection term to account for anomalous advection. Boundedness, existence and uniqueness for the developed advection-focused transport equation is presented. In addition, a semi-discretisation analysis is performed to demonstrate the equation stability in time. Augmented upwind schemes are investigated as they have been found to address stability problems when solute transport is advection-dominated. The upwind-based schemes are developed, and stability analysis conducted, to facilitate the solution of the complex equation. The numerical stability analysis found the upwind Crank–Nicolson to be the most stable, and is thus recommended for use with the ABC fractional advection-dispersion equation.

Keywords Fractional calculus · Atangana–Baleanu fractional derivative · Fractional advection-dispersion equation

---

A. Allwright · A. Atangana (✉)
Faculty of Agricultural and Natural Sciences, Institute for Groundwater Studies, University of the Free State, Bloemfontein 9301, Free State, South Africa
e-mail: AtanganaA@ufs.ac.za

J. F. Gómez et al. (eds.), *Fractional Derivatives with Mittag-Leffler Kernel*, Studies in Systems, Decision and Control 194,
https://doi.org/10.1007/978-3-030-11662-0_18

1 Introduction

Real-world systems are complex. By definition, a complexity by which any one method is not able to capture all the nuances of that system. It is this that compels science to improve and continuously strive for new methods and approaches, ever endeavouring to reconcile the difference between modelled and observed.

Simulating the transport of particles using the advection-dispersion equation is a real-world problem, where the general discrepancy between modelled and observed is particularly large. This discrepancy lead to the development of the term anomalous diffusion (non-Fickian diffusion), especially when using linear or traditional methods. For this reason, numerous nonlocal approaches have been applied to the advection-dispersion equation to reduce this divergence, ranging from multiple-rate mass transfer method and rate-limited mass transfer, stochastic averaging, continuous-time random walk, to fractal and fractional differential equations [1–9].

Complexity from the perspective of fractional calculus is explored in [10], where fractional differential equations are one method to improve the simulation of real world problems. Fractional calculus is not a new topic, having its original inception in the late 1600s, but the application of fractional derivations to practical problems has steadily increased since the 1970s. With the endeavour to continually improve simulation methods, a progression of fractional derivative definitions have been developed over the years, with definitions including Riemann-Liouville, Caputo, Caputo-Fabrizio, and the latest Atangana–Baleanu [11–21].

The newest fractional derivative definition Atangana–Baleanu, is used to develop an advection-focused fractional transport equation. The suitability of this particular formulation and fractional derivative definition is investigated for the specific real-world system of groundwater transport within fractured aquifers. Modelling groundwater transport in fractured aquifer systems is complex due to the uncertainty associated with demarcating the specific fractures along which water and potential contaminants could be transported along. A result in this uncertainty is the misrepresentation of the expected movement of a potential contaminant in the groundwater system. This potentially increases the impact on the environment because the misrepresented transport could result in inadequate mitigation or remediation measures [22–28].

A faster than expected transport along unknown fractures is referred to as super-advection associated with anomalous advection by [29], and a fractional derivative was applied to the advection term of the advection-dispersion equation to better simulate this phenomena. A fractional derivative was also applied to the time component of the advection-dispersion equation to activate the waiting-time distribution properties as discussed by [30, 31]. A similar approach is followed in developing the Atangana–Baleanu in Caputo sense (ABC) fractional advection-dispersion equation.

2 Advection-Focused Space-Time Fractional Transport Equation with Atangana–Baleanu in Caputo Sense (ABC) Derivative

The one-dimensional, non-reactive fractional advection-dispersion equation with the ABC fractional derivative definition is given by

$$ {}_{0}^{ABC}D_{t}^{\alpha}\left(c\left(x,t\right)\right)=-v_{0}^{ABC}D_{x}^{\alpha}\left(c\left(x,t\right)\right)+D_{L}\frac{\partial^{2}}{\partial x^{2}}\left(c\left(x,t\right)\right). \tag{1} $$

The boundedness, existence and uniqueness of the ABC fractional advection-dispersion equation is first determined, using the Picard–Lindelöf theorem, before the numerical approximation in the following sections.

2.1 *Picard–Lindelöf Theorem for Existence and Uniqueness*

Applying the AB integral to both sides of the ABC fractional advection-dispersion equation, we get

$$ \begin{aligned} c\left(x,t\right)-c\left(x,0\right)&={}_{0}^{AB}I_{x}^{\alpha}\left(-v_{0}^{ABC}D_{x}^{\alpha}\left(c\left(x,\tau\right)\right)+D_{L}\frac{\partial^{2}}{\partial x^{2}}\left(c\left(x,\tau\right)\right)\right)d\tau\\ &=\frac{1-\alpha}{AB\left(\alpha\right)}\left(-v_{0}^{ABC}D_{x}^{\alpha}\left(c\left(x,\tau\right)\right)+D_{L}\frac{\partial^{2}}{\partial x^{2}}\left(c\left(x,\tau\right)\right)\right)\\ +\frac{\alpha}{AB\left(\alpha\right)\Gamma\left(\alpha\right)}&\int_{0}^{t}\left(-v_{0}^{ABC}D_{x}^{\alpha}\left(c\left(x,\tau\right)\right)+D_{L}\frac{\partial^{2}}{\partial x^{2}}\left(c\left(x,\tau\right)\right)\right)\left(t-\tau\right)^{\alpha-1}d\tau. \end{aligned} \tag{2} $$

Now, we consider a new function $F\left(x,t,c\right)$ to simplify:

$$ F\left(x,t,c\right)=-v_{0}^{ABC}D_{x}^{\alpha}\left(c\left(x,t\right)\right)+D_{L}\frac{\partial^{2}}{\partial x^{2}}\left(c\left(x,t\right)\right). \tag{3} $$

Thus

$$ c\left(x,t\right)-c\left(x,0\right)=\frac{1-\alpha}{AB\left(\alpha\right)}F\left(x,t,c\right)+\frac{\alpha}{AB\left(\alpha\right)\Gamma\left(\alpha\right)}\int_{0}^{t}F\left(x,t,c\right)\left(t-\tau\right)^{\alpha-1}d\tau. \tag{4} $$

Let

$$ C_{\lambda,\beta}=I_{\lambda}\left(t_{0}\right)\times B_{\beta}\left(x_{0}\right), $$

where

$$I_\lambda(t_0) = [t_0 - \lambda, t_0 + \lambda],$$

$$B_\beta(x_0) = [x_0 - \beta, x_0 + \beta].$$

The Banach fixed-point theorem is applied by introducing the norm of the supremum (statistical limit of a set) for I_λ,

$$M = \|\varphi\|_\infty = \begin{matrix} sup|\varphi(t)| \\ t \in I_\lambda(t_0). \end{matrix} \quad (5)$$

Considering the practical meaning of $c(x,t)$, it can be assumed that the initial concentration (c_0) will always be greater than subsequent concentrations (c_n) due to advection, dispersion and diffusion processes which reduce the concentration over time and space,

$$\|c\|_\infty < c_0. \quad (6)$$

Considering the max norm for the function $F(x,t,c)$,

$$\|F\|_\infty = \left\| -v\frac{AB(\alpha)}{(1-\alpha)} \int_0^x \frac{d}{d\tau} c(\tau,x) E_\alpha\left[-\frac{\alpha}{1-\alpha}(x-\tau)^\alpha\right] d\tau + D_L \frac{\partial^2}{\partial x^2}(c(x,t)) \right\|_\infty. \quad (7)$$

Thus

$$\|F\|_\infty \le v\frac{AB(\alpha)}{(1-\alpha)} \left\| \int_0^x \frac{d}{d\tau} c(\tau,x) E_\alpha\left[-\frac{\alpha}{1-\alpha}(x-\tau)^\alpha\right] d\tau \right\|_\infty + D_L \left\| \frac{\partial^2}{\partial x^2}(c(x,t)) \right\|_\infty. \quad (8)$$

Applying the proven theorem for partial differential Lipschitz condition in [9], the second order derivative is bounded (M_1), thus

$$\|F\|_\infty \le v\frac{AB(\alpha)}{(1-\alpha)} \int_0^x \left\| \frac{d}{d\tau} C(\tau,x) \right\|_\infty \left\| E_\alpha\left[-\frac{\alpha}{1-\alpha}(x-\tau)^\alpha\right] \right\|_\infty d\tau + D_L M_1. \quad (9)$$

The Mittag-Leffler function is bounded because $1 > \alpha > 0$. The first-order derivative is bounded due to its physical meaning being related to the spread of a particle defined by its concentration (M_2). Thus, the derivative is considered at the maximum physical time that is applicable to the existence of the concentration (T_{max}),

$$\|F\|_\infty \le v\frac{AB(\alpha)}{(1-\alpha)} M_2 T_{max} + D_L M_1 < \infty. \quad (10)$$

Therefore, the solution is bounded because we obtain a positive constant, such that

$$A = \|F\|_\infty = \begin{matrix} sup|F(x,t,c)| \\ t \in C_{\lambda,\beta}. \end{matrix} \tag{11}$$

Let $C_{\lambda,\beta}$ be a set where, $F : C_{\lambda,\beta} \to C_{\lambda,\beta}$, such that

$$\Gamma\phi(x,t) = c(x,0) + \frac{1-\alpha}{AB(\alpha)}F(x,t,\phi) + \frac{\alpha}{AB(\alpha)\Gamma(\alpha)}\int_0^t F(x,\tau,\phi)(t-\tau)^{\alpha-1}d\tau. \tag{12}$$

Thus

$$\begin{aligned} \|\Gamma\phi(x,t) - c(x,0)\|_\infty &= \left\| \frac{1-\alpha}{AB(\alpha)}F(x,t,\phi) \right. \\ &\left. + \frac{\alpha}{AB(\alpha)\Gamma(\alpha)}\int_0^t F(x,\tau,\phi)(t-\tau)^{\alpha-1}d\tau \right\|_\infty \|\Gamma\phi(x,t) - c(x,0)\|_\infty \\ &\le \frac{1-\alpha}{AB(\alpha)}\|F(x,t,\phi)\|_\infty + \frac{\alpha}{AB(\alpha)\Gamma(\alpha)}\int_0^t (t-\tau)^{\alpha-1}\|F(x,\tau,\phi)\|_\infty d\tau. \end{aligned} \tag{13}$$

The function $F(x,t,c)$ has been shown to be bounded (Eqs. (7)–(11))

$$\left[\|\Gamma\phi(x,t) - c(x,0)\|_\infty \le \frac{1-\alpha}{AB(\alpha)}A + \frac{\alpha}{AB(\alpha)\Gamma(\alpha)}A\int_0^t (t-\tau)^{\alpha-1}d\tau\right].$$

The integral is considered at the maximum physical time that is applicable to the existence of the concentration (T_{max})

$$\begin{aligned} \left[\|\Gamma\phi(x,t) - c(x,0)\|_\infty \le \frac{1-\alpha}{AB(\alpha)}A + \frac{\alpha}{AB(\alpha)\Gamma(\alpha)}A\frac{T_{max}^\alpha}{\alpha}\right] \\ \le \frac{1-\alpha}{AB(\alpha)}A + \frac{AT_{max}^\alpha}{AB(\alpha)\Gamma(\alpha)} < \infty. \end{aligned}$$

Therefore Γ is well-posed because we obtain a positive constant.
We want to prove that Γ is Lipschitz

$$\begin{aligned} \|\Gamma\phi_1 - \Gamma\phi_2\|_\infty &= \left\| \frac{1-\alpha}{AB(\alpha)}(F(x,t,\phi_1) - F(x,t,\phi_2)) \right. \\ &\left. + \frac{\alpha}{AB(\alpha)\Gamma(\alpha)}\int_0^t (F(x,\tau,\phi_1) - F(x,\tau,\phi_2))(t-\tau)^{\alpha-1}d\tau \right\|_\infty \|\Gamma\phi_1 - \Gamma\phi_2\|_\infty \\ &\le \frac{1-\alpha}{AB(\alpha)}\|(F(x,t,\phi_1) - F(x,t,\phi_2))\|_\infty \\ &+ \frac{\alpha}{AB(\alpha)\Gamma(\alpha)}\int_0^t (t-\tau)^{\alpha-1}\|F(x,\tau,\phi_1) - F(x,\tau,\phi_2)\|_\infty d\tau. \end{aligned} \tag{14}$$

To achieve this, we first evaluate

$$\|F(x,t,\phi_1) - F(x,t,\phi_2)\|_\infty = \left\| -v_0^{ABC} D_x^\alpha (\phi_1-\phi_2) + D_L \frac{\partial^2}{\partial x^2}(\phi_1-\phi_2) \right\|_\infty \|F(x,t,\phi_1) - F(x,t,\phi_2)\|_\infty \leq v \|_0^{ABC} D_x^\alpha (\phi_1-\phi_2)\|_\infty + D_L \left\| \frac{\partial^2}{\partial x^2}(\phi_1-\phi_2) \right\|_\infty . \tag{15}$$

Applying the ABC fractional derivative, and the proven theorem for partial differential Lipschitz condition in [9], the second order derivative is bounded (ρ_2^2)

$$\|F(x,t,\phi_1) - F(x,t,\phi_2)\|_\infty \leq v \frac{AB(\alpha)}{(1-\alpha)} \int_0^x \left\| \frac{d}{d\tau}(\phi_1-\phi_2) \right\|_\infty \left\| E_\alpha \left[-\frac{\alpha}{1-\alpha}(x-\tau)^\alpha \right] \right\|_\infty d\tau + D_L \rho^2 \|(\phi_1-\phi_2)\|_\infty . \tag{16}$$

Similarly as in Eq. (9), the Mittag-Leffler function is bounded due to the constrain $1 > \alpha > 0$, and the first order derivative is bounded as explained (ρ_1). Thus, the derivative is considered at the maximum physical space that is applicable to the concentration (X_{max})

$$\|F(x,t,\phi_1) - F(x,t,\phi_2)\|_\infty \leq v \frac{AB(\alpha)}{(1-\alpha)} X_{max} \rho_1 \|(\phi_1-\phi_2)\|_\infty + D_L \rho_2^2 \|(\phi_1-\phi_2)\|_\infty . \tag{17}$$

Simplifying

$$\|F(x,t,\phi_1) - F(x,t,\phi_2)\|_\infty \leq \left(v \frac{AB(\alpha)}{(1-\alpha)} X_{max} \rho_1 + D_L \rho_2^2 \right) \|(\phi_1-\phi_2)\|_\infty < K_\alpha \|(\phi_1-\phi_2)\|_\infty . \tag{18}$$

Applying to Eq. (14), we get

$$\|\Gamma\phi_1 - \Gamma\phi_2\|_\infty \leq \frac{1-\alpha}{AB(\alpha)} K_\alpha \|(\phi_1-\phi_2)\|_\infty + \frac{\alpha}{AB(\alpha)\Gamma(\alpha)} K_\alpha \|(\phi_1-\phi_2)\|_\infty \int_0^t (t-\tau)^{\alpha-1} d\tau . \tag{19}$$

Applying a similar process as previously, we obtain

$$\begin{aligned}\|\Gamma\phi_1 - \Gamma\phi_2\|_\infty &\le \frac{1-\alpha}{AB(\alpha)} K_\alpha \| (\phi_1 - \phi_2) \|_\infty + \frac{\alpha}{AB(\alpha)\,\Gamma(\alpha)} K_\alpha \| (\phi_1 - \phi_2) \|_\infty \frac{T_{max}^\alpha}{\alpha} \\ &\le \left(\frac{1-\alpha}{AB(\alpha)} K_\alpha + \frac{T_{max}^\alpha}{AB(\alpha)\,\Gamma(\alpha)} K_\alpha \right) \| (\phi_1 - \phi_2) \|_\infty \le V \| (\phi_1 - \phi_2) \|_\infty . \end{aligned} \tag{20}$$

Therefore, Γ is a contraction when $V < 1$, which translates to a condition

$$K_\alpha < \frac{1}{\frac{1-\alpha}{AB(\alpha)} + \frac{T_{max}^\alpha}{AB(\alpha)\Gamma(\alpha)}}. \tag{21}$$

Then $F(x, t, c)$ has a fixed point using the Banach fixed-point theorem and the ABC fractional advection-dispersion equation is bounded and has a unique solution under this condition.

2.2 *Semi-discretisation Stability*

The stability of the defined fractional advection-dispersion equation is evaluated in time, and thus discretised in time while the concentration in space is considered constant. The forward finite difference approximation in time is applied to the ABC fractional derivative, considered for a specific time (t_n), and the numerical integration of the Mittag-Leffler function is performed in [32]

$$ {}_0^{ABC}D_t^\alpha \left(c\,(x, t_n) \right) = \frac{AB(\alpha)}{(1-\alpha)} \sum_{k=0}^{n-1} \left(c_x^{k+1} - c_x^k \right) \delta_{n,k}^\alpha ,$$

where,

$$\delta_{n,k}^\alpha = (n-k)\, E_{\alpha,2} \left[-\frac{\alpha \Delta t}{1-\alpha} (n-k) \right] - (n-k-1)\, E_{\alpha,2} \left[-\frac{\alpha \Delta t}{1-\alpha} (n-k-1) \right]. \tag{22}$$

Now, substituting back into fractional advection-dispersion equation with the ABC derivative, and applying the assumption of discretisation in time only

$$\frac{AB(\alpha)}{(1-\alpha)} \sum_{k=0}^{n-1} \left(c_x^{k+1} - c_x^k \right) \delta_{n,k}^\alpha = -v\,{}_0^{ABC}D_x^\alpha \left(c_x^n \right) + D_L \frac{\partial^2}{\partial x^2} \left(c_x^{n+1} \right). \tag{23}$$

A function $A^{\alpha}_{n,k}$ is applied to simplify

$$\sum_{k=0}^{n-1}\left(c_x^{k+1}-c_x^{k}\right)A^{\alpha}_{n,k}=-v_0^{ABC}D_x^{\alpha}\left(c_x^{n}\right)+D_L\frac{\partial^2}{\partial x^2}\left(c_x^{n+1}\right). \tag{24}$$

Reformulating to obtain,

$$\left(c_x^{n+1}-c_x^{n}\right)A_n^{\alpha}+\sum_{k=0}^{n-1}\left(c_x^{k+1}-c_x^{k}\right)A^{\alpha}_{n,k}=-v_0^{ABC}D_x^{\alpha}\left(c_x^{n}\right)+D_L\frac{\partial^2}{\partial x^2}\left(c_x^{n+1}\right). \tag{25}$$

Rearranging

$$c_x^{n+1}=c_x^{n}-\frac{v}{A^{\alpha}_{n}}{}_0^{ABC}D_x^{\alpha}\left(c_x^{n}\right)+\frac{D_L}{A_n^{\alpha}}\frac{\partial^2}{\partial x^2}\left(c_x^{n+1}\right)-\frac{1}{A_n^{\alpha}}\sum_{k=0}^{n-1}\left(c_x^{k+1}-c_x^{k}\right)A^{\alpha}_{n,k}. \tag{26}$$

Equation (26) is the numerical approximation of the ABC fractional advection-dispersion equation with respect to time. Now, the semi-stability can be evaluated defining the following norms

$$(f,g)=\int_{\Omega}(f\cdot g)(x)\,dx,$$

where,

$$\|g\|_0=\sqrt{(g\cdot g)},$$

$$\|g\|_1=\sqrt{\|g\|_0+\varepsilon\|\frac{d^2}{dx^2}g\|_0}.$$

When $n=0$, Eq. (26) becomes

$$c_x^{1}=c_x^{0}-\frac{v}{A^{\alpha}_{n}}{}_0^{ABC}D_x^{\alpha}\left(c_x^{0}\right)+\frac{D_L}{A_n^{\alpha}}\frac{\partial^2}{\partial x^2}\left(c_x^{1}\right). \tag{27}$$

Simplifying using functions λ_1 and λ_2,

$$c_x^{1}=c_x^{0}-\lambda_{1}{}_0^{ABC}D_x^{\alpha}\left(c_x^{0}\right)+\lambda_2\frac{\partial^2}{\partial x^2}\left(c_x^{1}\right). \tag{28}$$

Applying the norm with respect to g,

$$\left(c_x^{1},g\right)=\left(c_x^{0},g\right)-\lambda_1\left({}_0^{ABC}D_x^{\alpha}c_x^{0},{}_0^{ABC}D_x^{\alpha}g\right)+\lambda_2\left(\frac{\partial^2}{\partial x^2}c_x^{1},\frac{\partial^2}{\partial x^2}g\right). \tag{29}$$

Let $\forall g \in H^1(\Omega), \ g = c_x^1$

$$\left(c_x^1, c_x^1\right) = \left(c_x^0, c_x^1\right) - \lambda_1 \left({}_0^{ABC}D_x^\alpha c_x^0, {}_0^{ABC} D_x^\alpha c_x^1\right) + \lambda_2 \left(\frac{\partial^2}{\partial x^2} c_x^1, \frac{\partial^2}{\partial x^2} c_x^1\right). \tag{30}$$

From the defined norms, the following statement is to be proven,

$$\|c_x^1\|_1 \leq \|c_x^0\|_0.$$

Reformulating in terms of the defined norms

$$\begin{aligned}\left(c_x^1, c_x^1\right) - \lambda_2 \left(\frac{\partial^2 c_x^1}{\partial x^2}, \frac{\partial^2 c_x^1}{\partial x^2}\right) &= \left(c_x^0, c_x^1\right) - \lambda_1 \left({}_0^{ABC}D_x^\alpha c_x^0, {}_0^{ABC} D_x^\alpha c_x^1\right) \|c_x^1\|_1^2 \\ &= \|c_x^0\|_0 \|c_x^1\|_0 - \lambda_1 \|{}_0^{ABC}D_x^\alpha c_x^0\|_0 \|{}_0^{ABC}D_x^\alpha c_x^1\|_0,\end{aligned} \tag{31}$$

where,

$$\left[\left\|{}_0^{ABC}D_x^\alpha c_x^0\right\|_0 = \left\| \frac{AB(\alpha)}{(1-\alpha)} \int_0^x \frac{dc_\tau^0}{d\tau} E_\alpha \left[-\frac{\alpha}{1-\alpha}(x-\tau)^\alpha\right] d\tau \right\|_0\right].$$

Thus,

$$\left\|{}_0^{ABC}D_x^\alpha c_x^0\right\|_0 \leq \frac{AB(\alpha)}{(1-\alpha)} \int_0^x \left\|\frac{dc_\tau^0}{d\tau}\right\|_0 \left\|E_\alpha \left[-\frac{\alpha}{1-\alpha}(x-\tau)^\alpha\right]\right\|_0 d\tau. \tag{32}$$

As before, the Mittag-Leffler function is bounded due to the limited range of α, where $1 > \alpha > 0$. The first-order derivative represents the spread of a particle, defined by its concentration, and is thus bound due to its inherent physical meaning. The derivative is considered at the maximum physical space that is applicable to the concentration (X_{max})

$$\|{}_0^{ABC}D_x^\alpha c_x^0\|_0 \leq \frac{AB(\alpha)}{(1-\alpha)} \int_0^x \left\|\frac{dc_\tau^0}{d\tau}\right\|_0 d\tau \leq \frac{AB(\alpha)}{(1-\alpha)} \theta \|c_x^0\|_0 \int_0^{X_{max}} d\tau \leq \frac{AB(\alpha)}{(1-\alpha)} \theta \|c_x^0\|_0 X_{max}. \tag{33}$$

Substituting back into Eq. (31)

$$\|c_x^1\|_1^2 < \|c_x^0\|_0 \|c_x^1\|_0 - \lambda_1 \left(\frac{AB(\alpha)}{(1-\alpha)} \theta \|c_x^0\|_0 X_{max}\right) \left(\frac{AB(\alpha)}{(1-\alpha)} \theta \|c_x^1\|_0 X_{max}\right). \tag{34}$$

Rearranging

$$\|c_x^1\|_1^2 < \|c_x^0\|_0\|c_x^1\|_0 - \lambda_1\left(\frac{AB(\alpha)\,\theta X_{max}}{(1-\alpha)}\right)^2\|c_x^0\|_0\|c_x^1\|_0$$
$$< \left(1-\lambda_1\left(\frac{AB(\alpha)\,\theta X_{max}}{(1-\alpha)}\right)^2\right)\|c_x^0\|_0\|c_x^1\|_0. \quad (35)$$

Applying the assumption that

$$\|c_x^1\|_0 \le \|c_x^1\|_1.$$

Simplifying, the stability condition becomes

$$\|c_x^1\|_1^2 < \left(1-\lambda_1\left(\frac{AB(\alpha)\,\theta X_{max}}{(1-\alpha)}\right)^2\right)\|c_x^0\|_0\|c_x^1\|_1\|c_x^1\|_1$$
$$< \left(1-\lambda_1\left(\frac{AB(\alpha)\,\theta X_{max}}{(1-\alpha)}\right)^2\right)\|c_x^0\|_0\frac{\|c_x^1\|_1}{\|c_x^0\|_0} < 1-\lambda_1\left(\frac{AB(\alpha)\,\theta X_{max}}{(1-\alpha)}\right)^2, \quad (36)$$

where,

$$1-\lambda_1\left(\frac{AB(\alpha)\,\theta X_{max}}{(1-\alpha)}\right)^2 < 1\lambda_1\left(\frac{AB(\alpha)\,\theta X_{max}}{(1-\alpha)}\right)^2 > 0.$$

The first condition is thus upheld and unconditionally stable.
Secondly, Let $\forall g \in H^1(\Omega)$, $g = c_x^{n+1}$

$$\left(c_x^{n+1}, c_x^{n+1}\right) = \left(c_x^n, c_x^{n+1}\right) - \lambda_1\left({}_0^{ABC}D_x^\alpha c_{x,0}^n\, {}^{ABC}D_x^\alpha c_x^{n+1}\right) + \lambda_2\left(\frac{\partial^2}{\partial x^2}c_x^{n+1}, \frac{\partial^2}{\partial x^2}c_x^{n+1}\right)$$
$$- \lambda_3\sum_{k=0}^{n-1}\left(\left(c_x^{k+1}, c_x^{n+1}\right) - \left(c_x^k, c_x^{n+1}\right)\right), \quad (37)$$

where,

$$\lambda_3 = \frac{1}{A_n^\alpha}A_{n,k}^\alpha.$$

From the defined norms, the following statement is to be proven

$$\|c_x^{n+1}\|_1 \le \|c_x^0\|_0.$$

Reformulating Eq. (37) in terms of the defined norms, we have

$$\left(c_x^{n+1}, c_x^{n+1}\right) - \lambda_2 \left(\frac{\partial^2 c_x^{n+1}}{\partial x^2}, \frac{\partial^2 c_x^{n+1}}{\partial x^2}\right) = \left(c_x^n, c_x^{n+1}\right) - \lambda_1 \left({}_0^{ABC}D_x^\alpha c_x^n, {}_0^{ABC} D_x^\alpha c_x^{n+1}\right)$$
$$- \lambda_3 \sum_{k=0}^{n-1} \left(\left(c_x^{k+1}, c_x^{n+1}\right) - \left(c_x^k, c_x^{n+1}\right)\right)$$
$$\|c_x^{n+1}\|_1^2 = \|c_x^n\|_0 \|c_x^{n+1}\|_0 - \lambda_1 \|{}_0^{ABC}D_x^\alpha c_x^n\|_0 \|{}_0^{ABC}D_x^\alpha c_x^{n+1}\|_0$$
$$- \lambda_3 \sum_{k=0}^{n-1} \left(\left(\|c_x^{k+1}\|_0 \|c_x^{n+1}\|_0\right) - \left(\|c_x^k\|_0 \|c_x^{n+1}\|_0\right)\right). \quad (38)$$

Applying Eq. (33), we get

$$\|c_x^{n+1}\|_1^2 \leq \|c_x^n\|_0 \|c_x^{n+1}\|_0 - \lambda_1 A \|c_x^n\|_0 \|c_x^{n+1}\|_0 - \lambda_3 \sum_{k=0}^{n-1} \left(\left(\|c_x^{k+1}\|_0 \|c_x^{n+1}\|_0\right) - \left(\|c_x^k\|_0 \|c_x^{n+1}\|_0\right)\right), \quad (39)$$

where,

$$A = \left(\frac{AB(\alpha)\,\theta X_{max}}{(1-\alpha)}\right)^2.$$

Using the inductive method for

$$\|c_x^n\|_0 \leq \|c_x^0\|_0,$$

Equation (39) becomes

$$\|c_x^{n+1}\|_1^2 \leq \|c_x^0\|_0 \|c_x^{n+1}\|_0 - \lambda_1 A \|c_x^0\|_0 \|c_x^{n+1}\|_0 - \lambda_3 \sum_{k=0}^{n-1} \left(\left(\|c_x^0\|_0 \|c_x^{n+1}\|_0\right) - \left(\|c_x^0\|_0 \|c_x^{n+1}\|_0\right)\right). \quad (40)$$

Reformulating in terms of the defined norms, we have

$$\|c_x^{n+1}\|_1^2 \leq \|c_x^0\|_0 \|c_x^{n+1}\|_1 - \lambda_1 A \|c_x^0\|_0 \|c_x^{n+1}\|_1. \quad (41)$$

Rearranging and simplifying

$$\|c_x^{n+1}\|_1^2 \leq (1 - \lambda_1 A)\, \|c_x^0\|_0 \|c_x^{n+1}\|_1 \|c_x^{n+1}\|_1 \leq (1 - \lambda_1 A)\, \|c_x^0\|_0, \quad (42)$$

where,

$$[1 - \lambda_1 A < 1],$$

$$\lambda_1 A > 0.$$

The second condition is thus unconditionally stable. This concludes the semi-discretisation analysis for an evolution equation, where the proposed ABC fractional advection-dispersion equation has been found to be stable in time.

3 Upwind Numerical Approximation Schemes

Boundedness, existence and uniqueness has been established for the ABC fractional advection-dispersion equation. Furthermore, the stability in time has been demonstrated for the equation. Numerical schemes for this equation are now explored to facilitate the solution of the complex equation. Upwind-based finite difference schemes are investigated as motivated in [33], where upwind schemes aim to improve the stability of advection-dominated transport [34].

3.1 *First-Order Upwind Explicit*

The numerical approximation of the ABC fractional derivative with respect to time is considered, with the resulting scheme as [32]:

$$ {}_{0}^{ABC}D_t^{\alpha}\left(c\left(x_m,t_k\right)\right)=\frac{AB\left(\alpha\right)}{\left(1-\alpha\right)}\sum_{k=0}^{n-1}\left(c_m^{k+1}-c_m^k\right)\left(\begin{array}{c}\left(n-k\right)E_{\alpha,2}\left[-\frac{\alpha\Delta t}{1-\alpha}\left(n-k\right)\right]\\ -\left(n-k-1\right)E_{\alpha,2}\left[-\frac{\alpha\Delta t}{1-\alpha}\left(n-k-1\right)\right]\end{array}\right), \tag{43} $$

where, a function $\delta_{n,k}^{\alpha}$ is applied to simplify,

$$ \delta_{n,k}^{\alpha}=\left(n-k\right)E_{\alpha,2}\left[-\frac{\alpha\Delta t}{1-\alpha}\left(n-k\right)\right]-\left(n-k-1\right)E_{\alpha,2}\left[-\frac{\alpha\Delta t}{1-\alpha}\left(n-k-1\right)\right]. $$

Thus,

$$ {}_{0}^{ABC}D_t^{\alpha}\left(c\left(x_m,t_k\right)\right)=\frac{AB\left(\alpha\right)}{\left(1-\alpha\right)}\sum_{k=0}^{n-1}\left(c_m^{k+1}-c_m^k\right)\delta_{n,k}^{\alpha}. \tag{44} $$

The upwind finite difference scheme uses a one-sided finite difference in the upstream direction to approximate the advection term of the advection-dispersion equation (assuming $v>0$). Applying the upwind scheme and the ABC fractional derivative with respect to space (explicit) becomes [32]

$$ {}_{0}^{ABC}D_x^{\alpha}\left(c\left(x_m,t_k\right)\right)=\frac{AB\left(\alpha\right)}{\left(1-\alpha\right)}\sum_{i=0}^{m}\left(c_i^{n-1}-c_{i-1}^{n-1}\right)\left(\begin{array}{c}\left(m-i\right)E_{\alpha,2}\left[-\frac{\alpha\Delta x}{1-\alpha}\left(m-i\right)\right]\\ -\left(m-i-1\right)E_{\alpha,2}\left[-\frac{\alpha\Delta x}{1-\alpha}\left(m-i-1\right)\right]\end{array}\right), \tag{45} $$

where, a function $\delta_{m,i}^{\alpha}$ is applied to simplify

$$
{}_{0}^{ABC}D_{x}^{\alpha}\left(c\left(x_{m},t_{k}\right)\right)=\frac{AB\left(\alpha\right)}{\left(1-\alpha\right)}\sum_{i=0}^{m}\left(c_{i}^{n-1}-c_{i-1}^{n-1}\right)\delta_{m,i}^{\alpha}. \tag{46}
$$

Substituting into the ABC fractional advection-dispersion equation and using the traditional finite difference approach for the local second-order derivative, we have

$$
\begin{aligned}
&\frac{AB\left(\alpha\right)}{\left(1-\alpha\right)}\sum_{k=0}^{n-1}\left(c_{m}^{k+1}-c_{m}^{k}\right)\delta_{n,k}^{\alpha}+v\frac{AB\left(\alpha\right)}{\left(1-\alpha\right)}\sum_{i=0}^{m}\left(c_{i}^{n-1}-c_{i-1}^{n-1}\right)\delta_{m,i}^{\alpha}\\
&\qquad-D_{L}\left(\frac{c_{m+1}^{n-1}-2c_{m}^{n-1}+c_{m-1}^{n-1}}{\left(\Delta x\right)^{2}}\right)=0.
\end{aligned} \tag{47}
$$

Reformulating the following can be obtained

$$
\begin{aligned}
&\frac{AB\left(\alpha\right)}{\left(1-\alpha\right)}\left(c_{m}^{n}-c_{m}^{n-1}\right)\delta_{n,n-1}^{\alpha}+\frac{AB\left(\alpha\right)}{\left(1-\alpha\right)}\sum_{k=0}^{n-2}\left(c_{m}^{k+1}-c_{m}^{k}\right)\delta_{n,k}^{\alpha}\\
&+v\frac{AB\left(\alpha\right)}{\left(1-\alpha\right)}\left(c_{m}^{n-1}-c_{m-1}^{n-1}\right)\delta_{m,i}^{\alpha}+v\frac{AB\left(\alpha\right)}{\left(1-\alpha\right)}\sum_{i=0}^{m}\left(c_{i}^{n-1}-c_{i-1}^{n-1}\right)\delta_{m,i}^{\alpha}\\
&-D_{L}\left(\frac{c_{m+1}^{n-1}-2c_{m}^{n-1}+c_{m-1}^{n-1}}{\left(\Delta x\right)^{2}}\right)=0.
\end{aligned} \tag{48}
$$

Rearranging

$$
\begin{aligned}
&c_{m}^{n}\frac{AB\left(\alpha\right)}{\left(1-\alpha\right)}\delta_{n,n-1}^{\alpha}=c_{m}^{n-1}\left(\frac{AB\left(\alpha\right)}{\left(1-\alpha\right)}\delta_{n,n-1}^{\alpha}-v\frac{AB\left(\alpha\right)}{\left(1-\alpha\right)}\delta_{m,i}^{\alpha}-\frac{2D_{L}}{\left(\Delta x\right)^{2}}\right)\\
&+c_{m-1}^{n-1}\left(v\frac{AB\left(\alpha\right)}{\left(1-\alpha\right)}\delta_{m,i}^{\alpha}+\frac{D_{L}}{\left(\Delta x\right)^{2}}\right)+c_{m+1}^{n-1}\left(\frac{D_{L}}{\left(\Delta x\right)^{2}}\right)-\frac{AB\left(\alpha\right)}{\left(1-\alpha\right)}\sum_{k=0}^{n-2}\left(c_{m}^{k+1}-c_{m}^{k}\right)\delta_{n,k}^{\alpha}\\
&\qquad-v\frac{AB\left(\alpha\right)}{\left(1-\alpha\right)}\sum_{i=0}^{m}\left(c_{i}^{n-1}-c_{i-1}^{n-1}\right)\delta_{m,i}^{\alpha}.
\end{aligned} \tag{49}
$$

The numerical scheme can be simplified using place-keeper functions as follows

$$
ac_{m}^{n}=bc_{m}^{n-1}+dc_{m-1}^{n-1}+fc_{m+1}^{n-1}-g\sum_{k=0}^{n-2}\left(c_{m}^{k+1}-c_{m}^{k}\right)\delta_{n,k}^{\alpha}-vg\sum_{i=0}^{m}\left(c_{i}^{n-1}-c_{i-1}^{n-1}\right)\delta_{m,i}^{\alpha}, \tag{50}
$$

where,

$$a = \frac{AB(\alpha)}{(1-\alpha)}\delta^{\alpha}_{n,n-1};\ b = \frac{AB(\alpha)}{(1-\alpha)}\delta^{\alpha}_{n,n-1} - v\frac{AB(\alpha)}{(1-\alpha)}\delta^{\alpha}_{m,i} - \frac{2D_L}{(\Delta x)^2};\ d = v\frac{AB(\alpha)}{(1-\alpha)}\delta^{\alpha}_{m,i} + \frac{D_L}{(\Delta x)^2},$$

$$f = \frac{D_L}{(\Delta x)^2};\ g = \frac{AB(\alpha)}{(1-\alpha)}.$$

3.2 First-Order Upwind Implicit

Following the same approach as the explicit upwind numerical approximation, the following is obtained for the implicit upwind scheme

$$\frac{AB(\alpha)}{(1-\alpha)}\sum_{k=0}^{n-1}\left(c_m^{k+1} - c_m^k\right)\delta^{\alpha}_{n,k} + v\frac{AB(\alpha)}{(1-\alpha)}\sum_{i=0}^{m}\left(c_i^n - c_{i-1}^n\right)\delta^{\alpha}_{m,i} - D_L\left(\frac{c_{m+1}^n - 2c_m^n + c_{m-1}^n}{(\Delta x)^2}\right) = 0. \tag{51}$$

Reformulating and rearranging the following can be obtained

$$c_m^n\left(\frac{AB(\alpha)}{(1-\alpha)}\delta^{\alpha}_{n,n-1} + v\frac{AB(\alpha)}{(1-\alpha)}\delta^{\alpha}_{m,i} + \frac{2D_L}{(\Delta x)^2}\right) = c_{m-1}^n\left(v\frac{AB(\alpha)}{(1-\alpha)}\delta^{\alpha}_{m,i} - \frac{D_L}{(\Delta x)^2}\right)$$
$$+c_{m+1}^n\frac{D_L}{(\Delta x)^2} - \frac{AB(\alpha)}{(1-\alpha)}\sum_{k=0}^{n-2}\left(c_m^{k+1} - c_m^k\right)\delta^{\alpha}_{n,k} - v\frac{AB(\alpha)}{(1-\alpha)}\sum_{i=0}^{m}\left(c_i^n - c_{i-1}^n\right)\delta^{\alpha}_{m,i}$$
$$+ c_m^{n-1}\frac{AB(\alpha)}{(1-\alpha)}\delta^{\alpha}_{n,n-1}. \tag{52}$$

The numerical scheme is simplified by substituting functions as followings

$$hc_m^n = jc_{m-1}^n + fc_{m+1}^n - g\sum_{k=0}^{n-2}\left(c_m^{k+1} - c_m^k\right)\delta^{\alpha}_{n,k} - vg\sum_{i=0}^{m}\left(c_i^n - c_{i-1}^n\right)\delta^{\alpha}_{m,i} + ac_m^{n-1}, \tag{53}$$

where,

$$h = \frac{AB(\alpha)}{(1-\alpha)}\delta^{\alpha}_{n,n-1} + v\frac{AB(\alpha)}{(1-\alpha)}\delta^{\alpha}_{m,i} + \frac{2D_L}{(\Delta x)^2};\ j = v\frac{AB(\alpha)}{(1-\alpha)}\delta^{\alpha}_{m,i} - \frac{D_L}{(\Delta x)^2}.$$

3.3 First-Order Upwind Crank Nicolson Scheme

The upwind Crank Nicolson finite difference scheme for ABC fractional advection-dispersion equation is now considered [33]. The time component remains the same as with the implicit/explicit upwind schemes, but the space components change to

$$ {}_{0}^{ABC}D_{x}^{\alpha}\left(c\left(x_m, t_k\right)\right) = \frac{AB\left(\alpha\right)}{\left(1-\alpha\right)}\sum_{i=0}^{m}\left[0.5\left(c_i^{n-1} - c_{i-1}^{n-1}\right) + 0.5\left(c_i^{n} - c_{i-1}^{n}\right)\right)\delta_{m,i}^{\alpha}. \quad (54) $$

Substituting

$$ \frac{AB\left(\alpha\right)}{\left(1-\alpha\right)}\sum_{k=0}^{n-1}\left(c_m^{k+1} - c_m^{k}\right)\delta_{n,k}^{\alpha} + v\frac{AB\left(\alpha\right)}{\left(1-\alpha\right)}\sum_{i=0}^{m}\left[0.5\left(c_i^{n-1} - c_{i-1}^{n-1}\right) + 0.5\left(c_i^{n} - c_{i-1}^{n}\right)\right]\delta_{m,i}^{\alpha} - D_L\left(\frac{c_{m+1}^{n} - 2c_m^{n} + c_{m-1}^{n}}{\left(\Delta x\right)^2}\right) = 0. \quad (55) $$

Reformulating and rearranging, the following can be obtained

$$ \begin{aligned} c_m^{n}&\left(\frac{AB\left(\alpha\right)}{\left(1-\alpha\right)}\delta_{n,n-1}^{\alpha} + 0.5v\frac{AB\left(\alpha\right)}{\left(1-\alpha\right)}\delta_{m,i}^{\alpha} + \frac{2D_L}{\left(\Delta x\right)^2}\right) = c_m^{n-1} \\ &\times\left(\frac{AB\left(\alpha\right)}{\left(1-\alpha\right)}\delta_{n,n-1}^{\alpha} + 0.5v\frac{AB\left(\alpha\right)}{\left(1-\alpha\right)}\delta_{m,i}^{\alpha}\right) \\ &+ c_{m-1}^{n}\left(0.5v\frac{AB\left(\alpha\right)}{\left(1-\alpha\right)}\delta_{m,i}^{\alpha} - \frac{D_L}{\left(\Delta x\right)^2}\right) + c_{m+1}^{n}\frac{D_L}{\left(\Delta x\right)^2} + c_{m-1}^{n-1}0.5v\frac{AB\left(\alpha\right)}{\left(1-\alpha\right)}\delta_{m,i}^{\alpha} \\ &-\frac{AB\left(\alpha\right)}{\left(1-\alpha\right)}\sum_{k=0}^{n-2}\left(c_m^{k+1} - c_m^{k}\right)\delta_{n,k}^{\alpha} - v\frac{AB\left(\alpha\right)}{\left(1-\alpha\right)}\sum_{i=0}^{m}\left[0.5\left(c_i^{n-1} - c_{i-1}^{n-1}\right) + 0.5\left(c_i^{n} - c_{i-1}^{n}\right)\right]\delta_{m,i}^{\alpha}. \end{aligned} \quad (56) $$

Simplifying by substituting place-keeper functions

$$ lc_m^{n} = mc_m^{n-1} + oc_{m-1}^{n} + fc_{m+1}^{n} + pc_{m-1}^{n-1} - g\sum_{k=0}^{n-2}\left(c_m^{k+1} - c_m^{k}\right)\delta_{n,k}^{\alpha} - vg\sum_{i=0}^{m}\left[0.5\left(c_i^{n-1} - c_{i-1}^{n-1}\right) + 0.5\left(c_i^{n} - c_{i-1}^{n}\right)\right]\delta_{m,i}^{\alpha}, \quad (57) $$

where,

$$l = \frac{AB(\alpha)}{(1-\alpha)}\delta^{\alpha}_{n,n-1} + 0.5v\frac{AB(\alpha)}{(1-\alpha)}\delta^{\alpha}_{m,i} + \frac{2D_L}{(\Delta x)^2};\, m = \frac{AB(\alpha)}{(1-\alpha)}\delta^{\alpha}_{n,n-1} + 0.5v\frac{AB(\alpha)}{(1-\alpha)}\delta^{\alpha}_{m,i},$$

$$o = 0.5v\frac{AB(\alpha)}{(1-\alpha)}\delta^{\alpha}_{m,i} - \frac{D_L}{(\Delta x)^2};\, p = 0.5v\frac{AB(\alpha)}{(1-\alpha)}\delta^{\alpha}_{m,i}.$$

3.4 *First-Order Upwind-Downwind Weighted Scheme (Explicit)*

For the upwind-downwind weighted scheme, the upwind and downwind direction for the advection term are both integrated using a weighting factor. The weighting factor of upwind to downwind is defined as θ, where $0 \le \theta \le 1$ [33]. Thus, the advection component is approximated as

$$ {}_{0}^{ABC}D_x^{\alpha}\left(c\left(x_m, t_k\right)\right) = \frac{AB(\alpha)}{(1-\alpha)}\sum_{i=0}^{m}\left[\theta\left(c_i^{n-1} - c_{i-1}^{n-1}\right) + (1-\theta)\left(c_{i+1}^{n-1} - c_i^{n-1}\right)\right]\delta^{\alpha}_{m,i}. \tag{58}$$

Substituting this back into the advection-dispersion equation

$$\frac{AB(\alpha)}{(1-\alpha)}\sum_{k=0}^{n-1}\left(c_m^{k+1} - c_m^k\right)\delta^{\alpha}_{n,k} + v\frac{AB(\alpha)}{(1-\alpha)}\sum_{i=0}^{m}\left[\theta\left(c_i^{n-1} - c_{i-1}^{n-1}\right) + (1-\theta)\left(c_{i+1}^{n-1} - c_i^{n-1}\right)\right]\delta^{\alpha}_{m,i}$$
$$-D_L\left(\frac{c_{m+1}^{n-1} - 2c_m^{n-1} + c_{m-1}^{n-1}}{(\Delta x)^2}\right) = 0. \tag{59}$$

Reformulating and rearranging, the following can be obtained

$$c_m^n\frac{AB(\alpha)}{(1-\alpha)}\delta^{\alpha}_{n,n-1} = c_m^{n-1}\left(\frac{AB(\alpha)}{(1-\alpha)}\delta^{\alpha}_{n,n-1}\right.$$
$$\left.-v\theta\frac{AB(\alpha)}{(1-\alpha)}\delta^{\alpha}_{m,i} + v(1-\theta)\frac{AB(\alpha)}{(1-\alpha)}\delta^{\alpha}_{m,i} - \frac{2D_L}{(\Delta x)^2}\right)$$
$$+c_{m-1}^{n-1}\left(v\theta\frac{AB(\alpha)}{(1-\alpha)}\delta^{\alpha}_{m,i} + \frac{D_L}{(\Delta x)^2}\right) - c_{m+1}^{n-1}\left(v(1-\theta)\frac{AB(\alpha)}{(1-\alpha)}\delta^{\alpha}_{m,i} + \frac{D_L}{(\Delta x)^2}\right)$$
$$-\frac{AB(\alpha)}{(1-\alpha)}\sum_{k=0}^{n-2}\left(c_m^{k+1} - c_m^k\right)\delta^{\alpha}_{n,k} - v\frac{AB(\alpha)}{(1-\alpha)}\sum_{i=0}^{m}\left[\theta\left(c_i^{n-1} - c_{i-1}^{n-1}\right) + (1-\theta)\left(c_{i+1}^{n-1} - c_i^{n-1}\right)\right]\delta^{\alpha}_{m,i}. \tag{60}$$

Place-keeper functions are used to simplify the explicit upwind-downwind weighted scheme as followings

$$ac_m^n = qc_m^{n-1} + rc_{m-1}^{n-1} - sc_{m+1}^{n-1} - g\sum_{k=0}^{n-2}\left(c_m^{k+1} - c_m^k\right)\delta_{n,k}^{\alpha}$$
$$-vg\sum_{i=0}^{m}\left[\theta\left(c_i^{n-1} - c_{i-1}^{n-1}\right) + (1-\theta)\left(c_{i+1}^{n-1} - c_i^{n-1}\right)\right]\delta_{m,i}^{\alpha}, \quad (61)$$

where,

$$q = \frac{AB(\alpha)}{(1-\alpha)}\delta_{n,n-1}^{\alpha} - v\theta\frac{AB(\alpha)}{(1-\alpha)}\delta_{m,i}^{\alpha} + v(1-\theta)\frac{AB(\alpha)}{(1-\alpha)}\delta_{m,i}^{\alpha} - \frac{2D_L}{(\Delta x)^2};$$
$$r = v\theta\frac{AB(\alpha)}{(1-\alpha)}\delta_{m,i}^{\alpha} + \frac{D_L}{(\Delta x)^2},$$

$$s = v(1-\theta)\frac{AB(\alpha)}{(1-\alpha)}\delta_{m,i}^{\alpha} + \frac{D_L}{(\Delta x)^2}.$$

3.5 *First-Order Upwind-Downwind Weighted Scheme (Implicit)*

Correspondingly, both the upwind and downwind directions are considered for the advection term in the implicit upwind-downwind weighted scheme and the space advection component becomes

$${}_0^{ABC}D_x^{\alpha}\left(c\left(x_m, t_k\right)\right) = \frac{AB(\alpha)}{(1-\alpha)}\sum_{i=0}^{m}\left[\theta\left(c_i^n - c_{i-1}^n\right) + (1-\theta)\left(c_{i+1}^n - c_i^n\right)\right]\delta_{m,i}^{\alpha}. \quad (62)$$

Substituting

$$\frac{AB(\alpha)}{(1-\alpha)}\sum_{k=0}^{n-1}\left(c_m^{k+1} - c_m^k\right)\delta_{n,k}^{\alpha} + v\frac{AB(\alpha)}{(1-\alpha)}\sum_{i=0}^{m}\left[\theta\left(c_i^n - c_{i-1}^n\right) + (1-\theta)\left(c_{i+1}^n - c_i^n\right)\right]\delta_{m,i}^{\alpha}$$
$$- D_L\left(\frac{c_{m+1}^n - 2c_m^n + c_{m-1}^n}{(\Delta x)^2}\right) = 0. \quad (63)$$

Rearranging

$$c_m^n\left(\frac{AB(\alpha)}{(1-\alpha)}\delta_{n,n-1}^{\alpha} + v\theta\frac{AB(\alpha)}{(1-\alpha)}\delta_{m,i}^{\alpha} - v(1-\theta)\frac{AB(\alpha)}{(1-\alpha)}\delta_{m,i}^{\alpha} + \frac{2D_L}{(\Delta x)^2}\right)$$
$$= c_{m+1}^n\left(\frac{D_L}{(\Delta x)^2} - v(1-\theta)\frac{AB(\alpha)}{(1-\alpha)}\delta_{m,i}^{\alpha}\right) + c_{m-1}^n\left(\frac{D_L}{(\Delta x)^2} + v\theta\frac{AB(\alpha)}{(1-\alpha)}\delta_{m,i}^{\alpha}\right)$$

$$+ c_m^{n-1} \frac{AB(\alpha)}{(1-\alpha)} \delta_{n,n-1}^{\alpha} - \frac{AB(\alpha)}{(1-\alpha)} \sum_{k=0}^{n-2} \left(c_m^{k+1} - c_m^k\right) \delta_{n,k}^{\alpha}$$

$$- v \frac{AB(\alpha)}{(1-\alpha)} \sum_{i=0}^{m} \left[\theta \left(c_i^n - c_{i-1}^n\right) + (1-\theta) \left(c_{i+1}^n - c_i^n\right)\right] \delta_{m,i}^{\alpha}. \quad (64)$$

Simplifying

$$uc_m^n = vc_{m+1}^n + rc_{m-1}^n + ac_m^{n-1} - g \sum_{k=0}^{n-2} \left(c_m^{k+1} - c_m^k\right) \delta_{n,k}^{\alpha}$$

$$- vg \sum_{i=0}^{m} \left[\theta \left(c_i^n - c_{i-1}^n\right) + (1-\theta) \left(c_{i+1}^n - c_i^n\right)\right] \delta_{m,i}^{\alpha}, \quad (65)$$

where,

$$u = \frac{AB(\alpha)}{(1-\alpha)} \delta_{n,n-1}^{\alpha} + v\theta \frac{AB(\alpha)}{(1-\alpha)} \delta_{m,i}^{\alpha} - v(1-\theta) \frac{AB(\alpha)}{(1-\alpha)} \delta_{m,i}^{\alpha} + \frac{2D_L}{(\Delta x)^2},$$

$$v = \frac{D_L}{(\Delta x)^2} - v(1-\theta) \frac{AB(\alpha)}{(1-\alpha)} \delta_{m,i}^{\alpha}.$$

This concludes the formulation of the numerical approximations schemes to be investigated for the ABC fractional advection-dispersion equation. In the following section, the numerical stability of each scheme will be assessed.

4 Numerical Stability Analysis

The numerical stability analysis is performed using the recursive numerical stability method [33, 35, 36]. The numerical stability for the upwind schemes are evaluated to validate their use in solving the ABC fractional advection-dispersion equation for fracture flow in groundwater systems.

4.1 First-Order Upwind Implicit

Substituting the induction method terms for the developed finite difference first-order upwind (implicit) numerical scheme discussed in Sect. 3.2

$$hc_n e^{jk_i m} = jc_n e^{jk_i(m-\Delta m)} + fc_n e^{jk_i x(m+\Delta m)} - g\sum_{k=0}^{n-2}\left(c_{k+1}e^{jk_i m} - c_k e^{jk_i m}\right)\delta^{\alpha}_{n,k}$$
$$-vg\sum_{i=0}^{m}\left(c_n e^{jk_i m} - c_n e^{jk_i(m-\Delta m)}\right)\delta^{\alpha}_{m,i} + ac_{n-1}e^{jk_i m}. \quad (66)$$

Expand and simplify

$$hc_n = jc_n e^{-jk_i\Delta m} + fc_n e^{jk_i\Delta m} - g\sum_{k=0}^{n-2}\left(c_{k+1} - c_k\right)\delta^{\alpha}_{n,k}$$
$$-vg\sum_{i=0}^{m}\left(c_n - c_n e^{-jk_i\Delta m}\right)\delta^{\alpha}_{m,i} + ac_{n-1}. \quad (67)$$

The first procedure for the induction numerical stability analysis requires proving for a set $\forall\, n > 1$, that

$$|c_n| < |c_o|. \quad (68)$$

If $n = 1$, then

$$hc_1 = jc_1 e^{-jk_i\Delta m} + fc_1 e^{jk_i\Delta m} - vg\sum_{i=0}^{m}\left(c_1 - c_1 e^{-jk_i\Delta m}\right)\delta^{\alpha}_{m,i} + ac_0. \quad (69)$$

A subset for m is now considered, where $m = 0$

$$hc_1 = jc_1 e^{-jk_i\Delta m} + fc_1 e^{jk_i\Delta m} + ac_0. \quad (70)$$

Simplifying and rearranging

$$\frac{c_1}{c_0} = \frac{a}{h - je^{-jk_i\Delta m} - fe^{jk_i\Delta m}}. \quad (71)$$

Taking the norm, the condition for the first induction requirement becomes

$$\frac{|a|}{|h| + |j| + |f|} < 1. \quad (72)$$

The term is expanded using the simplification terms associated with Eq. (53)

$$\frac{\left|\frac{AB(\alpha)}{(1-\alpha)}\delta^{\alpha}_{n,n-1}\right|}{\left|\frac{AB(\alpha)}{(1-\alpha)}\delta^{\alpha}_{n,n-1} + v\frac{AB(\alpha)}{(1-\alpha)}\delta^{\alpha}_{m,i} + \frac{2D_L}{(\Delta x)^2}\right| + \left|v\frac{AB(\alpha)}{(1-\alpha)}\delta^{\alpha}_{m,i} - \frac{D_L}{(\Delta x)^2}\right| + \left|\frac{D_L}{(\Delta x)^2}\right|} < 1. \quad (73)$$

The following assumption is made

$$\left[v\frac{AB(\alpha)}{(1-\alpha)}\delta_{m,i}^{\alpha} > \frac{D_L}{(\Delta x)^2}\right].$$

Then, the condition is

$$\frac{\frac{AB(\alpha)}{(1-\alpha)}\delta_{n,n-1}^{\alpha}}{\frac{AB(\alpha)}{(1-\alpha)}\delta_{n,n-1}^{\alpha} + v\frac{AB(\alpha)}{(1-\alpha)}\delta_{m,i}^{\alpha} + \frac{2D_L}{(\Delta x)^2} + v\frac{AB(\alpha)}{(1-\alpha)}\delta_{m,i}^{\alpha} - \frac{D_L}{(\Delta x)^2} + \frac{D_L}{(\Delta x)^2}} < 1. \tag{74}$$

Simplifying

$$2v\frac{AB(\alpha)}{(1-\alpha)}\delta_{m,i}^{\alpha} + \frac{2D_L}{(\Delta x)^2} > 0. \tag{75}$$

Therefore, under this assumption, the first inductive stability condition for this subset is upheld and unconditionally stable.

The complementary assumption is made, and the condition becomes

$$\frac{\frac{AB(\alpha)}{(1-\alpha)}\delta_{n,n-1}^{\alpha}}{\frac{AB(\alpha)}{(1-\alpha)}\delta_{n,n-1}^{\alpha} + v\frac{AB(\alpha)}{(1-\alpha)}\delta_{m,i}^{\alpha} + \frac{2D_L}{(\Delta x)^2} - v\frac{AB(\alpha)}{(1-\alpha)}\delta_{m,i}^{\alpha} + \frac{D_L}{(\Delta x)^2} + \frac{D_L}{(\Delta x)^2}} < 1. \tag{76}$$

Simplifying

$$\frac{4D_L}{(\Delta x)^2} > 0. \tag{77}$$

Thus, the first inductive stability condition for this subset is supported and unconditionally stable under this assumption as well.

A subset for (m) is now considered for all $(m \geq 1)$

$$hc_1 = jc_1e^{-jk_i\Delta m} + fc_1e^{jk_i\Delta m} - vg\sum_{i=0}^{m}\left(c_1 - c_1e^{-jk_i\Delta m}\right)\delta_{m,i}^{\alpha} + ac_0 \tag{78}$$

Simplifying

$$\left(h - je^{-jk_i\Delta m} - fe^{jk_i\Delta m} + vg\left(1 - e^{-jk_i\Delta m}\right)\sum_{i=0}^{m}\delta_{m,i}^{\alpha}\right)c_1 = ac_0. \tag{79}$$

Expanding the summation, and simplifying

$$\sum_{i=0}^{m}\delta_{m,i}^{\alpha} = \left((m)\,E_{\alpha,2}\left[-\frac{\alpha\Delta x}{1-\alpha}(m)\right]\right) + \left(E_{\alpha,2}\left[-\frac{\alpha\Delta x}{1-\alpha}\right] - (-1)\,E_{\alpha,2}\left[-\frac{\alpha\Delta x}{1-\alpha}(-1)\right]\right). \tag{80}$$

Substituting back into Eq. (77), we get

$$\begin{pmatrix} h - je^{-jk_i\Delta m} - fe^{jk_i\Delta m} + vg\left(1 - e^{-jk_i\Delta m}\right)\cdot \\ \left((m)\, E_{\alpha,2}\left[-\frac{\alpha\Delta x}{1-\alpha}(m)\right] + E_{\alpha,2}\left[-\frac{\alpha\Delta x}{1-\alpha}\right] - (-1)\, E_{\alpha,2}\left[-\frac{\alpha\Delta x}{1-\alpha}(-1)\right]\right) \end{pmatrix} c_1 = ac_0, \tag{81}$$

where the function $\beta_{m,E_{\alpha,2}}$ is used to simplify as follows

$$\left(h - je^{-jk_i\Delta m} - fe^{jk_i\Delta m} + vg\left(1 - e^{-jk_i\Delta m}\right)\beta_{m,E_{\alpha,2}}\right) c_1 = ac_0. \tag{82}$$

Let a function simplify to

$$[\phi = k_i \Delta x],$$

where,

$$[e^{-j\phi} = e^{-jk_i\Delta x}].$$

Remembering Euler's formula for complex numbers, and substituting back into Eq. (80)

$$\left(h - je^{-jk_i\Delta m} - fe^{jk_i\Delta m} + vg\left(1 - cos\phi + isin\phi\right)\beta_{m,E_{\alpha,2}}\right) c_1 = ac_0. \tag{83}$$

Applying a norm and simplifying

$$\left(|h| + |j| + |f| + v|g|\left(2 - 2cos\phi\right)|\beta_{m,E_{\alpha,2}}|\right)|c_1| = |a||c_0|. \tag{84}$$

Rearranging

$$\frac{|c_1|}{|c_0|} = \frac{|a|}{\left(|h| + |j| + |f| + v|g|\left(2 - 2cos\phi\right)|\beta_{m,E_{\alpha,2}}|\right)}. \tag{85}$$

Thus, the condition becomes

$$\frac{|a|}{\left(|h| + |j| + |f| + v|g|\left(2 - 2cos\phi\right)|\beta_{m,E_{\alpha,2}}|\right)} < 1. \tag{86}$$

The term is expanded using the simplification terms associated with Eq. (53)

$$\frac{\left|\frac{AB(\alpha)}{(1-\alpha)}\delta^{\alpha}_{n,n-1}\right|}{\begin{pmatrix} \left|\frac{AB(\alpha)}{(1-\alpha)}\delta^{\alpha}_{n,n-1} + v\frac{AB(\alpha)}{(1-\alpha)}\delta^{\alpha}_{m,i} + \frac{2D_L}{(\Delta x)^2}\right| + \left|v\frac{AB(\alpha)}{(1-\alpha)}\delta^{\alpha}_{m,i} - \frac{D_L}{(\Delta x)^2}\right| + \left|\frac{D_L}{(\Delta x)^2}\right| \\ +v\left|\frac{AB(\alpha)}{(1-\alpha)}\right|\left(2 - 2cos\phi\right)|\beta_{m,E_{\alpha,2}}| \end{pmatrix}} < 1. \tag{87}$$

The assumption is made where

$$\left[v\frac{AB(\alpha)}{(1-\alpha)}\delta^{\alpha}_{m,i} > \frac{D_L}{(\Delta x)^2}\right].$$

Then, the conditions is

$$2v\frac{AB(\alpha)}{(1-\alpha)}\left(\delta_{m,i}^{\alpha}+(1-cos\phi)\,\beta_{m,E_{\alpha,2}}\right)+\frac{2D_L}{(\Delta x)^2}>0. \tag{88}$$

Under this assumption, the first inductive stability condition for the second subset of m is sustained and unconditionally stable.

The opposite assumption is made, and the condition becomes

$$\frac{4D_L}{(\Delta x)^2}+v\frac{AB(\alpha)}{(1-\alpha)}(2-2cos\phi)\,\beta_{m,E_{\alpha,2}}>0. \tag{89}$$

The first inductive stability condition for the second subset of m is upheld and unconditionally stable under this assumption as well.

The second procedure for the induction numerical stability analysis requires proving for a set $\forall\, n\geq 1$

$$|c_n|<|c_o|. \tag{90}$$

Rearranging Eq. (66) for c_n

$$\left(h+e^{-jk_i\Delta m}-fe^{jk_i\Delta m}+vg\left(1-e^{-jk_i\Delta m}\right)\sum_{i=0}^{m}\delta_{m,i}^{\alpha}\right)c_n=ac_{n-1}-g\sum_{k=0}^{n-2}(c_{k+1}-c_k)\,\delta_{n,k}^{\alpha}. \tag{91}$$

Following a similar simplification process as previously performed

$$\begin{aligned}&\left(h+je^{-jk_i\Delta m}-fe^{jk_i\Delta m}+vg\,(1-cos\phi+isin\phi)\,\beta_{m,E_{\alpha,2}}\right)c_n\\&=ac_{n-1}-g\sum_{k=0}^{n-2}(c_{k+1}-c_k)\,\delta_{n,k}^{\alpha}.\end{aligned} \tag{92}$$

Applying a norm

$$\begin{aligned}&\left|\left(h+je^{-jk_i\Delta m}-fe^{jk_i\Delta m}+vg\,(1-cos\phi+isin\phi)\,\beta_{m,E_{\alpha,2}}\right)c_n\right|\\&=\left|ac_{n-1}-g\sum_{k=0}^{n-2}(c_{k+1}-c_k)\,\delta_{n,k}^{\alpha}\right|.\end{aligned} \tag{93}$$

Therefore,

$$\begin{aligned}&|h+je^{-jk_i\Delta m}-fe^{jk_i\Delta m}+vg\,(1-cos\phi+isin\phi)\,\beta_{m,E_{\alpha,2}}||c_n|\\&<|a||c_{n-1}|+|g|\left|\sum_{k=0}^{n-2}(c_{k+1}-c_k)\,\delta_{n,k}^{\alpha}\right|.\end{aligned} \tag{94}$$

Remembering that it has been proved that for a set $\forall\, n > 1$

$$[|c_{n-1}| < |c_o|].$$

Thus,

$$|h + je^{-jk_i\Delta m} - fe^{jk_i\Delta m} + vg\,(1 - cos\phi + isin\phi)\,\beta_{m,E_{\alpha,2}}||c_n| < |a||c_{n-1}| \\ +|g|\left|\sum_{k=0}^{n-2}(c_{k+1} - c_k)\,\delta_{n,k}^{\alpha}\right| < |a||c_o| + |g|\left|\sum_{k=0}^{n-2}(c_{k+1} - c_k)\,\delta_{n,k}^{\alpha}\right|. \tag{95}$$

Therefore, it can be inferred that

$$|h + je^{-jk_i\Delta m} - fe^{jk_i\Delta m} + vg\,(1 - cos\phi + isin\phi)\,\beta_{m,E_{\alpha,2}}||c_n| \\ < |a||c_o| + |g|\left|\sum_{k=0}^{n-2}(c_{k+1} - c_k)\,\delta_{n,k}^{\alpha}\right|. \tag{96}$$

The remaining summation is considered at the upper limit

$$\left|\sum_{k=0}^{n-2}(c_{k+1} - c_k)\,\delta_{n,k}^{\alpha}\right| < \sum_{k=0}^{n-2}|c_{k+1}|\left(\left|1 - \frac{c_k}{c_{k+1}}\right|\right)\delta_{n,k}^{\alpha}. \tag{97}$$

Subset (k) will follow the same assumption made for a set $(\forall\, n \geq 1)$, where

$$\sum_{k=0}^{n-2} c_{k+1}\left(1 - \frac{c_k}{c_{k+1}}\right)\delta_{n,k}^{\alpha} < |c_0|\sum_{k=0}^{n-2}\delta_{n,k}^{\alpha}.$$

Substituting back into Eq. (97)

$$|h + je^{-jk_i\Delta m} - fe^{jk_i\Delta m} + vg\,(1 - cos\phi + isin\phi)\,\beta_{m,E_{\alpha,2}}||c_n| < |a||c_o| + |g||c_0|\sum_{k=0}^{n-2}\delta_{n,k}^{\alpha}. \tag{98}$$

Expanding the summation

$$\sum_{k=0}^{n-2}\delta_{n,k}^{\alpha} = \left((n)\,E_{\alpha,2}\left[-\frac{\alpha\Delta t}{1-\alpha}\,(n)\right]\right) + \left(2E_{\alpha,2}\left[-2\frac{\alpha\Delta t}{1-\alpha}\right] - E_{\alpha,2}\left[-\frac{\alpha\Delta t}{1-\alpha}\right]\right).$$

Substituting back into Eq. (98) and simplifying

$$|h + je^{-jk_i \Delta m} - fe^{jk_i \Delta m} + vg\left(1 - cos\phi + isin\phi\right)\beta_{m,E_{\alpha,2}}||c_n| < |a||c_o| + |g||c_0|\beta_{n,E_{\alpha,2}}, \quad (99)$$

where,

$$\beta_{n,E_{\alpha,2}} = nE_{\alpha,2}\left[-\frac{\alpha\Delta tn}{1-\alpha}\right] + 2E_{\alpha,2}\left[-\frac{2\alpha\Delta t}{1-\alpha}\right] - E_{\alpha,2}\left[-\frac{\alpha\Delta t}{1-\alpha}\right].$$

Simplifying and rearranging

$$\frac{|c_n|}{|c_o|} < \frac{|a| + |g|\beta_{n,E_{\alpha,2}}}{|h| + |j| + |f| + v|g|\left(|1 - cos\phi| + i|sin\phi|\right)|\beta_{m,E_{\alpha,2}}|}. \quad (100)$$

Thus, the condition becomes

$$\frac{|a| + |g|\beta_{n,E_{\alpha,2}}}{|h| + |j| + |f| + v|g|\left(|1 - cos\phi| + i|sin\phi|\right)|\beta_{m,E_{\alpha,2}}|} < 1. \quad (101)$$

The condition is expanded using the simplification terms associated with Eq. (53)

$$\frac{|\frac{AB(\alpha)}{(1-\alpha)}\delta^{\alpha}_{n,n-1}| + |\frac{AB(\alpha)}{(1-\alpha)}|\beta_{n,E_{\alpha,2}}}{|\frac{AB(\alpha)}{(1-\alpha)}\delta^{\alpha}_{n,n-1} + v\frac{AB(\alpha)}{(1-\alpha)}\delta^{\alpha}_{m,i} + \frac{2D_L}{(\Delta x)^2}| + |v\frac{AB(\alpha)}{(1-\alpha)}\delta^{\alpha}_{m,i} - \frac{D_L}{(\Delta x)^2}| + |\frac{D_L}{(\Delta x)^2}| \\ +v|\frac{AB(\alpha)}{(1-\alpha)}|\left(2 - 2cos\phi\right)|\beta_{m,E_{\alpha,2}}|} < 1. \quad (102)$$

The following assumption is made

$$\left[v\frac{AB(\alpha)}{(1-\alpha)}\delta^{\alpha}_{m,i} > \frac{D_L}{(\Delta x)^2}\right].$$

Simplifying

$$2v\left(\delta^{\alpha}_{m,i} + (1 - cos\phi)\,\beta_{m,E_{\alpha,2}}\right) + \frac{2D_L}{(\Delta x)^2} > \beta_{n,E_{\alpha,2}}. \quad (103)$$

Under this assumption, the second inductive stability requirement is found to be true. And, the numerical scheme is conditionally stable, under this condition.

The opposite assumption is made, and the condition becomes

$$2v\left(1 - cos\phi\right)\beta_{m,E_{\alpha,2}} + \frac{4D_L}{(\Delta x)^2} > \beta_{n,E_{\alpha,2}}. \quad (104)$$

Thus, under the opposite assumption, the second inductive stability condition is also found to be appropriate. Moreover, under this condition the numerical scheme is conditionally stable.

This concludes the stability analysis for the implicit upwind scheme for the ABC advection-dispersion equation, where the scheme is stable under the condition stated in Eqs. (103) and (104). These conditions can be simplified to an overall condition

$$min\left(\gamma,\eta\right)>\beta_{n,E_{\alpha,2}},$$

where,

$$\gamma=2v\left(\delta^{\alpha}_{m,i}+\left(1-cos\phi\right)\beta_{m,E_{\alpha,2}}\right)+\frac{2D_L}{\left(\Delta x\right)^2},$$

$$\eta=2v\left(1-cos\phi\right)\beta_{m,E_{\alpha,2}}+\frac{4D_L}{\left(\Delta x\right)^2}.$$

Under these conditions, the error of the approximation is not propagated throughout the solution, but rather decreases with each time step, as according to the induction method, where for all values of n, $|c_{n+1}|<|c_o|$.

4.2 First-Order Upwind Explicit

The induction method terms for the developed explicit upwind numerical scheme (Sect. 3.1) are substituted as follows

$$\begin{aligned}ac_ne^{jk_im}&=bc_{n-1}e^{jk_im}+dc_{n-1}e^{jk_ix(m-\Delta m)}+fc_{n-1}e^{jk_i(m+\Delta m)}-g\sum_{k=0}^{n-2}\left(c_{k+1}e^{jk_im}-c_ke^{jk_im}\right)\delta^{\alpha}_{n,k}\\&\quad-vg\sum_{i=0}^{m}\left(c_{n-1}e^{jk_im}-c_{n-1}e^{jk_i(m-\Delta m)}\right)\delta^{\alpha}_{m,i}.\end{aligned}\tag{105}$$

The same procedure applied in Sect. 4.1 is followed for the explicit upwind numerical scheme. When $n=1$, and a subset for m is considered where $m=0$, the explicit upwind numerical scheme for the ABC fractional advection-dispersion equation is conditionally stable, under the assumption

$$\frac{AB\left(\alpha\right)}{\left(1-\alpha\right)}\delta^{\alpha}_{n,n-1}<v\frac{AB\left(\alpha\right)}{\left(1-\alpha\right)}\delta^{\alpha}_{m,i}+\frac{2D_L}{\left(\Delta x\right)^2}.\tag{106}$$

Following the condition

$$2v\frac{AB\left(\alpha\right)}{\left(1-\alpha\right)}\delta^{\alpha}_{m,i}+\frac{4D_L}{\left(\Delta x\right)^2}<\frac{2AB\left(\alpha\right)}{\left(1-\alpha\right)}\delta^{\alpha}_{n,n-1}.\tag{107}$$

When a subset for m is considered where $m\geq 1$, the explicit upwind numerical scheme is also conditionally stable under the same assumption, with the condition

$$2v\frac{AB(\alpha)}{(1-\alpha)}\left(\delta^{\alpha}_{m,i}+(1-cos\phi)\,\beta_{m,E_{\alpha,2}}\right)+\frac{4D_L}{(\Delta x)^2}<\frac{2AB(\alpha)}{(1-\alpha)}\delta^{\alpha}_{n,n-1}. \tag{108}$$

Therefore, the assumption has been established where $|c_{n-1}|<|c_o|$. The next step for the recursive stability analysis is to use this assumption to demonstrate that for a set $\forall\, n\geq 1$, that

$$|c_n|<|c_o|.$$

Following the equivalent technique as shown in Sect. 4.1, and making the same assumption, the scheme is conditionally stable, with the following condition

$$\frac{AB(\alpha)}{(1-\alpha)}\left(2v\delta^{\alpha}_{m,i}-v\,(2-2cos\phi)\,\beta_{m,E_{\alpha,2}}+\beta_{n,E_{\alpha,2}}\right)+\frac{4D_L}{(\Delta x)^2}<\frac{2AB(\alpha)}{(1-\alpha)}\delta^{\alpha}_{n,n-1}. \tag{109}$$

This concludes the stability analysis for the explicit upwind scheme for the ABC advection-dispersion equation, where the scheme is found to be conditionally stable under the assumption made Eq. (106), with conditions Eqs. (107)–(109). The conditions can be simplified into

$$max\,(\lambda,\mu,\rho)<\frac{2AB(\alpha)}{(1-\alpha)}\delta^{\alpha}_{n,n-1},$$

where,

$$\lambda=2v\frac{AB(\alpha)}{(1-\alpha)}\delta^{\alpha}_{m,i}+\frac{4D_L}{(\Delta x)^2};\;\mu=2v\frac{AB(\alpha)}{(1-\alpha)}\left(\delta^{\alpha}_{m,i}+(1-cos\phi)\,\beta_{m,E_{\alpha,2}}\right)+\frac{4D_L}{(\Delta x)^2},$$

$$\rho=\frac{AB(\alpha)}{(1-\alpha)}\left(2v\delta^{\alpha}_{m,i}-v\,(2-2cos\phi)\,\beta_{m,E_{\alpha,2}}+\beta_{n,E_{\alpha,2}}\right)+\frac{4D_L}{(\Delta x)^2}.$$

4.3 First-Order Upwind Crank–Nicolson Scheme

The recursive induction method terms are substituted into the upwind Crank–Nicolson numerical scheme presented in Sect. 3.3

$$\begin{aligned}lc_ne^{jk_im}&=mc_{n-1}e^{jk_im}+oc_ne^{jk_i(m-\Delta m)}+fc_ne^{jk_ix(m+\Delta m)}+pc_{n-1}e^{jk_i(m-\Delta m)}\\&-g\sum_{k=0}^{n-2}\left(c_{k+1}e^{jk_im}-c_ke^{jk_im}\right)\delta^{\alpha}_{n,k}-vg\sum_{i=0}^{m}\left[\begin{array}{l}0.5\left(c_{n-1}e^{jk_im}-c_{n-1}e^{jk_i(m-\Delta m)}\right)\\+0.5\left(c_ne^{jk_im}-c_ne^{jk_i(m-\Delta m)}\right)\end{array}\right]\delta^{\alpha}_{m,i}.\end{aligned} \tag{110}$$

In the same way, the method outlined in Sect. 4.1 is employed to determine the numerical stability of the upwind Crank–Nicolson numerical scheme. If $n=1$, and a subset for m is considered where $m=0$, then the scheme is unconditionally stable, under the following assumption

$$0.5v\frac{AB\left(\alpha\right)}{\left(1-\alpha\right)}\delta_{m,i}^{\alpha} > \frac{D_L}{(\Delta x)^2}. \tag{111}$$

With the following condition resulting in the unconditional stability

$$\frac{2D_L}{(\Delta x)^2} > 0. \tag{112}$$

If the complementary assumption is made

$$0.5v\frac{AB\left(\alpha\right)}{\left(1-\alpha\right)}\delta_{m,i}^{\alpha} < \frac{D_L}{(\Delta x)^2}. \tag{113}$$

The scheme is conditionally stable, with the following condition

$$v\frac{AB\left(\alpha\right)}{\left(1-\alpha\right)}\delta_{m,i}^{\alpha} < \frac{4D_L}{(\Delta x)^2}. \tag{114}$$

When a subset for m is considered for all $m \geq 1$, under the assumption made in Eq. (111), the scheme is unconditionally stable. Similarly, when the complementary assumption is made, the scheme is conditionally stable.

The next step for the recursive stability analysis was performed as described in Sects. 4.1 and 4.2. The same assumption is made as in Eq. (111), and once more the same unconditionally stable condition was found. Likewise, the opposite condition resulting in the same condition as in Eq. (114).

This completes the stability analysis for the upwind Crank–Nicolson scheme for the ABC advection-dispersion equation, where the scheme is stable under the following condition stated in Eq. (114). Under this single condition, the error of the approximation is not propagated throughout the solution, but rather decreases with each time step, as according to the induction method, where for all values of n, $|c_{n+1}| < |c_o|$.

4.4 Upwind-Downwind Weighted Scheme (Implicit)

Substituting the induction method terms for the developed finite difference implicit upwind-downwind weighted numerical scheme discussed in Sect. 3.5

$$uc_n e^{jk_i m} = vc_n e^{jk_i x(m+\Delta m)} + rc_n e^{jk_i(m-\Delta m)} + ac_{n-1}e^{jk_i m} - g\sum_{k=0}^{n-2}\left(c_{k+1}e^{jk_i m} - c_k e^{jk_i m}\right)\delta^{\alpha}_{n,k}$$

$$-vg\sum_{i=0}^{m}\left[\theta\left(c_n e^{jk_i m} - c_n e^{jk_i(m-\Delta m)}\right) + (1-\theta)\left(c_n e^{jk_i x(m+\Delta m)} - c_n e^{jk_i m}\right)\right]\delta^{\alpha}_{m,i}. \quad (115)$$

A parallel procedure as applied in Sect. 4.1 is followed for the implicit upwind-downwind weighted numerical scheme. When $n = 1$, and a subset for m is considered where $m = 0$, the scheme for the ABC fractional advection-dispersion equation is unconditionally stable when $0.5 \leq \theta \leq 1$, but conditionally stable when $0 \leq \theta < 0.5$, under the assumption

$$\frac{D_L}{(\Delta x)^2} + v\theta\frac{AB(\alpha)}{(1-\alpha)}\delta^{\alpha}_{m,i} > v\frac{AB(\alpha)}{(1-\alpha)}\delta^{\alpha}_{m,i}. \quad (116)$$

With the condition being

$$(2\theta - 1)v\frac{AB(\alpha)}{(1-\alpha)}\delta^{\alpha}_{m,i} + \frac{2D_L}{(\Delta x)^2} > 0. \quad (117)$$

If the opposed assumption is made

$$\frac{D_L}{(\Delta x)^2} + v\theta\frac{AB(\alpha)}{(1-\alpha)}\delta^{\alpha}_{m,i} < v\frac{AB(\alpha)}{(1-\alpha)}\delta^{\alpha}_{m,i}. \quad (118)$$

The implicit upwind-downwind weighted numerical scheme is conditionally stable, with the following condition

$$v\frac{AB(\alpha)}{(1-\alpha)}\delta^{\alpha}_{m,i} > \left(\delta^{\alpha}_{n,n-1} + v\theta\delta^{\alpha}_{m,i}\right)\frac{AB(\alpha)}{(1-\alpha)} + \frac{D_L}{(\Delta x)^2}. \quad (119)$$

A subset for m is then considered for all $m \geq 1$. The assumption in Eq. (112) is made again, under which the scheme is unconditionally stable when $0.5 \leq \theta \leq 1$; and conditionally stable when $0 \leq \theta < 0.5$, with the condition

$$(2\theta - 1)\frac{AB(\alpha)\,v}{(1-\alpha)}\delta^{\alpha}_{m,i} + (1 - cos\phi)\frac{AB(\alpha)\,v}{(1-\alpha)}\beta_{m,E_{\alpha,2}} + \frac{2D_L}{(\Delta x)^2} > 0. \quad (120)$$

If complementary assumption is made, and the scheme is conditionally stable under the following condition

$$v\frac{AB(\alpha)}{(1-\alpha)}\left(\beta_{m,E_{\alpha,2}} + \delta^{\alpha}_{m,i}\right) > \frac{AB(\alpha)}{(1-\alpha)}\left(\delta^{\alpha}_{n,n-1} + v\theta\delta^{\alpha}_{m,i} + vcos\phi\beta_{m,E_{\alpha,2}}\right) + \frac{D_L}{(\Delta x)^2}. \quad (121)$$

The next step for the recursive stability analysis was performed as described in Sects. 4.1 and 4.2, and first the same assumption is made as in Eq. (116), and once more the scheme is unconditionally stable when $0.5 \leq \theta \leq 1$; and conditionally stable when $0 \leq \theta < 0.5$, with the condition now being

$$2v\frac{AB(\alpha)}{(1-\alpha)}\left(\delta^{\alpha}_{m,i}(2\theta-1)+(1-cos\phi)\,\beta_{m,E_{\alpha,2}}\right)+\frac{4D_L}{(\Delta x)^2}>\frac{AB(\alpha)}{(1-\alpha)}\beta_{n,E_{\alpha,2}}. \quad (122)$$

In the same way, when the complementary assumption is made, the scheme is conditionally stable, with the condition now being

$$2v\frac{AB(\alpha)}{(1-\alpha)}\left((1-\theta)\,\delta^{\alpha}_{m,i}+(1-cos\phi)\,\beta_{m,E_{\alpha,2}}\right)-\frac{2D_L}{(\Delta x)^2}>\frac{AB(\alpha)}{(1-\alpha)}\left(2\delta^{\alpha}_{n,n-1}+\beta_{n,E_{\alpha,2}}\right). \quad (123)$$

The stability analysis has established that the implicit upwind-downwind weighted scheme is stable under the conditions in Eqs. (119)–(122). Additional conditions are activated when the weighting factor is $0 \leq \theta < 0.5$, as stated in Eqs. (117) and (120). Only under these conditions, the error of the approximation made by the implicit upwind-downwind weighted numerical scheme is not propagated throughout the solution.

4.5 Upwind-Downwind Weighted Scheme (Explicit)

For the stability analysis of the explicit upwind-downwind weighted numerical scheme (Sect. 3.4), the recursive induction terms are substituted as follows

$$\begin{aligned} ac_n e^{jk_i m} &= qc_{n-1}e^{jk_i m}+rc_{n-1}e^{jk_i(m-\Delta m)}-sc_{n-1}e^{jk_i(m+\Delta m)}-g\sum_{k=0}^{n-2}\left(c_{k+1}e^{jk_i m}-c_k e^{jk_i m}\right)\delta^{\alpha}_{n,k} \\ &\quad -vg\sum_{i=0}^{m}\left[\begin{array}{l}\theta\left(c_{n-1}e^{jk_i m}-c_{n-1}e^{jk_i(m-\Delta m)}\right)\\ +(1-\theta)\left(c_{n-1}e^{jk_i x(m+\Delta m)}-c_{n-1}e^{jk_i m}\right)\end{array}\right]\delta^{\alpha}_{m,i}. \end{aligned} \quad (124)$$

Again, the process outlined in Sect. 4.1 is used to define the numerical stability of the explicit upwind-downwind weighted numerical scheme. When $n = 1$, and a subset for m is considered where $m = 0$, then the scheme is conditionally stable, under the following assumption

$$\frac{AB(\alpha)}{(1-\alpha)}\delta^{\alpha}_{n,n-1}+v\frac{AB(\alpha)}{(1-\alpha)}\delta^{\alpha}_{m,i}<2v\theta\frac{AB(\alpha)}{(1-\alpha)}\delta^{\alpha}_{m,i}+\frac{2D_L}{(\Delta x)^2}, \quad (125)$$

with the condition

$$2v(\theta-1)\frac{AB(\alpha)}{(1-\alpha)}\delta_{m,i}^{\alpha}+\frac{2D_L}{(\Delta x)^2}>\frac{2AB(\alpha)}{(1-\alpha)}\delta_{n,n-1}^{\alpha}. \tag{126}$$

A subset for m is then considered for all $m \geq 1$, and under the same assumption as in Eq. (121), the scheme is conditionally stable, under the following condition

$$\frac{2AB(\alpha)}{(1-\alpha)}\left(\delta_{n,n-1}^{\alpha}+v\delta_{m,i}^{\alpha}+vcos\phi\beta_{m,E_{\alpha,2}}\right)>\frac{2AB(\alpha)}{(1-\alpha)}\left(2v\theta\delta_{m,i}^{\alpha}+v\beta_{m,E_{\alpha,2}}\right)+\frac{4D_L}{(\Delta x)^2}. \tag{127}$$

The subsequent stage of the recursive stability analysis was performed as described in Sects. 4.1 and 4.2. The same assumption is made as in Eq. (125), and again the scheme is conditionally stable, with the following stability condition

$$\frac{2AB(\alpha)}{(1-\alpha)}\left(\delta_{n,n-1}^{\alpha}+v\delta_{m,i}^{\alpha}+vcos\phi\beta_{m,E_{\alpha,2}}\right)>\frac{AB(\alpha)}{(1-\alpha)}\left(2v\beta_{m,E_{\alpha,2}}+4v\theta\delta_{m,i}^{\alpha}+\beta_{n,E_{\alpha,2}}\right)+\frac{4D_L}{(\Delta x)^2}. \tag{128}$$

The stability analysis for the explicit upwind-downwind weighted scheme is concluded, where the scheme is conditionally stable under the assumption stated in Eq. (125), with the conditions stated in Eqs. (126)–(128). Only under these conditions, the error of the approximation made by the explicit upwind-downwind weighted scheme is not proliferated throughout the solution.

5 Comparison of Numerical Stability

The stability conditions for the traditional upwind (implicit/explicit), and the new upwind Crank-Nicholson and weighted upwind-downwind (implicit/explicit) numerical schemes are tabulated in Appendix A. The traditional implicit upwind scheme applied to the ABC fractional advection-dispersion equation is conditionally stable under both assumptions made with a single condition for each assumption. There is only one practically applicable assumption for the customary explicit upwind scheme, which has three sub-conditions. The upwind Crank–Nicolson numerical scheme applied to the ABC fractional advection-dispersion equation is unconditionally stable under the first assumption, and has a single condition under the second assumption made. The implicit upwind-downwind weighted numerical scheme is conditionally stable under both assumptions made, but unconditionally stable for the first assumption when the weighting factor is $0.5 \leq \theta \leq 1$. Similar to the explicit upwind scheme, the explicit upwind-downwind weighted numerical scheme has one practically applicable assumption, which has three conditions for stability. Table 1 show the summary of numerical schemes.

Table 1 Summary of numerical schemes

Summary		
Scheme	Assumption	Stability conditions
Implicit upwind	$v\frac{AB(\alpha)}{(1-\alpha)}\delta^{\alpha}_{m,i} > \frac{D_L}{(\Delta x)^2}$	$2v\left(\delta^{\alpha}_{m,i} + (1-cos\phi)\,\beta_{m,E_{\alpha,2}}\right) + \frac{2D_L}{(\Delta x)^2} > \beta_{n,E_{\alpha,2}}$
	$v\frac{AB(\alpha)}{(1-\alpha)}\delta^{\alpha}_{m,i} < \frac{D_L}{(\Delta x)^2}$	$2v\,(1-cos\phi)\,\beta_{m,E_{\alpha,2}} + \frac{4D_L}{(\Delta x)^2} > \beta_{n,E_{\alpha,2}}$
Explicit upwind	Assumption not made $\frac{AB(\alpha)}{(1-\alpha)}\delta^{\alpha}_{n,n-1} < v\frac{AB(\alpha)}{(1-\alpha)}\delta^{\alpha}_{m,i} + \frac{2D_L}{(\Delta x)^2}$	$max\,(\lambda,\mu,\rho) < \frac{2AB(\alpha)}{(1-\alpha)}\delta^{\alpha}_{n,n-1}$
Upwind Crank-Nicholson	$0.5v\frac{AB(\alpha)}{(1-\alpha)}\delta^{\alpha}_{m,i} > \frac{D_L}{(\Delta x)^2}$ $0.5v\frac{AB(\alpha)}{(1-\alpha)}\delta^{\alpha}_{m,i} < \frac{D_L}{(\Delta x)^2}$	Unconditionally stable $v\frac{AB(\alpha)}{(1-\alpha)}\delta^{\alpha}_{m,i} < \frac{4D_L}{(\Delta x)^2}$
Upwind-downwind weighted scheme (implicit)	$\frac{D_L}{(\Delta x)^2} + v\theta\frac{AB(\alpha)}{(1-\alpha)}\delta^{\alpha}_{m,i} > v\frac{AB(\alpha)}{(1-\alpha)}\delta^{\alpha}_{m,i}$	$(2\theta-1)\,v\frac{AB(\alpha)}{(1-\alpha)}\delta^{\alpha}_{m,i} + \frac{2D_L}{(\Delta x)^2} > 0$
		$(2\theta-1)\,\frac{AB(\alpha)v}{(1-\alpha)}\delta^{\alpha}_{m,i} + (1-cos\phi)\,\frac{AB(\alpha)v}{(1-\alpha)}\beta_{m,E_{\alpha,2}} + \frac{2D_L}{(\Delta x)^2} > 0$
		$2v\frac{AB(\alpha)}{(1-\alpha)}\left(\delta^{\alpha}_{m,i}(2\theta-1) + (1-cos\phi)\,\beta_{m,E_{\alpha,2}}\right) + \frac{4D_L}{(\Delta x)^2} > \frac{AB(\alpha)}{(1-\alpha)}\beta_{n,E_{\alpha,2}}$
		Unconditionally stable when $0.5 \le \theta \le 1$
		Conditionally stable when $0 \le \theta < 0.5$
	$\frac{D_L}{(\Delta x)^2} + v\theta\frac{AB(\alpha)}{(1-\alpha)}\delta^{\alpha}_{m,i} < v\frac{AB(\alpha)}{(1-\alpha)}\delta^{\alpha}_{m,i}$	$v\frac{AB(\alpha)}{(1-\alpha)}\delta^{\alpha}_{m,i} > \left(\delta^{\alpha}_{n,n-1} + v\theta\delta^{\alpha}_{m,i}\right)\frac{AB(\alpha)}{(1-\alpha)} + \frac{D_L}{(\Delta x)^2}$
		$v\left(\beta_{m,E_{\alpha,2}} + \delta^{\alpha}_{m,i}\right)\frac{AB(\alpha)}{(1-\alpha)} > 2v\left((1-\theta)\,\delta^{\alpha}_{m,i} + (1-cos\phi)\,\beta_{m,E_{\alpha,2}}\right)\frac{AB(\alpha)}{(1-\alpha)}$
		$-\frac{2D_L}{(\Delta x)^2} > \left(2\delta^{\alpha}_{n,n-1} + \beta_{n,E_{\alpha,2}}\right)\frac{AB(\alpha)}{(1-\alpha)}$
	Assumption not made	
Upwind-downwind weighted scheme (explicit)	$\frac{AB(\alpha)}{(1-\alpha)}\delta^{\alpha}_{n,n-1} + v\frac{AB(\alpha)}{(1-\alpha)}\delta^{\alpha}_{m,i} < 2v\theta\frac{AB(\alpha)}{(1-\alpha)}\delta^{\alpha}_{m,i} + \frac{2D_L}{(\Delta x)^2}$	$2v\,(\theta-1)\,\frac{AB(\alpha)}{(1-\alpha)}\delta^{\alpha}_{m,i} + \frac{2D_L}{(\Delta x)^2} > \frac{2AB(\alpha)}{(1-\alpha)}\delta^{\alpha}_{n,n-1}$
		$\left(\delta^{\alpha}_{n,n-1} + v\delta^{\alpha}_{m,i} + vcos\phi\beta_{m,E_{\alpha,2}}\right)\frac{2AB(\alpha)}{(1-\alpha)} > \left(2v\theta\delta^{\alpha}_{m,i} + v\beta_{m,E_{\alpha,2}}\right)\frac{2AB(\alpha)}{(1-\alpha)} + \frac{4D_L}{(\Delta x)^2}$
		$\left(\delta^{\alpha}_{n,n-1} + v\delta^{\alpha}_{m,i} + vcos\phi\beta_{m,E_{\alpha,2}}\right)\frac{2AB(\alpha)}{(1-\alpha)} > \left(2v\beta_{m,E_{\alpha,2}} + 4v\theta\delta^{\alpha}_{m,i} + \beta_{n,E_{\alpha,2}}\right)\frac{AB(\alpha)}{(1-\alpha)} + \frac{4D_L}{(\Delta x)^2}$

6 Conclusions

An advection-focused fractional transport equation is developed using the ABC fractional derivative. The boundedness, existence and uniqueness is determined using the Picard–Lindelöf theorem. The semi-discretisation stability is evaluated in time, and demonstrated that the developed equation is stable in time. Upwind-based finite difference approximations were developed, and the stability of each was determined. The implicit upwind formulations are found to be more stable than their comparable explicit formulations. The proposed implicit weighted upwind-downwind scheme is more stable than the traditional upwind scheme when the weighting factor is $0.5 \leq \theta \leq 1$, which denotes at least half upwind-weighted or more, and the downwind influence less than half. Of the numerical schemes analysed, the upwind Crank–Nicolson is the most stable numerical scheme, and would be suggested for use with the ABC fractional advection-dispersion equation.

References

1. Koch, D.L., Brady, J.F.: Anomalous diffusion in heterogeneous porous media. Phys. Fluids **31**(5), 965–973 (1988)
2. Schumer, R., Benson, D.A., Meerschaert, M.M., Baeumer, B.: Multiscaling fractional advection-dispersion equations and their solutions. Water Resour. Res. **39**(1), 1–11 (2003)
3. Berkowitz, B., Cortis, A., Dentz, M., Scher, H.: Modeling non-Fickian transport in geological formations as a continuous time random walk. Rev. Geophys. **44**(2), 1–49 (2006)
4. Singha, K., Day-Lewis, F.D., Lane, J.W.: Geoelectrical evidence of bicontinuum transport in groundwater. Geophys. Res. Lett. **34**(12), 1–14 (2007)
5. Zhang, Y., Papelis, C., Young, M.H., Berli, M.: Challenges in the application of fractional derivative models in capturing solute transport in porous media: Darcy-scale fractional dispersion and the influence of medium properties. Math. Probl. Eng. **1**, 1–21 (2013)
6. Neuman, S.P., Tartakovsky, D.M.: Perspective on theories of non-Fickian transport in heterogeneous media. Adv. Water Resour. **32**(5), 670–680 (2009)
7. Zhang, Y., Benson, D.A., Reeves, D.M.: Time and space nonlocalities underlying fractional-derivative models: distinction and literature review of field applications. Adv. Water Resour. **32**(4), 561–581 (2009)
8. Sun, H., Zhang, Y., Chen, W., Reeves, D.M.: Use of a variable-index fractional-derivative model to capture transient dispersion in heterogeneous media. J. Contam. Hydrol. **157**, 47–58 (2014)
9. Allwright, A., Atangana, A.: Fractal advection-dispersion equation for groundwater transport in fractured aquifers with self-similarities. Eur. Phys. J. Plus **133**(2), 1–14 (2018)
10. West, B.J.: Fractional Calculus View of Complexity: Tomorrow's Science. CRC Press, Florida (2016)
11. Oldham, K.B., Spanier, J.: The Fractional Calculus: Theory and Applications of Differentiation and Integration to Arbitrary Order. Academic, New York (1974)
12. Herrmann, R.: Fractional Calculus: An Introduction for Physicists. World Scientific Publishing, Singapore (2011)
13. Li, C., Qian, D., Chen, Y.: On riemann-liouville and caputo derivatives. Discret. Dyn. Nat. Soc. **1**, 1–15 (2011)
14. Caputo, M., Fabrizio, M.: A new definition of fractional derivative without singular kernel. Prog. Fract. Differ. Appl. **1**(2), 73–85 (2015)
15. Caputo, M., Fabrizio, M.: Applications of new time and spatial fractional derivatives with exponential kernels. Prog. Fract. Differ. Appl. **2**(1), 1–11 (2016)
16. Yépez-Martínez, H., Gómez-Aguilar, J.F.: A new modified definition of Caputo-Fabrizio fractional-order derivative and their applications to the Multi Step Homotopy Analysis Method (MHAM). J. Comput. Appl. Math. **346**, 247–260 (2019)
17. Atangana, A., Baleanu, D.: New fractional derivatives with nonlocal and non-singular kernel: theory and application to heat transfer model. Therm. Sci. **20**(2), 763–769 (2016)
18. Atangana, A., Gómez-Aguilar, J.F.: A new derivative with normal distribution kernel: theory, methods and applications. Phys. A Stat. Mech. Appl. **476**, 1–14 (2017)

19. Morales-Delgado, V.F., Gómez-Aguilar, J.F., Escobar-Jiménez, R.F., Taneco-Hernández, M.A.: Fractional conformable derivatives of Liouville-Caputo type with low-fractionality. Phys. A Stat. Mech. Appl. **503**, 424–438 (2018)
20. Atangana, A., Gómez-Aguilar, J.F.: Hyperchaotic behaviour obtained via a nonlocal operator with exponential decay and Mittag-Leffler laws. Chaos Solitons Fractals **102**, 285–294 (2017)
21. Sun, H., Hao, X., Zhang, Y., Baleanu, D.: Relaxation and diffusion models with non-singular kernels. Phys. A **468**, 590–596 (2017)
22. Schmelling, S.G., Ross, RR.: Contaminant transport in fractured media: models for decision makers. (EPA Superfund) Issue Paper, Groundwater **28**(2), 272–279 (1989)
23. Zimmerman, D.A., De Marsily, G., Gotway, C.A., Marietta, M.G., Axness, C.L., Beauheim, R.L., Gallegos, D.P.: A comparison of seven geostatistically based inverse approaches to estimate transmissivities for modeling advective transport by groundwater flow. Water Resour. Res. **34**(6), 1373–1413 (1998)
24. Fomin, S., Chugunov, V., Hashida, T.: The effect of non-Fickian diffusion into surrounding rocks on contaminant transport in a fractured porous aquifer. Proc. R. Soc. Lond. A Math. Phys. Eng. Sci. **461**(2061), 2923–2939 (2005)
25. Goode, D.J., Tiedeman, C.R., Lacombe, P.J., Imbrigiotta, T.E., Shapiro, A.M., Chapelle, F.H.: Contamination in fractured-rock aquifers: research at the former naval air warfare center, West Trenton, New Jersey, p. 3074 (2007)
26. Cello, P.A., Walker, D.D., Valocchi, A.J., Loftis, B.: Flow dimension and anomalous diffusion of aquifer tests in fracture networks. Vadose Zone J. **8**(1), 258–268 (2009)
27. Shapiro, A.M.: The challenge of interpreting environmental tracer concentrations in fractured rock and carbonate aquifers. Hydrogeol. J. **19**(1), 9–12 (2011)
28. Masciopinto, C., Palmiotta, D.: Flow and transport in fractured aquifers: new conceptual models based on field measurements. Transp. Porous Media **96**(1), 117–133 (2013)
29. Allwright, A., Atangana, A.: Augmented upwind numerical schemes for a fractional advection-dispersion equation in fractured groundwater systems. Discret. Contin. Dyn. Syst.-Ser. S **1**, 1–14 (2018)
30. Tateishi, A.A., Ribeiro, H.V., Lenzi, E.K.: The role of fractional time-derivative operators on anomalous diffusion. Front. Phys **5**(52), 1–17 (2017)
31. Atangana, A., Gómez-Aguilar, J.F.: Decolonisation of fractional calculus rules: breaking commutativity and associativity to capture more natural phenomena. Eur. Phys. J. Plus **133**, 1–22 (2018)
32. Alkahtani, B.S.T., Koca, I., Atangana, A.: New numerical analysis of Riemann-Liouville time-fractional Schrödinger with power, exponential decay, and Mittag-Leffler laws. J. Nonlinear Sci. Appl. **10**(8), 4231–4243 (2017)
33. Allwright, A., Atangana, A.: Augmented upwind numerical schemes for the groundwater transport advection-dispersion equation with local operators. Int. J. Numer. Methods Fluids **87**, 437–462 (2018)
34. Ewing, R.E., Wang, H.: A summary of numerical methods for time-dependent advection-dominated partial differential equations. J. Comput. Appl. Math. **128**(1), 423–445 (2001)
35. Atangana, A.: On the stability and convergence of the time-fractional variable order telegraph equation. J. Comput. Phys. **293**, 104–114 (2015)
36. Gnitchogna, R., Atangana, A.: New two-step Laplace Adam-Bashforth method for integer a noninteger order partial differential equations. Numer. Methods Partial. Differ. Equ. **1**, 1–20 (2017)

FSC
www.fsc.org
MIX
Papier aus verantwortungsvollen Quellen
Paper from responsible sources
FSC® C105338

If you have any concerns about our products,
you can contact us on
ProductSafety@springernature.com

In case Publisher is established outside the EU,
the EU authorized representative is:
Springer Nature Customer Service Center GmbH
Europaplatz 3, 69115 Heidelberg, Germany

Printed by Libri Plureos GmbH
in Hamburg, Germany